Functionalized Cardiovascular Stents

Woodhead publishing series in biomaterials

Functionalized Cardiovascular Stents

Edited by

J. Gerard Wall
Halina Podbielska
Magdalena Wawrzyńska

ELSEVIER

WP
WOODHEAD
PUBLISHING
An imprint of Elsevier

Woodhead Publishing is an imprint of Elsevier
The Officers' Mess Business Centre, Royston Road, Duxford, CB22 4QH, United Kingdom
50 Hampshire Street, 5th Floor, Cambridge, MA 02139, United States
The Boulevard, Langford Lane, Kidlington, OX5 1GB, United Kingdom

Library of Congress Cataloging-in-Publication Data
A catalog record for this book is available from the Library of Congress

British Library Cataloguing-in-Publication Data
A catalogue record for this book is available from the British Library

ISBN: 978-0-08-100496-8 (print)
ISBN: 978-0-08-100498-2 (online)

Publisher: Mathew Deans
Acquisition Editor: Laura Overend
Editorial Project Manager: Natasha Welford
Production Project Manager: Surya Narayanan Jayachandran
Cover Designer: Miles Hitchen

Typeset by SPi Global, India

Contents

List of contributors

H.Y. Ang National Heart Centre Singapore, Singapore

J. Arkowski Wrocław Medical University, Wrocław, Poland

A. Arunkumar Sri Shakthi Institute of Engineering and Technology, Coimbatore, India

D. Biały Wrocław Medical University, Wrocław, Poland

M.M.M. Bilek University of Sydney, Camperdown, NSW, Australia

F. Boccafoschi University of Oriental Piedmont (UPO), Novara, Italy

H. Bulluck National Heart Centre Singapore, Singapore

M. Cannas University of Oriental Piedmont (UPO), Novara, Italy

A.H.P. Chan The Heart Research Institute, Sydney; University of Sydney, Camperdown, NSW, Australia

P. Cools Ghent University, Gent, Belgium

A. de Belder Brighton and Sussex University Hospital, Brighton, United Kingdom

N. De Geyter Ghent University, Gent, Belgium

M. Duda Wroclaw University of Science and Technology, Wroclaw, Poland

E.C. Filipe The Heart Research Institute, Sydney; University of Sydney, Camperdown, NSW, Australia

A. Foerster Wroclaw University of Science and Technology, Wroclaw, Poland

N. Foin National Heart Centre Singapore; Duke-NUS Medical School, Singapore

L. Fusaro University of Oriental Piedmont (UPO), Novara, Italy

P. Gopinath Indian Institute of Technology Roorkee, Roorkee, India

Y. Huang Nanyang Technological University, Singapore

J. Hung The Heart Research Institute, Sydney; University of Sydney, Camperdown, NSW, Australia

S.K. Jaganathan Ton Duc Thang University, Ho Chi Minh City, Vietnam; Universiti Teknologi Malaysia, Johor, Malaysia

M. Kopaczyńska Wrocław University of Technology, Wrocław, Poland

H. Kraśkiewicz Wroclaw University of Science and Technology, Wroclaw, Poland

V. Kumar Indian Institute of Technology Roorkee, Roorkee, India

B.S.L. Lee The Heart Research Institute, Sydney; University of Sydney, Camperdown, NSW, Australia

C.H. Lee University of Missouri-Kansas City, Kansas City, MO, United States

D. Mantovani Laval University, Quebec, QC, Canada

C. McCormick University of Strathclyde, Glasgow, United Kingdom

P.L. Michael The Heart Research Institute, Sydney; University of Sydney, Camperdown, NSW, Australia

E.C. Michel Laval University, Quebec, QC, Canada

V. Montaño-Machado Laval University, Quebec, QC, Canada

R. Morent Ghent University, Gent, Belgium

J. Ng National Heart Centre Singapore, Singapore

I.B. O'Connor National University of Ireland, Galway, Ireland

B. Oh University of Missouri-Kansas City, Kansas City, MO, United States

S.A. Omar Brighton and Sussex University Hospital, Brighton, United Kingdom

K.N. Pandiyaraj Sri Shakthi Institute of Engineering and Technology, Coimbatore, India

H. Podbielska Wroclaw University of Science and Technology, Wroclaw, Poland

M.C. Ramkumar Sri Shakthi Institute of Engineering and Technology, Coimbatore, India

M. Santos The Heart Research Institute, Sydney; University of Sydney, Camperdown, NSW, Australia

A.R. Saraf Dr. Babasaheb Ambedkar Technological University, Lonere, India

P. Sojitra Concept Medical Research Private Limited, Surat, India

A. Srivastava National Institute of Pharmaceutical Education and Research, Ahmedabad, India

R.P. Tan The Heart Research Institute, Sydney; University of Sydney, Camperdown, NSW, Australia

S. Udaykumar Indian Institute of Technology Roorkee, Roorkee, India

S. Venkatraman Nanyang Technological University, Singapore

J.G. Wall National University of Ireland, Galway, Ireland

A. Waterhouse The Heart Research Institute, Sydney, NSW, Australia

M. Wawrzyńska Wrocław Medical University, Wrocław, Poland

S.G. Wise The Heart Research Institute, Sydney; University of Sydney, Camperdown, NSW, Australia

A. Włodarczak Copper Health Centre, Legnica, Poland

P. Wong National Heart Centre Singapore; Duke-NUS Medical School, Singapore

S.P. Yadav Dr. Babasaheb Ambedkar Technological University, Lonere, India

Y.F. Zheng Peking University, Beijing, China

Preface

Cardiovascular disease is the main cause of mortality globally. About half of all deaths from cardiovascular diseases are due to coronary artery disease, which results from narrowing of arteries due to the build-up of plaques, thereby restricting blood flow. The long-term prognosis for patients is also poor, with up to a third re-hospitalized within 1 year of initial presentation, at enormous economic cost to healthcare systems worldwide.

The development of angioplasty to re-open vessels, coupled with stent deployment to avoid restenosis, was a major innovation in improving patient outcomes in the 1980s. Engineering cardiovascular stents further to improve physical properties such as radial strength or corrosion resistance, enhance biocompatibility, or delay thrombosis while allowing healing to proceed, has been a major focus of cardiology research over the past three decades. As well as the resultant collection of routine bare metal stents (BMS) and drug-eluting stents (DES) available on the market, commercially available and in-pipeline devices now also deliver NO and antirestenotic drugs, are entirely bioabsorbable or biodegradable, or exhibit endothelialization-promoting coatings.

The aim of this book is to chart the evolution of cardiovascular stents from their initial emergence as simple, relatively crude bare metal cage structures, through drug-eluting stents, to multifunctional devices assembled using cutting-edge surface coating, cell stimulating and biofunctionalization technologies. The book arises from the European Union Industry-Academia Partnerships and Pathways "EPiCSTENT" grant which brought together a team of industry- and academia-based partners from throughout Europe, with synergistic competencies in materials science, protein engineering, cardiology, and manufacturing, to engineer a cardiovascular stent with improved clinical performance for better patient outcomes.

We hope that the book will be of interest and benefit to students studying biomedicine, biomedical engineering, materials science, and cardiology in particular, as well as researchers in the broad fields of interventional cardiology, medical devices, materials science, and surface engineering. Feedback from readers is welcome.

J.G. Wall

Acknowledgments

The editors wish to acknowledge the financial support of the European Union through Industry-Academia Partnerships and Pathways "EPiCSTENT" grant FP7-PEOPLE-2012-IAPP-324514. We acknowledge also the support of colleagues during preparation of the book, particularly Dr. Silvia Maretto, NUI Galway, Project Manager of the EPiCSTENT grant, and thank the Elsevier team for their help throughout this project.

Part One

Fundamentals of cardiovascular stents

Overview of cardiovascular stent designs

1

C. McCormick
University of Strathclyde, Glasgow, United Kingdom

1.1 Introduction

Coronary heart disease (CHD) is the leading cause of death globally. In Europe, each year it is responsible for around 1.8 million deaths and it costs the European Union economies around €60 billion, of which around one third is due to direct treatment costs, with the remainder associated with informal care costs and productivity losses [1]. It is characterized by the development of plaque material within the artery wall, due to atherosclerosis. Ultimately, this leads to a progressive reduction in the lumen available for blood flow, thereby increasing the risk of ischemia in the surrounding tissue and producing painful episodes of angina. If left untreated, the plaque can become vulnerable to rupture, with the resultant acute vessel occlusion leading to myocardial infarction and potentially stroke. Research over recent decades has greatly increased understanding of the pathophysiology of this condition and revealed potential targets for therapeutic intervention. Despite the new pharmacological-based treatments and preventative strategies that such progress has produced, CHD remains a major health challenge [1].

Depending on the severity and extent of the disease, it may be appropriate to manage the condition with conventional drug treatments. Such treatment strategies are designed to lower the risk of myocardial infarction through improved control of blood pressure and cholesterol levels. Lifestyle modifications are also recommended as a means of managing this condition. However, in more advanced forms of the disease then a revascularization procedure will be performed in order to restore blood flow to the affected tissue. For many years, this was achieved through the use of a coronary artery bypass graft (CABG) operation and this remains the gold standard treatment, particularly in those patients with multivessel disease or complex lesions in major vessels [2,3]. While CABG therefore remains a key revascularization treatment; in recent decades, it has been overtaken in many countries by the use of percutaneous coronary intervention (PCI). In the United Kingdom, for example, around 92,000 coronary artery revascularizations are currently carried out annually by PCI, compared to around 17,000 by CABG [4]. There have been various technology developments that have helped drive this increased use of PCI. This chapter will provide an overview of the most important of these developments, with particular focus given to innovations in coronary stent design.

Functionalized Cardiovascular Stents. https://doi.org/10.1016/B978-0-08-100496-8.00001-9

1.2 Percutaneous coronary interventions

1.2.1 Percutaneous transluminal coronary angioplasty (PTCA)

The use of a balloon for percutaneous transluminal coronary angioplasty, PTCA, was pioneered in the 1970s by Andreas Gruentzig, a German physician. The process involves delivering a balloon to the site of vessel narrowing via a catheter inserted in a peripheral artery, commonly the femoral or radial artery. The balloon is then inflated under pressure, thus expanding the vessel, restoring lumen size, and hence improving blood flow. Gruentzig performed the first coronary balloon angioplasty on an awake human in September 1977, successfully revascularizing a stenotic lesion of the left anterior descending artery. In 1979, he reported the results from the first 50 patients to receive this treatment [5], with various technical advances in catheter design and materials helping the technique become increasingly widespread in the decade that followed [6]. However, while such developments increased the effectiveness of the procedure, it remained limited by acute vessel closure and elastic recoil in the short term and restenosis in the longer term. The process of restenosis is thought to be the artery's response to the severe injury induced by balloon dilatation. It is characterized by increased smooth muscle cell proliferation and extracellular matrix deposition, leading to progressive luminal narrowing, and is observed in around one third of patients [7]. In addition to improved catheter design and deployment techniques, significant research efforts focussed on the investigation of various pharmacological and radiation-based therapies to reduce restenosis rates [8]. However, such approaches have had limited long-term clinical impact, and high restenosis rates remained a major limitation of PTCA up until the introduction of coronary stents.

1.3 Bare metal stents

It was found that expansion and permanent placement of a mechanical support device, known as a stent, into the stenosed lesion immediately following balloon angioplasty, could reduce acute vessel occlusion and restenosis. This technique was first performed in man by Sigwart and Puel in 1986, using a self-expanding stainless steel stent. Shortly afterwards, a study of 25 patients with occlusions of the coronary or iliac arteries was completed, with no evidence of restenosis observed in either lesion group observed at 9-month follow-up, thus demonstrating the potential utility of this device [9]. Since then, improvements in stent design and materials, deployment technique, delivery methods, and the use of dual antiplatelet therapies have led to a rapid growth in the use of stenting. Such growth was stimulated by results from a number of clinical trials, which consistently demonstrated that stent placement leads to reduced restenosis rates in comparison to balloon angioplasty alone [10–12]. However, while such reductions in restenosis were welcome, stents did not completely remove this problem. In fact, it became clear that the use of these devices gave rise to a new healing response, so-called in-stent restenosis (ISR). While overlapping mechanisms exist, in-stent restenosis is thought to be a more cellular and proliferative response than balloon angioplasty [13]. Although ISR rates have been found to be highly dependent on patient and lesion characteristics, average rates of around 25% of cases following bare metal stent placement are typically reported [14].

1.3.1 In-stent restenosis

Repeat revascularization of a previously stented atherosclerotic lesion following ISR is a complex procedure. This complexity and the high number of repeat revascularizations required meant that ISR represented a significant clinical problem. Consequently, attention became focussed on the development of strategies to inhibit ISR and thus reduce the need for repeat revascularizations [15]. Although the mechanisms underlying ISR are still not completely understood, it has generally been viewed as a distortion of the natural healing response to the injury invoked by arterial expansion and stent placement (Fig. 1.1). This healing process can be viewed as a cascade of events, where the ultimate product is the development of a neointima consisting of smooth muscle

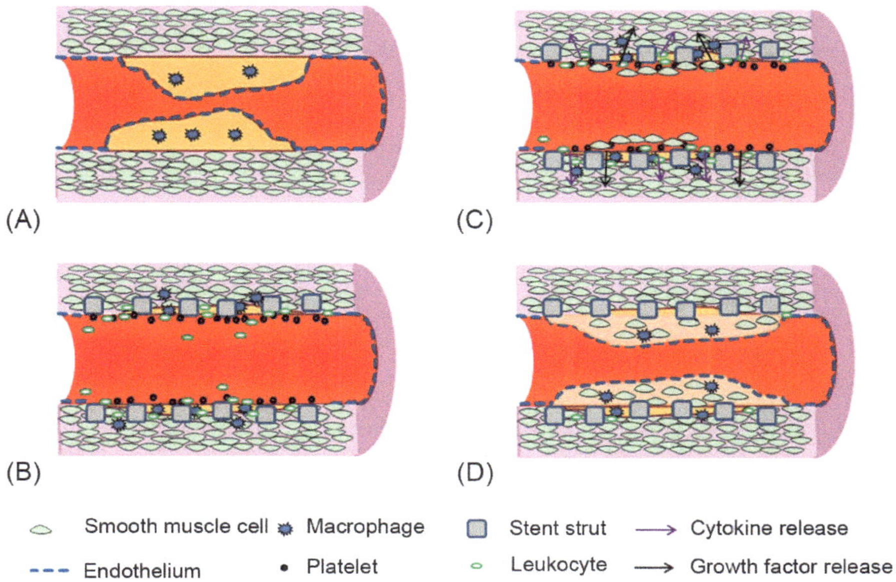

Fig. 1.1 Schematic overview of in-stent restenosis, showing key responses to stent implantation within a diseased coronary artery. (A) Diseased coronary artery with lumen narrowed by presence of atherosclerotic plaque, (B) stent implementation causes severe injury to the vessel wall, including endothelial denudation, resulting in platelet adhesion/activation with recruitment and activation of neutrophils, monocytes, and macrophages, (C) release of cytokines and growth factors triggers smooth muscle cells to proliferate and migrate towards the lumen, and (D) mature neointima is formed, comprising smooth muscle cells and extracellular matrix, covered by a restored endothelium.
Modified from F.G.P. Welt, C. Rogers, Inflammation and restenosis in the stent era, Arterioscler. Thromb. Vasc. Biol. 22 (2002) 1769–1776.

cells and extracellular matrix components [16,17]. The individual processes leading to neointima formation were well-summarized by Babapulle and Eisenberg [18]. Briefly, the immediate response to stent insertion is characterized by platelet adhesion and activation, with recruitment and activation of neutrophils, monocytes, and macrophages within the emergent thrombus occurring over the first week or so. This acute phase response releases a cocktail of cytokines and growth factors, which in combination with the vessel injury induced by stent deployment leads to activation of medial smooth muscle cells. The resulting proliferation and migration of smooth muscle cells towards the lumen, together with the deposition of extracellular matrix, results in the neointima formation that is characteristic of ISR. This process of neointima formation stabilizes after around 4 weeks in animal models [15], with recovery of the endothelium normally being completed over a similar period. In humans, the responses are thought to be similar, but occur over a more prolonged period of 3–6 months following the procedure [19].

There is strong evidence from histological observations in the pig [20] and human coronary artery [21] that the extent of ISR is closely related to both the severity and the characteristic features of the balloon and stent-induced injury [22]. Vessel stretch is thought to be a key determinant of this injury response [23]. Moreover, strut-induced laceration of the lamina that separate the individual tunica elements of the artery has also been associated with increased restenosis, which may be a factor driving an increased inflammatory response [24].

In general, strategies designed to overcome ISR have attempted to target many of the processes within the cascade outlined above. These have ranged from strategies aimed at reducing the extent of vessel injury, to pharmacological and radiation-based therapies to minimize the inflammatory and proliferative responses. Our focus in this chapter is on those developments in stent design that have had the most significant long-term clinical impact.

1.3.2 Stent platform design

There are numerous design criteria that a coronary stent must meet if it is to fulfill its function. Paramount among these is that the stent platform must provide mechanical support to the vessel wall, such that it readily conforms to the curvature of the vessel following expansion and maintains sufficient radial strength to withstand the forces being exerted by the arterial wall. In addition, the stent platform should be radio-opaque to provide the required visibility to ensure optimal sizing and deployment. The platform must be easily deliverable and have a narrow profile when crimped, such that it can readily pass through narrow vessels with significant stenoses. Finally, the stent material should, as far as is possible, be compatible with the blood and surrounding vascular wall tissue. A wide range of quite distinct stents were developed to meet these criteria, and through the evaluation of these early-generation devices, it became quickly apparent that stent platform design and construction were key factors driving clinical performance [25,26]. Great attention has therefore subsequently been paid to the optimization of stent materials and platform designs. We will look at some of the most important developments within this area. The reader is also directed towards a number of more comprehensive sources for detailed information on this aspect of stent design [25,27–29].

1.3.2.1 Stent construction

The first stent implanted in humans was the WALLSTENT [9]. This was a self-expanding platform, comprising a stainless steel metal mesh structure. Although the positive results from early clinical studies of this stent provided the stimulus for the widespread use of stenting we see today, its delivery mechanism was rather cumbersome in comparison to other systems that quickly became available. The Palmaz-Schatz stent overcame some of the practical limitations of the original WALLSTENT by using an alternative deployment mechanism, through the use of balloon expansion. Today, nearly all coronary stents are balloon-mounted in this way. Similarly, the slotted tube design of the Palmaz-Schatz stent, which demonstrated greater radial strength and superior clinical outcomes over the more flexible mesh and coil structures of competitor stents, is reflected in the platforms used in current stents. The further development of modular stent platforms has also offered a balance between the flexibility of coils and the radial strength of slotted-tube designs.

1.3.2.2 Stent geometry

While stent flexibility and radial strength are key features of successful stents, it is important to recognize that the different geometric configurations employed to achieve these properties can impact on the clinical performance of the device. This was perhaps most clearly demonstrated by Rogers and Edelman [30], who showed that stent geometry was an important determinant of restenosis. In particular, they demonstrated that increasing the number of strut-strut intersections, while maintaining the stent material and surface area constant, led to a proportionate increase in neointimal area. The reduction in vascular injury score also achieved with the reduced number of strut intersections provided a potential mechanistic explanation for the difference in efficacy between the two different stent geometries. While the study authors subsequently went on to show that stent geometry can impact neointima formation independent of vascular injury [31], several subsequent investigators have demonstrated a link between the extent of stent-induced injury and the level of restenosis [22,26].

1.3.2.3 Stent strut thickness

Strut thickness has been a key element of stent design, with thinner struts associated with greater deliverability. However, results from a series of animal and clinical studies also suggested that thinner struts produced lower rates of restenosis [26]. The first clinical study to demonstrate this link was the ISAR-STEREO [32], a study of 651 patients randomized to receive one of two bare metal stent types (ACS RX Multilink with strut thickness of 50 μm or ACS Multilink RX Duet with strut thickness of 140 μm). It was found that angiographic restenosis was observed in 15% of patients who had received the thin strut stent, while this figure increased to 26% in the thick strut group. Importantly, the relative similarity in geometry between the two stent types used in the study suggested that the difference observed was due to strut thickness. The same group went on to show that the benefits of reduced strut thickness were maintained when the geometry of the two stents test was different, further demonstrating the important role of strut thickness [33]. Consequently, there has been a move to the use of stents with minimal strut thickness, with current stents typically employing metal struts ranging in thickness from around 60 to 100 μm. Any further reductions in stent strut size will have to be balanced by the need for visibility of the stent location, and ideally the platform itself, during delivery and deployment in the affected artery lesion.

1.3.2.4 Stent platform materials

The improvements in stent platform design described above have been greatly facilitated by developments in materials science. Perhaps of greatest clinical impact has been the use of metal alloys with greater mechanical strength than the stainless steel 316L that was used in the majority of early bare metal stent platforms. This enhanced material strength has been key to achieving the reductions in stent strut thickness which have helped improve delivery and efficacy of stents in the last decade [28]. A great many materials have been examined with different mechanical properties and degrees of biocompatibility [34,35]. These include alloys of nickel-titanium, cobalt-chromium, magnesium, platinum-iridium, as well as single metal designs of tantalum, nitinol, iron, and variable grades of stainless steel. Although even some of the most recently developed stents still use stainless steel, the superior strength and biocompatibility of cobalt-chromium alloys meant that it is increasingly the preferred stent platform material. The Multi-Link Vision stent (Abbott Vascular) was the first such stent to receive FDA approval in the United States and other devices quickly followed, notably the Driver stent (Medtronic). More recently, Boston Scientific developed the OMEGA stent, which employs Platinum-Chromium. Generally, the transition from stainless steel to the use of these more advanced metal alloys has helped reduce strut thickness and improved deliverability, while maintaining radial strength.

1.4 Drug-eluting stents

While it is important to recognize the contribution that the above innovations in stent platforms have had, the key technology that has had by far the greatest impact on reducing ISR has been the development of drug-eluting stents (DES). Indeed, in a direct comparison between the first-generation Cypher DES (Cordis Corp., Johnson and Johnson) and a bare metal stent with thinner struts, the Cypher had significantly lower restenosis [36]. The modifications to stent platforms may therefore be viewed as incremental evolution, while the development of DES can be more accurately described as a revolution in stent design.

It was clear from the multifactorial nature of ISR that there were a great many biological targets that could be potentially addressed by pharmacological treatment. However, it was found that pharmacological therapies delivered systemically largely failed to deliver on the initial promise shown in animal studies. Aside from generic difficulties in the interpretation and clinical applicability of results from animal models [22,37], a significant reason for this failure was that the systemic concentrations found to be efficacious in animals could not be tolerated in humans. In this context, extending the function of the stent platform such that it could be used to provide localized drug delivery, directly to the affected lesion in high concentrations, became a particularly attractive concept. Positive findings during animal studies were quickly followed by early clinical trials that demonstrated quite dramatic reductions in restenosis. These new devices, the so-called DES, have since gone on to revolutionize interventional cardiology. Currently, they are used in the majority of PCI revascularization procedures [38]. However, their performance is limited in certain patient groups and lesion types. They have not removed the problem of stent thrombosis. For these reasons, extensive research budgets across academia and industry remain targeted at the development of enhanced DES. In this section, we will describe the key features of DES design and examine how it has evolved since the introduction of the first generation of these devices.

1.4.1 DES design

There are three key components of a DES that together contribute to the overall safety and efficacy of the device. These are the stent platform, the drug, and the drug delivery coating technology (Fig. 1.2). The performance of the device is dependent on optimization of each of these important aspects in what is a complex multidisciplinary process [39,40]. As an example of this complexity and how the design of one element can impact on the performance of another aspect of the device, it has been shown that there is a relationship between the stent platform configuration and strut thickness and the drug delivery characteristics of the device [41–43]. Indeed, DES design has been described as a multidisciplinary success story [39,40].

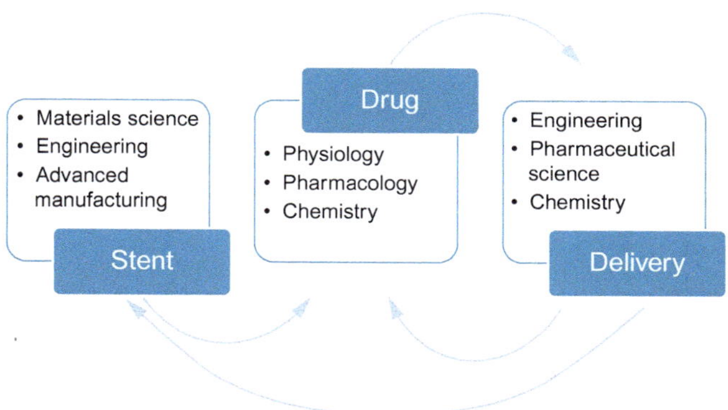

Fig. 1.2 Overview of the key features of a DES. Device performance is dependent on optimization of the relationship between the stent platform, the drug, and the choice of drug delivery technology. Successful DES design requires contributions from across many disciplines.

1.4.2 DES stent platforms

We have shown above how critical certain aspects of the stent platform materials and design are to the performance of bare metal stents. The same is true for DES. Generally, the different generations of these devices have made use of the most advanced bare metal stent platforms available at the time of their development. So, the first-generation DES, Cypher (Cordis Corp.) and Taxus (Boston Scientific) devices, employed stainless steel stent platforms with strut thicknesses of around 130–140 μm, while the development of platforms such as the Driver (Medtronic), Multi-Link Vision (Abbott Vascular), and OMEGA (Boston Scientific) meant that later-generation DES had much thinner struts, typically around 80–90 μm. While this trend towards thinner struts was consistent across conventional DES [27], more recent bioresorbable platforms have required a return to relatively thick struts to ensure sufficient radial strength [44]. The shape of stent struts has also undergone significant changes as DES design has developed over the years, with a move away from the rectangular-like struts used in first-generation devices, which have more angular edges than the more circular shapes and rounded edges characteristic of latest DES [45].

1.4.3 DES drugs

Our attention now turns towards the drug component of DES. We have seen that the mechanisms driving ISR and the accompanying neointima formation are numerous. It was therefore perhaps natural that the search for suitable drugs for use within DES has yielded a wide and varied range of candidates. It is beyond the scope of this chapter to detail all of these, and the reader is therefore referred to the following sources for further information on this aspect of DES design [46,47]. Briefly, notable strategies included the local release of antioxidants to inhibit the oxidative stress response to stenting [48]. A particularly promising approach appeared to be the release of endogenous molecules, such as prostacyclin [49,50] and nitric oxide [51], which would help compensate for the absence of the endothelium in the immediate aftermath of stent implantation. However, the real clinical breakthrough came with the use of the antiproliferative drugs, sirolimus and paclitaxel.

1.4.3.1 Sirolimus

The first DES to be implanted in humans was the Cypher stent (Cordis Corp.). The pharmacological agent in this stent is the macrocyclic lactone, sirolimus. This drug acts on the mammalian target of rapamycin (mTOR) and has a varied profile of action, with antibiotic, immunosuppressive, and antiproliferative effects. Its beneficial effects on restenosis are thought to arise from a combination of its inhibition of the immune response, through inhibition of T-lymphocyte activation, and perhaps more importantly through an antiproliferative effect on smooth muscle cells. In a randomized, double-blind controlled trial, target vessel revascularization rates were 17% in the bare metal control group, compared to 4% in those patients treated with the Cypher stent [52]. Such positive outcomes were replicated in subsequent trials, helping make Cypher the market leader in the early days of DES use. Cordis Corp. went on to develop a second-generation sirolimus-eluting stent (NEVO), using a different stent platform that allowed a more sophisticated drug delivery strategy to be adopted [53], but they were unable to maintain their market position and withdrew from the stent market in 2011. Although sirolimus is still used in a small number of DES from other manufacturers, it has been replaced with the use of limus analogs in the majority of currently used DES [27,45].

1.4.3.2 Paclitaxel

Paclitaxel has been used as a chemotherapy drug in the treatment of a range of different cancers. Its stabilization of microtubules means that the normal dynamic reorganization of the microtubule network during the interphase and mitosis stages of the cell cycle is inhibited. Although the precise mechanism(s) by which this drug inhibits neointima formation is still not completely clear, it is thought that this microtubule stabilization effect prevents the proliferation of smooth muscles cells, thereby reducing restenosis. This drug was first used in the Taxus stent (Boston Scientific), one of the first-generation DES devices. In a key early multicenter, randomized, double-blind controlled trial of this stent against bare metal controls, the target-lesion revascularization rate was reduced from 11% to 3% [54]. The dramatically improved clinical outcomes observed in such early trials saw a rapid increase in the use of the Taxus stent. Paclitaxel is now rarely used within the latest generation of DES, although there has been increased interest in its use within the drug-coated balloons that are now emerging.

1.4.4 Late-stent thrombosis and the search for better drugs

While the first generation of DES were a revolutionary step, the development of subsequent generations of these devices is probably best described as a more incremental process. There were clear commercial factors driving medical device manufacturers, most notably Abbott Vascular and Medtronic at the time, to develop competitor DES to challenge the early market dominance of Boston Scientific and Cordis Corp. However, the challenge of late-stent thrombosis (LST) that emerged following the use of first-generation DES also provided a crucial clinical stimulus to such development. LST is often fatal [55] and therefore early reports that there was an increased risk of this condition following the use of first-generation DES raised considerable concern within the cardiology community [56,57]. Although the precise causes of LST were not well-defined and to some extent uncertainty on this persists, a cessation of antiplatelet therapy was linked with its occurrence, suggesting that the endothelium had not completely recovered [58]. Indeed, postmortem studies revealed evidence of delayed re-endothelialization in patients who had been treated with first-generation DES [59,60]. Such evidence of delayed recovery of the endothelium following use of first-generation DES caused the research community to revisit all aspects of device design, with greatest attention focused on the use of alternative drugs and more biocompatible drug delivery technologies. It was known from in vitro investigations that both sirolimus and paclitaxel inhibited human endothelial cell proliferation at the same concentrations required to inhibit human smooth muscle cell proliferation [61].

There was also evidence that exposure to both drugs could impair endothelial cell function. Pig coronary arteries, preincubated ex vivo in paclitaxel, subsequently produced impaired in vitro relaxation responses to the endothelium-dependent relaxant, calcimycin [62]. Of more relevance clinically were findings from a series of relatively small studies, which revealed that patients with an implanted Cypher or Taxus stent exhibited coronary artery vascular dysfunction [63–65]. As a result, the search began for alternative drugs that could inhibit smooth muscle cell proliferation while having no detrimental effect on recovery of a functional endothelium.

1.4.4.1 Limus analogs

Several alternative drug types were investigated and research in these areas continues to this day. Particular emphasis has been placed on the identification of novel molecular pathways to allow more targeted inhibition of smooth muscle cells [66,67]. However, limus analogs have emerged as the most common drug type used in the second-generation DES [27] and they remain dominant in the latest DES designs that are emerging. Zotarolimus is used in the Endeavour, Resolute and Resolute Integrity DES produced by Medtronic. This compound has a similar pharmacological profile to sirolimus, with inhibition of mTOR thought to give rise to its suppressive effect on neointima formation. Although structurally similar to sirolimus, it is reported that the incorporation of a tetrazole ring within Zotarolimus provides significantly greater lipophilicity, which may provide increased partitioning and prolonged retention within the artery [68]. Like Zotarolimus, Everolimus is structurally similar to sirolimus and shares the same broad mechanism of action [69]. It is used in the Xience-V DES from Abbott Vascular and the Promus and Promus Element DES (Boston Scientific). Biolimus A9 has been used in a range of different DES devices, including the BioMatrix and BioFreedom stents from Biosensors International. Although both drugs have similar potency, the greatly enhanced lipophilicity of Biolimus A9 may provide advanced pharmacokinetics over sirolimus [70]. While it is therefore clear that each limus analog has potential advantages over sirolimus, it is difficult to precisely determine whether the improved performance of these latest DES arises from such mechanistic and pharmacokinetic benefits, or simply reflects other improvements made to the stent platform and the drug delivery method. We will now go on to consider the last of these factors in more detail.

1.4.5 DES drug delivery technologies

1.4.5.1 Drug release profile

The drug release profile is a key factor in the performance of a DES [71]. Release too little drug and there will be no inhibition of restenosis, while too much drug may result in unwanted effects, including delayed re-endothelialization. Put simply, the optimal drug release profile will be one that ensures that the arterial wall concentrations of the drug are maintained at the therapeutic dose within the target cells for the period required. In vitro cell culture studies have traditionally been used to identify a therapeutic dose, although in vivo dose-ranging studies are ultimately still required [72]. The optimal release period for a given drug is likely to be dependent upon the specific process(es) within the ISR cascade that it targets [73]. Animal studies have provided an indication of the temporal nature of the key features within this cascade [15,74]. The design of DES has involved optimization of the drug release profile based on the proposed mechanism(s) of action of the candidate drug. Therefore, for drugs acting primarily through an antiproliferative effect on smooth muscle cells, prolonged release profiles that maintain effective drug levels within the artery wall over several weeks and months were developed. The first generation of stents incorporated drug coatings to provide such release, with the Cypher Stent releasing sirolimus for up to around 90 days and the Taxus stent releasing paclitaxel for a similar period [73]. Given the similarity in drug type and mechanism of action, it is therefore not surprising that the subsequent generations of DES have targeted similar release profiles, although faster release formulations have also been found to be effective [75]. Recent insights from computational modeling suggest that there are opportunities for further optimization of the drug release profile for existing drug types and most certainly for new drugs currently under investigation [76].

In order to achieve the desired release profile, the biological agent must be bound to the surface of the stent in some way. Several approaches to this challenge have been investigated, including polymer- and nonpolymer-based systems and the discussion below is focused on those that have reached most widespread clinical application. For more detailed information, the reader is referred to Venkatraman and Boey [73] and to other chapters within this book that are dedicated to this topic.

1.4.5.2 Polymer-controlled drug release

Permanent polymers

A wide range of polymers have been studied as potential stent coatings, with initial focus on their potential to confer enhanced biocompatibility before their exploitation as a means of providing sustained drug delivery was fully realized [77]. The two first-generation DES utilized permanent polymer coatings, with release of the drug dominated by diffusion [78]. In the Taxus stent, a coating containing paclitaxel and the copolymer, poly(styrene-B-isobutylene-B-styrene) (Translute) is applied to the stent to provide sustained paclitaxel release [73]. It is important to note that the majority of drug loaded on to the Taxus stent is in fact permanently retained within the polymer and is therefore not released from the stent. In contrast, near-complete release of sirolimus is achieved with the Cypher stent. The Cypher stent employs a three-layer coating, comprising a base layer of Parylene C which is first applied to the metal stent to promote adhesion of the sirolimus containing middle layer [73]. This layer contains sirolimus mixed within a combination of two co-polymers, polyethylene-co-vinyl acetate and poly-n-butyl methylacrylate. An outer layer of these two co-polymers, in the absence of drug initially, is then added to provide an additional means of retarding the initial release.

In the same way that we have seen that commercial considerations and emerging clinical data provided important stimuli in the search for more effective drugs, these same factors drove the development of enhanced stent drug delivery approaches.

Reports of hypersensitivity responses to the polymers used in first-generation devices were particularly significant in this context [59,79]. Consequently, research efforts were focused on the development of technologies that would not only provide the optimal drug release, the overriding criteria that had motivated the design of the earliest DES, but would also provide enhanced biocompatibility. Much of the resultant activity was built on the results of earlier research that sought to improve the biocompatibility of stents in the period immediately preceding the introduction of DES [28]. In particular, the use of phoshorylcholine had generated promising results. The biomimetic nature of this polymer made it an excellent candidate and it was used to provide release of Zotarolimus from the Endeavour stent (Medtronic) [80]. The fairly rapid release of Zotarolimus from the phoshorylcholine coating, with drug release occurring over the first 2 weeks, was thought to be a limitation of this device [81] and so Medtronic went on to develop a slower release version, the Endeavour Resolute DES. The key difference being that a new polymer blend, BioLinx, incorporating polyvinyl pyrrolidone as a carrier layer for the drug, was used to provide the majority of release over a 2-month period [82]. Poly(vinylidene fluoride-co-hexafluoropropylene (PVDF-HFP)) has also been used as the outer polymer carrier layer for Everolimus in order to enhance biocompatibility, with this combination being used in Abbott Vascular's Xience V [69] and Boston Scientific's Promus Element DES products [83]. In addition to the use of alternative permanent polymers, a more fundamental shift in approach quickly emerged. Many manufacturers increasingly moved towards the use of coatings that provide controlled drug release over the initial month(s) followed by a return to a bare metal stent in the longer term. In large part, such strategies were based on the concerns that prolonged exposure to polymer-coated stents contributed to delayed vessel healing [84]. Two main approaches have been attempted in this pursuit, with the use of degradable polymer coatings and polymer-free strategies proving particularly successful.

Degradable polymers

Polymers of lactide and glycolide undergo degradation to produce lactic and glycolic acid, respectively, which can be readily metabolized to nontoxic products within the body, and therefore represented an attractive alternative to existing nondegradable polymers [77]. An early investigation by Finkelstein et al. [85] characterized in vitro release pharmacokinetics of paclitaxel from poly (lactide-*co*-glycolide) (PLGA)-coated stents. They were able to produce a variety of release profiles, with variations in the rate and duration of drug release being controlled by varying the number of layers and ratios of lactide to glycolide. Moreover, the lead paclitaxel-PLGA coating identified was found to inhibit neointima formation in an in vivo porcine coronary model of stent injury. Comparable results were achieved using poly(D,L-lactide-*co*-glycolide) to release either paclitaxel or sirolimus over 1 month in vitro [85a]. Polymers of this family have since made an important contribution to the evolution of DES design and are now used in many of the latest generation devices. The reader is referred to the following articles for a more comprehensive description of such developments [28,86]. Briefly, Poly(D,L)lactide (PDLLA) is the polymer coating used in the BioMatrix (Biosensors International) and Nobori (Terumo) Biolimus A9 eluting stents. It is also worth noting that these stents were among the first to apply the polymer- drug coating to only the abluminal facing side of the stent, an innovation intended to better target drug delivery to the artery tissue and achieve more rapid endothelialization. These stents had a relatively large strut thickness of 120 μm in comparison to the thinner struts used by competitors. For example, the Synergy stent (Boston Scientific), which uses a very thin layer of PLGA to provide controlled release of Everolimus, has a platinum-chromium platform with strut thickness ranging from 74 to 81 μm. Although the use of such thinner struts may yield additional clinical benefit beyond that potentially conferred by the use of biodegradable coatings, the second-generation devices described thus far have been based on the use of conventional bare metal stent platforms and broadly traditional coating deposition techniques. Other devices have used novel stent platforms that facilitate the use of alternative coating techniques. The NEVO stent (Cordis Corp.) is an example of such approach, where the stent platform design allows precise loading of the drug and rate-controlling polymer layers within specially created reservoirs within the stent struts [53]. A variation on this type of hybrid approach is the Yukon Choice PC (Translumina GmbH). This device is based on the application of a sirolimus-containing PLA polymer layer to a microporous stainless steel stent platform, with the application of a Shellac resin topcoat [87]. In contrast to conventional DES, once degradation of this polymer coating is complete, the stent presents a microporous surface that is believed to encourage more rapid endothelialization than comparable smooth stent surfaces [88]. Subsequent trials indicated that this device is associated with similar clinical outcomes as Cypher and Xience stents at 3 years [89]. Although subtle variations from device to device make teasing out the relative effect of biodegradable coatings on clinical outcomes difficult to quantify precisely, a pooled analysis of several large randomized clinical trials found that biodegradable polymer-coated DES had improved safety and efficacy when compared with first-generation durable polymer DES [90]. The extent to which the introduction of such biodegradable polymers yielded an improvement on other second-generation durable polymer DES appears to be less clear.

1.4.5.3 Polymer-free DES

While the biodegradable polymer coatings described above represented one approach for removing concerns over the use of permanent polymer coatings on stents, an important alternative strategy has saw the development of completely polymer-free DES. Several such DES have since been developed, with varying degrees of clinical success. The absence of the polymer means that all such devices must include some other mechanism to control drug release from the stent surface. This class of DES will be described in greater detail later on in this book and so only a brief summary is presented here. Ultimately, the introduction of surface porosity, at macro-, micro-, or nanoscale, has proved a popular approach to controlling drug release from polymer-free stent systems [91,92]. Of these approaches, it has been macro- and microporous stent surfaces that have seen the greatest use clinically. Examples of stents with pores at the macro level, essentially holes or slotted grooves, include the Janus Tacrolimus-eluting Carbostent (Sorin Group) and a drug-filled stent system from Medtronic [86]. One of the first microporous stents to be approved for use was the Yukon stent (Translumina

GmbH) [75,93]. In this stent, a conventional stainless steel stent platform is sandblasted to produce a rough surface finish characterized by the presence of microporous pits of the order of 1–2 μm in size. These pits are then loaded with drug, usually sirolimus, by spray-coating to provide controlled drug release. Upon completion of drug release, it has been suggested by the manufacturers that the microporous rough stent surface will accelerate re-endothelialization and help reduce restenosis. Certainly, a trend towards reduced restenosis through the use of microporous surfaces has been observed in one clinical trial of the bare metal version of this device, although the overall effect was not found to be statistically significant [88]. Another potential advantage of the Yukon stent is that the drug coating is applied on-site within the catheterization lab at the time of the stent procedure, thereby allowing for the drug type(s) and dose to be selected by the cardiologist. However, this approach has had relatively low clinical uptake, perhaps because there remains a limited understanding of how drug dose and type should be tailored towards particular lesion types or patients [94]. The drug release from this stent was also found to be fairly rapid, and in fact, this manufacturer's more recent DES now use a more conventional biodegradable polymer coating to slow drug release [87]. A more recent microporous DES that is showing particularly impressive clinical results is the BioFreedom stent from Biosensors International. Here, a rough stent surface is created through the use of proprietary micro-abrasion process. Although this produces a surface finish that appears similar to the Yukon stents, in the BioFreedom platform the rough finish is applied only to the abluminal surface, thereby providing potentially more targeted drug delivery to the affected lesion. Whereas the Yukon stent has generally been used to provide release of sirolimus, the BioFreedom stent is coated with Biolimus A9. The greater lipophilicity of Biolimus A9 may be an important factor in the impressive clinical results that have been reported for this stent. It has demonstrated noninferiority against the Taxus stent, in terms of late lumen loss measured at year, with all clinical event rates found to be similar between the two stent types after 5 years [95]. It has recently been shown to be superior to bare metal stents in the treatment of complex lesions in patients at high risk of bleeding (LEADERS FREE) [96]. An important aspect of the LEADERS FREE trial was that dual antiplatelet therapy was withdrawn from both the bare metal and BioFreedom stent groups after just 4 weeks. The VESTAsync stent (MIV Therapeutics) is a further example of a microporous stent, with porosity in this case being provided through the use of a hydroxyapatite surface finish [97]. Our recent modeling studies have shown that the induction of microporosity alone may in some cases be insufficient to slow drug release to levels seen with conventional polymer coatings [91]. Indeed, the induction of nanoporosity and/or changes in the physical characteristics of the drug coating, most notably modifications to drug solubility, are likely required and we will describe recent developments in these two areas in more detail in a separate chapter within this book. Ultimately, research into the development of alternative approaches to polymer-free drug coatings continues at pace, as evidenced by the continued registration of a variety of patented technologies in this area [98]. The ongoing interest in the crucial role that the stent surface plays in the performance of the device is also reflected in the return to drug-free approaches to stent design that we are now increasingly seeing [86]. It is also likely that the development of more biocompatible surfaces that promote rapid endothelialization will have an important role to play. Stents that can successfully inhibit neointima formation and promote recovery of a healthy endothelium remain particularly attractive. Early clinical experience with the Combo stent (OrbusNeich Medical), which combines conventional sirolimus release with an endothelial progenitor cell coating technology, has shown the potential of such approaches [99].

1.5 Bioresorbable stents

The final and latest category of coronary stents that we will now consider is bioresorbable stents. These stents have the potential to revolutionize cardiology in a manner similar to that seen with the original bare metal stents and then DES [100]. While it would be easy to categorize these devices as a natural response to clinical experience with more conventional stents, it is important to note that the concept of bioresorbable stents has actually been around since introduction of the first DES in the early 2000s [101]. The basic rationale for such stents set out at this time, namely that the stent should disappear once its main function had been served, remains the same today. That this concept has clear potential clinical benefits has recently been nicely summarized by Sotomi et al. [44]. Firstly, there are potential reductions in adverse clinical events associated with the ongoing presence of polymer and metal materials within the artery wall. Secondly, there is the possibility that the vessel can recover important functions, such as the unimpeded vasomotion that is so critical to control of blood flow and pressure. Thirdly, the treated patient can freely undergo future diagnostic investigations involving MRI without concern. Finally, the stented vessel will remain suitable for the full range of revascularization strategies should this be required in future, which is particularly important for younger adult patients and children in particular. There has therefore been intense research activity across industry and academia into the development of bioresorbable stent scaffolds that will provide these benefits. The leading biodegradable polymer platform for which there is most clinical data is Abbott Vascular's Absorb bioresorbable vascular scaffold (BVS) stent. The main stent platform comprises PLLA, with the grade chosen to provide a polymer degradation time in excess of 2 years. Everolimus is incorporated into a PDLA surface coating in the same way as the Xience V, to provide diffusion-controlled release. The initial enthusiasm that greeted the generally positive results from early clinical trials of this stent has now given way to increasing concern that the long-term results may be inferior to equivalent DES [44]. Indeed, in a recent metaanalysis of seven randomized controlled trials, comprising 5583 patients, the Absorb BVS had lower efficacy and higher thrombosis risk than the conventional Everolimus-eluting stent [102]. These safety concerns call for further research and it is likely that improvements in device design and deployment procedure will be required if the true potential of these devices is to be realized [103]. To this end, investigations into the impact of deployment methods on clinical outcomes with the Absorb BVS have already begun [100]. On the device design side, several alternative devices are in development, with some already undergoing clinical trials [44]. A Novolimus-eluting DESolve stent (Elixir Medical Corp.), which uses a similar polymer and strut thickness to the Absorb BVS, has shown promising clinical results at 2-year follow-up [104]. Degradable metallic platforms have also been developed, with magnesium being one of the most promising materials investigated thus far. The Magmaris stent (Biotronik AG) comprises a magnesium alloy stent platform, with elution of sirolimus from an outer coating of biodegradable PLLA [44]. Initial 12-month clinical data on this stent look promising [105], although further longer-term studies are required. In all approaches described thus far, the desired degradation characteristics are achieved at a potential cost to performance, with the stent strut thickness needing to be increased up to 150–170 μm in order to provide the required mechanical strength. This strut thickness is similar to the level used within early BMS and DES designs and some of the clinical benefits seen with thinner struts may therefore have been sacrificed within existing bioresorbable stents. Consequently, current research is focused on the investigation of new materials, processing techniques, and stent configurations that can allow strut thickness to be reduced. Such advances are also needed if deliverability is to be improved, such that it is more comparable to conventional DES. Current developments in this area are described in more detail by Sotomi et al. [44].

1.6 Summary of current state of the art and future perspective

We have seen that there have been a great many innovations in stent design in the last two to three decades, such that the devices now available clinically are quite distinct from the very early-generation stents. These innovations have helped make stents a key treatment for advanced forms of coronary heart disease. Globally, it is estimated that over 6 million stents are implanted into patients each year [38]. Current state of the art stents that are now most commonly used in clinical practice incorporate advanced stent platforms, drugs, and coating technologies. It is clear that these developments have improved clinical outcomes, with the introduction of bare metal stents and then first-generation DES yielding the greatest positive impacts. However, the effect of more recent advances appears smaller. Indeed, a recent analysis by Byrne et al. [38], based on a systematic review of large scale randomized clinical trial data for CE-approved coronary stents between 2002 and 2013, has shown that median 9- to 12-month target-lesion revascularization rates have fallen from 12.32% with bare metal stents to 4.34% with so-called early-generation DES (Cypher, Taxus, Taxus Element and Endeavour), representing a considerable improvement in clinical performance. However, revascularization rates with new-generation DES (Xience, Promus and Promus Element, Resolute, Biomatrix, Nobori, Yukon Choice PC and Yukon Choice (polymer-free)) were found to be 2.91%, a relatively modest improvement considering the vast research budgets that have been spent on the introduction of these newer devices. The relatively low rates of revascularization with DES do of course make it difficult to demonstrate further improvement and it is therefore tempting to conclude that we have reached a plateau in device performance. However, to do so would be to accept that around 6% of patients will have to return to hospital for a repeat revascularization within 2 years, with this rate known to be more than double for patients with multivessel disease and in the more complex lesions that are being treated more commonly now [2,3]. Similarly, although the latest DES and dual antiplatelet strategies have reduced stent thrombosis rates to around 0.5% [38], they have not completely removed this problem and the high mortality associated with this condition means that this remains an important clinical challenge. With such a wide array of advanced stents now available to clinicians and more in development, there will likely be greater opportunity for clinicians to increasingly tailor their treatments to individual patients. The development of more advanced in silico models has already enhanced basic understanding of the performance of first-generation DES that have helped shape the design of existing devices [106]. Further advances in such modeling may reveal further insights into how stent performance is related to patient and lesion-specific differences [94] and therefore open up the more personalized treatment approaches that leading clinicians are increasingly calling for [107,108]. Further, advances in technology will also continue to have an important role to play, with fierce competition within the stent market providing ongoing stimulus for such developments. Although recent clinical results from the use of first-generation bioresorbable stents have been somewhat disappointing [103], current research into the development of second-generation versions and better deployment techniques may yield the improvements originally envisaged [44,100]. Enhanced understanding of the interactions between the stent surface and the surrounding biological tissue will also be crucial to the development of better devices, with modified surfaces and more advanced biologically active coatings likely to be increasingly important features of stent design in future.

References

[1] M. Nichols, N. Townsend, R. Luengo-Fernandez, J. Leal, A. Gray, P. Scarborough, M. Rayner, European Cardiovascular Disease Statistics 2012, European Heart Network/European Society of Cardiology, Brussels/Sophia Antipolis, 2012.

[2] M.E. Farkouh, M. Domanski, L.A. Sleeper, F.S. Siami, G. Dangas, M. Mack, M. Yang, D.J. Cohen, Y. Rosenberg, S.D. Solomon, A.S. Desai, B.J. Gersh, E.A. Magnuson, A. Lansky, R. Boineau, J. Weinberger, K. Ramanathan, J.E. Sousa, J. Rankin, B. Bhargava, J. Buse, W. Hueb, C.R. Smith, V. Muratov, S. Bansilal, S. King 3rd, M. Bertrand, V. Fuster, Strategies for multivessel revascularization in patients with diabetes, N. Engl. J. Med. 367 (25) (2012) 2375–2384.

[3] P.W. Serruys, M.C. Morice, A.P. Kappetein, A. Colombo, D.R. Holmes, M.J. Mack, E. Stahle, T.E. Feldman, M. van den Brand, E.J. Bass, N. Van Dyck, K. Leadley, K.D. Dawkins, F.W. Mohr, Percutaneous coronary intervention versus coronary-artery bypass grafting for severe coronary artery disease, N. Engl. J. Med. 360 (10) (2009) 961–972.

[4] N. Townsend, P. Bhatnagar, E. Wilkins, K. Wickramasinghe, M. Rayner, Cardiovascular Disease Statistics, British Heart Foundation, London, 2015.

[5] A.R. Gruntzig, A. Senning, W.E. Siegenthaler, Nonoperative dilatation of coronary-artery stenosis: percutaneous transluminal coronary angioplasty, N. Engl. J. Med. 301 (2) (1979) 61–68.

[6] R.L. Mueller, T.A. Sanborn, The history of interventional cardiology: cardiac catheterization, angioplasty, and related interventions, Am. Heart J. 129 (1) (1995) 146–172.

[7] M. Hamon, C. Bauters, E.P. McFadden, N. Wernert, J.M. LaBlanche, B. Dupuis, M.E. Bertrand, Restenosis after coronary angioplasty, Eur. Heart J. 16 (Suppl. I) (1995) 33–48.
[8] S.M. Garas, P. Huber, N.A. Scott, Overview of therapies for prevention of restenosis after coronary interventions, Pharmacol. Ther. 92 (2–3) (2001) 165–178.
[9] U. Sigwart, J. Puel, V. Mirkovitch, F. Joffre, L. Kappenberger, Intravascular stents to prevent occlusion and restenosis after transluminal angioplasty, N. Engl. J Med. 316 (12) (1987) 701–706.
[10] A. Betriu, M. Masotti, A. Serra, J. Alonso, F. Fernández-Avilés, F. Gimeno, T. Colman, J. Zueco, J.L. Delcan, E. García, J. Calabuig, Randomized comparison of coronary stent implantation and balloon angioplasty in the treatment of de novo coronary artery lesions (START): a four-year follow-up, J. Am. Coll. Cardiol. 34 (5) (1999) 1498–1506.
[11] D.L. Fischman, M.B. Leon, D.S. Baim, R.A. Schatz, M.P. Savage, I. Penn, K. Detre, L. Veltri, D. Ricci, M. Nobuyoshi, et al., A randomized comparison of coronary-stent placement and balloon angioplasty in the treatment of coronary artery disease. Stent restenosis study investigators, N. Engl. J. Med. 331 (8) (1994) 496–501.
[12] P.W. Serruys, P. de Jaegere, F. Kiemeneij, C. Macaya, W. Rutsch, G. Heyndrickx, H. Emanuelsson, J. Marco, V. Legrand, P. Materne, et al., A comparison of balloon-expandable-stent implantation with balloon angioplasty in patients with coronary artery disease. Benestent study group, N. Engl. J. Med. 331 (8) (1994) 489–495.
[13] P.R. Moreno, I.F. Palacios, M.N. Leon, J. Rhodes, V.N. Fuster, J.T. Fallon, Histopathologic comparison of human coronary in-stent and post-balloon angioplasty restenotic tissue, Am. J. Cardiol. 84 (4) (1999) 462–466.
[14] E. Van Belle, C. Bauters, E. Hubert, J.C. Bodart, K. Abolmaali, T. Meurice, E.P. McFadden, J.M. Lablanche, M.E. Bertrand, Restenosis rates in diabetic patients: a comparison of coronary stenting and balloon angioplasty in native coronary vessels, Circulation 96 (5) (1997) 1454–1460.
[15] H.C. Lowe, S.N. Oesterle, L.M. Khachigian, Coronary in-stent restenosis: current status and future strategies, J. Am. Coll. Cardiol. 39 (2) (2002) 183–193.
[16] P.H. Grewe, T. Deneke, A. Machraoui, J. Barmeyer, K.M. Muller, Acute and chronic tissue response to coronary stent implantation: pathologic findings in human specimen, J. Am. Coll. Cardiol. 35 (1) (2000) 157–163.
[17] A.K. Mitra, D.K. Agrawal, In stent restenosis: bane of the stent era, J. Clin. Pathol. 59 (3) (2006) 232–239.
[18] M.N. Babapulle, M.J. Eisenberg, Coated stents for the prevention of restenosis: part I, Circulation 106 (21) (2002) 2734–2740.
[19] R. Hoffmann, G.S. Mintz, Coronary in-stent restenosis—predictors, treatment and prevention, Eur. Heart J. 21 (21) (2000) 1739–1749.
[20] R.S. Schwartz, K.C. Huber, J.G. Murphy, W.D. Edwards, A.R. Camrud, R.E. Vlietstra, D.R. Holmes, Restenosis and the proportional neointimal response to coronary artery injury: results in a porcine model, J. Am. Coll. Cardiol. 19 (2) (1992) 267–274.
[21] A. Farb, G. Sangiorgi, A.J. Carter, V.M. Walley, W.D. Edwards, R.S. Schwartz, R. Virmani, Pathology of acute and chronic coronary stenting in humans, Circulation 99 (1) (1999) 44–52.
[22] R.S. Schwartz, N.A. Chronos, R. Virmani, Preclinical restenosis models and drug-eluting stents: still important, still much to learn, J. Am. Coll. Cardiol. 44 (7) (2004) 1373–1385.
[23] J. Gunn, N. Arnold, K.H. Chan, L. Shepherd, D.C. Cumberland, D.C. Crossman, Coronary artery stretch versus deep injury in the development of in-stent neointima, Heart 88 (4) (2002) 401–405.

[24] R. Kornowski, M.K. Hong, F.O. Tio, O. Bramwell, H. Wu, M.B. Leon, In-stent restenosis: contributions of inflammatory responses and arterial injury to neointimal hyperplasia, J. Am. Coll. Cardiol. 31 (1) (1998) 224–230.

[25] K.W. Lau, A. Johan, U. Sigwart, J.S. Hung, A stent is not just a stent: stent construction and design do matter in its clinical performance, Singapore Med. J. 45 (7) (2004) 305–311. quiz 312.

[26] A.C. Morton, D. Crossman, J. Gunn, The influence of physical stent parameters upon restenosis, Pathol. Biol. 52 (4) (2004) 196–205.

[27] J. Iqbal, J. Gunn, P.W. Serruys, Coronary stents: historical development, current status and future directions, Br. Med. Bull. 106 (2013) 193–211.

[28] B. O'Brien, W. Carroll, The evolution of cardiovascular stent materials and surfaces in response to clinical drivers: a review, Acta Biomater. 5 (4) (2009) 945–958.

[29] P. Serruys, B. Rensing, Handbook of Coronary Stents, fourth ed., Martin Dunitz Ltd, London, 2002.

[30] C. Rogers, E.R. Edelman, Endovascular stent design dictates experimental restenosis and thrombosis, Circulation 91 (12) (1995) 2995–3001.

[31] J.M. Garasic, E.R. Edelman, J.C. Squire, P. Seifert, M.S. Williams, C. Rogers, Stent and artery geometry determine intimal thickening independent of arterial injury, Circulation 101 (7) (2000) 812–818.

[32] A. Kastrati, J. Mehilli, J. Dirschinger, F. Dotzer, H. Schühlen, F.-J. Neumann, M. Fleckenstein, C. Pfafferott, M. Seyfarth, A. Schömig, Intracoronary stenting and angiographic results. Strut thickness effect on restenosis outcome (ISAR-STEREO) trial, Circulation 103 (23) (2001) 2816–2821.

[33] J.ü. Pache, A. Kastrati, J. Mehilli, H. Schühlen, F. Dotzer, J.ö. Hausleiter, M. Fleckenstein, F.-J. Neumann, U. Sattelberger, C. Schmitt, M. Müller, J. Dirschinger, A. Schömig, Intracoronary stenting and angiographic results: strut thickness effect on restenosis outcome (ISAR-STEREO-2) trial, J. Am. Coll. Cardiol. 41 (8) (2003) 1283–1288.

[34] O.F. Bertrand, R. Sipehia, R. Mongrain, J. Rodés, J.-C. Tardif, L. Bilodeau, G. Côté, M.G. Bourassa, Biocompatibility aspects of new stent technology, J. Am. Coll. Cardiol. 32 (3) (1998) 562–571.

[35] G. Mani, M.D. Feldman, D. Patel, C.M. Agrawal, Coronary stents: a materials perspective, Biomaterials 28 (9) (2007) 1689–1710.

[36] J. Pache, A. Dibra, J. Mehilli, J. Dirschinger, A. Schomig, A. Kastrati, Drug-eluting stents compared with thin-strut bare stents for the reduction of restenosis: a prospective, randomized trial, Eur. Heart J. 26 (13) (2005) 1262–1268.

[37] J. Iqbal, J. Chamberlain, S.E. Francis, J. Gunn, Role of animal models in coronary stenting, Ann. Biomed. Eng. 44 (2) (2016) 453–465.

[38] R.A. Byrne, P.W. Serruys, A. Baumbach, J. Escaned, J. Fajadet, S. James, M. Joner, S. Oktay, P. Juni, A. Kastrati, G. Sianos, G.G. Stefanini, W. Wijns, S. Windecker, Report of a european society of cardiology-european association of percutaneous cardiovascular interventions task force on the evaluation of coronary stents in europe: executive summary, Eur. Heart J. 36 (38) (2015) 2608–2620.

[39] H.M. Burt, W.L. Hunter, Drug-eluting stents: a multidisciplinary success story, Adv. Drug Deliv. Rev. 58 (3) (2006) 350–357.

[40] H.M. Burt, W.L. Hunter, Drug-eluting stents: an innovative multidisciplinary drug delivery platform, Adv. Drug Deliv. Rev. 58 (3) (2006) 345–346.

[41] T. Htay, M.W. Liu, Drug-eluting stent: a review and update, Vasc. Health Risk Manag. 1 (4) (2005) 263–276.

[42] C.W. Hwang, D. Wu, E.R. Edelman, Physiological transport forces govern drug distribution for stent-based delivery, Circulation 104 (5) (2001) 600–605.

[43] C. Yang, H.M. Burt, Drug-eluting stents: factors governing local pharmacokinetics, Adv. Drug Deliv. Rev. 58 (3) (2006) 402–411.
[44] Y. Sotomi, Y. Onuma, C. Collet, E. Tenekecioglu, R. Virmani, N.S. Kleiman, P.W. Serruys, Bioresorbable scaffold: the emerging reality and future directions, Circ. Res. 120 (8) (2017) 1341–1352.
[45] M.G. Mennuni, P.A. Pagnotta, G.G. Stefanini, Coronary stents: the impact of technological advances on clinical outcomes, Ann. Biomed. Eng. 44 (2) (2016) 488–496.
[46] W. Khan, S. Farah, A.J. Domb, Drug eluting stents: developments and current status, J. Control. Release 161 (2) (2012) 703–712.
[47] C. McCormick, The search for endothelium friendly stents, Med. Device Technol. 18 (3) (2007). 30, 32, 33.
[48] J. Watt, R. Wadsworth, S. Kennedy, K.G. Oldroyd, Pro-healing drug-eluting stents: a role for antioxidants? Clin. Sci. (Lond.) 114 (4) (2008) 265–273.
[49] E. Alt, I. Haehnel, C. Beilharz, K. Prietzel, D. Preter, A. Stemberger, T. Fliedner, W. Erhardt, A. Schomig, Inhibition of neointima formation after experimental coronary artery stenting: a new biodegradable stent coating releasing hirudin and the prostacyclin analogue iloprost, Circulation 101 (12) (2000) 1453–1458.
[50] C. McCormick, R.M. Wadsworth, R.L. Jones, S. Kennedy, Prostacyclin analogues: the next drug-eluting stent? Biochem. Soc. Trans. 35 (Pt 5) (2007) 910–911.
[51] D. Hou, H. Narciso, K. Kamdar, P. Zhang, B. Barclay, K.L. March, Stent-based nitric oxide delivery reducing neointimal proliferation in a porcine carotid overstretch injury model, Cardiovasc. Intervent. Radiol. 28 (1) (2005) 60–65.
[52] J.W. Moses, M.B. Leon, J.J. Popma, P.J. Fitzgerald, D.R. Holmes, C. O'Shaughnessy, R.P. Caputo, D.J. Kereiakes, D.O. Williams, P.S. Teirstein, J.L. Jaeger, R.E. Kuntz, Sirolimus-eluting stents versus standard stents in patients with stenosis in a native coronary artery, N. Engl. J. Med. 349 (14) (2003) 1315–1323.
[53] R. Falotico, T. Parker, R. Grishaber, S. Price, S.A. Cohen, C. Rogers, NEVO: a new generation of sirolimus-eluting coronary stent, EuroIntervention 5 (Suppl. F) (2009) F88–F93.
[54] G.W. Stone, S.G. Ellis, D.A. Cox, J. Hermiller, C. O'Shaughnessy, J.T. Mann, M. Turco, R. Caputo, P. Bergin, J. Greenberg, J.J. Popma, M.E. Russell, A polymer-based, paclitaxel-eluting stent in patients with coronary artery disease, N. Engl. J. Med. 350 (3) (2004) 221–231.
[55] I. Iakovou, Thrombosis after stent implantation: how much of a problem is there? Future Cardiol. 4 (3) (2008) 261–267.
[56] D.R. Holmes Jr., D.J. Kereiakes, S. Garg, P.W. Serruys, G.J. Dehmer, S.G. Ellis, D.O. Williams, T. Kimura, D.J. Moliterno, Stent thrombosis, J. Am. Coll. Cardiol. 56 (17) (2010) 1357–1365.
[57] G. Nakazawa, Stent thrombosis of drug eluting stent: pathological perspective, J. Cardiol. 58 (2) (2011) 84–91.
[58] T.F. Lüscher, J. Steffel, F.R. Eberli, M. Joner, G. Nakazawa, F.C. Tanner, R. Virmani, Drug-eluting stent and coronary thrombosis, Biol. Mech. Clin. Implic. 115 (8) (2007) 1051–1058.
[59] M. Joner, A.V. Finn, A. Farb, E.K. Mont, F.D. Kolodgie, E. Ladich, R. Kutys, K. Skorija, H.K. Gold, R. Virmani, Pathology of drug-eluting stents in humans: delayed healing and late thrombotic risk, J. Am. Coll. Cardiol. 48 (1) (2006) 193–202.
[60] J. Kotani, M. Awata, S. Nanto, M. Uematsu, F. Oshima, H. Minamiguchi, G.S. Mintz, S. Nagata, Incomplete neointimal coverage of sirolimus-eluting stents: angioscopic findings, J. Am. Coll. Cardiol. 47 (10) (2006) 2108–2111.

[61] T.J. Parry, R. Brosius, R. Thyagarajan, D. Carter, D. Argentieri, R. Falotico, J. Siekierka, Drug-eluting stents: sirolimus and paclitaxel differentially affect cultured cells and injured arteries, Eur. J. Pharmacol. 524 (1–3) (2005) 19–29.

[62] S. Kennedy, R.M. Wadsworth, C.L. Wainwright, Effect of antiproliferative agents on vascular function in normal and in vitro balloon-injured porcine coronary arteries, Eur. J. Pharmacol. 481 (1) (2003) 101–107.

[63] S.H. Hofma, W.J. van der Giessen, B.M. van Dalen, P.A. Lemos, E.P. McFadden, G. Sianos, J.M. Ligthart, D. van Essen, P.J. de Feyter, P.W. Serruys, Indication of long-term endothelial dysfunction after sirolimus-eluting stent implantation, Eur. Heart J. 27 (2) (2006) 166–170.

[64] M. Togni, L. Raber, R. Cocchia, P. Wenaweser, S. Cook, S. Windecker, B. Meier, O.M. Hess, Local vascular dysfunction after coronary paclitaxel-eluting stent implantation, Int. J. Cardiol. 120 (2) (2007) 212–220.

[65] M. Togni, S. Windecker, R. Cocchia, P. Wenaweser, S. Cook, M. Billinger, B. Meier, O.M. Hess, Sirolimus-eluting stents associated with paradoxic coronary vasoconstriction, J. Am. Coll. Cardiol. 46 (2) (2005) 231–236.

[66] S.O. Marx, H. Totary-Jain, A.R. Marks, Vascular smooth muscle cell proliferation in restenosis, Circ. Cardiovasc. Interv. 4 (1) (2011) 104–111.

[67] R. Tang, S.Y. Chen, Smooth muscle-specific drug targets for next-generation drug-eluting stent, Expert Rev. Cardiovasc. Ther. 12 (1) (2014) 21–23.

[68] D.E. Kandzari, M.B. Leon, Overview of pharmacology and clinical trials program with the zotarolimus-eluting endeavor stent, J. Interv. Cardiol. 19 (5) (2006) 405–413.

[69] I. Sheiban, G. Villata, M. Bollati, D. Sillano, M. Lotrionte, G. Biondi-Zoccai, Next-generation drug-eluting stents in coronary artery disease: focus on everolimus-eluting stent (Xience V(®)), Vasc. Health Risk Manag. 4 (1) (2008) 31–38.

[70] P.W. Serruys, V. Farooq, B. Kalesan, T. de Vries, P. Buszman, A. Linke, T. Ischinger, V. Klauss, F. Eberli, W. Wijns, M.C. Morice, C. Di Mario, R. Corti, D. Antoni, H.Y. Sohn, P. Eerdmans, T. Rademaker-Havinga, G.-A. van Es, B. Meier, P. Jüni, S. Windecker, Improved safety and reduction in stent thrombosis associated with biodegradable polymer-based biolimus-eluting stents versus durable polymer-based sirolimus-eluting stents in patients with coronary artery disease: final 5-year report of the LEADERS (Limus Eluted From A Durable Versus ERodable Stent Coating) randomized, noninferiority trial, J. Am. Coll. Cardiol. Interv. 6 (8) (2013) 777–789.

[71] M.I. Papafaklis, Y.S. Chatzizisis, K.K. Naka, G.D. Giannoglou, L.K. Michalis, Drug-eluting stent restenosis: effect of drug type, release kinetics, hemodynamics and coating strategy, Pharmacol. Ther. 134 (1) (2012) 43–53.

[72] R.S. Schwartz, E. Edelman, R. Virmani, A. Carter, J.F. Granada, G.L. Kaluza, N.A.F. Chronos, K.A. Robinson, R. Waksman, J. Weinberger, G.J. Wilson, R.L. Wilensky, Drug-eluting stents in preclinical studies: updated consensus recommendations for preclinical evaluation, Circ. Cardiovasc. Inter. 1 (2) (2008) 143–153.

[73] S. Venkatraman, F. Boey, Release profiles in drug-eluting stents: Issues and uncertainties, J. Control. Release 120 (3) (2007) 149–160.

[74] R. Virmani, F.D. Kolodgie, A. Farb, A. Lafont, Drug eluting stents: are human and animal studies comparable? Heart 89 (2) (2003) 133–138.

[75] J. Hausleiter, A. Kastrati, R. Wessely, A. Dibra, J. Mehilli, T. Schratzenstaller, I. Graf, M. Renke-Gluszko, B. Behnisch, J. Dirschinger, E. Wintermantel, A. Schömig, Prevention of restenosis by a novel drug-eluting stent system with a dose-adjustable, polymer-free, on-site stent coating, Eur. Heart J. 26 (15) (2005) 1475–1481.

[76] F. Bozsak, J.M. Chomaz, A.I. Barakat, Modeling the transport of drugs eluted from stents: physical phenomena driving drug distribution in the arterial wall, Biomech. Model. Mechanobiol. 13 (2) (2014) 327–347.
[77] S. Commandeur, H.M.M. Van Beusekom, W.J. Van Der Giessen, Polymers, drug release, and drug-eluting stents, J. Interv. Cardiol. 19 (6) (2006) 500–506.
[78] G. Acharya, K. Park, Mechanisms of controlled drug release from drug-eluting stents, Adv. Drug Deliv. Rev. 58 (3) (2006) 387–401.
[79] R. Virmani, G. Guagliumi, A. Farb, G. Musumeci, N. Grieco, T. Motta, L. Mihalcsik, M. Tespili, O. Valsecchi, F.D. Kolodgie, Localized hypersensitivity and late coronary thrombosis secondary to a sirolimus-eluting stent, Should we be cautious? Circulation 109 (6) (2004) 701–705.
[80] A. Garcia-Touchard, S.E. Burke, J.L. Toner, K. Cromack, R.S. Schwartz, Zotarolimus-eluting stents reduce experimental coronary artery neointimal hyperplasia after 4 weeks, Eur. Heart J. 27 (8) (2006) 988–993.
[81] N. Nikam, T.B. Steinberg, D.H. Steinberg, Advances in stent technologies and their effect on clinical efficacy and safety, Med. Devices (Auckl.) 7 (2014) 165–178.
[82] S. Brugaletta, F. Burzotta, M. Sabate, Zotarolimus for the treatment of coronary artery disease: pathophysiology, DES design, clinical evaluation and future perspective, Expert Opin. Pharmacother. 10 (6) (2009) 1047–1058.
[83] J. Bennett, C. Dubois, A novel platinum chromium everolimus-eluting stent for the treatment of coronary artery disease, Biol. Targets Ther. 7 (2013) 149–159.
[84] R.A. Byrne, M. Joner, A. Kastrati, Polymer coatings and delayed arterial healing following drug-eluting stent implantation, Minerva Cardioangiol. 57 (5) (2009) 567–584.
[85] A. Finkelstein, D. McClean, S. Kar, K. Takizawa, K. Varghese, N. Baek, K. Park, M.C. Fishbein, R. Makkar, F. Litvack, N.L. Eigler, Local drug delivery via a coronary stent with programmable release pharmacokinetics, Circulation 107 (5) (2003) 777–784.
[85a] F. Alexis, S.S. Venkatraman, S.K. Rath, F. Boey, In vitro study of release mechanisms of paclitaxel and rapamycin from drug-incorporated biodegradable stent matrices, J. Control. Release 98 (1) (2004) 67–74.
[86] B. O'Brien, H. Zafar, A. Ibrahim, J. Zafar, F. Sharif, Coronary stent materials and coatings: a technology and performance update, Ann. Biomed. Eng. 44 (2) (2016) 523–535.
[87] K. Steigerwald, S. Merl, A. Kastrati, A. Wieczorek, M. Vorpahl, R. Mannhold, M. Vogeser, J. Hausleiter, M. Joner, A. Schomig, R. Wessely, The pre-clinical assessment of rapamycin-eluting, durable polymer-free stent coating concepts, Biomaterials 30 (4) (2009) 632–637.
[88] A. Dibra, A. Kastrati, J. Mehilli, J. Pache, R. von Oepen, J. Dirschinger, A. Schomig, Influence of stent surface topography on the outcomes of patients undergoing coronary stenting: a randomized double-blind controlled trial, Catheter. Cardiovasc. Interv. 65 (3) (2005) 374–380.
[89] R.A. Byrne, A. Kastrati, S. Massberg, A. Wieczorek, K.-L. Laugwitz, M. Hadamitzky, S. Schulz, J. Pache, M. Fusaro, J. Hausleiter, A. Schömig, J. Mehilli, Biodegradable polymer versus permanent polymer drug-eluting stents and everolimus- versus sirolimus-eluting stents in patients with coronary artery disease: 3-year outcomes from a randomized clinical trial, J. Am. Coll. Cardiol. 58 (13) (2011) 1325–1331.
[90] G.G. Stefanini, R.A. Byrne, P.W. Serruys, A. de Waha, B. Meier, S. Massberg, P. Juni, A. Schomig, S. Windecker, A. Kastrati, Biodegradable polymer drug-eluting stents reduce the risk of stent thrombosis at 4 years in patients undergoing percutaneous coronary intervention: a pooled analysis of individual patient data from the ISAR-TEST 3, ISAR-TEST 4, and LEADERS randomized trials, Eur. Heart J. 33 (10) (2012) 1214–1222.

[91] S. McGinty, T.T.N. Vo, M. Meere, S. McKee, C. McCormick, Some design considerations for polymer-free drug-eluting stents: a mathematical approach, Acta Biomater. 18 (2015) 213–225.
[92] I. Tsujino, J. Ako, Y. Honda, P.J. Fitzgerald, Drug delivery via nano-, micro and macroporous coronary stent surfaces, Exp. Opin. Drug Deliv. 4 (3) (2007) 287–295.
[93] R. Wessely, J. Hausleiter, C. Michaelis, B. Jaschke, M. Vogeser, S. Milz, B. Behnisch, T. Schratzenstaller, M. Renke-Gluszko, M. Stover, E. Wintermantel, A. Kastrati, A. Schomig, Inhibition of neointima formation by a novel drug-eluting stent system that allows for dose-adjustable, multiple, and on-site stent coating, Arterioscler. Thromb. Vasc. Biol. 25 (4) (2005) 748–753.
[94] C.M. McKittrick, S. Kennedy, K.G. Oldroyd, S. McGinty, C. McCormick, Modelling the impact of atherosclerosis on drug release and distribution from coronary stents, Ann. Biomed. Eng. 44 (2) (2016) 477–487.
[95] R.A. Costa, A. Abizaid, R. Mehran, J. Schofer, G.C. Schuler, K.E. Hauptmann, M.A. Magalhães, H. Parise, E. Grube, Polymer-free biolimus a9-coated stents in the treatment of de novo coronary lesions: 4- and 12-month angiographic follow-up and final 5-year clinical outcomes of the prospective, multicenter biofreedom FIM clinical trial, J. Am. Coll. Cardiol. Interv. 9 (1) (2016) 51–64.
[96] P. Urban, I.T. Meredith, A. Abizaid, S.J. Pocock, D. Carrié, C. Naber, J. Lipiecki, G. Richardt, A. Iñiguez, P. Brunel, M. Valdes-Chavarri, P. Garot, S. Talwar, J. Berland, M. Abdellaoui, F. Eberli, K. Oldroyd, R. Zambahari, J. Gregson, S. Greene, H.-P. Stoll, M.-C. Morice, Polymer-free drug-coated coronary stents in patients at high bleeding risk, N. Engl. J. Med. 373 (21) (2015) 2038–2047.
[97] J.R. Costa Jr., A. Abizaid, R. Costa, F. Feres, L.F. Tanajura, A. Abizaid, G. Maldonado, R. Staico, D. Siqueira, A.G.M.R. Sousa, R. Bonan, J.E. Sousa, 1-year results of the hydroxyapatite polymer-free sirolimus-eluting stent for the treatment of single de novo coronary lesions: the VESTASYNC I trial, J. Am. Coll. Cardiol. Interv. 2 (5) (2009) 422–427.
[98] V. Demidov, D. Currie, J. Wen, Patent watch: patent insight into polymer-free drug-eluting stents, Nat. Rev. Drug Discov. 16 (4) (2017) 230–231.
[99] M. Haude, S.W.L. Lee, S.G. Worthley, S. Silber, S. Verheye, S. Erbs, M.A. Rosli, R. Botelho, I. Meredith, K.H. Sim, P.R. Stella, H.-C. Tan, R. Whitbourn, S. Thambar, A. Abizaid, T.H. Koh, P. Den Heijer, H. Parise, E. Cristea, A. Maehara, R. Mehran, The REMEDEE trial: a randomized comparison of a combination sirolimus-eluting endothelial progenitor cell capture stent with a paclitaxel-eluting stent, J. Am. Coll. Cardiol. Interv. 6 (4) (2013) 334–343.
[100] C. Indolfi, S. De Rosa, A. Colombo, Bioresorbable vascular scaffolds - basic concepts and clinical outcome, Nat. Rev. Cardiol. 13 (12) (2016) 719–729.
[101] A. Colombo, E. Karvouni, Biodegradable stents : "fulfilling the mission and stepping away", Circulation 102 (4) (2000) 371–373.
[102] S. Sorrentino, G. Giustino, R. Mehran, A.S. Kini, S.K. Sharma, M. Faggioni, S. Farhan, B. Vogel, C. Indolfi, G.D. Dangas, Everolimus-eluting bioresorbable scaffolds versus metallic everolimus-eluting stents: meta-analysis of randomized controlled trials. J. Am. Coll. Cardiol. (2017), https://doi.org/10.1016/j.jacc.2017.04.011 (in press).
[103] S. Bangalore, E.R. Edelman, D.L. Bhatt, First-generation bioresorbable vascular scaffolds: disappearing stents or disappearing evidence? J. Am. Coll. Cardiol. (2017), https://doi.org/10.1016/j.jacc.2017.04.012 (in press).
[104] A. Abizaid, R.A. Costa, J. Schofer, J. Ormiston, M. Maeng, B. Witzenbichler, R.V. Botelho, J.R. Costa Jr., D. Chamie, A.S. Abizaid, J.P. Castro, L. Morrison,

S. Toyloy, V. Bhat, J. Yan, S. Verheye, Serial multimodality imaging and 2-year clinical outcomes of the novel desolve novolimus-eluting bioresorbable coronary scaffold system for the treatment of single de novo coronary lesions, JACC Cardiovasc. Interv. 9 (6) (2016) 565–574.

[105] M. Haude, H. Ince, A. Abizaid, R. Toelg, P.A. Lemos, C. von Birgelen, E.H. Christiansen, W. Wijns, F.J. Neumann, C. Kaiser, E. Eeckhout, S.T. Lim, J. Escaned, Y. Onuma, H.M. Garcia-Garcia, R. Waksman, Sustained safety and performance of the second-generation drug-eluting absorbable metal scaffold in patients with de novo coronary lesions: 12-month clinical results and angiographic findings of the BIOSOLVE-II first-in-man trial, Eur. Heart J. 37 (35) (2016) 2701–2709.

[106] S. McGinty, A decade of modelling drug release from arterial stents, Math. Biosci. 257 (2014) 80–90.

[107] S. Garg, C. Bourantas, P.W. Serruys, New concepts in the design of drug-eluting coronary stents, Nat. Rev. Cardiol. 10 (5) (2013) 248–260.

[108] K. Kolandaivelu, B.B. Leiden, E.R. Edelman, Predicting response to endovascular therapies: dissecting the roles of local lesion complexity, systemic comorbidity, and clinical uncertainty, J. Biomech. 47 (4) (2014) 908–921.

Further Reading

[1] F.G.P. Welt, C. Rogers, Inflammation and restenosis in the stent era, Arterioscler. Thromb. Vasc. Biol. 22 (2002) 1769–1776.

Fundamentals of bare-metal stents

2

A.R. Saraf, S.P. Yadav
Dr. Babasaheb Ambedkar Technological University, Lonere, India

2.1 Clinical study of bare-metal stents

The historical study of stents into clinical practice suggests that the balloon angioplasty method of treatment was the only percutaneous available for atherosclerotic cardiovascular disease. The primary objective for the introduction of bare-metal stents (BMS) was to overcome the risk of acute restenosis that is prevailed by the balloon angioplasty that causes recoil, thrombosis, neointimal (NI) hyperplasia, or constructive remodeling. In 1987, the stent technology was first introduced and since then due to its high medical impact has constantly progressed from the fundamental bare-metal stents to the ongoing models of biodegradable stents. The porous and meshed structure of the stents employs resistance to corrosion materials like Co-Cr alloys, Ni-Ti alloys, and SS316L. However, while designing the tubular space, it is necessary to consider the typical diameter of large coronary artery that is generally about 2–5 mm with a vessel thickness of 0.5–1 mm. The key properties of the designed stents are the following: They inhibit positive remodeling and vasomotor and also prevent acute recoil due to porous construction, but simultaneously, it is prone to late thrombosis. The invention of drug-eluting stents (DES) reduced NI proliferation and also clinical restenosis, simultaneously suppressing the smooth muscle cell migration leading to less NI proliferation and restenosis. The better performance of the drug-eluting stents in terms of increase in the frequency of late stent thrombosis and degradation of drug-containing polymer in comparison with bare-metal stents (BMS) was also accompanied by the delayed healing [1].

The principle of biodegradable stents (BDS) is that they lose their scaffolding ability followed by complete degradation of the stent material making it possible for late positive remodeling, biological vasomotion, and its growth.

For a particular time instant, the mechanical varsity is unknown of a stent, whereas the mechanical scaffolding is required for the first 6 months only as negative remodeling is limited to this period. It consumes another 6 months for the healing that contains the vessel to be remodeled, which provides the desired degradation rate. The mechanical strength of the stent should remain constant for the initial 6 months, after which the gradual degradation and loss in its scaffolding ability is preferred. The remodeling process terminates at the end of 12 months that concludes with small traces of stent leftovers [2]. The arterial walls are predicted to be supported by the biodegradable stents during the process of remodeling beyond which degradation is preferred due to which the damage of late stent thrombosis is avoided. Also, it is not obligatory to consider the extended antiplatelet therapy. Igaki-Tamai stent was the first degradable polymer stent implanted in humans that was raised from poly-L-lactic acid.

Functionalized Cardiovascular Stents. https:/doi.org/10.1016/B978-0-08-100496-8.00002-0

Magnesium (Mg) stents that are degradable are applied for critical limb ischemia, tracheal, esophageal, etc. Support to mechanical scaffolding and early recoil are the limitations of balloon angioplasty [3].

2.2 Complimentary manufacturing of bare-metal stents

Manufacturing of biodegradable stents involves either polymers or metals. Polymeric materials composed of polylactic and polyglycolic acid copolymers or polycaprolactone were investigated initially. There are many limitations in the application of polymers that include high response to inflammation and poor mechanical properties like lower ultimate tensile strength (UTS), lower Young's modulus, or higher elastic recoil. In order to attain similar strength from polymer stents, thicker polymeric walls would be required that in turn increases the risk of side-branch occlusion; also, the polymer stents consume a longer period for its complete degradation. Iron (Fe)- or magnesium (Mg)-based stents or bare-metal biodegradable stents have the preferred mechanical properties for manufacturing of stents [4]. The prominent materials for stents include iron (Fe) having Young's modulus of 200 GPa with elongation ability of 40% and also SS316L with Young's modulus of 193 GPa and elongation ability of 40%. The degradation of pure iron stents requires comparatively more time while alloying the stent with manganese which lead to induction of new lesions due to degraded products in the form of voluminous flakes at the degradation region [5]. Almost 300 cell reactions involve manganese in trace amount that exhibits noncarcinogenic property, and the content of physiological plasma ranges between 0.70 and 1.05 mmol/L, whereas the threshold of its toxicity lies as low between 2.5 and 3.5 mmol/L. Considering a large stent with a length of 30 mm with a diameter of 6 mm, the content of Mg would not exceed 450 mg, which is close to the recommended daily oral intake. Magnesium has many medical applications, while its high-dose infusion causes recruitment of collaterals and vasodilation during ischemia [6]. However, its antiarrhythmic properties considering its depletion lead to the development of atherosclerosis coronary, vasoconstriction, cardiac arrhythmia, and increase in blood pressure. The mechanical properties of Mg have similar trends but lower in magnitude as compared with Fe or SS316L, having the advantage that the complete degradation of its alloys in physiological environment is possible [7]. Additive to the physiological and biocompatibility advantages, a successful background of Mg in biomedical implants presents it as a genuine replacement to the traditional stents. Recently, testing of Mg stents in animals and clinical trials has yielded capable results for replacement [8].

Research on reproducible and controllable time required for degradation has been worked upon to make it sustainable for regular clinical application. Also, improvements in the production technology and quality of Mg were improved by alloying it. The microstructure and crystal structure plays an important role while deciding the corrosive and mechanical properties of a metal. The influence of alloy composition drastically affects the crystal structure, whereas the process of production and its post treatment results in its microstructure. It can be concluded that the decisive parameters for the property of the stents are alloying and its processing [9]. The applications of bare-metal stents in coronaries are typically used in small children and newborn, whereas for adults and adolescence children, peripheral stents are employed. Cobalt-chromium, stainless steel, and platinum iridium are the different stent materials, while at the site of the stent implantation, polytetrafluoroethylene (e-PTFE) is used for sealing the vessel and treating aneurysms [10]. Coronary stent, typically for the small vessels, is made up of stainless steel that extrudes an external diameter of 1.8 mm, which is crimped in the section of 1.05 mm of the balloon catheter, and its expansion is set at 3.0 mm. Pressure of 10–14 atm is required for a positioned stainless-steel stent that induces 20% plastic deformation to the strut of the stent [11].

Fig. 2.1 concludes the above effects in the literature of the various properties of stents both desirable and undesirable in clinical trials in a schematic. After the deployment of the stents, remodeling of the arterial wall causes mechanical stresses to be induced on its walls. The time for the process of remodeling consumes 6–12 months and a desired equilibrium position attained after which the presence of stent is not required. The presence of stents for a long term leads to chronic inflammation and late thrombosis and also causes problem of in-stent restenosis. The vessel growth is motorized by the degradation and the disappearance of the stents avoiding the necessity of adulthood stent dilatation. The MRI quality is most contributed by the paramagnetic properties of Mg stent without severe artifacts. In pediatric patients, established treatment for stenotic vessels and vascular malformations is caused by stent implantation. Most often, they are indicated in patients where balloon dilatation alone fails. Recurrent implantations of stent in pulmonary artery and the aorta of patients are observed in congenital heart disease that employs different material and sizes depending upon the patient's size and age. The indications are not given for most of the stents like the efficiency of working of an implanted stent in pediatric patient and an adult. The stented lesion stenosis with continuing growth of the patient does not occur in a bare-metal stent. Many stents exhibit constrained potentials for further expansion; hence, redilation of balloon expandable stents can only be considered as a possibility [12]. For a patient who is currently in the process of growth along with the vessel, removal of stents with the help of surgery is very important as it may resemble long-term effect on the patient specifically in younger-aged patients. The flow pattern of

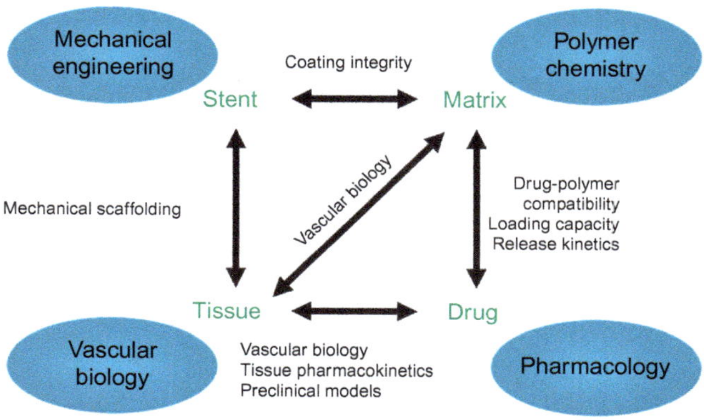

Fig. 2.1 Collaboration of clinical trials with mechanical properties.

the blood up- and downstream is changed by the induction of stents in the vessel due to the straightening of the vessel, and hence, the vessel growth is impacted by the resulting wall shear stress [13].

Implantation of stents in the heart valves or tissue patches for the treatment of stenosis would behave as a carrier of prostheses. Some implants have the possibility to grow that include decellularized homologous, decellularized xenologous, or living tissue-engineered autologous heart. The degradation rate of a degradable stent grows with the patient, and the implantation of these prostheses can be accomplished with minimum vessel invasion. The basic advantage of a bare metal over biodegradable stent is its potential to assist growth in pediatrics. Computed tomography (CT) and magnetic resonance imaging (MRI) are the vital indicators for congenital heart disease. Mg stents are not highlighted during fluoroscopy because of its low atomic weight [14]. Cardiac imaging of the stents is easily possible without causing artifact in CT or X-ray although the implantation process is complicated, which is a huge advantage to the bare-metal and biodegradable stents.

2.3 Validation of mechanical properties of metals for bare-metal stent

SS316L or cobalt-chromium (Co-Cr) poses superior mechanical properties as compared with Mg and its alloys although the capacity of plastic deformation is of 2% and Young's modulus is sufficient about 44 GPa; it appears to restrict the implantation of stent made up of Mg alloys. The hexagonal close-packed (hcp) crystals leads to less formability of Mg and its alloys when considered at room temperature as the active slip system of the crystal lies in the basal plane at room temperature. The normal of the basal plane is the only axis along which the deformation mechanism of twinning takes place. However, twinning is subjected to restricted deformation capability that is contrary to slip. The activation of mechanism is not possible once the whole crystallite undergoes twinning. The crystal structure is completely transformed from hcp to body-centered cubic (bcc) when alloying Mg with lithium (Li), which results in higher formability as the slip systems are more activate in bcc materials, and also, less energy is employed when compared with crystals packed with hcp arrangement.

The design requirements and properties must be satisfied in order to successfully transfer the stent above the catheter, reducing the restenosis and placing the stent at appropriate place and the deployment of stent:

1. High radial strength
   Radial strength plays a key role in preventing the recoil of stent by providing radial or structural support to the vessel.
2. Low elastic radial recoil
   In order to attain fixed final diameter of the stent to be manufactured when compared with the host artery diameter, the property of low elastic radial recoil is of importance.
3. Good flexibility
   For the proper placement of the stent in the tortuous geometry of blood vessels, good flexibility of the designed stent is essential in order to place it with the help of a catheter.

4. Excellent trackability

   The degree of ability of guiding catheter to follow a curved route to place the stent at its final destination is called its trackability.

   It depends on the following:

   (i) Low friction between adjacent environment and stent surface
   (ii) High flexibility of shaft

   Axial twisting of the catheter is minimized due to high axial deformation that is possible by balancing the above parameters, which are dependent variables on each other, and hence, a good trackability can be achieved.

5. Minimal stent profile

   During implantation to avoid the unnecessary disturbance of blood flow, it is desirable to have minimal profile of the stent.

6. Minimal foreshortening

   During the expansion of the vessel, precise placing of the stent is important; hence, it should possess minimum foreshortening.

7. Low longitudinal elastic recoil

   The undesirable effects of foreshortening and longitudinal recoil cause the shearing of stent. Denudation of the endothelial cells from the lumen of the vessel is caused during the expansion of the stent. Hence, low longitudinal elastic recoil is preferred.

8. Optimum scaffolding

   Rupture-free optimal coverage to the vessel to ensure original position is not regained by plaque. In order to reduce aggressive thrombotic response of the stent material, there should be minimum contact area between the surface of stent and vessel.

9. Requirements of stent material are the following:

   i. Biocompatibility: Human body must not be caused any adverse reaction or injury by the material of the stents, and hence, biocompatibility is very essential.
   ii. Radiopacity: For delivering the stent at the appropriate position, the property of radiopacity is very essential when viewed under fluoroscopic imagining.
   iii. Corrosion-resistant: Growth of passive oxide layers on the material of stent must be critically avoided, for which selected stent material must be corrosion-resistant.
   iv. Excellent fatigue properties: The blood flow induces cyclic stresses, and hence, due to the application of this cyclic load, the possibility of fatigue failure in the material drastically increases. The selection of the stent material is such that it can withstand a minimum of 380 million cyclic load means up to 10 years.

   The above properties depend upon the following variables:

1. Selection of material
2. Cross section and dimensions of the strut
3. The fabrication method used for stent
4. Number of axial and circumferential iterating elements and their geometry

2.4 Material selection

Grain size and orientation along with the microstructure and crystal structure have a major influence on the mechanical properties of metals. The grain size of the metal decides the stress required to deform it. Grain boundaries in large quantities present more obstacles for slip; hence, smaller grains result in higher stress requirement and more obstacles. The slip rises in materials with hcp structure when the number of grains increases also dilating the orientation of the crystals. The strength of the material increases with loss in its plasticity. The possibility of grain boundary slip is more likely in case of smaller grain sizes. The plasticity of materials with hcp structure is mainly caused due to the basal plane orientation, and the preferred deformation directions can accommodate the required orientation of the microstructures. The preparation of stent precursors employs diverse procedures for its production that include drawing and extrusion to form wires or tubes [15]. Desired microstructure orientation can be achieved by controlling the parameters. Along the grain boundaries, the presence of line-shaped particles can be noticed. Both strength and resistance to corrosion can be increased by alloying rare earth metal with Mg.

High modulus of elasticity for minimum recoil is considered as the ideal material for balloon expandable stent accompanied by high strength and allows yield stress.

The ideal material for self-expandable stent should possess a high yield stress and low elasticity. Thus, characteristics of materials required for stent vary for different types of stent.

(I) Material for balloon expandable stent are the following:
  1. Stainless steel 316L
  2. Tantalum
  3. Martensitic nitinol
  4. Polymers
  5. Cobalt alloy

(II) Material for self-expanding stent are the following:
  1. Nickel-titanium
  2. Nitinol
  3. Cobalt alloy
  4. Full hard (stainless steel)

2.4.1 Iron and its alloys

The decomposition of proteins, lipids, and DNA mutilation in human body takes place due to the presence of enzymes in large quantity that primarily consists of iron and proteins. The process of decomposition occurs due to the reactivity of iron and proteins to the molecules of oxygen-producing species that are reactive Fenton reaction. Also the transport, reduction of ribonucleotides and dinitrogen, storage and activation of molecular oxygen, etc. in the human body are assisted by the presence of iron in human body. As the inhibition of metabolic activities is beneficial in proliferation of endothelial cell in humans, it is very necessary for the involvement of pure Fe for the elution medium with concentration higher than 50 g/mL regardless of incubation time. Smooth muscle cell growth seizes with the excessive presence of Fe ions that was reported by an in vitro study. However, in terms of preventing restenosis in stent application, this seizure can be viewed as positive. A case study up to 18 months on the in vivo implantation of pure Fe stents on the rabbits of New Zealand with a descending aorta has revealed optimum results where significant neointimal proliferation, thromboembolic complications, pronounced inflammatory response, and systemic toxicity were not observed.

Modulus of elasticity of pure iron that is 211.4 GPa is the highest, which is five times as large as pure magnesium having modulus of elasticity of 41 GPa, whereas magnesium alloy shows slight improvement in elasticity going up to 44 GPa, while the modulus of elasticity of SS316L is 190 GPa. However, in terms of in vivo degradation rate, Fe has been rated very low, which means it consumes considerably large amount of time for degradation. Implanted device are ideally required to be nonferromagnetic; however, iron being ferromagnetic in nature constitutes a problem for its prime motive. Improvements in degradation rates of 0.44 mm/year and superior mechanical properties as compared with SS316L have been observed in newly developed Fe-based alloys. The technique to convert Fe into antiferromagnetic material by alloying it with 30–35 wt% of manganese gave it suitable austenitic alloy properties that made it compatible in the presence of magnetic field, which is produced during MRI acting as a diagnostic tool leading to the rapid growth and considered to be noninvasive in the prosthetics of medical imaging.

2.4.2 Magnesium and its alloys

There can be significant increase in the growth and the strength of bones as the bones have a substantial amount of magnesium present in them that contributes to its composition. Stabilization of RNA and DNA is also contributed by the presence of Mg that is a cofactor for several metabolic enzymes. For a normal adult, the average daily intake of magnesium is in the range of 300–400 mg that constitutes as the fourth most abundant cation in human body. In blood plasma, the amount of magnesium can be tolerated up to a comparatively elevated level of 85–121 mg/L. However, cardiac arrest, muscular paralysis, and hypotension respiratory distress can be caused due to the presence or increase in the level of Mg excessively. Contrarily, its deficiency leads to increased incidence of heart disease, susceptibility to oxidative stress, cell membrane dysfunction, and sometimes cancer. Although these problems due to its overconsumption prove unlikely as efficient, filtration of excretion in the urine with the assistance of the kidney takes place. Considering the feasibility of implantation in human bodies, the fabrication process of magnesium and its alloys is considered to be safe and feasible for the purpose [16]. The involvement of sterilized pure Mg and Mg-Ca alloys has proved its cytocompatibility in works that includes an Mg-hydroxyapatite composite test on its application by using derived cells of human bones, indirect contact cytotoxicity test using L-929 cells, and MG-63 plus RAW 264.7 cells. Different studies on cytotoxicity studies for Mg alloys were also reported [17].

As the density of magnesium, which ranges between 1.8 and 2 g/cm^3, is very near to that of human bones, it proves to be of extreme importance in orthopedic implant application that is the most targeted zone of utilizing Mg and its alloys in a supportive nature to the progressing field of biomedics. The degradation of pins implanted into the rabbit femoral shaft that were manufactured with Mg-Ca alloy was accomplished within 90 days, which terms the huge success of the implants made of this alloy and has delivered results in the formation of new bone tissue [18]. The bone cell activation in human body can also have been said to be supported by the implantation of magnesium stents. Since the rate of corrosion of magnesium is very swift, it provides a limitation to its usage in pure form in orthopedic patients that resembles a passive layer formation rate of 10–200 mm/year having a purity of 99.9% in 3% NaCl. The enhancement of the corrosion resistance has been the current most eluding research, and many efforts have been showcased, which include by incorporating coating of dicalcium phosphate dehydrate (DCPD), by alkali-heat treatment, and also by alloying.

2.4.3 Stainless steel 316L

The application of SS316L has been noticed as an advantageous material in the manufacturing of stents. SS316L is being prominently used for both bare metals and with the forecast of a coating material on the metal. The significance of this material lies under its underlining properties of outstanding resistance to corrosion and optimum mechanical properties, which assigns its applicability in the stated application. The most desired quality of fatigue strength is to provide subsistence strength of the application of cyclic load; also, the biocompatibility of SS316L biomaterial that offers good resistance to degradation has preferential selection when compared with the array of materials to be used for the fabrication of stents [19]. Being a prominent and a very enormously developed section of bare-metal stent contributing largely and successfully in biomedical applications in the past, stent heart valve, knee cups, fracture fixation, surgical instruments, etc. have found typical applications of SS316L. Hence, many manufacturing advances in the process of machining of biocompatible SS316L material have been and are consistently been explored recently. SS316L seamless tubes have been used in the fabrication of 3-D PCM of stent. Ultrasonic cleaning of the seamless tubes before the manufacturing of the stents removes contaminants like dirt, oil, and grease, which would affect the precision of machining of the stents [20]. For the manufacturing of stents by the process of photochemical machining, $FeCl_3$ has been considered as the most commonly used etchant for the material removal for precision applications.

Table 2.1 considers various metal composition for biodegradable metals taking into account their mechanical properties with SS316L revealing that the mechanical properties of SS316L is considered to be as a standard reference in developing new alloys, which consists of metallic biomaterials for most of the used medical grade alloy. It can be concluded that SS316L has proximity in its mechanical properties with pure Fe when compared with Mg alloys. This prefers Fe in high-strength applications, and also, ductility requirements are high such as coronary stents. But the property of ferromagnetism enables SS316L material to be preferred when compared with Fe, as the property of ferromagnetism is reduced considerably in SS316L material, which enables MRI in the implantation of the stents. Alloying Mg affects and improves the ductility when compared with most Fe alloys that can be accomplished by engaging techniques for advanced processing. A typical example of alloying Mg with slight amount of Li has the ability to modify the crystal structure from hexagonal to body-centered cubic, producing a significant rise in ductility [21]. However, when Mg was alloyed with lithium with weight percentage of 8.7, elongation observed was 52% due to which the drop in tensile strength brought it to 132 MPa. Human body can tolerate the alloy of Mg to a very limited set of metals that include Zn, Ca, and Mn, and

Table 2.1 Mechanical properties of bare-metal and biodegradable stents

Composition of metal (wt%)	Metall-urgy	Density (g/cm^3)	Magneti-city	YS (MPa)	UTS (MPa)	YM (GPA)	e (%)
WE43 Mg alloy	Hot extruded bar	1.84	NF	150	250	44	4
Mg, 3.7–4.3 Y, 2.4–4.4 Nd, 0.4–1 Zr, pure FEA 99.8 Fe	Annealed plate	7.87	FM	150	210	200	40
SS316L	Annealed plate	8	NF	190	490	193	40

e, maximum elongation; FM, ferromagnetic; NF, nonferromagnetic; UTS, ultimate tensile strength; YM, Young's modulus; YS, yield strength.

rare earth elements such as Zr and Y in very small traces can also be utilized. For easy and optimum regulation of OH and H_2 gas, a very slow degradation rate is preferred to allow its generation during degradation. Due to the toxic nature of effects of Zr, Mn, and Al on human body, caution must be taken while designing Mg alloy to avoid the release of alloying elements that might prove fatal for the patient. The deficiency of phosphate source in the human body can easily take place by deportation of aluminum ions that induce dementia, which binds with the inorganic phosphate consuming the phosphate source from the human body.

Table 2.2 presents the implantation of stents in animals, and the following considerations were made. Magnesium alloy stents possessed accelerated degradation, low thrombogenicity, and low-grade response to inflammation and lacked local or systemic toxicity, which intimate promising future. However, when comparing it with SS316L stent, the desired exhibition of prolonged degradation is desirable for neointimal proliferation that cannot be observed in Mg and its alloys stents, while continued studies (3 months) were also required to demonstrate the positive remodeling of VBT as assistant to stenting [22]. Also supplementary decrease in intimal hyperplasia and better lumen area was considered to be safe when compared with pure Fe stents. There is a risk of Parkinsonian syndrome due to alloying of Fe stents with manganese that can lead to neurotoxicity. The possibility of breast, liver, nasopharyngeal, and lung cancers is more prone in the presence of Zr as an alloying substance in Fe stents.

As derived from Fig. 2.2, manufacturing of biodegradable stent involves the determination of three design criteria for developing new alloys that include the following:

(1) Mechanical properties have similar trend and peaks of SS316L.
(2) There should be a fixed balance between the vessel remodeling period of 6–12 months which must be in close contrast to the degradation rate, and a complete degradation or disappearance of the implant during a reasonable period of time ranging between 12 and 24 months.
(3) During the degradation process, there must be no release of any toxic substances.

Table 2.2 Biodegradable metallic stents implantation in animal

Metal	Site of implantation	Conclusions
Fe	Rabbits of New Zealand implanted at descending aorta for 18 months	The lack of systemic or local toxicity, thrombogenicity was low, response to inflammation was mild, warranty in the acceleration of degradation
AE21	Pigs implanted at coronary artery for 56 days	Promising results were exhibited by Mg alloy stents; prolonged degradation for a prolonged period was anticipated
Fe	Minipig implanted at descending aorta for 12 months	In terms of stents of SS316L, similarity was observed in neointimal proliferation; toxicity of local or systemic kind was not observed; increase in the rate of degradation was required
WE43	Minipig or domestic implanted at coronary artery for 3 months	Safety accompanied with less formation of neointima was observed in Mg alloy stents; in order to verify positive remodeling, there is a need of long-term studies (13 months)
WE43	Domestic pig implanted at coronary artery for 28 days with VBT	Intimal hyperplasia was reduced by employing VBT as assistant to stenting; also, lumen area was improved in comparison with unaided stenting
Fe	Porcine implanted at coronary arteries for 28 days	Safe consideration of Fe stents

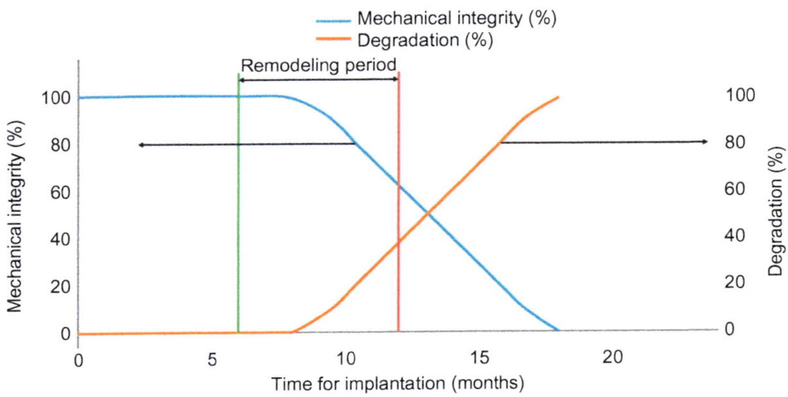

Fig. 2.2 Design criteria for stent material.

The mechanical properties of iron when compared with Mg and its alloys and SS316L were found to be superior in most of its regards. Enhancement in the properties of iron by thermomechanical treatment and proper alloying is also possible. Transformation of the magnetic properties of iron, that is, from ferromagnetic to nonmagnetic, is possible by considering the appropriated alloying element, thus assisting a better compatibility toward magnetic field produced during the procedure of MRI [23].

The austenite formation of the element that has the caliber to convert iron into nonmagnetic material was concluded upon extensive choice of manganese as most of other materials that could be employed to alloy with Fe to reduce its magnetic properties were toxic in nature, which may have hazardous effects in the degradation process. The phase diagram of Fe-Mn has been represented in Fig. 2.3. Alloying of iron and nickel has also been possible; however, the formation of carcinogenic compounds for human is quite possible with it. The presence of manganese in trace elemental form is very essential for the survival and proper functioning of the bodies of all mammals. Also any toxic effect to the cardiovascular system was not reported due to extensive binding counteract of plasma protein and due to the excessive presence of manganese [24].

The trend shows that the formation of austenitic phase represented by c initializes when manganese with 15 wt% is mixed with martensite (e) that is ~27 wt% of manganese and after which the presence of c should only be prevalent. The phases that contribute extensively to the nonmagnetic nature are represented by e and c phases that present the antiferromagnetic properties of Fe-Mn alloys. The analysis was performed for alloy consisting of 20–35 wt% Mn keeping into view its potential to be evaluated as biodegradable stent material.

The production of four Fe-Mn alloys with high purity of 99.98% of its elemental powders was observed that were, namely, Fe-20Mn, Fe-25Mn, Fe-30Mn, and Fe-35Mn. The fabrication of the high-purity homogenous material that provides porosity to the alloy that assists in the degradation control of the manufactured component was prepared with the process of powder metallurgy where the mixtures of powder were sintered and compacted and in a mixture of Ar 5% H_2 gas flowing at 1200°C for 3 h.

Application of two-step thermomechanical treatment, so as to obtain aligned porosity accompanied by high density, of sintered pellets included the following:

(a) Resintering cold at 1200°C for 2 h after rolling it to reduction in 50% of its thickness followed by furnace cooling.

(b) Resintering cold at 1200°C for 2 h after rolling it to reduction in 50% of its thickness followed by water quenching. The degradation rate was controlled by the aligned porosity that acted as a physical barrier.

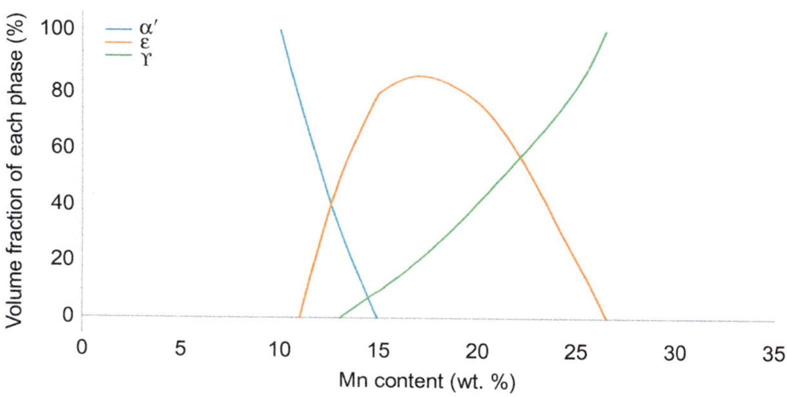

Fig. 2.3 Phase diagram of binary Fe-Mn system.

2.5 Finite element analysis of stents

In the finite element analysis of the component, whose prototype model is created consisting of the desired material and design, analysis of stress is performed to obtain specific results. Product designing and refinement of the existing design are possible quite easily with the observation of stress concentration and possible failure points during fatigue loading and during defined number of cycles. The redefinition of service condition leads to modification in existing products, and its structure is quite possible for the qualification of a product in its desired service conditions. Modification and optimization of the design parameters can be made with the help of finite element analysis of the design before the manufacturing and prediction in the failure of the final product can be made and altered. Figs. 2.4 and 2.5 show the design and meshing of the cardiovascular stent, respectively.

The feasibility in the simulated service conditions of the designed stent can be verified easily in the FE tool. The von Mises stresses, radial displacement, and changed blood flow velocity are critically evaluated, and results are obtained for the same that are the possible components of failures.

On the deployment pressure of 2 atm that in the average pressure acting on the vessel, in the FEA tool, the results of the static analysis of the stent evaluating the von Mises stresses were found to be more concentrated in the longitudinal direction at the central part of the stent. The results obtained are complimentary to the increased surface area contact of the stent with the arterial at its central part. At the ends of the stent are observed to have von Mises stresses, while there is an exponential decay of stresses from the central part toward the edges. Calibration of the radial displacement is also performed in a similar fashion. During the deployed condition, the maximum radial displacement is observed to be within the range for the standard deformation values of the artery. Simulation results of the designed stent for the evaluation of von Mises stresses and displacement counter are shown in Figs. 2.6 and 2.7.

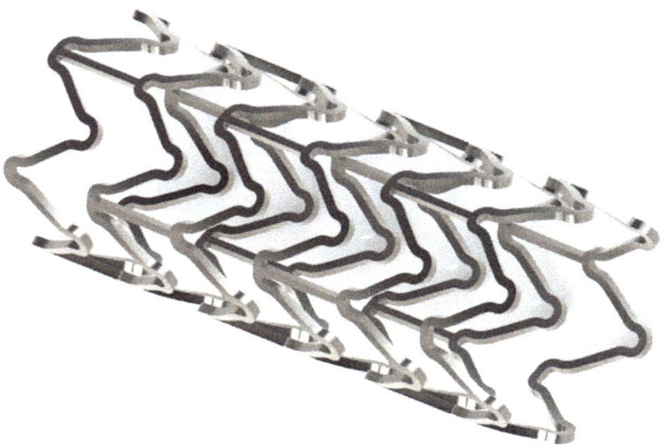

Fig. 2.4 Designed model of cardiovascular stent.

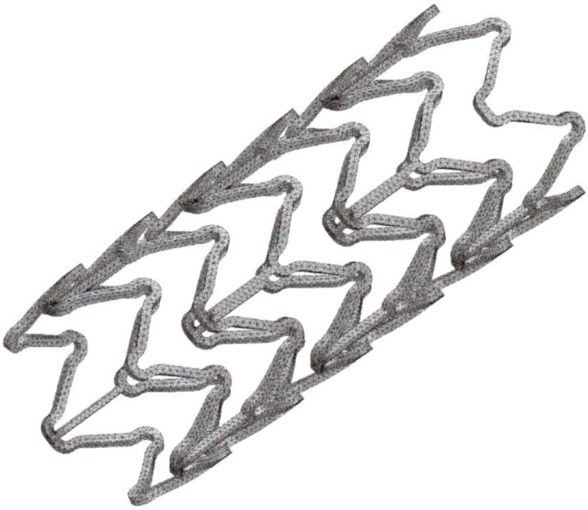

Fig. 2.5 Meshing of cardiovascular stent.

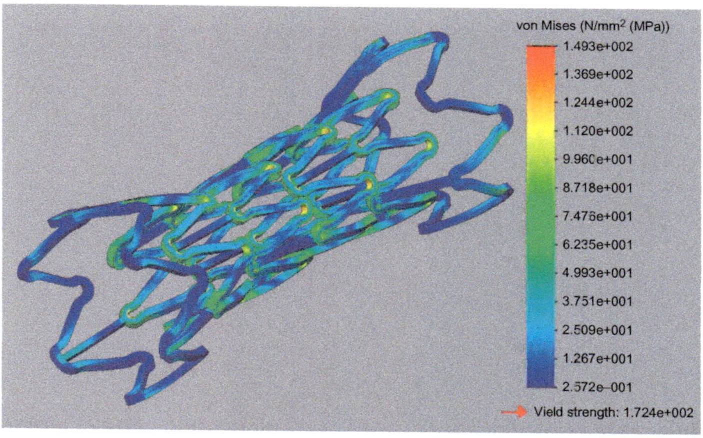

Fig. 2.6 FE simulation of von Mises stress distribution in the cardiovascular stent.

Blood flow velocity and buckling displacement are the two most relevant parameter preferred for the various stenting considerations while performing the dynamic analysis of the stent. Buckling analysis simulation is presented in Fig. 2.8. Circumferential pressure in the radial direction on the surface of the stent by the arterial wall is exerted simultaneously to the blood flow within the artery that exerts pressure on the internal surface of the stents. The pressure exerted at the internal and external wall of the stent combined with the posture of body in different conditions leads to the exhibition of buckling of the stents. For the simulation results of the finite element analysis, it was observed that maximum displacement due to buckling at the interior surface of the stent occurs where blood flow starts and gradually goes on decreasing in the blood flow direction along its length and attains a minimum value at the extreme ends.

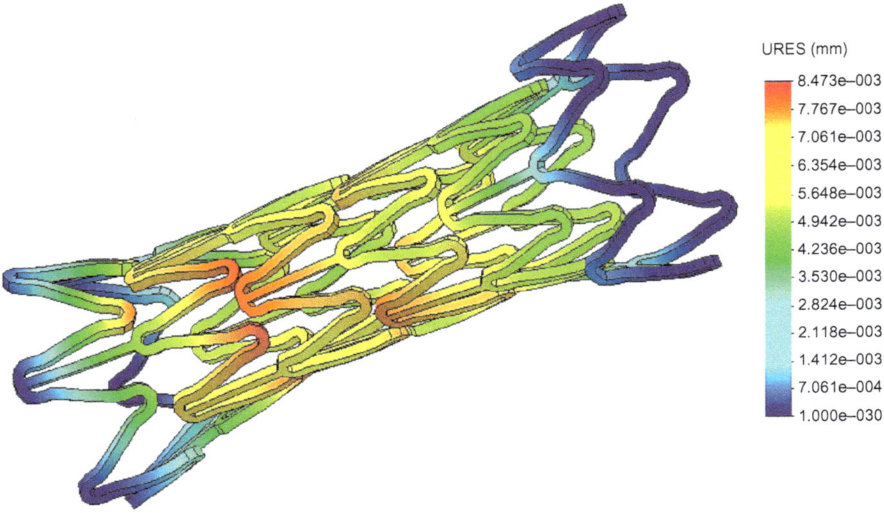

Fig. 2.7 FE simulation of radial displacement in the cardiovascular stent.

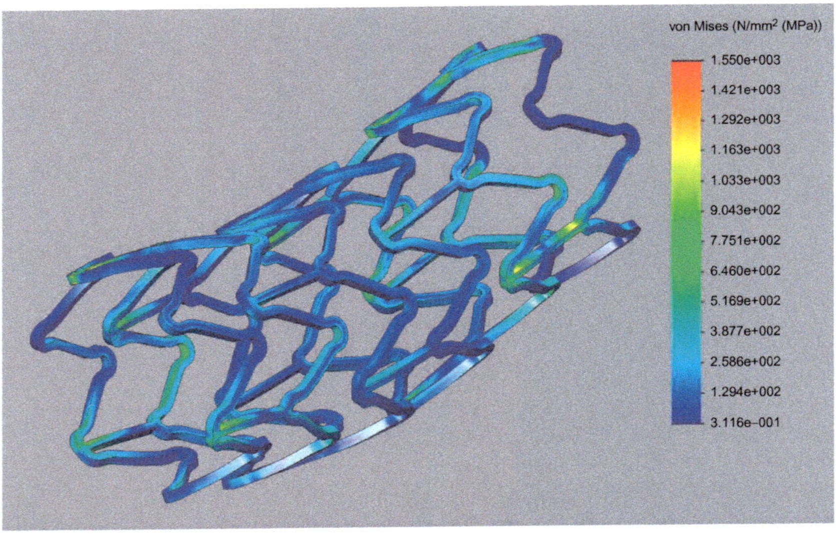

Fig. 2.8 FE buckling analysis in the cardiovascular stent.

Analysis report of the changes in the pattern in the blood flow reveals that the maximum blood flow is at the entry to the stent while a longitudinal directional decay is observed along the length of the stent. As there is friction along the internal surface of the stent that obstructs the blood flow, there is significant buckling that is presented in the simulation results.

The simulation analysis of model is implicated by the designing followed by the material selection procedure and later stressing the model to ultimate values with the help of finite element analysis and collaborating its results with the desired properties of the component. Thus, all these processes are computational having the feasibility to be redefined and modified design for any failure iterations. The designed stent needs to be checked with its mechanical behavior that presents the primary aim of this work. Length, width, thickness, and geometric distributions of each strut, number of struts, radius and length of stent, Poison's ratio, and Young's modulus of the material are the features that have been considered for the described study.

The simulated service condition is defined, and the feasibility of the stent designed has been implicated with finite element analysis in its dynamic conditions. Radial displacement and von Mises stresses are the desired results obtained from FEA.

The compression process has been studied during the static analysis with the help of von Mises stress and radial displacement. It can be observed from the illustration of the analysis represented in Fig. 2.6 that at the von Mises stresses are greater in magnitude at the curved portion, and it decays when moved away from the curved portion. The occurrence of such a trend can be easily be compensated at the struts whose compression is the major for the generation of stress in higher magnitude at the connectors and the curved portion. From Fig. 2.7, it is revealed that the radial displacement is minimum or can be said to be negligible at the ends of the stent while it increases exponentially as we move toward the central portion from the edges. The maximum radial displacement is observed at the center of the stent along its length however the value of deformation being well within the range for the deployed conditions of the standard deformation values of the artery. Radial displacement counter of the FE simulation for the designed stent is shown in Fig. 2.7.

The buckling analysis is very important to be considered when simulating the model in its dynamic conditions. Buckling causes the production of large deformation and also leads to failure of structures that is consequence of mechanical instability. Due to body movements during bending, there is an induction of twist, axial tension, and outer pressure leading to complex mechanical loading conditions on the veins and arteries by the application of lumen pressure and tissue tethering. Fig. 2.8 shows the simulation of buckling analysis. The outer wall of the stents has been acted upon the radially acting load due to pressure that is exerted upon it. However, this load is applied to the contact between the walls of the artery and the outer surface of stent. Also the stent walls are acted upon by an internal pressure in the radially outward direction due to the flow of blood in the artery. However, buckling is caused due to the deliverance of varied postures of human body alongside the internal and external pressure being acted that ultimately leads to the buckling of the stents. Lifestyle, diseases, genetics, diet (high-cholesterol diet), sex, and habits (smoking, drinking, etc.) are the various factors that have influential effect on the buckling of stents. As the strength of blood vessels decreases with the increasing age of the patient that causes atherosclerosis, calcium deposition with arterial wall, blood pressures, and decrease in strength of blood vessels in old age (after age of 30), it also leads to buckling of the stents. It also depends upon postangioplasty treatment. The results of the buckling analysis reveal that the human body condition can be sustained by the stents without any collapse. The details of buckling conditions in human body are twisting buckling (vertebral artery-head rotation, vein grafts, anastomoses, and coronary vessel cardiac contraction), lumen collapse (arteries stenosis, veins-surface compression, internal jugular vein catheter insertion, and coronary artery cardiac contraction), kinking (iliac arteries and internal carotid arteries), helical buckling(corkscrew collaterals and vein grafts), and bend buckling (vein retinopathy and vein grafts over length, hypertension, retinal arteries, and carotid artery aging). From the analysis, it is clear that the stent does not undergo any type of buckling.

2.6 Conclusions

The design and development of bare-metal stents have led to the revolution toward true biological solutions. The artery needs a mechanical support for its healing and remodeling that is provided by the bare-metal stents for over a period of time up to its complete cure. Stainless steel is considered as the elements for such applications as its mechanical properties and biocompatibility in human body possess ultimate assets for the application of stents in load bearing. The results of implantation of all the bare-metal stents including SS316L, iron, and magnesium demonstrate that even though the degradation rate of these stents is yet to be modified, the simulation and FEA analysis performed on its modeled structure of these materials have the adequate strength to withstand the load that is dispersed on its walls, which is a major setback leading to a clear obstruction in the development of alternative material like polymer. The degradation rate of magnesium was observed to be extremely rapid in comparison with SS316L eluting it to be used as a biodegradable stent. By altering the microstructure and composition of the alloying elements, there is a possibility of controlling the degradation rate of SS316L-, Fe-, and Mg-based stents. However, while designing the stents and making the righteous choice for its materials factors such as mechanical stability, reduction on neointimal hyperplasia and degradation rate must be considered. Modifications in the design of its strut and providing coatings of noncorrosive materials are the current advancements toward which the present-day research is concentrated upon. While designing the stents, it was observed that the radial profiles should be minimum along with adequate deliverability and minimum tissue contact to acknowledge the sophistication of new stent design. Also altering the material and its composition has proved to be beneficial by many researchers. The present-day research is also focused on endothelial repair by inductive scaffolding. The magnification of flexibility of stents by revolutionized design employing micro- and nanotechnology for its manufacturing has enhanced the ability of deploying the stent to its targeted site. Improvisation in blood-contacting properties and acceleration of endothelialization are the capacities under which the recent modifications take place. However, the scope of further building and modification in the bare-metal stents is likely to take place, but at the same time, they provide the benchmark for the various properties of the stents that a designer must keep in mind while designing the new era stents incorporating all the modifications.

References

[1] M. Bartosch, S. Schubert, F. Berger, Magnesium stents—fundamentals, biological implications and applications beyond coronary arteries, BioNanoMaterials 16 (1) (2015) 3–17.
[2] B. Sripal, S. Kumar, M. Fusaro, N. Amoroso, A.J. Kirtane, R.A. Byrne, D.O. Williams, J. Slater, D.E. Cutlip, F. Feit, Outcomes with various drug eluting or bare metal stents in patients with diabetes mellitus: mixed treatment comparison analysis of 22 844 patient years of follow-up from randomised trials, Br. Med. J. 345 (2012) 1–14.
[3] B. Sripal, B. Toklu, N. Amoroso, M. Fusaro, S. Kumar, E.L. Hannan, D.P. Faxon, F. Feit, Bare metal stents, durable polymer drug eluting stents, and biodegradable polymer drug eluting stents for coronary artery disease: mixed treatment comparison meta-analysis, Br. Med. J. 347 (2013) 1–20.
[4] P.K. Bowen, J. Drelich, J. Goldman, Zinc exhibits ideal physiological corrosion behavior for bioabsorbable stents, Adv. Mater. 25 (2013) 2577–2582.
[5] Y. Chen, Z. Xu, C. Smith, J. Sankar, Recent advances on the development of magnesium alloys for biodegradable implants, Acta Biomater. 10 (2014) 4561–4573.
[6] D.E. Cutlip, S. Windecker, R. Mehran, A. Boam, D.J. Cohen, G.A. van Es, P.G. Steg, M.A. Morel, L. Mauri, P. Vranckx, E. McFadden, A. Lansky, M. Hamon, M.W. Krucoff, P.W. Serruys, Clinical end points in coronary stent trials: a case for standardized definitions, J. Am. Heart Assoc. 115 (2007) 2344–2351.
[7] A. Drynda, N. Deinet, N. Braun, M. Peuster, Rare earth metals used in biodegradable magnesium-based stents do not interfere with proliferation of smooth muscle cells but do induce the upregulation of inflammatory genes, J. Biomed. Mater. Res. 91A (2009) 360–369.
[8] M.E. Hogg, B.G. Peterson, W.H. Pearce, M.D. Morasch, M.R. Kibbe, Bare metal stent infections: case report and review of the literature, J. Vasc. Surg. 46 (4) (2007) 813–820.
[9] M. Mantovani, D. Mantovani, Biodegradable metals for cardiovascular stent application: interests and new opportunities, Int. J. Mol. Sci. 12 (2011) 4250–4270.
[10] F. Jung, C. Wischke, A. Lendlein, Degradable, multifunctional cardiovascular implants: challenges and hurdles, MRS Bull. 35 (2010) 607–613.
[11] S.-H. Kang, K.W. Park, D.Y. Kang, W.H. Lim, K.T. Park, J.K. Han, H.J. Kang, B.K. Koo, B.H. Oh, Y.B. Park, D.E. Kandzari, D.J. Cohen, S.S. Hwang, H.S. Kim, Biodegradable-polymer drug-eluting stents vs. bare metal stents vs. durable-polymer drug-eluting stents: a systematic review and Bayesian approach network meta-analysis, Eur. Heart J. 35 (2014) 1147–1158.
[12] B. Lagerqvist, S.K. James, U. Stenestrand, J. Lindbäck, T. Nilsson, L. Wallentin, Long-term outcomes with drug-eluting stents versus bare-metal stents in Sweden, N. Engl. J. Med. 356 (10) (2007) 1009–1019.
[13] T.M. Jeewandara, S.G. Wise, M.K.C. Ng, Biocompatibility of coronary stents, Materials 7 (2014) 769–786.
[14] M. Moravej, F. Prima, M. Fiset, D. Mantovani, Electroformed iron as new biomaterial for degradable stents: development process and structure–properties relationship, Acta Biomater. 6 (2010) 1726–1735.
[15] A.J. Nordmann, M. Briel, H.C. Bucher, Mortality in randomized controlled trials comparing drug-eluting vs. bare metal stents in coronary artery disease: a meta-analysis, Eur. Heart J. 27 (2006) 2784–2814.
[16] C. Spaulding, J. Daemen, E. Boersma, D.E. Cutlip, P.W. Serruys, A pooled analysis of data comparing sirolimus-eluting stents with bare-metal stents, N. Engl. J. Med. 356 (10) (2007) 989–998.

[17] G.G. Stefanini, R.A. Byrne, P.W. Serruys, A. de Waha, B. Meier, S. Massberg, P. Jüni, A. Schömig, S. Windecker, A. Kastrati, Biodegradable polymer drug-eluting stents reduce the risk of stent thrombosis at 4 years in patients undergoing percutaneous coronary intervention: a pooled analysis of individual patient data from the ISAR-TEST 3, ISAR-TEST 4, and LEADERS randomized trials, Eur. Heart J. 33 (2012) 1214–1222.

[18] S. Christoph, S. Allemann, S. Wandel, A. Kastrati, M.C. Morice, A. Schomig, M.E. Pfisterer, G.W. Stone, M.B. Leon, J.S. de Lezo, J.J. Goy, S.J. Park, M. Sabate, M.J. Suttorp, H. Kelbaek, Drug eluting and bare metal stents in people with and without diabetes: collaborative network meta-analysis, Br. Med. J. 337 (2008) 1–11.

[19] G.W. Stone, S.G. Ellis, L. Cannon, J.T. Mann, J.D. Greenberg, D. Spriggs, C.D. O'Shaughnessy, S. DeMaio, P. Hall, J.J. Popma, J. Koglin, M.E. Russell, Comparison of a polymer-based paclitaxel-eluting stent with a bare metal stent in patients with complex coronary artery disease: a randomized controlled trial, J. Am. Med. Assoc. 294 (10) (2005) 1215–1223.

[20] F. Vogta, A. Stein, G. Rettemeier, N. Krott, R. Hoffmann, J. vom Dahl, A.K. Bosserhoff, W. Michaeli, P. Hanrath, C. Weber, R. Blindt, Long-term assessment of a novel biodegradable paclitaxel-eluting coronary polylactide stent, Eur. Heart J. 25 (2004) 1330–1340.

[21] Z.H. Wang, N. Li, R. Li, Y.W. Li, L.Q. Ruan, Biodegradable intestinal stents: a review, Prog. Nat. Sci. Mater. Int. 24 (2014) 423–432.

[22] P. Zartner, M. Buettner, H. Singer, M. Sigler, First biodegradable metal stent in a child with congenital heart disease: evaluation of macro and histopathology, Catheter. Cardiovasc. Interv. 69 (2007) 443–446.

[23] B. Zberg, P.J. Uggowitzer, J.F. Löffler, MgZnCa glasses without clinically observable hydrogen evolution for biodegradable implants, Nat. Mater. 8 (2009) 887–892.

[24] R. Zeng, W. Dietzel, F. Witte, N. Hort, C. Blawert, Progress and challenge for magnesium alloys as biomaterials, Adv. Biomater. 10 (2008) B3–B14.

Development of drug-eluting stents (DES)

3

M. Wawrzyńska, J. Arkowski*, A. Włodarczak†, M. Kopaczyńska‡, D. Biały**
*Wrocław Medical University, Wrocław, Poland, †Copper Health Centre, Legnica, Poland,
‡Wrocław University of Technology, Wrocław, Poland

3.1 First coronary intervention and development of stents

Coronary angiography was introduced as a diagnostic method for coronary artery disease in the late 1950s. Although balloon angioplasty was successfully applied to larger, "peripheral" arteries (mainly femoral and renal) since 1960s, only the development of more sophisticated equipment and technique allowed Andreas Grunzig to perform the first successful balloon angioplasty of a coronary artery in 1977. Thus, a new (and by far more convenient for the patient) method of coronary revascularization was introduced, adding another option to already well-established surgical coronary artery bypass grafting. At first, balloon angioplasty BA was associated with significant rate of procedural failure, mainly due to atherosclerotic plaque disruption which resulted in vessel thrombosis and acute closure. Another problem, more mechanical in nature, was elastic recoil of the vessel wall shortly after initial dilation. In addition to that short-term effects, it became clearly visible that in many cases, several months after the intervention, a new narrowing (restenosis) develops inside the dilated artery. Restenosis was reported to occur after 20%–40% of cases of initially successful BA [1]. Several additional techniques: rotational atherectomy (rotablation), Directional Coronary Atherectomy (DCA), and Excimer Laser Coronary Angioplasty (ELCA), were developed in 1980s and early 1990s. All of them, however, were not successful in reducing incidence of restenosis [2,3].

Another innovation was the development of metallic scaffolding stents. Stent implantation definitely reduced the incidence of acute periprocedural complications and initial failures of angioplasty (acute occlusion and elastic recoil). In most initial studies, stents also significantly reduced the incidence of restenosis, for example, from 42.1% to 31.6% [4] or even from 32% to 22% (P=.02) [5] when compared to BA. The first-generation stents, which presently are called bare metal stents (BMS) to distinct them from more sophisticated modern designs, were, however, associated with somewhat higher incidence of subacute thrombosis and late in-stent restenosis (ISR). The former has been successfully overcome with more aggressive antiplatelet therapy. ISR remained an issue especially in some high-risk settings such as narrow vessels, long lesions, bifurcations, and diabetic patients.

Functionalized Cardiovascular Stents. https://doi.org/10.1016/B978-0-08-100496-8.00003-2

Initially, to prevent ISR, brachytherapy was used. The method, although moderately effective, had many limitations. It required sophisticated and expensive equipment and special precautions since potentially harmful radioactive materials were used [6]. Therefore, it was not a treatment modality suited for routine clinical practice.

3.2 Pathophysiology of restenosis

Three different mechanisms are involved in the development of restenosis. The first one, elastic recoil of arterial wall, occurs usually within hours of the procedure. The second, negative remodeling, consists of arterial wall contraction and narrowing of the lumen. It is usually understood as a reaction to injury, related to healing process and endothelial cells activity [7]. Those processes are successfully eliminated by stent implantation as stiff metal scaffolding makes it impossible for the vessel wall to contract.

In contrast to the above, the third process, neointimal hyperplasia, is the main mechanism of ISR. Smooth muscle cells migrate from the media into the intima forming a multicellular layer that grows into the vessel lumen. In the first stage of the process, platelet aggregation and activation occur. In the second stage, inflammatory cells (macrophages and leukocytes) are attracted to the injury site. They in turn release growth factors and cytokines which activate smooth muscle cells in the media, causing them to migrate to the intima and to proliferate. Even when their cell divisions stop (usually several weeks after injury), the cells continue to produce extracellular matrix which grows into the vessel lumen causing further narrowing [8].

3.3 Methods of testing stent performance and their limitations

There are several ways to measure stent performance and to compare different designs. The most obvious laboratory tests (mainly mechanical) may give useful information on flexibility, radial force, possible deformation, and cell opening for side branch access. In the next step, animal testing, important information on thrombosis, vascular healing, and endothelialization can be elicited. These data are relatively easy to obtain (as large number of stents can be implanted, animals may be sacrificed and tissue cross sections may be examined). It must be remembered though that there is no prefect animal model for human atherosclerosis (and subsequently for coronary intervention, vascular healing, etc.). Consequently, the results of animal test may be used only with caution to draw conclusions about stent performance in humans.

Finally, clinical trials when stents are implanted into human coronary arteries give the most useful data. There are, however, several parameters that can be measured and used to judge how well a particular type of stent performs. The most obvious is to measure (by means of angiography or intravascular modalities such as IVUS) the degree of restenosis, sometimes also expressed as late lumen loss. Restenosis may be expressed in per cent of lumen area or as "binary restenosis"—whether less or more than 50% of the lumen is taken by the tissue. Even at this stage, the same date may lead to different interpretations (as, e.g., 30% of lumen loss is twice as much as 15%, but still 0 in binary terms and probably clinically insignificant). Clinical events are of more value. The most often used MACE term (major adverse cardiac event) usually comprises cardiac death, myocardial infarction and ischemia-driven revascularization, although exact criteria may differ between investigators. Most clinicians agree that MACE is the best single parameter to be considered when measuring stent performance. If no difference in MACE is found, sometimes data for only one specific type of event (but usually less clinically significant than MACE) is used. As stents and intravascular techniques are constantly improved, adverse clinical events and large differences in measured parameters become less frequent and very large series of procedures are needed for enough statistical power. Therefore, more and more often so-called "surrogate endpoints" are used—events or parameters that are not clinically significant by themselves, but their incidence may reflect the incidence of clinically important events. One needs also to bear in mind that the largest clinical trials are sponsored by stent manufacturers and understandably the data most likely to be published are those showing superiority or at least noninferiority of the latest design.

3.4 First-generation drug-eluting stents

A logical solution to the problem of restenosis was to locally (i.e., at the site of the injury) deliver a drug that stops smooth muscle cell proliferation. Thus, a typical drug-eluting stent (DES) design consists of metallic scaffolding (BMS), a layer of drug carrier substance, and antiproliferative agent itself.

Several studies showed that drug distribution into the arterial wall depends on stent strut configuration. Open strut cells, although often preferred by operators due to better conformability with vessel wall curvature and side branch access, usually result in much less homogenous drug distribution than closed cell designs, which in turn cover larger areas of the vessel wall in a more homogenous fashion.

In some DES designs, direct bonding of antiproliferative drug to metal was used. Most manufacturers, however, use matrix polymer that allows drug retention during stent deployment and homogenous drug distribution on the stent surface. The polymers can be divided into organic and inorganic and also alternatively into bioerodable and nonbioerodable. Most often, nonbioerodable polymers are used to prevent inflammatory response. Usually, synthetic organic polymers are used, but to some extent naturally occurring organic substances such as fibrin, cellulose, albumin, and phospahorylcholine were also tested.

Theoretically, an ideal drug for DES should have an antirestenotic effect, but also allow re-endothelialization and vessel healing. It should also have no adverse systemic effects

Many antiproliferative and antiinflammatory substances were used as active agents in drug-eluting stents. Most of them act by inhibiting DNA synthesis. Most important examples include paclitaxel, everolimus, tacrolimus sirolimus, interferon, dexamethasone, and cyclosporine. Migration inhibitors, such as batimastat or halofuginone, were thought to prevent migration of smooth muscle cells into the intima. However, in practice such stents releasing those substances failed to reduce restenosis [9]. So-called healing factors aiming to reduce platelet activation and promote reendothelization were also tested. Other compounds such as estradiol have also been used.

Initially, two antirestenotic drugs showed significant efficacy in clinical trials: sirolimus and paclitaxel. Sirolimus (rapamycin) was discovered in 1977 as an antifungal macrolide antibiotic with strong immunosuppressive effect. As a lipophilic molecule, it easily diffuses across the cell membranes of vascular smooth muscle cells and leukocytes. In the cytoplasm, it forms a complex with an intracellular protein FKBP12. This compound molecule inhibits in a turn a regulatory enzyme TOR (target of rapamycin). It blocks cell cycle progression from G1 to S phase, thus inhibiting smooth muscle replication and proliferation [10–12].

Paclitaxel is an antineoplastic drug originally isolated from the Pacific yew tree, *Taxus brevifolia*. It was initially used for treatment of breast and ovarian cancer. It is also a lipophilic molecule easily diffusing through cell membranes. Its main effect consists of stabilizing microtubules [13,14], making it impossible for the cells to pass from the G2 to M phase of the mitotic cycle.

The initial feasibility study of sirolimus-eluting stent was carried out on a small cohort in 2001 showing very good angiographic and clinical result—only minimal intimal hyperplasia and low level of target lesion revascularization. A large randomized trial RAVEL followed soon. It showed very low level of major adverse cardiac events (MACE) in the sirolimus-eluting (SES) arm (5.8% vs 28.8% for BMS $P<.001$) at 1-year follow-up [15,16].

A larger American trial SIRIUS included more high-risk patients and lesions (diabetes, narrow vessels, and long stented segments). It also showed much lower restenosis rate of 8.9% for SES than 36.3% for BMS ($P<.001$) at 9-month follow-up [17]. Other trials consistently showed much lower rates of restenosis, mace, and target vessel revascularization (TVR).

For paclitaxel and its derivatives, different stent designs and different coatings have been studied. Two types of coatings (polyacrylate sleeves and nonpolymer coating) proved to be not suitable for stent use. In third type of coating, polymer was used in TAXUS trials. The first feasibility study was carried out in 2003. Randomized trials comparing paclitaxel-eluting stents (PES) with BMS followed soon. They showed significant reduction in the target lesion revascularization (TLR) rate (4.4% vs 15.1%, $P<.0001$), TVR rate (7.1% vs 17.1%, $P<.0001$), and composite MACE rate (10.8% vs 20.0%, $P<.0001$) [18]. In TAXUS VI trial where more complex lesions (long plaques, small vessels) were stented, similarly lower incidences of restenosis (12.4% vs 35.7%; $P<.0001$) and TLR (6.8% vs 18.9%; $P=.0001$) were observed when DES were compared with BMS at 9 months [19].

3.5 Second-generation DES

Since much evidence suggested a correlation between stent strut thickness and the inflammatory process, subsequent restenosis second-generation stent platforms had thinner struts, but conserved the radial strength of the first-generation design.

This innovation was made possible by employing cobalt chromium alloy. Another advantage of the new platform was increased radiopacity and conformability to vessel curvature. In a large clinical trial (SPIRIT III), one of the second-generation DES everolimus- eluting stents (EES) showed 43% relative reduction in MACE when compared to paclitaxel-eluting stent. These outcomes persisted for 3 years proving long-term safety and efficacy. Similar results were obtained in SPIRIT IV trial; this time in more complex coronary lesions [18]. Understandably, questions were raised whether better clinical results are due to platform material (CoCr vs stainless steel), strut thickness, or the drug itself (-limus vs paclitaxel) or some combination of those factors.

In COMPARE trial, second-generation EES (thin struts, CoCr) was clinically more efficient than slightly improved first-generation paclitaxel-eluting stent (stainless steel, thin struts) [20]. Several trials comparing first-generation SES (thicker stainless steel struts) with second-generation EES (thin CoCr struts) did not show any clear advantage of new design in terms of overall MACE incidence. They did show, however, noninferiority of the second-generation DES and also significant reduction of some more specific endpoints such as TVR, myocardial infarction (MI), and stent thrombosis.

Another type of second-generation DES—zotarolimus-eluting stent (ZES)—with thin CoCr struts used different biopolymer formulations which resulted in more rapid drug release in the first 2 weeks. Both possible kinetic types (rapid vs slow drug release) may offer some advantages and drawbacks. Short release times may reduce the time period when dual antiplatelet therapy is necessary (this treatment has potential adverse effects of its own). On the other hand, in some cases, the restenotic process is still active several months after implantation and only stents with long drug release could prevent this phenomenon. The latter kinetics type does also have its drawbacks—it leaves stent struts unendothelized for longer time and late stent thrombosis may occur. In ENDAVOR III trial, significantly higher rates of late lumen loss were observed in second-generation ZES (compared to first-generation SES) [21]. In SORT OUT III trial, some composite endpoints (not as clinically important as MACE) were also more frequent in ZES group compared to first-generation SES [22].

Another innovation in DES design was to use a different PtCr platform. Theoretically, it was supposed to be more flexible and thus offer better deliverability. The so-called newer second-generation DES demonstrated noninferiority in clinical trials when compared to CoCrEES. There were, however, some data suggesting stent deformation after deployment, most probably due to less rigid structure (less connectors per ring). In some cases, more flexible stent structure may offer advantages (better deliverability and conformability in tortuous vessels, better side branch access); in other situation when more radial strength is required (e.g., calcified lesions), it is probably less appropriate. Periprocedural stent deformation has been described as risk factor of stent thrombosis. To address these issues, some minor modifications were made (more connectors resulting in more rigid structure) [23]. Somewhat similarly, newer second-generation ZES was designed to allow more flexibility, but the alignment of helical struts resulted in more longitudinal stability than new EES. In addition, the drug elution kinetic was changed, allowing a more prolonged release. It resulted in a lower incidence of late lumen loss than the previous ZES design.

There have been some trials comparing the newer EES with the newer ZES: they showed no difference in terms of primary composite endpoints. There was, however, a significant difference in incidence of definite stent thrombosis (more in ZES group, although it was a rarely occurring event) [24].

3.5.1 *Synthesis of data on currently approved DES*

Large number of data from clinical trials on the DES make it sometimes difficult to elicit clear conclusions on actual differences between various DES designs (in terms of major clinical events). Network metaanalysis of more than 50,000 patients in 49 randomized controlled trials demonstrated lower incidence of stent thrombosis for CoCr EES compared to BMS, SES, PES, and ZES [25]. The data also prove that newer second-generation DES perform very well, with perhaps slight favor towards EES-CoCr (Table 3.1).

3.6 Next-generation DES

3.6.1 *Abluminal coating*

A further development of DES is to cover only outer (abluminal) surface of the stent with polymer and drug. This design would allow the drug to be delivered where it is most needed (vessel wall) leaving the inner surface with bare metal only, making endothelialization easier. Such stents are already commercially available and first clinical trials showed promising results [26,27].

3.6.2 *Bioresorbable polymers*

As organic polymers may provoke immune response and hypersensitivity and, after the drug has been completely released and the polymer is no longer necessary, it was logical to try to design a stent with resorbable polymer. An example of such design is Nobori stent (stainless steel platform, bioresorbable polylactic acid polymer, and biolimus drug). The biolimus-eluting stent (BES) proved to be noninferior to SES in terms of MACE. However, BES seemed to perform better in cases of more complex anatomy and in terms of very late stent thrombosis. In other trials, Nobori BES performed equally well as CoCrEES [28,29].

Other designs included SES with bioresorbable polymer and EES with bioresorbable polymer. In clinical trials, no significant difference was shown between SES with biodegradable polymer and CoCrEES [30]. EES with bioresorbable polymer was noninferior to PtCr EES. On the other hand, there is some evidence that stents with bioresorbable polymers may be associated with slightly higher incidence of stent thrombosis. This trend was further confirmed by some more metaanalyses. Currently, most experts believe that CoCrEES is the best combination in terms of efficacy and long-term safety and the already approved newer second-generation DES can only hardly be improved.

Table 3.1 Overview of current DES designs

Overview of Most Important DES Designs						
Commercial Name	**Strut Material**	**Drug**	**Strut Thickness (μm)**	**Manufacturer**	**Drug Release Profile**	**On the Market Since**
Cypher	Stainless steel	Sirolmus	140	Cordis	1 month 80%	2003
Taxus Express/Liberte	Stainless steel	Paclitaxel	132	Boston Scientific	10% in 10 days	2004
Endavor	CoCr	Zotarolimus	91	Medtronic	14d >95%	2008
Resolute Integrity	CoCr	Zotarolimus	91	Medtronic	2 m 85%	2012
Xcience V Prime xpedition	CoCr	Everolimus	81	Abbot Vascular	1 m 80%	2008
Promus element	CoCr	Everolimus	81	Boston Scientific	1 m 80%	2008
Promus Premier	PtCr	Everolimus	81	Boston Scientific	1 m 80%	2013
SYNERGY	PtCr	Everolimus	74	Boston Scientific	2 m 80%	2012
Absorb	Poly-L-lactic acid	Everolimus	150	Abbot Vascular	1 m 90%	2011

3.6.3 Pro-healing stents

The drugs used in DES affect different cell types. They inhibit smooth muscle cell proliferation, but they also slow down endothelialization of stent surface. The idea to promote endothelial growth on stent surface seems, therefore, an attractive solution. Stents-eluting vascular endothelial growth factor was tested, but they increased proliferation of neointima [31]. There is also a commercially available stent covered with CD34 antibodies aiming to capture endothelial cells. The results of preclinical trials were favorable. In a clinical trial, no significant difference was seen between this type of stent and PES. It was speculated that CD34 antibodies attract not only endothelial progenitor cells, but also other hematopoietic cells, for example, smooth muscle progenitor cell [32]. Consequently, a new generation of prohealing stents was developed: with CD34 antibodies on luminal and sirolimus-eluting biodegradable polymer on abluminal surface. It performed well on porcine model and was safe and noninferior to ZES in clinical trial [27].

3.6.4 Bioresorbable stents

As incomplete endothelialization of DES may persist even for many years, research on bioresorbable stents started already in 1990s. An ideal bioresorbable scaffold would provide sufficient mechanical support in the days and weeks after the angioplasty procedure. It would then start to gradually dissolve and finally leave only healed vessel wall behind. Thus, many problems associated with stenting such as leaving a rigid mechanical caging or a need for a prolonged dual antiplatelet therapy could be overcome. Several other problems: covering of side branches ostia and inaccessibility of the stented segment for surgical grafting would also be solved. Another advantage of such design would be a completely restored vasomotion once the stent is dissolved.

Some manufacturers came up with bioresorbable metallic materials of which magnesium seems to be the most interesting. Unfortunately, pure magnesium bioresorbable stents were associated with poor radial strength and high rates of restenosis. With platform design improvement and addition of paclitaxel, the clinical results were slightly better, although not superior to current state of art DES [33].

Among the polymers, most frequently used for bioresorbable stents are poly-L-lactic acid (PLLA) and poly-DL-lactic acid (PDLLA). Other substances were also tested. Polymeric scaffolds have usually much less radial strength and therefore require much thicker struts than steel stents. First, small series of bioresorbable stents (Igaki-Tamai, made of PLLA with no drug coating) was implanted between 1998 and 2000. It has shown very good safety and efficacy profile in 10 years follow-up [34]. Nevertheless, the design has not been developed further, mainly due to its rather complicated expansion mode (use of heated contrast) and lack of drug coating which made it obsolete in the DES era.

Currently, there are several bioresorbable stent designs (usually called bioresorbable scaffolds to make clear distinction from permanent metallic stents) of which Absorb is the most developed and was the first to become commercially available. It is made of poly(L-lactide) 150 μm thick struts and is covered with bioresorbable polymer that releases everolimus. The scaffold starts to disintegrate around 1 year after implantation via a series of chemical reactions that break the chains down to smaller molecules, leading eventually to carbon dioxide and water. It dissolves completely after 3–5 years. The data from randomized trials and registries available so far suggest no safety concerns [35]. Initially, due to low radial strength of the scaffold, a meticulous lesion preparation (via conventional balloon angioplasty) was required. The new version of Absorb scaffold was developed to increase radial support. The data from trials suggest there is no scaffold thrombosis, but the device is characterized by more pronounced lumen loss than current DES designs. Nevertheless, this phenomenon does not lead to significant clinical events (Fig. 3.1).

The most important advantages of bioresorbable stents include:

- Restoration of vessel anatomy (after resorption there is no segment straightening, no edges that could alter blood flow)
- Restoration of vessel physiology—as there is no metal cage full vasomotion is possible—both in terms of cell signals and vessel relaxation/contraction
- No thrombogenity when the scaffold is resorbed
- Possibly, no neoatherosclerosis as normal endothelium and no residual scaffold constitute a normal vessel environment

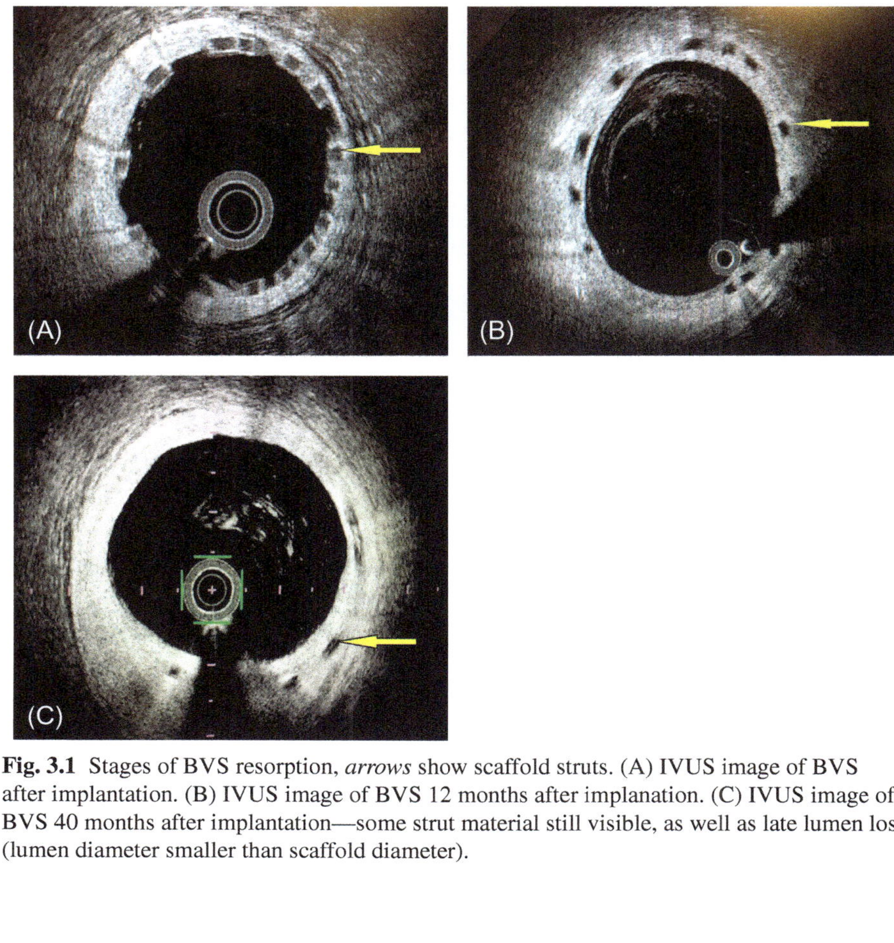

Fig. 3.1 Stages of BVS resorption, *arrows* show scaffold struts. (A) IVUS image of BVS after implantation. (B) IVUS image of BVS 12 months after implanation. (C) IVUS image of BVS 40 months after implantation—some strut material still visible, as well as late lumen loss (lumen diameter smaller than scaffold diameter).

The important limitations of BVS are: strut thickness larger than in conventional metallic stents and somewhat less perfect performance during the procedure (deliverability, radial strength).

3.7 Conclusion

The developments in cardiovascular stent design over the last 30 years have been truly remarkable. It seems the newest generation of DES is hard to beat in terms of low risk of restenosis and stent thrombosis. Nevertheless, some minor improvements in platform design, polymers, drug release kinetics, and drugs itself are still possible. On the other hand, the modern bioresorbable stents are getting close to clinical performance of current best metallic DES designs. Many experts believe that there will probably never be an ideal DES. But as more and more information is gathered about advantages and drawbacks of different solutions, clinicians in the future will be able to choose an ideal stent for risk profile and lesion characteristic of every particular patient.

References

[1] J.M. Miller, E.M. Ohman, D.J. Moliterno, et al., Restenosis: the clinical issues, in: E.J. Topol (Ed.), Textbook of Interventional Cardiology, third ed., WB Saunders, Philadelphia, PA, 1999, pp. 379–415.

[2] R.L. Mueller, T.A. Sanborn, The history of interventional cardiology: cardiac catheterization, angioplasty, and related interventions, Am. Heart J. 129 (1995) 146–172.

[3] G. Karthikeyan, B. Bhargava, Prevention of restenosis after coronary angioplasty, Curr. Opin. Cardiol. 19 (2004) 500–509.

[4] D.L. Fischman, M.B. Leon, D.S. Baim, et al., A randomized comparison of coronary-stent lacement and balloon angioplasty in the treatment of coronary artery disease. Stent restenosis study investigators, N. Engl. J. Med. 331 (1994) 496–501.

[5] P.W. Serruys, P. de Jaegere, F. Kiemeneij, et al., A comparison of balloon-expandable-stent implantation with balloon angioplasty in patients with coronary artery disease. BENESTENT Study Group, N. Engl. J. Med. 331 (1994) 489–495.

[6] A.E. Raizner, S.N. Oesterle, R. Waksman, et al., Inhibition of restenosis with beta-emitting radiotherapy: report of the proliferation reduction with vascular energy trial (PREVENT), Circulation 102 (2000) 951–958.

[7] M.W. Liu, G.S. Roubin, S.B. King III, Restenosis following coronary angioplasty potential biological determinants and role of intimal hyperplasia, Circulation 79 (1989) 1374–1387.

[8] R.S. Schwartz, T.D. Henry, Pathophysiology of coronary artery restenosis, Rev. Cardiovasc. Med. 3 (Suppl. 5) (2003) S4–S9.

[9] H.M. van Beusekom, M.J. Post, D.M. Whelan, et al., Metalloproteinase inhibition by batimastat does not reduce neointimal thickening in stented atherosclerotic porcine femoral arteries, Cardiovasc. Radiat. Med. 4 (4) (2003) 186–191.

[10] M. Poon, S.O. Marx, R. Gallo, et al., Rapamycin inhibits vascular smooth muscle cell migration, J. Clin. Invest. 98 (1996) 2277–2283.

[11] S.O. Marx, A.R. Marks, Bench to bedside: development of rapamycin and its application to stent restenosis, Circulation 104 (2001) 852–855.

[12] J.E. Sousa, A.G. Sousa, M.A. Costa, et al., Use of rapamycin-impregnated stents in coronary arteries, Transplant. Proc. 35 (2003) 5165–5170.
[13] P.B. Schiff, J. Fant, S.B. Horwitz, Promotion of microtubule assembly in vitro by taxol, Nature 277 (1979) 665–667.
[14] S.J. Sollott, L. Cheng, R.R. Pauly, et al., Taxol inhibits neointimal smooth muscle cell accumulation after angioplasty in the rat, J. Clin. Invest. 95 (1995) 1869–1876.
[15] M.C. Morice, P.W. Serruys, J.E. Sousa, et al., A randomized comparison of a sirolimus-eluting stent with a standard stent for coronary revascularization, N. Engl. J. Med. 346 (2002) 1773–1780.
[16] P.W. Serruys, M. Degertekin, K. Tanabe, et al., Intravascular ultrasound findings in the multicenter, randomized, double-blind RAVEL (randomized study with the sirolimus-eluting velocity balloonexpandable stent in the treatment of patients with de novo native coronary artery lesions) trial, Circulation 106 (2002) 798–803.
[17] J.W. Moses, M.B. Leon, J.J. Popma, et al., Sirolimus-eluting stents versus standard stents in patients with stenosis in a native coronary artery, N. Engl. J. Med. 349 (2003) 1315–1323.
[18] G.W. Stone, A. Rizvi, K. Sudhir, et al., Randomized comparison of everolimus- and paclitaxeleluting stents. 2-year follow-up from the SPIRIT (clinical evaluation of the XIENCE V everolimus eluting coronary stent system) IV trial, J. Am. Coll. Cardiol. 58 (2011) 19–25.
[19] K.D. Dawkins, E. Grube, G. Guagliumi, et al., Clinical efficacy of polymer-based paclitaxel-eluting stents in the treatment of complex, long coronary artery lesions from a multicenter, randomized trial: support for the use of drug-eluting stents in contemporary clinical practice, Circulation 112 (21) (2005) 3306–3313.
[20] E. Kedhi, K.S. Joesoef, E. McFadden, et al., Second-generation everolimus-eluting and paclitaxel-eluting stents in real-life practice (COMPARE): a randomised trial, Lancet 375 (2010) 201–209.
[21] D.E. Kandzari, M.B. Leon, J.J. Popma, et al., Comparison of zotarolimus-eluting and sirolimus-eluting stents in patients with native coronary artery disease, J. Am. Coll. Cardiol. 48 (2006) 2440–2447.
[22] M. Maeng, H.H. Tilsted, L.O. Jensen, et al., 3-year clinical outcomes in the randomized SORT OUT III superiority trial comparing zotarolimus- and sirolimus-eluting coronary stents, JACC Cardiovasc. Interv. 5 (2012) 812–818.
[23] P. Williams, M. Mamas, K. Morgan, et al., Longitudinal stent deformation: a retrospective analysis of frequency and mechanisms, EuroIntervention 8 (2012) 267–274.
[24] P.W. Serruys, S. Silber, S. Garg, et al., Comparison of zotarolimus-eluting and everolimuseluting coronary stents, N. Engl. J. Med. 363 (2010) 136–146.
[25] T. Palmerini, G. Biondi-Zoccai, D. Della Riva, et al., Stent thrombosis with drug-eluting and baremetal stents: evidence from a comprehensive network meta-analysis, Lancet 379 (2012) 1393–1402.
[26] A. Abizaid, J.R. Costa Jr., New drug-eluting stents: an overview on biodegradable and polymerfree next-generation stent systems, Circ. Cardiovasc. Interv. 3 (2010) 384–393.
[27] M. Haude, S.W. Lee, S.G. Worthley, et al., The REMEDEE trial: a randomized comparison of a combination sirolimus-eluting endothelial progenitor cell capture stent with a paclitaxel-eluting stent, JACC Cardiovasc. Interv. 6 (4) (2013) 334–343.
[28] P.C. Smits, S. Hofma, M. Togni, N. Vázquez, M. Valdés, V. Voudris, et al., Abluminal biodegradable polymer biolimus-eluting stent versus durable polymer everolimus-eluting stent (COMPARE II): a randomised, controlled, non-inferiority trial, Lancet 381 (2013) 651–660.

[29] M. Natsuaki, K. Kozuma, T. Morimoto, K. Kadota, T. Muramatsu, Y. Nakagawa, et al., Biodegradable polymer biolimus-eluting stent versus durable polymer everolimus-eluting stent: a randomized, controlled, noninferiority trial, J. Am. Coll. Cardiol. 62 (2013) 181–190.

[30] R.A. Byrne, A. Kastrati, S. Kufner, S. Massberg, K.A. Birkmeier, K.L. Laugwitz, et al., Randomized, non-inferiority trial of three limus agent-eluting stents with different polymer coatings: the intracoronary stenting and angiographic results: test efficacy of 3 limus-eluting stents (ISAR-TEST-4) trial, Eur. Heart J. 30 (2009) 2441–2449.

[31] N. Swanson, K. Hogrefe, Q. Javed, et al., Vascular endothelial growth factor (VEG-F)-eluting stents: in vivo effects on thrombosis, endothelialization and intimal hyperplasia, J. Invasive Cardiol. 15 (2003) 688–692.

[32] M.A. Beijk, M. Klomp, N.J. Verouden, et al., Genous endothelial progenitor cell capturing stent vs. The taxus liberte stent in patients with de novo coronary lesions with a high-risk of coronary restenosis: a randomized, single-centre, pilot study, Eur. Heart J. 31 (2010) 1055–1064.

[33] M. Haude, R. Erbel, P. Erne, et al., Safety and performance of the drug-eluting absorbable metal scaffold (dreams) in patients with de-novo coronary lesions: 12 month results of the prospective, multicentre, first-in-man biosolve-i trial, Lancet 381 (2013) 836–844.

[34] S. Nishio, K. Kosuga, K. Igaki, et al., Long-term (.10 years) clinical outcomes of first-in-human biodegradable poly-L-lactic acid coronary stents: Igaki-Tamai stents, Circulation 125 (2012) 2343–2353.

[35] D. Dudek, Y. Onuma, J.A. Ormiston, et al., Four-year clinical follow-up of the absorb everolimus-eluting bioresorbable vascular scaffold in patients with de novo coronary artery disease: the ABSORB trial, EuroIntervention 7 (2012) 1060–1061.

Polymer-free drug-eluting stents

4

C. McCormick
University of Strathclyde, Glasgow, United Kingdom

4.1 Introduction

Drug-eluting stents (DES) have revolutionized the way that advanced coronary heart disease is now treated. These devices are associated with much lower rates of restenosis than their bare metal stent counterparts and are now used in the vast majority of coronary artery revascularization procedures carried out [1]. As a result, fewer patients than even before now need to return to hospital for a repeat revascularization. However, clinical challenges remain, meaning that there are considerable opportunities to improve on the performance of current DES [2]. While there are a great many strategies currently being investigated in this pursuit, some with greater clinical potential than others, the development of polymer-free approaches to stent drug delivery appears particularly promising [3]. In this chapter, we will firstly describe the clinical rationale that stimulated the development of the first generation of polymer-free DES. We will then describe these devices and current state of the art in some detail before going on to explore the remaining challenges with respect to technology development in this area. The chapter will conclude with a brief consideration of future perspectives.

4.2 Moving beyond polymer controlled stent drug release

4.2.1 Rationale for polymer-free drug-eluting stents

The development of polymer-free DES has been viewed as a response to some of the negative clinical outcomes that were observed with first-generation DES [4,5]. In particular, the first significant reports of increased late stent thrombosis (LST) with the Cypher (Cordis Corp.) and Taxus (Boston Scientific) DES caused the clinical and industry communities to revisit all aspects of stent design [5,6]. In the investigations that followed, concern was raised that the ongoing presence of a polymer coating on the stent surface may be an important contributing factor in the occurrence of LST [7,8], and there is no doubt that such reports were a key driver in the development of polymer-free approaches to DES design. These clinical experiences with permanent polymer-coated DES added to existing evidence within the literature on the potential negative effects of certain polymer coatings within the vasculature. Specifically, a range of permanent and biodegradable polymer materials were shown to produce intense inflammatory responses and neointimal thickening within pig coronary arteries in vivo [9]. Although the methodological details of this particular study made direct

Functionalized Cardiovascular Stents. https://doi.org/10.1016/B978-0-08-100496-8.00004-4

extrapolation of the findings to coronary stent coatings difficult, and the clinical success of the first- and second-generation polymer-coated DES would appear to confirm these limitations, such studies nonetheless provided sufficient impetus for alternative approaches to polymer coatings to be investigated, even in the earliest days of DES development. Since then, a great many different polymer-free DES have been developed, and in this chapter, we will examine the key innovations that have helped drive increased clinical use of these devices.

4.2.2 Sustained drug release for clinical efficacy

The release profile of the drug is known to significantly impact on the overall performance of a DES [10]. It was known that the smooth muscle cell response, thought to be largely responsible for restenosis, is maintained for weeks and even months following stent implantation [11,12]. The first generation of DES were therefore designed to sustain drug release for periods up to around 3 months [13]. This was achieved through the use of permanent polymers, although more recent devices have increasingly featured biodegradable polymers [2]. The use of such polymers can provide great flexibility, allowing precise optimization of the drug release profile to maximize the therapeutic effect [14], although in reality, there is still similarity in the release profiles of many conventional polymer-coated DES, with a fairly rapid release within the first few days followed by sustained drug release over weeks and months [13]. The central challenge for polymer-free approaches was to ensure that the drug release profile could be sustained for a similar period to that achieved with conventional DES. Several approaches to addressing this challenge have since been pursued, which have been summarized recently by Chen et al. [3]. The optimal approach is ultimately dependent on the drug that is to be released, its physicochemical properties, the design of the stent surface, and the interaction between all these components. The following section will describe the key designs that have been developed, with particular emphasis given to those that have either reached clinical evaluation or have greatest future potential.

4.3 Direct coating of drug

Perhaps the simplest approach to achieving polymer-free drug elution is direct loading of the drug onto an unmodified stent surface. This can be achieved through dip coating of the stent into a drug solution, with evaporation of the solvent leaving a drug layer coating the stent surface. The release rate of the drug in this case is largely dependent on the physicochemical properties of the drug [15]. Such an approach is therefore unsuitable for hydrophilic drugs, such as heparin, since rapid dissolution would likely produce only short-term drug release [16]. However, such rapid release kinetics can potentially be avoided with the use of highly lipophilic drugs, such as paclitaxel and sirolimus, which have been used extensively in conventional DES [17]. In addition to lipophilicity, certain drugs may preferentially adhere to particular chemical groups present on the stent surface. For example, paclitaxel has been shown to bind preferentially to synthetic materials [18], and it is thought that this may provide an additional mechanism by which stent-based release can be slowed [3]. Ultimately, the method of drug deposition and the final physical form that the drug takes can significantly impact on the subsequent release characteristics for a given drug, but for now, we will focus on those approaches that have simply directly loaded unmodified drugs onto standard bare metal stents.

Heldman et al. [19] demonstrated that standard stainless steel stents (Palmaz-Schatz), coated directly with an ethanol-paclitaxel solution, inhibited neointima formation at 4 weeks in a porcine coronary artery model. One of the first polymer-free paclitaxel-eluting stents to reach clinical evaluation was the V-Flex Plus stent (Cook Inc.). In this device, a coating of paclitaxel is applied directly to the abluminal facing surface of the stent. In a pilot clinical trial (ELUTES), the high-dose version of this device (2.7 $\mu g/mm^2$) was found to reduce angiographic restenosis when compared with the bare metal control [20]. However, in a direct head-to-head clinical trial, it was shown that the Taxus stent was superior to the paclitaxel-coated V-Flex Plus stent [21]. Guidant Corp. applied the same coating approach developed by Cook Inc. to produce the RX ACHIEVE polymer-free paclitaxel-eluting stent. However, this device failed to achieve significantly improved clinical outcomes compared with the bare metal control in the DELIVER clinical trial [22]. This is in sharp contrast to the results from a trial of the TAXUS polymer-coated paclitaxel-eluting stent reported in the same year, which showed that it markedly reduced clinical and angiographic restenosis compared with bare metal stents [23]. The Amazonia PAX stent is a more recent addition to the family of paclitaxel-eluting polymer stents (MINVASYS). This device incorporates a cobalt-chromium platform, with a semicrystalline paclitaxel coating applied to the abluminal surface of the stent. Paclitaxel release was found to be rapid in the early phase, with up to 52% released within the first 8 h and 75% within 1 week, before the remaining drug is eluted more slowly up to 45 days [24]. In a small single-center clinical study (PAX A), it was shown to have similar rates of angiographic restenosis after 4 months and clinical events at 12 months as the Taxus stent [24]. However, less encouraging results were observed in a more recent prospective, nonrandomized, multicenter study (PAX B), where although it demonstrated safety after 1 year, target vessel revascularization rate was 21.3%, a performance more comparable with conventional BMS than DES [25].

It appeared from these early clinical evaluations that polymer-controlled approaches to paclitaxel release were superior to polymer-free approaches. Although the majority of research and product development across the major stent manufacturers has therefore focused on the use of polymer-controlled systems, with a gradual move away from the use of paclitaxel in recent years, the concept of direct paclitaxel coating has still been pursued by some groups. For example, it has recently been demonstrated that direct loading of paclitaxel onto cobalt-chromium alloy surfaces can provide release for up to 56 days [26], with this extended release offering a potential advantage over earlier approaches described above where release was thought to be fairly rapid. In addition, recent modeling studies have indicated that for paclitaxel, short-term release profiles may in fact be sufficient to generate therapeutic effects [27]. Although it is necessary to exercise caution in the interpretation of studies of this nature, the models do provide a potential explanation for some of the positive early findings observed with polymer-free paclitaxel elution from stents and more recent clinical results from paclitaxel-coated balloons [28]. It may therefore be that paclitaxel elution, perhaps from some of the more advanced metal alloys and through the use of some of the surface modification strategies discussed below, may still have a role to play in percutaneous coronary interventions.

One consistent finding from the early trials of the polymer-free paclitaxel-eluting DES described above was that there was a relationship between the dose of the drug loaded onto the stent and its efficacy. It was found that doses of between ~2.5 and 3.1 μg/mm^2 were clinically effective, which is considerably greater than the 1 μg/mm^2 used in the polymer-coated Taxus paclitaxel-eluting stent. The higher doses required for efficacy in polymer-free direct drug loading DES designs likely reflect rapid release and substantial drug loss to the blood, which represents a key limitation of such approaches and is a likely factor limiting the performance of these devices. Alternative means of optimizing drug release characteristics of polymer-free DES have therefore been investigated, and two broad strategies have emerged. One strategy involves modification of the stent platform in some way, such that that it can more effectively control the release of drug. Alternatively, the drug physical properties may be selected or tailored to ensure that release is slowed. Some of the most advanced approaches that have now reached clinical evaluation combine both strategies successfully. In keeping with the overall focus of this book, the following discussion centers on stent platform modifications.

4.4 Stent platform modifications

Stent platforms can be designed with particular surface features that help provide enhanced control over drug release compared with conventional smooth surfaces [29,30]. Our discussion will start with those stents that have introduced macroscale surface features in order to provide control over drug release, before going on to consider those devices that have introduced micro and nanoscale modifications.

4.4.1 Macroporous stents

It is clear that the paramount function of the stent platform is to provide radial support to the vessel wall, thereby helping to maintain long-term vessel patency. However, the introduction of DES has seen the function of the stent platform extended to include drug delivery. Polymer-coated DES have traditionally used conventional bare metal platforms, with polymer and drug coatings being carefully selected to ensure that the desired drug elution characteristics are not achieved at the expense of the mechanical performance of the stent. A potentially more advanced approach is to recognize that the platform itself could be specifically designed to not only provide the required mechanical support but also to provide optimal drug release kinetics.

4.4.1.1 NEVO stent

An early example of this type of approach to stent platform design was the NEVO stent (Cordis Corp.), where the stent platform incorporates a series of evenly distributed reservoirs embedded within the stent struts [31]. These reservoirs are then loaded with sirolimus and poly(lactide-*co*-glycolide) (PLGA) to provide sustained release of the drug over a period of 90 days. This novel platform also incorporated ductile hinges, which were designed to help ensure that the drug reservoirs were directly apposed to the vessel wall following expansion. This design allowed a similar dose of sirolimus to be loaded onto the stent and released over a similar period to the polymer-coated Cypher stent, but with a far greater percentage of the bare metal surface uncoated by drug/polymer. Strictly speaking, however, the incorporation of PLGA within the reservoirs means that the NEVO stent is something of a hybrid device that sits in the gap between polymer-coated and nonpolymer-coated designs. Despite the impressive level of technical innovation incorporated within the NEVO stent, it did not achieve the clinical uptake expected and was withdrawn from use when Cordis Corp. left the stent market in 2011.

4.4.1.2 Janus Carbostent

Despite its limitations, the NEVO stent represented an important innovation in stent design. There have been similar approaches that have sought to use macro-sized reservoirs to allow the incorporation of higher drug loads and potentially more targeted delivery than is possible with conventional platforms. The Janus tacrolimus-eluting Carbostent (Sorin Group) is a good example of such an approach [32]. This stent is based on the Tecnic Carbostent, which incorporates a Carbofilm coating to increase biocompatibility and thromboresistance. Tacrolimus, a sirolimus analogue with similar antiproliferative effects on vascular cells [33], is then loaded directly into reservoirs that have been created on the abluminal facing stent surface. Although it is possible to introduce polymers within the reservoirs to slow drug release, it was found in preclinical evaluation of the polymer-free version of this stent that around 50% of the drug remained on the stent after 1 month, with measurable levels of the drug within arterial tissue also maintained up to this point [32]. This suggests that the nature of the reservoirs and the relatively low solubility of the drug are sufficient to sustain drug release over periods comparable with polymer-controlled systems. However, although the Janus Carbostent has generally been shown to be safe, it did not yield any improvements in clinical outcomes at 6 months compared with the drug-free Tecnic Carbostent [34], suggesting that the tacrolimus release profile achieved had little therapeutic benefit. These results, which were followed by negative findings from clinical investigations in real-world patients [35,36], raised serious concerns about the clinical utility of the Janus Carbostent [37]. More encouraging results were obtained from a recent multicenter trial of the stent, which showed no significant difference in major adverse cardiac event or thrombosis rates between patients who had received either 2 or 6 months dual antiplatelet therapy [38]. An updated version of this stent, the OPTIMA TES, has also been shown to be safe with similarly low periods of dual antiplatelet therapy [39]. Although further larger, randomized, trials are clearly required, this suggests that such reservoir-based polymer-free approaches may allow reduced antiplatelet therapy regimes to be used, which may be particularly advantageous in patients at high risk of bleeding in the same way that microporous polymer-free DES have recently shown [40].

4.4.1.3 Cre8

Another stent that may be particularly useful for high bleeding risk patient groups is the DES Cre8 (Alvimedica). In this device, macroporous reservoirs are introduced into the struts of a cobalt-chromium stent platform. A passive carbon coating (i-Carbofilm, CID) is then applied, and a formulation containing sirolimus with long-chain fatty acids as an excipient is then loaded into the abluminally facing reservoirs. In a head-to-head comparison with the Taxus, the Cre8 was shown to have lower late lumen loss after 6 months and a trend toward enhanced safety and efficacy after 1 year [41]. These results have been followed by promising findings indicating that this device is noninferior to second-generation everolimus-eluting DES in patients with diabetes mellitus [42] and that it may enable reduced durations of dual antiplatelet therapy to be used in patients at high risk of bleeding [43].

4.4.1.4 Polymer-free drug-filled stent

Medtronic Inc. has recently developed a drug-filled stent (DFS), which has several innovative features that distinguish it from other polymer-free devices. The platform is made up of stent struts, which have been hollowed out to provide an internal lumen. Coating the lumen is a layer of tantalum to provide added radiopacity. A series of holes of around 20 μm are then drilled into the abluminal facing side of the stent struts. A coating of sirolimus is then applied through these holes, thereby targeting drug release to the artery wall. The first-in-man clinical trial (RevElution) results indicate that this device is safe, with efficacy that was noninferior to historical control data on the Resolute DES [44]. Optical coherence tomography (OCT) was used to assess strut coverage in separate subsets of patients within this trial and revealed the mean percentage strut coverage to be 89% after 1 month, increasing to 93% after 3 months and near complete coverage after 9 months (99%). Larger, randomized trials will clearly be required to fully evaluate this device. However, should such trials confirm the promise of these early findings, then this device will represent a very significant advance in the development of coronary stent technology [45].

4.4.2 Microporous stents

Of the various approaches to polymer-free stent drug elution that have been developed, it has been microporous stent surfaces that have perhaps received the most extensive evaluation thus far. The overall rationale behind the use of such devices is that the creation of microporous or textured stent surfaces enhances the drug loading capacity of the stent. One of the first such devices to be developed was the Yukon stent (Translumina GmbH) [46].

4.4.2.1 Yukon stent

Although the concerns around the use of polymers within the vasculature raised by van der Giessen et al. [9] would have provided supporting rationale for the development of this stent, it is important to note that the development of the Yukon stent preceded the height of the concerns around late stent thrombosis and delayed healing that were described at the start of this chapter. It is clear that the removal of the polymer was seen not only as a way of enhancing the biocompatibility of this stent but also as a way of introducing a new concept in stent design that would allow the clinician to select the drug type and dose to be coated onto the stent platform. This was achieved through the use of a dose-adjustable stent coating machine housed within the catheterization suite, which sprays a drug-ethanol solution onto the microporous stent surface, with the resultant drug-coated stent being available for use within minutes. The stent platform is made from stainless steel 316L, which is sandblasted to produce microporosity across the whole surface of the stent. This treatment generates microporous pits of the order of 1–2 μm in size, thereby increasing the drug loading capacity of the platform. To date, most clinical studies have examined the performance of sirolimus-coated versions of this stent, although dual-drug combinations have also been evaluated more recently. The sirolimus-coated version of this device was shown to be safe and effective in the initial dose-finding clinical study, with dose-dependent reductions in restenosis and repeat revascularization rates being observed in the drug-coated groups compared with the bare metal stent control group [47]. More recent data from the ISAR-TEST trial found no difference in clinical outcomes at 5 years in patients treated with either the sirolimus-eluting Yukon stent or the Taxus stent [48]. Although in vitro pharmacokinetic studies have demonstrated that sirolimus release is sustained for more than 21 days, there is a clear burst release with over two-thirds of the drug released within the first 6 days, and it is likely that in vivo release is even more rapid. Indeed, data from Watt et al. [49] have shown that the release of the lipophilic antioxidant drug, succinobucol, from the Yukon stent platform was sustained for only 14 days in vivo, with around 60% being released in the first 3 days. It is perhaps revealing that Translumina GmbH has now developed a polymer-coated version of this device in an attempt to provide more sustained sirolimus release and thus improve performance [50]. They have also gone on to develop a variety of polymer-free dual-drug versions of this device, with probucol-sirolimus being the most effective combination tested so far [51]. Although the dose-adjustable aspect of the Yukon stent has not found widespread use in clinical practice, perhaps because there remains a limited understanding of how drug dose and type should be tailored toward particular lesion types or patients [51a], it remains an interesting technology that may yet be useful in helping to realize some of the more personalized treatment approaches that are now being called for [52,53].

4.4.2.2 BioFreedom

The BioFreedom stent (Biosensors Int.) is similar to the Yukon device, although it does not currently allow on-site dose-adjustable coatings to be applied. There are also other differences that have potentially important impacts on performance. The microporous surface finish is applied only to the abluminal facing surface of the stent platform, thereby targeting drug release into the artery wall. The drug loaded onto this stent is Biolimus A9, which has greater lipophilicity and so is therefore believed to be taken up into the artery tissue more rapidly than sirolimus [54]. This may therefore help reduce drug loss to the circulation and improve retention of the drug within the artery wall, thereby overcoming key limitations of previous polymer-free approaches to stent drug release. The first-in-man trial compared the performance of the BioFreedom stent to the Taxus stent, with angiographic follow-up after 12 months and clinical outcome measures recorded at 5 years. There was no difference in performance observed between the two stent types, confirming the long-term safety and efficacy of the BioFreedom device [55]. This stent may also open up the benefits of DES treatments to patients who are at particularly high risk of bleeding. Such patients would normally be treated with a second-generation conventional DES and a reduced duration of dual antiplatelet therapy (DAPT) or with a bare metal stent and 1 month of DAPT. Both approaches are ultimately compromises that either prolong the risk of bleeding or increase the risk of restenosis. The LEADERS FREE trial investigated whether the BioFreedom stent could help remove this treatment dilemma for high bleeding risk patients by providing the efficacy benefits of a DES without increasing the duration of DAPT [40]. It was found that the BioFreedom stent reduced the incidence of target lesion revascularization compared with the bare metal stent. Crucially, this clinical benefit was achieved without increasing the rate of stent thrombosis, despite just 1 month of DAPT being used. The BioFreedom stent therefore has the potential to make a very significant impact on clinical treatment strategies in the years to come [56].

4.4.2.3 YINYI stent

The YINYI stent (Liaoning Biomedical Materials R&D Center Co., Ltd) is another microporous polymer-free DES that has recently undergone clinical evaluation. It comprises a stainless steel stent that is treated to produce a surface finish covered by micropores. The depth of the pores is reported to be <500 nm, with the pore diameter ranging between 1 and 2 μm (Yinyi Biotech). The surface porosity is therefore a similar order of magnitude to that used in the Yukon stent. However, in contrast to the Yukon stent that has mostly been used to release sirolimus alone or in combination with other drugs, the YINYI stent provides release of paclitaxel. The most recent clinical data on this device indicate that it is safe and effective, with low rates of stent thrombosis and target lesion revascularization being observed after 3 years [57], but although it is reported to have been implanted in over 100,000 patients now (Yinyi Biotech), further clinical evaluations of this device are required before its potential impact on clinical practice can be fully evaluated.

4.4.2.4 VESTASYNC

While the microporous and textured finishes described in the stents above were achieved through the use of processing treatments that modify the surface of the stent, in the VESTASYNC (MIV Therapeutics), a very thin layer of microporous hydroxyapatite is coated onto the stent platform, and sirolimus is then loaded into this surface structure. Hydroxyapatite has been used previously to improve the biocompatibility of medical implants, with the greatest use being found within orthopedic and related applications [58,59]. The pore sizes within hydroxyapatite surface of the VESTASYNC are between 100 and 500 nm in diameter. The mass of sirolimus released from this porous coating in the first hour is similar to that reported for the Cypher stent, although there is greatly reduced release thereafter, and it is estimated that all of the drug will have been released after 3–4 weeks [60]. One year data from the first-in-man trial in a small group of patients demonstrated the safety of the device (VESTASYNC I) [61]. A second generation of this device, with the stainless steel platform being replaced by a cobalt-chromium one, has been tested against the drug-free version of this stent and found to be effective, with reduced restenosis observed after 8 months [62].

4.4.3 Nanoporous stents

Although the use of microporous and microtextured surfaces described above provides a greater surface area, allowing drug loads to be increased, the drug release profiles generated have generally been characterized by rapid burst release, and it has been argued that this remains an important limitation of such devices in comparison with conventional DES [25]. Such rapid release characteristics are not surprising, since the pore dimensions are well in excess of the size of drug molecules typically used within DES. However, as the pore size is reduced down to nanoscale dimensions, their impact on drug transport is likely to become greater [63]. Indeed, McGinty et al. [29] modeled drug release from metal surfaces of varying surface topographies and identified nanoporosity and indeed tortuosity as important parameters, which can be tuned to provide enhanced control over stent drug release characteristics. There are a great many approaches to inducing such topographical features, ranging from the creation of highly ordered nanotubular structures to randomly distributed nanopores. Such approaches have been used in a wide range of drug-releasing materials, with applications in orthopedic, dental, and vascular implants. Gultepe et al. [63] has drawn together a very useful overview of these approaches and applications, which the reader is referred to for more detailed information.

A ceramic-coated tacrolimus-eluting stent was originally developed by Jomed International [64]. A two-step process is used to create a nanoporous ceramic surface coating on a standard stainless steel 316L stent platform. The first step involves the application of a thin layer of aluminum by a process of physical vapor deposition. This base layer is then electrochemically oxidized to induce nanoporosity [65]. This process produces around 10^9 pores per square centimeters with a pore size range of 5–15 nm. The stents are then immersed in a tacrolimus solution, with subsequent evaporation of the methanol solvent leaving drug-loaded pores. In a preclinical study using the rabbit common carotid artery model, both a low-dose (60 μg) and high-dose (120 μg) version of this device inhibited neointima formation at 28 days. This effect was achieved despite in vitro release studies indicating that there was little further drug release after 3 days [64]. However, a further preclinical study found evidence that the surface was prone to release particle debris [66], and disappointing clinical results mean that this device has not been taken forward.

An alternative to the electrochemical approach is to use sputter coating techniques, where nanoporosity is induced via deposition of various materials, including stainless steel and cobalt-chromium [30,67]. The Setagon stent is an example of such an approach, which comprises a metallic nanoporous surface layer capable of incorporating drugs [67]. Although Medtronic Inc. developed a zotarolimus-eluting version of this stent, it is not clear if this has been taken any further forward.

4.5 Role of stent surface in vessel healing

Up until now, our focus in this chapter has been on the drug delivery aspects of polymer-free DES technologies. However, given that the fundamental rationale driving the development of these devices was improved biocompatibility and reduced hypersensitivity, it is also important to briefly consider how some of the surface modifications described above may impact on vascular healing. The endothelium is crucial to vascular function, so any delayed healing of this layer following stenting leaves the vessel at risk of late stent thrombosis [68]. While the first generation of DES clearly inhibited restenosis, in some patients, this effect appeared to come at a cost of impaired recovery of the endothelium [4,69]. The impact of stent surfaces on the recovery of the endothelium has therefore been the subject of extensive research. It is beyond the scope of this chapter to comprehensively describe such investigations. The role of materials on stent performance has been comprehensively documented by other authors [70,71]. The following discussion will therefore focus on stent surface topography and describe some of the key findings that are of particular relevance to the subject matter at hand.

De Scheerder et al. [72] investigated the impact of electropolishing on stent thrombogenicity and neointima formation in a rat and pig model, respectively. They found that the polished stent surfaces reduced the extent of clot formation and fibrin deposition. Moreover, there was a significant reduction in neointima observed in the polished stent group. In a separate in vitro study, Tepe et al. [73] also demonstrated that electropolished stent surfaces are less thrombogenic. Such studies provide support for the widespread use of highly smooth surface finishes that have been a long-standing feature of most bare metal and drug-eluting stent platforms. However, it has become increasingly clear that cells can respond in quite different ways to variations in surface topography [74], and there are therefore opportunities to use modified surfaces to promote endothelialization. This has led to the investigation of a wide range of materials and stent surface topographies from the microscale down to the nanoscale [70,71,75,67]. Palmaz et al. [76] were pioneers in this area and revealed that the introduction of microscale grooves significantly increased the migration rate of endothelial cells on stent surfaces. This opened up the possibility of using vascular stents with topographical features specifically designed to aid vascular healing. In reality, very few such approaches have reached the market. However, research in this area continues, with a more recent study demonstrating that careful selection of the substrate groove dimensions can provide a means of encouraging targeted endothelial cell migration [77]. The potential of such approaches has also been highlighted by Sprague et al. [78], who have shown that the use of microengineered stent grooves reduces neointima formation in a pig coronary artery stent model. It is therefore clear that the use of grooves and other such features may have a positive role to play in future polymer-free DES development and indeed may have been a contributing factor in the performance of some of the macroporous stents described above.

We have seen that one of the leading examples of the use of microporous stent surfaces is the Yukon stent [46]. Although the micropits on the surface of this stent initially serve as drug reservoirs, following drug release, a rough surface is left behind, and it is claimed that this will promote reendothelialization (Translumina GmbH). A sirolimus-coated version of this stent significantly enhanced endothelialization compared with the Cypher stent in an in vivo rabbit study [79]. There is also some clinical evidence supporting the use of this surface, with the drug-free microporous Yukon stents displaying a trend toward less late lumen loss and lower restenosis, when compared with smooth surface bare metal stents [80]. While improved endothelialization may be the mechanism responsible for these beneficial clinical outcomes, it is very challenging to demonstrate this clinically. It may however have been expected that improved endothelialization with the Yukon stent would have led to reduced rates of stent thrombosis. However, the low rates of stent thrombosis achieved with current DES, typically around 0.5%–1%, make demonstrating such an improvement very difficult. Indeed, there was no difference in stent thrombosis observed between the Yukon and Taxus stents after 5 years [48]. Nonetheless, the equivalence that has been demonstrated in this trial and others that have preceded it provides confidence that stent surfaces need not necessarily be perfectly smooth to be safe and effective.

We have seen that the BioFreedom stent is now demonstrating great potential for use in patients at high bleeding risk [56]. It appears to have the benefits of a bare metal stent, faster reendothelialization allowing more rapid removal of dual antiplatelet therapy while also displaying the inhibitory effects on restenosis of conventional DES [40]. The fact that the microtextured finish is applied only to the abluminal side, an important feature distinguishing it from the Yukon stent appears to suggest that it is the selection of Biolimus A9 and its very rapid release profile that are the dominant factors behind the impressive results that have so far been achieved with this device.

4.6 Summary and future perspectives

Polymer-free DES technology has evolved considerably since the introduction of the first paclitaxel-eluting stents in the early 2000s. Direct loading of drug onto unmodified stent surfaces was generally found to be inferior to conventional polymer-coated DES. The use of surface modification strategies at the macro-, micro-, and nanoscale helped provide enhanced drug release profiles. The best of these strategies has produced impressive clinical results that are comparable with state-of-the art polymer-coated DES, although they still represent a small proportion of currently used DES. It has been a long-standing goal of manufacturers to develop a stent that will reduce restenosis while not impairing recovery of the endothelium. The most recent clinical data on the use of the Cre8, BioFreedom, and the latest drug-filled Stent from Medtronic Inc. suggest that these devices may have achieved a step change in performance, and we may therefore see an increase in their use. Although these devices have shown particular benefit in the treatment of high bleeding risk patients, the reduction in antiplatelet therapy duration required is likely to be an appealing feature that may see them make a wider impact more generally. So, just as polymer-coated DES represented a revolution in stent design, perhaps we are witnessing another such moment. In the same way that conventional DES design has continued to evolve from the introduction of the Cypher and Taxus DES, it is likely that we will see a similar level of technology innovation within polymer-free DES design. Indeed, a recent review of patent registrations demonstrates that there is still significant activity across industry in this area [81]. Such activity will be focused on the development of devices that perform better in particularly challenging lesion types and patient groups, thereby extending the use of stents across a greater number of patients in the future. It can be expected that further new stent platforms will be developed, incorporating modifications to the size and shape of the embedded reservoirs that may provide more effective drug release kinetics [82]. Recent developments within computational modeling are likely to have an increasingly important role to play in the optimization of such device designs [83]. It is interesting to note that the most advanced polymer-free DES that appear to have generated the most impressive recent results all incorporate a smooth lumen facing surface. However, as understanding of the relationship between stent surface and endothelialization continues to improve, this will likely reveal further opportunities for the development of enhanced materials with optimized surface characteristics. Although induction of nanoporosity for drug delivery has failed to make a large impact on the stent field thus far, the surface features at this scale may be of benefit even in the absence of drug effects. Continuing developments in advanced manufacturing, which enable the generation of highly defined topographies across a range of scales, will likely lead to further developments in these areas. Indeed, Liang et al. [84] recently used a femtosecond laser-based approach to generate a biomimetic surface pattern on stainless steel stents and found that this treatment accelerated reendothelialization in a rabbit model of stent injury. Similarly, new drugs are being investigated that may provide more selective targeting of smooth muscle cells than is possible with existing compounds [85]. The extent to which such advances in surface modification technologies and drugs, alone or in combination, will contribute toward continued evolution or lead to another revolution in stent design remains to be seen.

References

[1] R.A. Byrne, P.W. Serruys, A. Baumbach, J. Escaned, J. Fajadet, S. James, M. Joner, S. Oktay, P. Juni, A. Kastrati, G. Sianos, G.G. Stefanini, W. Wijns, S. Windecker, Report of a European Society of Cardiology-European Association of Percutaneous Cardiovascular Interventions task force on the evaluation of coronary stents in Europe: executive summary, Eur. Heart J. 36 (38) (2015) 2608–2620.

[2] M.G. Mennuni, P.A. Pagnotta, G.G. Stefanini, Coronary stents: the impact of technological advances on clinical outcomes, Ann. Biomed. Eng. 44 (2) (2016) 488–496.

[3] W. Chen, T.C. Habraken, W.E. Hennink, R.J. Kok, Polymer-free drug-eluting stents: an overview of coating strategies and comparison with polymer-coated drug-eluting stents, Bioconjug. Chem. 26 (7) (2015) 1277–1288.

[4] R.A. Byrne, M. Joner, A. Kastrati, Polymer coatings and delayed arterial healing following drug-eluting stent implantation, Minerva Cardioangiol. 57 (5) (2009) 567–584.
[5] D.R. Holmes Jr., D.J. Kereiakes, S. Garg, P.W. Serruys, G.J. Dehmer, S.G. Ellis, D.O. Williams, T. Kimura, D.J. Moliterno, Stent thrombosis, J. Am. Coll. Cardiol. 56 (17) (2010) 1357–1365.
[6] G. Nakazawa, Stent thrombosis of drug eluting stent: pathological perspective, J. Cardiol. 58 (2) (2011) 84–91.
[7] M. Joner, A.V. Finn, A. Farb, E.K. Mont, F.D. Kolodgie, E. Ladich, R. Kutys, K. Skorija, H.K. Gold, R. Virmani, Pathology of drug-eluting stents in humans: delayed healing and late thrombotic risk, J. Am. Coll. Cardiol. 48 (1) (2006) 193–202.
[8] R. Virmani, G. Guagliumi, A. Farb, G. Musumeci, N. Grieco, T. Motta, L. Mihalcsik, M. Tespili, O. Valsecchi, F.D. Kolodgie, Localized hypersensitivity and late coronary thrombosis secondary to a sirolimus-eluting stent: should we be cautious? Circulation 109 (6) (2004) 701–705.
[9] W.J. van der Giessen, A.M. Lincoff, R.S. Schwartz, H.M. van Beusekom, P.W. Serruys, D.R. Holmes Jr., S.G. Ellis, E.J. Topol, Marked inflammatory sequelae to implantation of biodegradable and nonbiodegradable polymers in porcine coronary arteries, Circulation 94 (7) (1996) 1690–1697.
[10] M.I. Papafaklis, Y.S. Chatzizisis, K.K. Naka, G.D. Giannoglou, L.K. Michalis, Drug-eluting stent restenosis: effect of drug type, release kinetics, hemodynamics and coating strategy, Pharmacol. Ther. 134 (1) (2012) 43–53.
[11] H.C. Lowe, S.N. Oesterle, L.M. Khachigian, Coronary in-stent restenosis: current status and future strategies, J. Am. Coll. Cardiol. 39 (2) (2002) 183–193.
[12] R. Virmani, F.D. Kolodgie, A. Farb, A. Lafont, Drug eluting stents: are human and animal studies comparable? Heart 89 (2) (2003) 133–138.
[13] S. Venkatraman, F. Boey, Release profiles in drug-eluting stents: issues and uncertainties, J. Control. Release 120 (3) (2007) 149–160.
[14] A. Finkelstein, D. McClean, S. Kar, K. Takizawa, K. Varghese, N. Baek, K. Park, M.C. Fishbein, R. Makkar, F. Litvack, N.L. Eigler, Local drug delivery via a coronary stent with programmable release pharmacokinetics, Circulation 107 (5) (2003) 777–784.
[15] C.W. Hwang, D. Wu, E.R. Edelman, Impact of transport and drug properties on the local pharmacology of drug-eluting stents, Int. J. Cardiovasc. Interv. 5 (1) (2003) 7–12.
[16] M.A. Lovich, E.R. Edelman, Mechanisms of transmural heparin transport in the rat abdominal aorta after local vascular delivery, Circ. Res. 77 (6) (1995) 1143–1150.
[17] A.D. Levin, N. Vukmirovic, C.W. Hwang, E.R. Edelman, Specific binding to intracellular proteins determines arterial transport properties for rapamycin and paclitaxel, Proc. Natl. Acad. Sci. U. S. A. 101 (25) (2004) 9463–9467.
[18] D. Song, L.F. Hsu, J.L. Au, Binding of taxol to plastic and glass containers and protein under in vitro conditions, J. Pharm. Sci. 85 (1) (1996) 29–31.
[19] A.W. Heldman, L. Cheng, G.M. Jenkins, P.F. Heller, D.-W. Kim, M. Ware, C. Nater, R.H. Hruban, B. Rezai, B.S. Abella, K.E. Bunge, J.L. Kinsella, S.J. Sollott, E.G. Lakatta, J.A. Brinker, W.L. Hunter, J.P. Froehlich, Paclitaxel stent coating inhibits neointimal hyperplasia at 4 weeks in a porcine model of coronary restenosis, Circulation 103 (18) (2001) 2289–2295.
[20] A. Gershlick, I. De Scheerder, B. Chevalier, A. Stephens-Lloyd, E. Camenzind, C. Vrints, N. Reifart, L. Missault, J.J. Goy, J.A. Brinker, A.E. Raizner, P. Urban, A.W. Heldman, Inhibition of restenosis with a paclitaxel-eluting, polymer-free coronary stent: the European evaLUation of pacliTaxel Eluting Stent (ELUTES) trial, Circulation 109 (4) (2004) 487–493.

[21] E. Iofina, R. Langenberg, R. Blindt, H. Kühl, M. Kelm, R. Hoffmann, Polymer-based paclitaxel-eluting stents are superior to nonpolymer-based paclitaxel-eluting stents in the treatment of De Novo Coronary lesions, Am. J. Cardiol. 98 (8) (2006) 1022–1027.
[22] A.J. Lansky, R.A. Costa, G.S. Mintz, Y. Tsuchiya, M. Midei, D.A. Cox, C. O'Shaughnessy, R.A. Applegate, L.A. Cannon, M. Mooney, A. Farah, M.A. Tannenbaum, S. Yakubov, D.J. Kereiakes, S.C. Wong, B. Kaplan, E. Cristea, G.W. Stone, M.B. Leon, W.D. Knopf, W.W. O'Neill, Non-polymer-based paclitaxel-coated coronary stents for the treatment of patients with de novo coronary lesions: angiographic follow-up of the DELIVER clinical trial, Circulation 109 (16) (2004) 1948–1954.
[23] G.W. Stone, S.G. Ellis, D.A. Cox, J. Hermiller, C. O'Shaughnessy, J.T. Mann, M. Turco, R. Caputo, P. Bergin, J. Greenberg, J.J. Popma, M.E. Russell, A polymer-based, paclitaxel-eluting stent in patients with coronary artery disease, N. Engl. J. Med. 350 (3) (2004) 221–231.
[24] D. Chamié, J.R. Costa Jr., A. Abizaid, R.A. Costa, F. Feres, R. Staico, D. Siqueira, L.F. Tanajura, A. Abizaid, A.G.M.R. Sousa, J.E. Sousa, Comparação randomizada entre o stent eluidor de paclitaxel de nova geração sem polímero e o stent eluidor de paclitaxel com polímero durável em pacientes com doença arterial coronária: resultados da análise angiográfica e ultrassonográfica seriada do estudo PAX-A, Rev. Bras. Cardiol. Invasiva 19 (2011) 379–391.
[25] E.O.d. Abreu-Silva, R.A. Costa, A. Abizaid, A. Ramondo, P. Brenot, H. Benamer, A. Desideri, J. Berland, B.O. Almeida, F. Digne, M.A. Perin, J.P.d. Castro, J.R. Costa Jr., R. Staico, L.F. Tanajura, A. Abizaid, Seguimento angiográfico e clínico tardio do novo stent farmacológico não-polimérico liberador de paclitaxel para o tratamento de lesões coronárias de novo: resultados do estudo PAX-B, Rev. Bras. Cardiol. Invasiva 20 (2012) 146–154.
[26] G. Mani, C.E. Macias, M.D. Feldman, D. Marton, S. Oh, C.M. Agrawal, Delivery of paclitaxel from cobalt–chromium alloy surfaces without polymeric carriers, Biomaterials 31 (20) (2010) 5372–5384.
[27] F. Bozsak, J.M. Chomaz, A.I. Barakat, Modeling the transport of drugs eluted from stents: physical phenomena driving drug distribution in the arterial wall, Biomech Model. Mechanobiol. 13 (2) (2014) 327–347.
[28] Y. Cheng, M.B. Leon, J.F. Granada, An update on the clinical use of drug-coated balloons in percutaneous coronary interventions, Expert Opin. Drug Deliv. 13 (6) (2016) 859–872.
[29] S. McGinty, T.T.N. Vo, M. Meere, S. McKee, C. McCormick, Some design considerations for polymer-free drug-eluting stents: a mathematical approach, Acta Biomater. 18 (2015) 213–225.
[30] I. Tsujino, J. Ako, Y. Honda, P.J. Fitzgerald, Drug delivery via nano-, micro and macroporous coronary stent surfaces, Expert Opin. Drug Deliv. 4 (3) (2007) 287–295.
[31] R. Falotico, T. Parker, R. Grishaber, S. Price, S.A. Cohen, C. Rogers, NEVO: a new generation of sirolimus-eluting coronary stent, EuroIntervention 5 (Suppl F) (2009) F88–F93.
[32] A.L. Bartorelli, D. Trabattoni, F. Fabbiocchi, P. Montorsi, S. de Martini, G. Calligaris, G. Teruzzi, S. Galli, P. Ravagnani, Synergy of passive coating and targeted drug delivery: the tacrolimus-eluting Janus CarboStent, J. Interv. Cardiol. 16 (6) (2003) 499–505.
[33] C.M. Matter, I. Rozenberg, A. Jaschko, H. Greutert, D.J. Kurz, S. Wnendt, B. Kuttler, H. Joch, J. Grunenfelder, G. Zund, F.C. Tanner, T.F. Luscher, Effects of tacrolimus or sirolimus on proliferation of vascular smooth muscle and endothelial cells, J. Cardiovasc. Pharmacol. 48 (6) (2006) 286–292.
[34] M.C. Morice, H.P. Bestehorn, D. Carrie, C. Macaya, W. Aengevaeren, W. Wijns, C. Dubois, R. de Winter, S. Verheye, S. Hoffmann, O. Pachinger, C. Di Mario, Direct

stenting of de novo coronary stenoses with tacrolimus-eluting versus carbon-coated carbostents. The randomized JUPITER II trial, EuroIntervention 2 (1) (2006) 45–52.
[35] J.M. Siller-Matula, I. Tentzeris, B. Vogel, S. Schacherl, R. Jarai, A. Geppert, G. Unger, K. Huber, Tacrolimus-eluting carbon-coated stents versus sirolimus-eluting stents for prevention of symptom-driven clinical end points, Clin. Res. Cardiol. 99 (10) (2010) 645–650.
[36] C. Tamburino, M.E. Di Salvo, D. Capodanno, P. Capranzano, R. Parisi, F. Mirabella, F. Scardaci, G. Ussia, A.R. Galassi, D. Fiscella, R. Mehran, G. Dangas, Real world safety and efficacy of the Janus tacrolimus-eluting stent: long-term clinical outcome and angiographic findings from the tacrolimus-eluting stent (TEST) registry, Catheter. Cardiovasc. Interv. 73 (2) (2009) 243–248.
[37] R. Kornowski, A critical appraisal of the Janus carbostent, Catheter. Cardiovasc. Interv. 73 (2) (2009) 249–250.
[38] S. Cassese, G. De Luca, B. Villari, S. Berti, P. Bellone, A. Alfieri, A. Montinaro, G. Quaranta, P. Marraccini, F. PiscioneM. S. I. on behalf of, Reduced antiplatelet therapy after drug-eluting stenting: Multicenter janus flex carbostent implantation with short dual antiplatelet treatment for 2 or 6 months—matrix study, Catheter. Cardiovasc. Interv. 80 (3) (2012) 408–416.
[39] N. Aslanabadi, A. Separham, R. Beheshti, S. Ghaffari, B. Sohrabi, OPTIMA tacrolimus-eluting stent: a twelve-month clinical follow up with two different periods of dual antiplatelet therapy; 2-month vs. 6-month approach, J. Cardiovasc. Thorac. Res. 4 (3) (2012) 81–84.
[40] P. Urban, Polymer-free drug-coated coronary stents in patients at high bleeding risk, N. Engl. J. Med. 373 (2015) 2038–2047.
[41] D. Carrie, J. Berland, S. Verheye, K.E. Hauptmann, M. Vrolix, R. Violini, A. Dibie, S. Berti, E. Maupas, D. Antoniucci, J. Schofer, A multicenter randomized trial comparing amphilimus- with paclitaxel-eluting stents in de novo native coronary artery lesions, J. Am. Coll. Cardiol. 59 (15) (2012) 1371–1376.
[42] R. Romaguera, J.A. Gómez-Hospital, J. Gomez-Lara, S. Brugaletta, E. Pinar, P. Jiménez-Quevedo, M. Gracida, G. Roura, J.L. Ferreiro, L. Teruel, E. Montanya, A. Fernandez-Ortiz, F. Alfonso, M. Valgimigli, M. Sabate, A. Cequier, A randomized comparison of reservoir-based polymer-free amphilimus-eluting stents versus everolimus-eluting stents with durable polymer in patients with diabetes mellitus the RESERVOIR clinical trial, J. Am. Coll. Cardiol. Intv. 9 (1) (2016) 42–50.
[43] C. Godino, M. Chiarito, M. Donahue, L. Testa, R. Colantonio, A. Cappelletti, A. Monello, V. Magni, D. Milazzo, R. Parisi, A. Nicolino, S. Moshiri, R. Fattori, G. Aprigliano, A. Palloshi, G. Caramanno, M. Montorfano, F. Bedogni, C. Briguori, A. Margonato, A. Colombo, Midterm and one-year outcome of amphilimus polymer free drug eluting stent in patients needing short dual antiplatelet therapy. Insight from the ASTUTE registry (AmphilimuS iTalian mUlticenTer rEgistry), Int. J. Cardiol. 231 (2017) 54–60.
[44] S.G. Worthley, A. Abizaid, A.J. Kirtane, D.I. Simon, S. Windecker, S. Brar, I.T. Meredith, S. Shetty, A. Sinhal, A.P. Almonacid, D. Chamié, A. Maehara, G.W. Stone, S. Worthley, S. Shetty, A. Sinhal, I. Meredith, A. Abizaid, N. Jepson, R. Bhindi, S.T. Lim, P. Stewart, P. Barlis, D. Walters, D. Muller, S. Cox, R. Bhagwandeen, First-in-human evaluation of a novel polymer-free drug-filled stent: angiographic, IVUS, OCT, and clinical outcomes from the RevElution study, J. Am. Coll. Cardiol. Intv. 10 (2) (2017) 147–156.
[45] R.L. Wilensky, A DES RevElution? J. Am. Coll. Cardiol. Intv. 10 (2) (2017) 157–159.
[46] R. Wessely, J. Hausleiter, C. Michaelis, B. Jaschke, M. Vogeser, S. Milz, B. Behnisch, T. Schratzenstaller, M. Renke-Gluszko, M. Stover, E. Wintermantel, A. Kastrati,

A. Schomig, Inhibition of neointima formation by a novel drug-eluting stent system that allows for dose-adjustable, multiple, and on-site stent coating, Arterioscler. Thromb. Vasc. Biol. 25 (4) (2005) 748–753.

[47] J. Hausleiter, A. Kastrati, R. Wessely, A. Dibra, J. Mehilli, T. Schratzenstaller, I. Graf, M. Renke-Gluszko, B. Behnisch, J. Dirschinger, E. Wintermantel, A. Schömig, Prevention of restenosis by a novel drug-eluting stent system with a dose-adjustable, polymer-free, on-site stent coating, Eur. Heart J. 26 (15) (2005) 1475–1481.

[48] L. King, R.A. Byrne, J. Mehilli, A. Schomig, A. Kastrati, J. Pache, Five-year clinical outcomes of a polymer-free sirolimus-eluting stent versus a permanent polymer paclitaxel-eluting stent: final results of the intracoronary stenting and angiographic restenosis - test equivalence between two drug-eluting stents (ISAR-TEST) trial, Catheter. Cardiovasc. Interv. 81 (1) (2013) E23–E28.

[49] J. Watt, S. Kennedy, C. McCormick, E.O. Agbani, A. McPhaden, A. Mullen, P. Czudaj, B. Behnisch, R.M. Wadsworth, K.G. Oldroyd, Succinobucol-eluting stents increase neointimal thickening and peri-strut inflammation in a porcine coronary model, Catheter. Cardiovasc. Interv. 81 (4) (2013) 698–708.

[50] K. Steigerwald, S. Merl, A. Kastrati, A. Wieczorek, M. Vorpahl, R. Mannhold, M. Vogeser, J. Hausleiter, M. Joner, A. Schomig, R. Wessely, The pre-clinical assessment of rapamycin-eluting, durable polymer-free stent coating concepts, Biomaterials 30 (4) (2009) 632–637.

[51] R.A. Byrne, J. Mehilli, R. Iijima, S. Schulz, J. Pache, M. Seyfarth, A. Schomig, A. Kastrati, A polymer-free dual drug-eluting stent in patients with coronary artery disease: a randomized trial vs. polymer-based drug-eluting stents, Eur. Heart J. 30 (8) (2009) 923–931.

[51a] C.M. McKittrick, S. Kennedy, K.G. Oldroyd, S. McGinty, C. McCormick, Modelling the impact of atherosclerosis on drug release and distribution from coronary stents, Ann. Biomed. Eng. 44 (2) (2016) 477–487.

[52] S. Garg, C. Bourantas, P.W. Serruys, New concepts in the design of drug-eluting coronary stents, Nat. Rev. Cardiol. 10 (5) (2013) 248–260.

[53] K. Kolandaivelu, B.B. Leiden, E.R. Edelman, Predicting response to endovascular therapies: dissecting the roles of local lesion complexity, systemic comorbidity, and clinical uncertainty, J. Biomech. 47 (4) (2014) 908–921.

[54] P.W. Serruys, V. Farooq, B. Kalesan, T. de Vries, P. Buszman, A. Linke, T. Ischinger, V. Klauss, F. Eberli, W. Wijns, M.C. Morice, C. Di Mario, R. Corti, D. Antoni, H.Y. Sohn, P. Eerdmans, T. Rademaker-Havinga, G.-A. van Es, B. Meier, P. Jüni, S. Windecker, Improved safety and reduction in stent thrombosis associated with biodegradable polymer-based biolimus-eluting stents versus durable polymer-based sirolimus-eluting stents in patients with coronary artery disease: final 5-year report of the LEADERS (limus eluted from a durable versus ERodable stent coating) randomized, noninferiority trial, J. Am. Coll. Cardiol. Intv. 6 (8) (2013) 777–789.

[55] R.A. Costa, A. Abizaid, R. Mehran, J. Schofer, G.C. Schuler, K.E. Hauptmann, M.A. Magalhães, H. Parise, E. Grube, Polymer-free biolimus a9-coated stents in the treatment of de novo coronary lesions: 4- and 12-month angiographic follow-up and final 5-year clinical outcomes of the prospective, multicenter biofreedom fim clinical trial, J. Am. Coll. Cardiol. Intv. 9 (1) (2016) 51–64.

[56] M.-C. Morice, F.J. Sawaya, The Quest for the Perfect Stent for a Given Patient: Drug-Coated Stents for the Treatment of Coronary Disease, J. Am. Coll. Cardiol. Intv. 9 (1) (2016) 65–67.

[57] J. Zhu, Q. Zhang, L. Chen, C. Zhang, X. Zhou, Y. Yuan, R. Zhang, Three-year clinical outcomes of a polymer-free paclitaxel-eluting microporous stent in real-world practice:

final results of the safety and efficacy registry of the Yinyi Stent (SERY-I), Acta Cardiol. Sin. 33 (1) (2017) 28–33.
[58] B.L. Eppley, W.S. Pietrzak, M.W. Blanton, Allograft and alloplastic bone substitutes: a review of science and technology for the craniomaxillofacial surgeon, J Craniofac Surg 16 (6) (2005) 981–989.
[59] J.A. Shepperd, H. Apthorp, A contemporary snapshot of the use of hydroxyapatite coating in orthopaedic surgery, J. Bone Joint Surg. (Br.) 87 (8) (2005) 1046–1049.
[60] J.R. Costa Jr., A. Abizaid, R. Costa, F. Feres, L.F. Tanajura, A. Abizaid, L.A. Mattos, R. Staico, D. Siqueira, A.G.M.R. Sousa, R. Bonan, J.E. Sousa, Preliminary results of the hydroxyapatite nonpolymer-based sirolimus-eluting stent for the treatment of single de novo coronary lesions: a first-in-human analysis of a third-generation drug-eluting stent system, J. Am. Coll. Cardiol. Intv. 1 (5) (2008) 545–551.
[61] J.R. Costa Jr., A. Abizaid, R. Costa, F. Feres, L.F. Tanajura, A. Abizaid, G. Maldonado, R. Staico, D. Siqueira, A.G.M.R. Sousa, R. Bonan, J.E. Sousa, 1-year results of the hydroxyapatite polymer-free sirolimus-eluting stent for the treatment of single de novo coronary lesions: the VESTASYNC I trial, J. Am. Coll. Cardiol. Intv. 2 (5) (2009) 422–427.
[62] J.R. Costa, B.A. Oliveira, A. Abizaid, R. Costa, M. Perin, A. Abizaid, D. Chamié, L. Fernando Tanajura, A. Sousa, J.E.M.R. Sousa, Clinical, angiographic, and intravascular ultrasound results of the VestSaync II trial, Catheter. Cardiovasc. Interv. 84 (7) (2014) 1073–1079.
[63] E. Gultepe, D. Nagesha, S. Sridhar, M. Amiji, Nanoporous inorganic membranes or coatings for sustained drug delivery in implantable devices, Adv. Drug Deliv. Rev. 62 (3) (2010) 305–315.
[64] H. Wieneke, O. Dirsch, T. Sawitowski, Y.L. Gu, H. Brauer, U. Dahmen, A. Fischer, S. Wnendt, R. Erbel, Synergistic effects of a novel nanoporous stent coating and tacrolimus on intima proliferation in rabbits, Catheter. Cardiovasc. Interv. 60 (3) (2003) 399–407.
[65] T.-A. Hanaoka, A. Heilmann, M. Kröll, H.-P. Kormann, T. Sawitowski, G. Schmid, P. Jutzi, A. Klipp, U. Kreibig, R. Neuendorf, Alumina membranes—templates for novel nanocomposites, Appl. Organomet. Chem. 12 (5) (1998) 367–373.
[66] M. Kollum, A. Farb, R. Schreiber, K. Terfera, A. Arab, A. Geist, J. Haberstroh, S. Wnendt, R. Virmani, C. Hehrlein, Particle debris from a nanoporous stent coating obscures potential antiproliferative effects of tacrolimus-eluting stents in a porcine model of restenosis, Catheter. Cardiovasc. Interv. 64 (1) (2005) 85–90.
[67] F. Variola, J. Brunski, G. Orsini, P.T. de Oliveira, R. Wazen, A. Nanci, Nanoscale surface modifications of medically-relevant metals: state-of-the art and perspectives, Nanoscale 3 (2) (2011) 335–353.
[68] T.F. Lüscher, J. Steffel, F.R. Eberli, M. Joner, G. Nakazawa, F.C. Tanner, R. Virmani, Drug-eluting stent and coronary thrombosis, biological mechanisms and clinical implications, Circulation 115 (8) (2007) 1051–1058.
[69] J. Kotani, M. Awata, S. Nanto, M. Uematsu, F. Oshima, H. Minamiguchi, G.S. Mintz, S. Nagata, Incomplete neointimal coverage of sirolimus-eluting stents: angioscopic findings, J. Am. Coll. Cardiol. 47 (10) (2006) 2108–2111.
[70] G. Mani, M.D. Feldman, D. Patel, C.M. Agrawal, Coronary stents: a materials perspective, Biomaterials 28 (9) (2007) 1689–1710.
[71] B. O'Brien, H. Zafar, A. Ibrahim, J. Zafar, F. Sharif, Coronary stent materials and coatings: a technology and performance update, Ann. Biomed. Eng. 44 (2) (2016) 523–535.
[72] I. De Scheerder, E. Verbeken, J. Van Humbeeck, Metallic surface modification, Semin. Interv. Cardiol. 3 (3-4) (1998) 139–144.

[73] G. Tepe, H.P. Wendel, S. Khorchidi, J. Schmehl, J. Wiskirchen, B. Pusich, C.D. Claussen, S.H. Duda, Thrombogenicity of various endovascular stent types: an in vitro evaluation, J. Vasc. Interv. Radiol. 13 (10) (2002) 1029–1035.

[74] A.S. Curtis, C.D. Wilkinson, Reactions of cells to topography, J. Biomater. Sci. Polym. Ed. 9 (12) (1998) 1313–1329.

[75] J.C. Palmaz, S. Bailey, D. Marton, E. Sprague, Influence of stent design and material composition on procedure outcome, J. Vasc. Surg. 36 (5) (2002) 1031–1039.

[76] J.C. Palmaz, A. Benson, E.A. Sprague, Influence of surface topography on endothelialization of intravascular metallic material, J. Vasc. Interv. Radiol. 10 (4) (1999) 439–444.

[77] S.A. Biela, Y. Su, J.P. Spatz, R. Kemkemer, Different sensitivity of human endothelial cells, smooth muscle cells and fibroblasts to topography in the nano-micro range, Acta Biomater. 5 (7) (2009) 2460–2466.

[78] E.A. Sprague, F. Tio, S.H. Ahmed, J.F. Granada, S.R. Bailey, Impact of parallel micro-engineered stent grooves on endothelial cell migration, proliferation, and function: an in vivo correlation study of the healing response in the coronary swine model, Circ. Cardiovasc. Interv. 5 (4) (2012) 499–507.

[79] M.C. John, R. Wessely, A. Kastrati, A. Schomig, M. Joner, M. Uchihashi, J. Crimins, S. Lajoie, F.D. Kolodgie, H.K. Gold, R. Virmani, A.V. Finn, Differential healing responses in polymer- and nonpolymer-based sirolimus-eluting stents, JACC Cardiovasc. Interv. 1 (5) (2008) 535–544.

[80] A. Dibra, A. Kastrati, J. Mehilli, J. Pache, R. von Oepen, J. Dirschinger, A. Schomig, Influence of stent surface topography on the outcomes of patients undergoing coronary stenting: a randomized double-blind controlled trial, Catheter. Cardiovasc. Interv. 65 (3) (2005) 374–380.

[81] V. Demidov, D. Currie, J. Wen, Patent watch: patent insight into polymer-free drug-eluting stents, Nat. Rev. Drug Discov. 16 (4) (2017) 230–231.

[82] Y.T. Wang, Y.P. Wang, T.Y. Lin, C.E. Lin, H.M. Hsiao, Drug-eluting stent with rhombic-shape reservoirs for drug delivery, in: 2016 International Conference on Applied System Innovation (ICASI), 2016.

[83] H.-M. Hsiao, Y.-H. Chiu, T.-Y. Wu, J.-K. Shen, T.-Y. Lee, Effects of through-hole drug reservoirs on key clinical attributes for drug-eluting depot stent, Med. Eng. Phys. 35 (7) (2013) 884–897.

[84] C. Liang, Y. Hu, H. Wang, D. Xia, Q. Li, J. Zhang, J. Yang, B. Li, H. Li, D. Han, M. Dong, Biomimetic cardiovascular stents for in vivo re-endothelialization, Biomaterials 103 (2016) 170–182.

[85] R. Tang, S.Y. Chen, Smooth muscle-specific drug targets for next-generation drug-eluting stent, Expert. Rev. Cardiovasc. Ther. 12 (1) (2014) 21–23.

Online sources

[1] Alvimedica Company Website. Product information page for DES—Cre8TM, 2017. http://www.alvimedica.com/product/cre8/ (Accessed 22 May 2017).

[2] Minvasys Company Website. Product information page for Amazonia Pax stent, 2017. http://www.minvasys.com/fr/stents/amazonia-pax.php (Accessed 22 May 2017).

[3] Translumina GmbH Company Website. Product Information Sheet for Yukon Choice 4 stent, 2017. http://www.translumina.de/products3/Yukon_Choice4_revision3.pdf. (Accessed 22 May 2017).

[4] Yinyi Biotech Company Website. Product information page for Yinyi stent, 2017. http://www.dlyinyi.com/yaowu-en.php (Accessed 22 May 2017).

Fundamentals of bioresorbable stents

5

H.Y. Ang, J. Ng*, H. Bulluck*, P. Wong*,†, S. Venkatraman‡, Y. Huang‡, N. Foin*,†*
*National Heart Centre Singapore, Singapore, †Duke-NUS Medical School, Singapore,
‡Nanyang Technological University, Singapore

5.1 Introduction

Coronary angioplasty was first conceptually described by Dotter and Judkins in 1964 and performed in 1977 by Gruntzig as a treatment for coronary artery disease (CAD). At the beginning, the angioplasty technique was performed only with a balloon to reopen the occluded vessel, but the clinical outcomes were compromised due to issues such as elastic recoil, acute closure secondary to dissection as well as neointimal hyperplasia [1,2]. Therefore, coronary stents were developed in mid-1980s to overcome these inherent limitations of balloon angioplasty. The bare metal stent (BMS) works by scaffolding the balloon-dilated artery, sealing the dissection flaps, and preventing late recoil. However, the efficacy of BMS was reduced by high incidence of in-stent restenosis (ISR), and neointimal hyperplasia was still prominent [3].

Therefore, to address the issue of in-stent restenosis, drug-eluting stents (DES) consisting of metallic stents coated with antiproliferative drugs such as sirolimus or paclitaxel were introduced in 1999 [4]. DES first received the CE mark in 2002 and was subsequently approved by the FDA in 2003. The first generation of DES was made from 316L stainless steel with a drug-eluting polymeric coating and has a strut thickness of 130–140 μm. Clinical evaluation of these DES had shown that they significantly reduced ISR and target lesion revascularization (TLR) compared to BMS, but might be associated with an increased risk of stent thrombosis [5,6].

Newer-generation DES using cobalt-chromium as stent material were then developed with thinner struts (80–90 μm), biocompatible or biodegradable polymer coating, and comparably improved safety profile. These DES have become the device of first choice for the treatment of CAD in interventional cardiology practice till this day [7]. Although the DES has shown positive clinical results, certain concerns still remain with the use of a metallic implant in the body permanently [8]. Caging the vessel constantly in a metallic stent is not ideal in the long term due to the risk of impaired endothelial function, reduced potential for positive lumen remodeling, interference with normal arterial healing process, and risk of occlusion of covered side branches by neointimal hyperplasia [9–13]. The placement of a metallic stent may also limit future treatment options to the same site [14,15]. Furthermore, although the use of DES has significantly reduced in-stent restenosis, apprehension about its efficacy (late "catch-up" phenomena) still persists. Clinical observations from several large-scale

Functionalized Cardiovascular Stents. https://doi.org/10.1016/B978-0-08-100496-8.00005-6

real-world DES registries have shown that late adverse events such as very late stent thrombosis or late target lesion revascularization beyond 1 year are cause for concerns of using DES [16].

5.1.1 Concept of bioresorbable scaffolds (BRS)

Thus, the concept of bioresorbable scaffolds (BRS) was introduced to address the limitation of using DES in coronary angioplasty. The theory of using a BRS is to provide a transient support to the vessel and subsequently be completely resorbed, allowing the vessel to heal and to return to a more natural state [17]. This decreases the risk of stent thrombosis (a concern with metallic DES) and late malapposition at long-term follow-up as the BRS would have already underwent complete resorption. Ideally, the BRS will retain sufficient radial strength after implantation to prevent vessel recoil and to release the antiproliferative drug. After the healing period, the BRS will degrade and be resorbed completely, leaving the vessel with a healthy endothelium, normal vasomotion, and free of caging [18–20]. BRS fabricated from bioresorbable polymeric material are, in theory, more flexible and conformable and would influence the shear stress pattern to a lesser degree compared to a metallic DES. Shear stress pattern has been shown to affect the development of neointimal hyperplasia after stent implantation [21]. By freeing the vessel from permanent caging, late positive remodeling in response to a physiological stimulus will be possible. The absence of any residual foreign material and restoration of endothelial coverage would also reduce the need for long-term dual antiplatelet treatment (DAPT), which decreases the risk of bleeding, especially in older patients [22–24]. For younger patients, the resorption of the BRS would allow future intervention at the same site and facilitate the access to side branches that were jailed by the original stent. A summary featuring the potential benefits of using BRS over BMS and DES (adapted from literature) [9,25] can be seen in Table 5.1.

Table 5.1 Comparison of BRS with another stent devices

	BMS	DES	BRS
Acute occlusion	+	+	+
Acute recoil	+	+	+
Acute thrombosis	−	+/−	+
Late thrombosis	−	−	+
Neointimal hyperplasia	−	+	+
Constrictive remodeling	+	+	+
Expansive remodeling	−	−	+
Restoration of vasomotion	−	−	+
Late lumen enlargement	−	−	+

"+" indicates prevented or not restricted and "−" indicates not prevented or restricted; *BMS*, bare metal stent; *BRS*, bioresorbable scaffold; *DES*, drug-eluting stent.

Modified from N. Gonzalo, C. Macaya, Absorbable stent: focus on clinical applications and benefits, Vasc. Health Risk Manag. 8 (2012) 125–132; Y. Onuma, P.W. Serruys, Bioresorbable scaffold: the advent of a new era in percutaneous coronary and peripheral revascularization?, Circulation 123 (2011) 779–797.

5.1.2 Current limitations of bioresorbable stents

5.1.2.1 Insufficient mechanical strength

Since the implantation of the first Igaki-Tamai BRS in human in 1999, BRS has gained significant interest and researches have focused on different methods to improve radial strength and making the device radiopaque [26]. Currently, the majority of the material use in manufacturing BRS are polymers, with poly-L-lactic acid (PLLA) being the main player in this field. Corrodible metals such as magnesium (Mg) have been gaining traction and examined as potential BRS material. Generally, the strength to weight ratio of polymers is lower than that of metals, thus making it a challenge to manufacture a degradable device with sufficient radial strength for an appropriate duration [27]. Table 5.2 highlights the gap between the mechanical properties of different biodegradable materials used presently for biomedical devices fabrication.

Polymeric materials such as PLLA will typically exhibit approximately 100-fold lower tensile modulus and Mg-based alloys are about 5-fold lower than conventional metallic DES material. As the tensile modulus is directly proportional to radial stiffness, stents fabricated from these bioresorbable materials require 240% and 50% thicker struts, respectively, in order to match up with the current DES [28]. Fig. 5.1 shows the difference in strut thickness between Abbott Vascular's ABSORB Bioresorbable Vascular Scaffold (BVS), which is fabricated from PLLA (Fig. 5.1A) and the metallic Xience DES (Fig. 5.1B).

Having struts almost twice the thickness of current metallic DES affected the profile and deliverability of the BRS [29,30]. Thicker struts can also result in more flow disturbance, which can potentially increase incidence of acute thrombotic events. PLLA devices have inherent limit of expansion (low ductility) and can fracture due to overdilation. It is important to improve the expandability of the BRS while maintaining the radial strength [31]. With lesser mechanical strength, the BRS require extensive vessel preparation and achieve, on average, lesser acute gain immediately postimplantation than metallic stent, particularly in more calcified lesions.

5.1.2.2 Lack of radiopacity

A cardiologist or interventional radiologist can track the delivery catheter through the patient's vasculature and accurately place the stent at the site of a lesion, which is achieved by fluoroscopy or similar X-ray visualization procedures. In order for a stent to be fluoroscopically visible, it must be more absorptive of X-rays than the surrounding tissue. However, the X-ray attenuation coefficient of most polymeric components is low, thus polymeric BRS suffer from a lack of radiopacity as compared to traditional metallic stents. Most of the current BRS have radiopaque metallic markers in the device for visibility under X-ray.

These radiopaque markers are made from metals of higher atomic weight (e.g., gold, platinum, and tantalum) and are usually placed at the distal and proximal ends of the scaffolds for visualization. The markers are usually secured to the structural element using techniques such as micro-welding and micro-riveting. Although the addition of radiopaque markers in the BRS can aid in accurate positioning of the

Table 5.2 Material properties of metallic alloys and biodegradable polymers

Material	Radiopacity	Tensile modulus (GPa)	Tensile strength (MPa)	Elongation at break (%)	Degradation (months)
Poly-L-lactic acid	–	2–4	60–70	2–6	18–36
Poly-D-lactic acid	–	1–3.5	40–55	2–6	12–16
Polyglycolic acid	–	6–7	90–110	1–2	4–6
Polycaprolactone	–	0.2–0.4	25–35	>300	24–36
Poly(lactic-*co*-glycolic acid) (85L/15G)	–	2–4	40–70	2–6	12–18
Poly(DL-lactide-*co*-glycolic acid) (50DL/50G)	–	2–4	40–50	1–4	1–2
Polycarbonate	–	2–2.4	55–75	80–150	>14
WE43 (Mg alloy)	–	40–50	220–330	2–20	~12
Fe-35MN	–	235	530	32	>12
Pure iron	–	150	210	40	>12
Stainless steel 316L	+	193	668	40	Biostable
Cobalt chromium	+	210–235	1449	40	Biostable

Data from Y. Onuma, P.W. Serruys, Bioresorbable scaffold: the advent of a new era in percutaneous coronary and peripheral revascularization?, Circulation 123 (2011) 779–797; H.M. Garcia-Garcia, P.W. Serruys, C.M. Campos, T. Muramatsu, S. Nakatani, Y.J. Zhang, et al., Assessing bioresorbable coronary devices: methods and parameters, JACC Cardiovasc. Imaging 7 (2014) 1130–1148; N. Foin, R.D. Lee, R. Torii, J.L. Guitierrez-Chico, A. Mattesini, S. Nijjer, et al., Impact of stent strut design in metallic stents and biodegradable scaffolds, Int. J. Cardiol. 177 (2014) 800–808; R.M. AL-Mangour Bandar, S. Yue, Coronary stents fracture: an engineering approach (review), Mater. Sci. Appl. 4 (2013) 606–621.

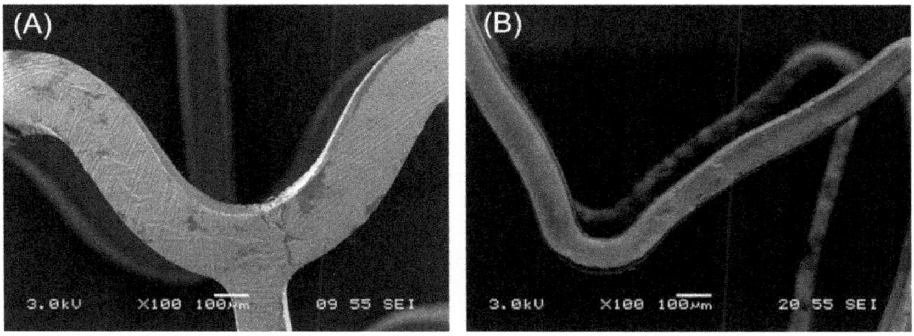

Fig. 5.1 Scanning electron micrographs of coronary stents fabricated from different materials at an ×100 magnification. (A) Abbott's BVS has a strut thickness of 150 μm (width 190 μm) + 6 μm polymer drug coating and (B) its Xience DES has a strut thickness of 81 μm + 7 μm polymer drug coating.

scaffold, assessment of scaffold expansion and lesion coverage remains a challenge [32]. This also complicates retrieval in the case of dislodgment of the BRS from the delivery catheter.

5.2 Current bioresorbable stents technology

This section will explore the different bioresorbable stents that are in different stages of development. Table 5.3 is a summary of different bioresorbable materials and their respective degradation process. The materials used, interaction of each material with tissue in terms of biocompatibility, and the bioresorption mechanism in the body will be discussed. Table 5.4 is a summary of current BRS technology based on PLLA and Table 5.5 will focus on bioresorbable stents that are using other polymers and metal alloys as the backbone material.

5.2.1 *PLLA-based scaffolds*

PLLA is a biodegradable, biocompatible, and biologically inert synthetic polymer, and has been used widely in biomedical application such as sutures and tissue engineering scaffolds. PLLA is semicrystalline polymer (maximum crystallinity = 70%) comprising a mixture of crystalline phase and a less dense amorphous phase. The crystallinity of the polymer defined by the degree of monomers' linear arrangement affects the mechanical strength and degradation rate of PLLA. PLLA with the highest Tg among the general biodegradable polymers has relatively high mechanical properties [33]. However, as mentioned previously, the mechanical strength of PLLA is still inferior as compared to the metals used traditionally in stent fabrication.

Therefore, material processing and stent design modifications are employed to maximize the performance of the BRS as compared to the metals used traditionally in stent fabrication. The processing technique employed can be used to orient individual polymer chains, which will in turn strengthen the material. In addition, annealing,

Table 5.3 **Biodegradable materials' mechanical properties and degradation mechanism**

Material	Absorption	Strength	Stiffness	Degradation mechanism	Degradation products
PLLA	Very slow	High	High	Hydrolysis of ester linkages; phagocytosis and Krebs cycle	Lactic acid to pyruvate to water and carbon dioxide
PDLA	Medium	Medium	Medium	Hydrolysis of ester linkages; phagocytosis and Krebs cycle	Lactic acid to pyruvate to water and carbon dioxide
PLGA	Fast	High	High	Hydrolysis of ester linkages; phagocytosis and Krebs cycle	Lactic acid, glycolic acid, water and carbon dioxide
PCL	Slow	Medium	Very low	Hydrolysis of ester linkages; enzymatic degradation and Krebs cycle	6-Hydroxyhexanoic acid to adipic acid, water, carbon dioxide, hydroxy caproic acid
Tyrosine PC	Very slow	High	Low	Hydrolysis of ester linkages	Water, carbon dioxide, ethanol, desaminotyrosyl-tyrosine
Magnesium	Fast	Very high	Very high	Biocorrosion in body fluid	Magnesium oxide and organic salts; hydrogen gas

PLLA, poly-L-lactic acid; *PDLA*, poly-D-lactic acid; *PLGA*, poly(lactic-*co*-glycolic acid); *PCL*, polycaprolactone; *PC*, polycarbonate.

Table 5.4 Summary of current BRS technology with PLLA as scaffold material

Company	BRS	Scaffold material	Strut thickness (μm)	Crossing profile (mm)	Radiopacity	Antiproliferative drug	Current status
Abbott Vascular	ABSORB BVS	PLLA	150	1.4	Platinum markers	Everolimus	CE mark FDA approved
Amaranth	FORTITUDE APTITUDE MAGNITUDE	PLLA	150 120 Sub 100	NA	NA	Sirolimus	Clinical trials
ART/Terumo	ART18Z	PDLA	170 (1st gen) 140–150 (2nd gen)	NA	NA	None (1st gen) Sirolimus (2nd gen)	Clinical trials CE mark
Arterius	ArterioSorb	PLLA	110–140	NA	RO markers	Sirolimus	Clinical trials
Boston Scientific	FAST	PLLA				Everolimus	Clinical trials
Elixir Medical Corp.	DeSolve DeSolve Cx DeSolve 100	PLLA	150 120 100	1.5 1.3	Metallic markers	Novolimus	CE mark
HuaAn Biotech	XINSORB	PLLA	150–160	NA	RO markers	Sirolimus	Clinical trials
Manli Cardiology	MIRAGE	PLLA fibers	125	1.12–1.47	RO microtubes	Sirolimus	Clinical trials
Meril Life Science	MeRes100	PLLA	100 ± 10	1.2	RO markers	Sirolimus	Clinical trials
OrbusNeich	ON-ABS	PLLA/PDLA/PCL	150	NA	RO markers	Sirolimus/CD34+	Preclinical

BVS, bioresorbable vascular scaffold; *PLLA*, poly-L-lactic acid; *PDLA*, poly-D-lactic acid; *PCL*, polycaprolactone; *RO*, radiopaque.

Table 5.5 Summary of bioresorbable stents that are using other polymers and metal alloys as stent material

Company	BRS	Scaffold material	Strut thickness (μm)	Crossing profile (mm)	Radiopacity	Antiproliferative drug	Current Status
REVA Medical	ReZolve	Desaminotyrosine	125	1.5	Iodine	Sirolimus	Clinical trials
	FANTOM	derived PC		<1.27			CE mark
Xenogenics	Ideal Biostent	PAE salicylic acid	175	1.5–1.7	NA	Sirolimus	Clinical trials
Biotronik AG	Magmaris	Mg alloy	120	1.75	Tantalum	Sirolimus	CE mark
Cardionovum	Renatural M	Mg alloy	NA	NA	NA	Sirolimus	Preclinical
Envision Scientific	Biolute	Mg alloy	120	NA	RO markers	Sirolimus	Preclinical
Life Tech Scientific	IBS	Nitrided Fe	70–80	NA	RO markers	Sirolimus	Preclinical

Fe, iron; *IBS*, iron-based bioresorbable scaffold; *Mg*, magnesium; *PC*, polycarbonate; *PAE*, poly(anhydride-ester).

which is the heating of the polymer above its Tg and then cooling it to relieve the internal stresses induced during the fabrication, can also be used.

5.2.1.1 Bioresorption process of PLLA

For PLLA, degradation and subsequent absorption of the polymer in vivo occurs in three stages via hydrolysis predominantly. The process is a bimolecular nucleophilic substitution reaction catalyzed by the presence of either acids or bases and a chain scission event usually takes place at an ester bond as seen in Fig. 5.2.

In the first stage after the hydration of the polymer, hydrolysis of the amorphous tie chains that are binding the crystal lamellae occurs, leading to a decrease in molecular weight with little effect on the mechanical properties. The amorphous segments are less packed and more hydrophilic due to the carboxylic acid end group, thus are more susceptible to hydration. In the second stage, mass loss happens and the polymer starts to fragment into segments of low-weight oligomers. (Fig. 5.3A) Subsequently, radial strength starts to decrease from the 6th month onwards as a result of the scission of amorphous tie chains connecting the crystalline regions and is completed at the 12th month after implantation [34]. The third stage is the dissolution of the monomers by phagocytes. The monomer (e.g., L-lactate) is changed into pyruvate, which enters the Krebs cycle and is further converted into carbon dioxide and water. These final products are excreted from the body through the kidneys or lungs, which results in complete bioresorption of the polymer [25,35].

5.2.1.2 Abbott vascular BVS

The first BRS to receive the CE mark and FDA approval is the Abbott Vascular's (California, United States) ABSORB Bioresorbable Vascular Scaffold (BVS). It is the most studied BRS on the market currently and it is fabricated from semicrystalline PLLA consisting of a coating of amorphous poly-D,L-lactide (PDLA) eluting Everolimus, with a strut thickness of 150 μm. Fig. 5.3B shows the optical coherence tomography (OCT) data of Abbott's BVS 1.0 after implantation at different time points. The results indicated that at the end of 2 years, the BVS struts were no longer discernible, showing resorption of the scaffold. The endoluminal surface appeared to be smooth and homogenous compared to the corrugated lining at 6 months, suggesting healing of the artery.

The second generation of Abbott's BVS (BVS 1.1) reported a 75% release of the coated Everolimus within 30 days, comparable to the Everolimus-coated Xience at the same dose density (100 μg/cm^2). It has been suggested that the release kinetics for Everolimus from the BVS follows a diffusion-controlled mechanism instead of erosion-controlled since PLLA experience little erosion within a 30-day period. The

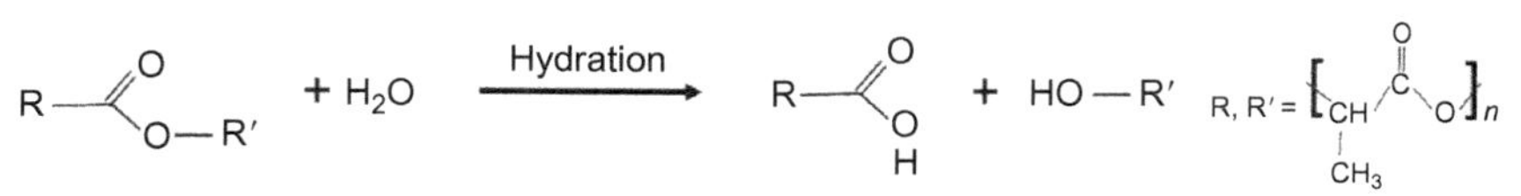

Fig. 5.2 Reaction pathway for hydrolytic degradation of PLLA.

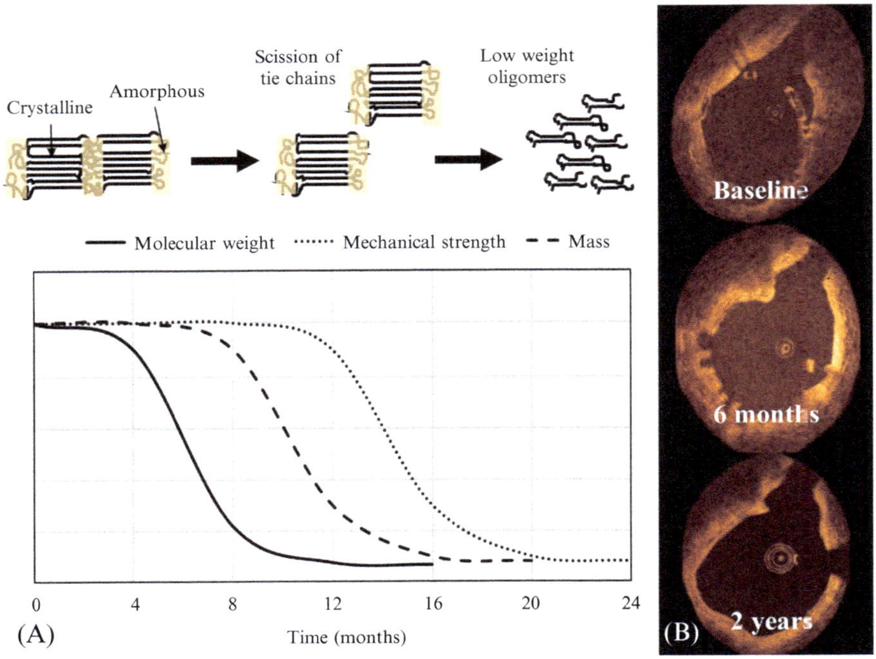

Fig. 5.3 The hydrolytic degradation profile of semicrystalline biodegradable polyesters. (A) During degradation, hydration of the polymer occurs, causing the susceptible amorphous tie chains to be hydrolyzed, decreasing the polymer's molecular weight. The decrease of mechanical support occurs approximately at around 6 months. Mass loss starts from the 12th month and is completed at 24 months. (B), OCT images after implantation of Abbott's BVS 1.0, showing that at 6 months, the endoluminal lining was corrugated and the struts were covered by neointimal tissues. At 24 months, most struts were no longer detected and the endoluminal surface was smooth and circular.
(A) Modified from Y. Onuma, P.W. Serruys, Bioresorbable scaffold: the advent of a new era in percutaneous coronary and peripheral revascularization? Circulation 123 (2011) 779–797. (B) Reproduced with permission from Elsevier (S. Verheye, J.A. Ormiston, J. Stewart, M. Webster, E. Sanidas, R. Costa, et al., A next-generation bioresorbable coronary scaffold system: from bench to first clinical evaluation: 6- and 12-month clinical and multimodality imaging results, JACC: Cardiovasc. Interv. 7 (1) (2014) 89–99).

clinical outcomes of the BVS has been encouraging; a 1-year metaanalysis of four ABSORB randomized controlled trials (involving 3889 patients) revealed that the patient-oriented composite endpoints did not differ significantly between the BVS and DES (Xience). However, patients treated with biodegradable scaffold appeared to have an overall higher risk of definite or probable stent thrombosis [36].

5.2.1.3 Elixir Medical Corp. DESolve

The second CE marked DESolve BRS (Elixir Medical Corporation, California, United States) is also fabricated from PLLA, but with Novolimus as the antiproliferative drug. The first generation of DESolve BRS had an initial strut thickness of 150 μm, but has since reduced it to 120 μm (DESolve Cx) and the DESolve 100 (with a strut thickness of 100 μm) has been introduced recently. The differentiating features of DESolve from the other BRS are: (i) inherent self-correcting properties of the device in case of minor strut malapposition, which is due to the proprietary processing technique and (ii) its relative ductility that allows DESolve a wide range of expansion without risk of strut fracture. DESolve has a wide safety margin for expansion and it was reported that the 3.0 mm scaffold can be expanded to 4.5 mm without strut fracture [37]. The initial DESolve Nx trial demonstrated the safety and efficacy of the scaffold (150 μm strut thickness) with good acute performance and low late lumen loss. The DESolve Cx and the DESolve 100 are currently being evaluated in clinical trials [38].

5.2.1.4 Amaranth Medical BRS

FORTITUDE (Amaranth Medical, California, United States) is a sirolimus-eluting BRS with a strut thickness of 150 μm, while APTITUDE has a strut thickness of 120 μm. Amaranth has recently revealed that a 90 μm strut thickness BRS is currently being developed [39]. FORTITUDE/APTITUDE is fabricated from a high molecular weight PLLA with a proprietary dip coating tube fabrication process. Preliminary results of the FORTITUDE/APTITUDE BRS revealed high radial strength, prolonged mechanical stability, and exhibited minimal recoil of the scaffold [3]. It has been reported that the FORTITUDE BRS can be dilated more than twice its initial diameter without compromising the mechanical strength [38]. FORTITUDE has been evaluated in clinical trials (MEND II, RENASCENT I) and has exhibited no scaffold restenosis and thrombosis to date. APTITUDE is currently in Phase 2 clinical trials (RENASCENT II). Amaranth has also introduced the MAGNITUDE BRS (strut thickness of sub 100 μm), which is presently undergoing clinical evaluation.

5.2.1.5 Manli Cardiology MIRAGE

The MIRAGE BRS (Manli Cardiology Ltd., Singapore) is a novel technology incorporating a PLLA microfiber helix coil design tagged on three backbones and a biodegradable PLA abluminal coating that releases sirolimus [40]. MIRAGE BRS has high flexibility (elongation at break: >20%) due to the helical design, radial strength with strut thickness ranging from 125 to 150 μm, and a relatively shorter bioresorption time of about 14 months compared to the BVS. The improved mechanical properties are attributed to the single microfiber structure processed by extrusion, drawing, and annealing. Other features of the MIRAGE BRS include: (i) no time limitation for staying in artery before deployment; (ii) re-entering artery allowed, and (iii) no gradual balloon inflation required during deployment [41]. The MIRAGE BRS was compared to the BVS in a prospective, randomized first-in-human evaluation on 68 patients (in two centers: Indonesia and Malaysia) and results showed comparable efficacy to the BVS group at 6 months [38,41].

5.2.1.6 Other PLLA-based scaffolds

Arterius Limited's (West Yorkshire, GB) ArterioSorb is fabricated by die drawing of the extruded PLLA tubing and is currently in the preclinical stage. ArterioSorb has a strut thickness of 110–140 μm and a sirolimus-eluting coating. Arterial Remodeling

Technologies (France)'s ART18Z is fabricated using a semicrystalline, amorphous PDLLA polymer with a strut thickness of 170 μm without any drug coating. Animal studies have shown that there was no MACE reported and acute recoil rates were similar with BMS [42]. The ARTDIVA first in man trial launched to evaluate the safety and efficacy of the ART18Z BRS has been completed. A second generation of the device with thinner struts (140–150 μm) and sirolimus coating is currently under preclinical evaluation [43].

The XINSORB BRS (HuaAn Biotech. Co. Ltd, China) has a strut thickness of 150 μm and a PDLLA/PLLA polymeric coating that elutes sirolimus. For the XINSORB BRS, the PDLLA/sirolimus coating was reported to release 80% of the drug in 28 days ex vivo [44]. Preclinical results have shown that the acute absolute/percent recoil of XINSORB was comparable to that of a metallic DES [45]. The MeRes100 BRS (Meril Life Science, India) has a hybrid geometry between an open cell design in mid-portion and a closed cell design at both ends, which can avoid overexpansion at the edges. The strut thickness of MeRes is 100 μm (±20%) with a sirolimus-eluting PDLLA coat. The results from animal studies indicate favorable healing responses at 30 and 60 days in a porcine model [46]. OrbusNeich's (Florida, United States) On-AVS BRS is fabricated from a polymeric blend of PLLA/PDLA/L-lactic-*co*-ε-caprolactone with CD34+ antibodies on the luminal surface for endothelial progenitor cell (EPC) capture and sirolimus on the abluminal surface. The antibodies draw circulating EPC to the site, achieving faster endothelialization with the aim to lower the risk of restenosis and potentially reduce stent thrombosis [38]. The On-AVS scaffold is currently in the developmental stage.

5.2.2 Other polymeric scaffolds

5.2.2.1 REVA Medical ReZolve and Fantom

Tyrosine-derived polycarbonates (PC) are a group of homologous carbonate-amide copolymers with differing length of their respective alkyl ester pendent chains. The difference in the structure of the pendent chains alters the mechanical and thermal properties, degradation rate, and cellulose response of the polymer system. Tyrosine-derived PC have three potential sites for hydrolysis to take place, namely the amide, carbonate, and ester bond. It has been shown that under physiological conditions (37°C, pH 7.4 and in the absence of enzymes) in vitro, the carbonate and ester bonds were susceptible to hydrolysis, but the amide bond remained stable. The hydrolysis of ester bonds results in the formation of carboxylic acid group leading to increased solubility of the degradation products and facilitate the complete resorption of the polymer [47].

REVA Medical's (California, United States) ReZolve and FANTOM BRS are made from a tyrosine-based PC polymer that degrades with water, carbon dioxide, and ethanol as the final products. The ReZolve series had a patented "slide and lock" design to mechanically open and lock into place at a range of diameter in order to preserve the acute lumen gain and give additional support [25]. The ReZolve BRS uses the same proprietary scaffold polymer as the coating for controlled release of sirolimus over 30 days, and it has been reported that majority of the drug was released within 90 days.

However, in 2014, the company introduced the latest FANTOM BRS, which moved away from the "slide and lock" mechanism, but has a reinforced PC material for better device performance. The Fantom BRS has full scaffold radiopacity like the ReZolve and has a reported strut thickness of 125 μm [48]. Fig. 5.4 shows the visibility of the FANTOM BRS under X-ray compared to Abbott's BVS and Xience. The FANTOM I trial showed good acute performance of the FANTOM BRS and REVA is presently evaluating its safety and performance in the FANTOM II trial. The FANTOM BRS received the CE mark in April 2017.

5.2.2.2 Xenogenics Corp. IDEAL (Xenogenics)

Polyanhydrides is a class of biocompatible and biodegradable polymers that has been researched as biomaterials for short-term drug delivery. The polyanhydride polymer has hydrophobic backbone with hydrolytically labile anhydride linkages that are more susceptible to hydrolysis than the ester bonds in PLA. Hence, hydrolytic degradation of the polymer blend can be tailored by altering the polymer composition and the polymer will degrade into their nonmutagenic and noncytotoxic acid counterparts [49]. The IDEAL BRS (Xenogenics Corp., Massachusetts, United States) contains two components: (i) polylactide anhydride mixed with a polymer of salicylic acid and sebacic acid linker forming the core polymer backbone and (ii) a salicylate layer containing sirolimus [50]. The salicylate component has shown to confer antiinflammatory and antiplatelet properties to the scaffold, thereby reducing restenosis and promoting vessel healing [3]. Anhydride polymers have high stiffness, which makes them unsuitable as stent material where material expansion without failure is required. In

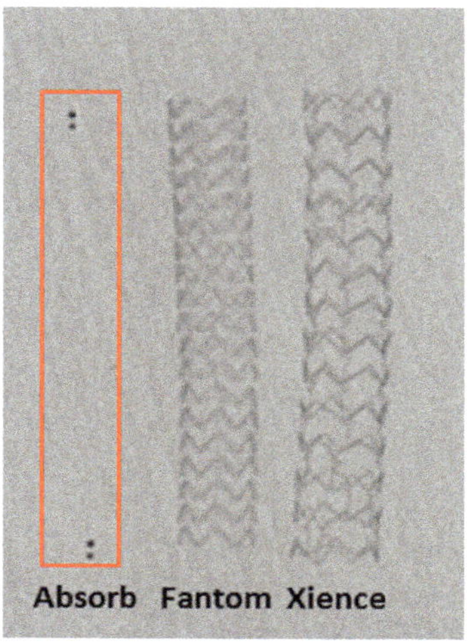

Fig. 5.4 Radiopacity of coronary stents made from PLLA (Absorb), tyrosine-derived PC (Fantom), and cobalt-chromium (Xience). Abbott Vascular's BVS has two platinum markers at the distal and proximal end of the BRS, but the whole scaffold cannot be visualized under X-ray. REVA Medical's FANTOM has comparable radiopacity with the metallic Xience. Reproduced with permission from REVA Medical, Inc.

the IDEAL BRS, the backbone polymer has a selected proportion of ester and anhydride linkages along the polymer. The concentration of anhydride linkages can be tailored to enhance the rate of surface degradation, which is preferred in order to promote continuous release of the drug. The combination of polyesters and polyanhydrides thus incorporates the release characteristics of both groups to allow the BRS to have the required deformability and shape retention, along with a suitable biodegradation rate over the desired stenting period [51].

5.2.3 Biodegradable metallic stents

Biodegradable metallic stents capitalize on the strength of metals while keeping the benefits of using a temporal scaffold to open up the narrowed vessel. Ideally, the degradation rate of the biodegradable metallic stents should be sufficient to prevent detrimental amount of degradation products around the implantation site and in systemic organs. The metals, the degradation, and its products should not induce adverse effect to the healing process too. The corrosion of the biodegradable metals occurs via electrochemical degradation of metals or alloys through their reaction with the local physiological environment.

5.2.3.1 Magnesium stents

Magnesium (Mg) is an element that is highly compatible with the body tissues, making Mg an attractive material with relatively low corrosion resistance and catering to the biocompatibility of both the metal itself and the corrosion reaction products. Pure Mg has a very high corrosion rate in physiological pH, leading to rapid degradation of the material. Therefore, alloying was employed to improve the corrosion resistance and mechanical properties of Mg. Alloying Mg with aluminum zinc or rare earth metals (e.g., cerium) forms a class of bioresorbable material known as the biocorrodible metals. The strength-to-weight ratio of precipitation-hardened Mg alloys has been reported to be comparable with that of strong aluminum alloys and steels [52]. This allows an Mg stent to have the potential of high radial strength for dilating atherosclerotic narrowing and thus higher acute gain of coronary lumen [18,52].

Therefore, bioresorbable stents made from biocorrodible metals are able to have thinner struts and lower profile compared to polymeric BRS [53]. Two phases of the resorption process of Mg alloy have been identified. Firstly, the Mg alloy undergoes degradation to give an Mg-rich compound containing a large amount of oxygen, possibly consisting of a mixture of Mg hydroxide and Mg carbonate. Secondly, after several weeks, these compounds are converted to amorphous calcium phosphate, taking the place of the space previously occupied by the dissolved scaffold struts [54].

BIOTRONIK drug-eluting absorbable magnesium scaffolds (DREAMS)

The DREAMS stent (Biotronik, Berlin, Germany) was an improvement from the original Absorbable Mg Scaffold (AMS) series with: (i) a Mg alloy with a higher collapse pressure and slower resorption, (ii) change from rectangular to square struts,

(iii) reduced strut thickness, and (iv) an antiproliferative drug-eluting polymeric coating [8,52]. The resorption time of DREAMS was also improved to 9–12 months compared to AMS's 2 months. Magnesium alloys have a rapid degradation rate, thus a polymeric coating was also added to slow down the stent degradation. DREAMS has improved mechanical properties due to grain refining of the Mg alloy. DREAMS BRS consisted of a 1 μm layer of poly(lactide-*co*-glycolide) and paclitaxel (0.07 μg/mm^2) coating.

The 2nd-generation DREAMS stent (DREAMS 2G), renamed Magmaris, received its CE mark in 2016. The device has a sirolimus coating and an improved design, which rendered greater radial stiffness and mechanical strength to the stent with the radial stiffness of this metallic BRS comparable to conventional metal stents [53]. In order to further reduce neointimal formation, Magmaris has a 7 μm PLA/sirolimus coating at a drug concentration of 1.4 μg/mm^2 [52]. Studies in a porcine coronary artery model revealed that the vessel tissue concentration of sirolimus was 2.6 ng/mg at 90 days. The recently presented BIOSOLVE II trial showed improved results of the Magmaris device compared with previous versions [55].

Envision Scientific BIOLUTE

The BIOLUTE BRS (Envision Scientific Pte. Ltd., Gujarat, India) has a sandwich layer coating between the fully resorbable magnesium alloy in order to maintain sufficient mechanical properties with a 120 μm strut thickness. The magnesium alloy used is reported to have three times higher radial strength than PLLA, thus allowing the Mg stent to have thinner struts compared to the polymeric scaffolds. BIOLUTE is coated first by a very thin layer of biocompatible proprietary polymer mixture with sirolimus before the next coating of nanocarriers (containing sirolimus) is added. The sandwich layer helps to maintain rigidity and strength of the device as the nanocarriers layer is degrading; the sandwich layer continues to release sirolimus [56,57].

5.2.3.2 Iron stents

Iron's mechanical properties, such as its elastic modulus, are similar to traditional stainless steel stents and it is radiopaque, making it an attractive candidate for bioresorbable stent fabrication. Furthermore, iron is less brittle than Mg; therefore, stents made from iron can potentially have a thinner strut. However, one major issue with iron-based bioresorbable stent will be its slow degradation rate and modifications need to be done to accelerate the resorption process [58].

Life Tech Scientific iron-based bioresorbable scaffold (IBS)

Life Tech Scientific's (Shenzhen, China) IBS makes use of nitrided pure iron as the backbone of the stent with a polymeric coating releasing sirolimus. Nitrided iron stents have shown to exhibit better radial strength and stent stiffness compared to pure iron stents due to the strengthening effect [59]. Extremely low nitrogen alloying into pure iron leads to solid solution strengthening of nitrogen that enters the Iron lattices as interstitial solute atoms. Dispersion strengthening also occurs as a result of the even dispersion of fine iron nitride particles. The strut thickness of the IBS was reported to be 70–80 μm at the 2015 TCT conference. The polymer layer of the IBS helps in creating a local environment that is favorable for iron corrosion, leading to faster corrosion and less solid corrosion products. This is employed to address the issue of iron's slow corrosion rate and the precipitation of corrosion products in tissue as the IBS degrades [60]. The results from the preclinical studies of the IBS showed good mechanical performance of this iron BRS, but corrosion products were found to precipitate in the tissue [60].

5.2.4 Clinical outcomes of the Absorb BVS

The most clinically studied BRS will be Abbott Vascular's Absorb BVS, which has been commonly compared with Xience, a cobalt-chromium everolimus-eluting stent (CoCr-EES). Currently, there have been several trials across the world comparing the short-term (<1 year) clinical outcomes of these two devices with the objective of establishing noninferiority of the BRS compared to metallic DES [61–63].

Presently, there are two studies that have summarized the results obtained from the various trials in a patient level, pooled metaanalysis to present an overall comparison of the two devices. Stone et al. have compiled four randomized trials to compare postprocedural angiographic results as well as 1-year clinical outcomes of both devices [64]. Another metaanalysis took into account the results from six trials, including the four from Stone's study and compared the 1-year follow-up angiographic results and clinical outcomes for both devices [65]. Fig. 5.5 is an outline highlighting the details of each trial in this metaanalysis.

Firstly, postprocedural angiographic results showed that the CoCr-EES performed better with larger in-device acute gain, larger in-device minimal luminal diameter as well as lower percentage diameter stenosis compared to the BVS. This outcome is likely attributed to the greater strut thickness of the BVS as well as its inability to overexpand without the risk of stent fracturing compared to the CoCr-EES.

Short-term follow-up angiographic results also favor CoCr-EES with lower in-device late lumen loss compared to BVS. With regard to success rate, the BVS showed about 4% lower device success rate and a 2% lower procedural rate compared to CoCr-EES. However, further optimization of the bioresorbable scaffold technology and future iterations is expected to improve the success rates of the scaffold in due time. Secondly, the 1-year clinical outcome of both devices showed similar overall patient- and device-orientated composite endpoints (Fig. 5.6). This shows the noninferiority of the BVS's short-term efficacy compared to the CoCr-EES, which is a promising outcome.

However, both metaanalyses show that the BVS had approximately double the risk of device thrombosis as compared to CoCr-EES, with the majority of it happening within 30 days of the procedure (Fig. 5.6B). Furthermore, both studies show that target vessel-related myocardial infarction is more common among BVS patients, although total myocardial infarction, cardiac mortality and all-cause mortality showed no statistical difference between the two devices. The increased risk of target vessel-related myocardial infarction could be partially due to the increased risk of device thrombosis.

	ABSORB China[11]	ABSORB II[7]	ABSORB III[12]	ABSORB Japan[10]	EVERBIO II[9]	TROFI II[8]
Patients						
Randomized, *n*	480	501	2008	400	158 (240*)	191
Age (years)	57.4 (10.5)	61.2 (10.0)	63.5 (10.5)	67.2 (9.5)	65 (11.0)	58.6 (10.1)
Men	343/475 (72%)	385 (77%)	1415/2006 (71%)	309 (77%)	125 (79%)	157 (82%)
Diabetes	115/475 (24%)	120 (24%)	640/2006 (32%)	144 (36%)	30 (19%)	32 (17%)
Insulin dependent	41/475 (9%)	36 (7%)	215/2006 (11%)	35 (9%)	5 (3%)	8 (4%)
Dyslipidaemia	192/475 (40%)	385 (77%)	1732 (86%)	328 (82%)	94 (59%)	115 (60%)
Acute coronary syndrome at admission	306/475 (64%)†	105 (21%)†	523/2007 (26%)†	48 (12%)†	55 (35%)	191 (100%)
Dual antiplatelet therapy						
New P2Y12 inhibitor	5/475 (1%)	NR	719/1990 (36%)	NA	91 (58%)	127 (66%)
Clopidogrel	470/475 (99%)	NR	1271/1990 (64%)	393 (98%)‡	67 (42%)	65 (34%)
Lesions						
Randomized, *n*	503	546	2098	412	208 (325*)	193
Diameter stenosis (%)	64.8 (12.8)	59.5 (11.5)	65.6 (12.1)	64.6 (11.0)	80.5 (15.7)	89.7 (15.2)
Reference vessel diameter (mm)	2.81 (0.44)	2.61(0.39)	2.66 (0.45)	2.75 (0.45)	2.58 (0.65)	2.81 (0.49)
Length (mm)	14 (4.93)	13.8 (6.55)	12.8 (5.6)	13.4 (10.8)	NA	13.1 (7.17)
Type B2/C	369/502 (74%)	254/543 (47%)	1462/2089 (70%)	313/409 (77%)	67/208 (32%)	192/192 (100%)

Data are mean (SD) or *n* (%), unless otherwise indicated. Denominators have been provided when they differ from the total number of patients or lesions. NR = not reported. NA = not available. *Totals in parentheses include patients or lesions in the biolimus-eluting stent group. †Unstable angina only. ‡Ten (2.5%) patients received ticlopidine.

***Table*: Baseline characteristics**

Fig. 5.5 Patient and lesion characteristics of each trial in Cassese et al.'s metaanalysis.
A total of 3738 patients were included in this study and the ratio between patients assigned with BRS and CoCr-EEs was 2:1.
Reproduced from S. Cassese, R.A. Byrne, G. Ndrepepa, S. Kufner, J. Wiebe, J. Repp, et al., Everolimus-eluting bioresorbable vascular scaffolds versus everolimus-eluting metallic stents: a meta-analysis of randomised controlled trials, Lancet 387 (2016) 537–544.

Recent 3-years follow-up from the ABSORB II trial reported that vasomotor reactivity was not statistically different between the BVS and CoCr-EES groups, while late lumen loss was significantly larger in the BVS arm. The BVS group also recorded eight definite scaffold thromboses and one late probable scaffold thrombosis, but no definite or probable stent thrombosis in CoCr-EES [66]. Two-years follow-up results from the ABSORB III trial, on the other hand, revealed an increase in major adverse cardiac events for BVS compared to Xience group (TLF: 11.0% vs 7.9% (P=.03), target-vessel MI: 7.3% vs 4.9% (P=.04)) [67]. This outcome has prompted the FDA to investigate the increased 2-year rate of MACE observed with the BVS. EU regulatory authority and Abbott Vascular have also decided to restrict commercial availability of the device in Europe from May 2017 onwards.

Recent clinical data has highlighted the importance of implantation procedure on clinical outcomes of the BVS; improvement was observed in the pooled ABSORB trials outcome with the adoption of BVS-specific strategy [68,69]. Overall, Abbott's BVS's short-term clinical efficacy is comparable to current DES, but the proposed long-term benefits remain to be seen. Currently, there is only one study that has reached the 5-years follow-up for BVS and the results appear promising, with stabilized luminal dimensions as well as no discernible struts detected by OCT or IVUS at 5-years follow-up [70]. However, this is only a small study and the clinical focus will

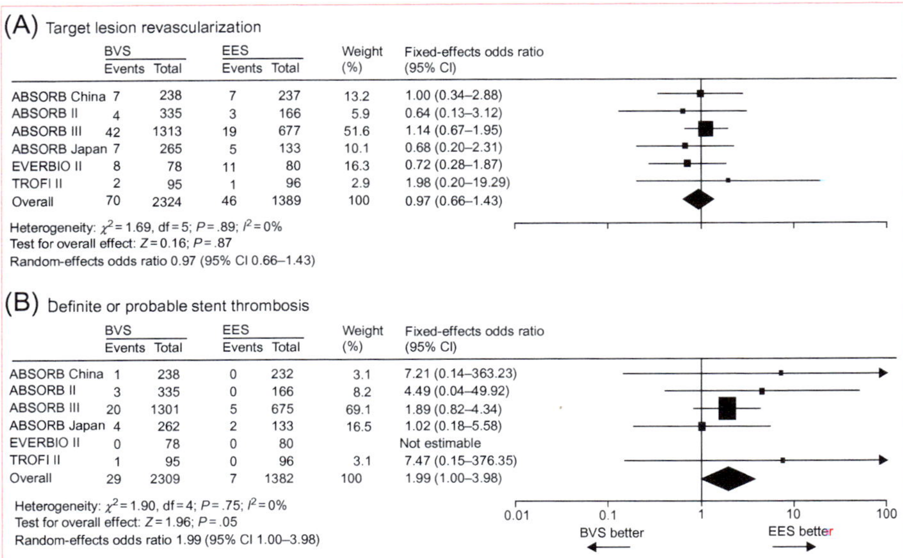

Fig. 5.6 One-year follow-up clinical outcome for target lesion revascularization and stent thrombosis. (A) While the clinical outcome (target lesion revascularization) of both devices were comparable, (B) patients who received the BVS had double the risk of definite and probable stent thrombosis compared to CoCr-EES.
Reproduced from S. Cassese, R.A. Byrne, G. Ndrepepa, S. Kufner, J. Wiebe, J. Repp, et al., Everolimus-eluting bioresorbable vascular scaffolds versus everolimus-eluting metallic stents: a meta-analysis of randomised controlled trials, Lancet 387 (2016) 537–544.

still be on the ongoing long-term clinical trials (most notably the Absorb IV trial) for their 5-years follow-up results and to evaluate if the BVS is truly superior to metallic DES in the long-term.

5.3 Future perspectives

While metallic DES have become the standard choice of device in coronary stenting, the long-term clinical trial outcomes of different DES have displayed a worrying trend of increased adverse events. The use of BRS can potentially address the shortcomings of DES by temporally opening up diseased vessel before completely being resorbed after the vessel heals. The BRS will have to match up with the DES's short-term performance with the potential advantage of longer-term benefits due to not caging the vessel permanently and restoring its vasomotion. The clinical outcomes of the ABSORB BVS have demonstrated noninferiority of the scaffolds compared to metallic DES, but the higher rate of scaffold thrombosis of the BVS remains a concern. The main limitation of the BRS is the weaker mechanical properties (e.g., low tensile modulus and elongation at break) of polymeric materials compared to cobalt-chromium used in the fabrication of metallic DES. Lower mechanical strength leads to thicker scaffold struts that might result in more flow disturbance, thus potentially increasing acute thrombotic events. Implantation technique is also crucial in optimizing clinical outcomes of the device.

The first generation of polymeric BRS has strut thickness of 150 μm and above and the newer generation of BRS is working towards scaffolds with thinner struts and adequate mechanical performance. Research has been done to address this technical challenge of fabricating a polymeric BRS with sufficient radial strength. Postprocessing methods of polymers such as extrusion, annealing, and microfiber technology have been employed to increase the crystallinity of the PLLA and enhanced the overall mechanical strength of the scaffold. For the biodegradable metallic stents, the degradation rate of Mg has to be prolonged, while iron-based stents need to address the issue of slow degradation. More randomized clinical trials will have to be conducted for the biodegradable metallic stents in order to assess its long-term performance. In conclusion, there is undoubtedly a lot of exciting development and research happening in the BRS field where further breakthroughs are expected to advance the PCI practice and improve clinical outcomes in patients' care.

References

[1] J. Iqbal, J. Gunn, P.W. Serruys, Coronary stents: historical development, current status and future directions, Br. Med. Bull. 106 (2013) 193–211.

[2] A. Gruntzig, A. Maresta, W. Gossler, M. Schlumpf, M. Turina, Percutaneous transluminal dilation by catheter of coronary—artery stenosis (author's transl), G. Ital. Cardiol. 10 (1980) 261–267.

[3] Y. Zhang, C.V. Bourantas, V. Farooq, T. Muramatsu, R. Diletti, Y. Onuma, et al., Bioresorbable scaffolds in the treatment of coronary artery disease, Med. Devices (Auckl.) 6 (2013) 37–48.

[4] R.P. Kraak, M.J. Grundeken, K.T. Koch, R.J. de Winter, J.J. Wykrzykowska, Bioresorbable scaffolds for the treatment of coronary artery disease: current status and future perspective, Expert Rev. Med. Devices 11 (2014) 467–480.

[5] A. Colombo, J. Drzewiecki, A. Banning, E. Grube, K. Hauptmann, S. Silber, et al., Randomized study to assess the effectiveness of slow- and moderate-release polymer-based paclitaxel-eluting stents for coronary artery lesions, Circulation 108 (2003) 788–794.

[6] P.W. Serruys, A.T. Ong, J.J. Piek, F.J. Neumann, W.J. van der Giessen, M. Wiemer, et al., A randomized comparison of a durable polymer Everolimus-eluting stent with a bare metal coronary stent: the SPIRIT first trial, EuroIntervention 1 (2005) 58–65.

[7] G.G. Stefanini, M. Taniwaki, S. Windecker, Coronary stents: novel developments, Heart 100 (2014) 1051–1061.

[8] T. Muramatsu, Y. Onuma, Y.J. Zhang, C.V. Bourantas, A. Kharlamov, R. Diletti, et al., Progress in treatment by percutaneous coronary intervention: the stent of the future, Rev. Esp. Cardiol. (Engl. Ed.) 66 (2013) 483–496.

[9] N. Gonzalo, C. Macaya, Absorbable stent: focus on clinical applications and benefits, Vasc. Health Risk Manag. 8 (2012) 125–132.

[10] J. Wiebe, H.M. Nef, C.W. Hamm, Current status of bioresorbable scaffolds in the treatment of coronary artery disease, J. Am. Coll. Cardiol. 64 (2014) 2541–2551.

[11] T. Takayama, T. Hiro, A. Hirayama, Stent thrombosis and drug-eluting stents, J. Cardiol. 58 (2011) 92–98.
[12] R. Kawaguchi, D.J. Angiolillo, H. Futamatsu, N. Suzuki, T.A. Bass, M.A. Costa, Stent thrombosis in the era of drug eluting stents, Minerva Cardioangiol. 55 (2007) 199–211.
[13] M. Joner, A.V. Finn, A. Farb, E.K. Mont, F.D. Kolodgie, E. Ladich, et al., Pathology of drug-eluting stents in humans: delayed healing and late thrombotic risk, J. Am. Coll. Cardiol. 48 (2006) 193–202.
[14] C. Felix, B. Everaert, R. Diletti, N. Van Mieghem, J. Daemen, M. Valgimigli, et al., Current status of clinically available bioresorbable scaffolds in percutaneous coronary interventions, Neth. Heart J. 23 (2015) 153–160.
[15] Y. Onuma, J. Ormiston, P.W. Serruys, Bioresorbable scaffold technologies, Circ. J. 75 (2011) 509–520.
[16] M. Natsuaki, T. Morimoto, Y. Furukawa, Y. Nakagawa, K. Kadota, K. Yamaji, et al., Late adverse events after implantation of sirolimus-eluting stent and bare-metal stent: long-term (5-7 years) follow-up of the Coronary Revascularization Demonstrating Outcome study-Kyoto registry Cohort-2, Circ. Cardiovasc. Interv. 7 (2014) 168–179.
[17] T. Sharkawi, F. Cornhill, A. Lafont, P. Sabaria, M. Vert, Intravascular bioresorbable polymeric stents: a potential alternative to current drug eluting metal stents, J. Pharm. Sci. 96 (2007) 2829–2837.
[18] M.T. Alfonso Lelasi, Current status and future perspectives on drug-eluting bioresorbable coronary scaffolds: Will the paradigm of PCI shfit? EMJ Int. Cardiol. 1 (2014) 81–90.
[19] M. Lesiak, A. Araszkiewicz, "Leaving nothing behind": is the bioresorbable vascular scaffold a new hope for patients with coronary artery disease? Postep. Kardiol Inter. 10 (2014) 283–288.
[20] S. Garg, P.W. Serruys, Coronary stents: looking forward, J. Am. Coll. Cardiol. 56 (2010) S43–S78.
[21] J. Gomez-Lara, H.M. Garcia-Garcia, Y. Onuma, S. Garg, E. Regar, B. De Bruyne, et al., A comparison of the conformability of everolimus-eluting bioresorbable vascular scaffolds to metal platform coronary stents, JACC Cardiovasc. Interv. 3 (2010) 1190–1198.
[22] I. Neamtu, A.P. Chiriac, A. Diaconu, L.E. Nita, V. Balan, M.T. Nistor, Current concepts on cardiovascular stent devices, Mini Rev. Med. Chem. 14 (2014) 505–536.
[23] T.L. Pinto Slottow, R. Waksman, Overview of the 2006 Food and Drug Administration Circulatory System Devices Panel meeting on drug-eluting stent thrombosis, Catheter. Cardiovasc. Interv. 69 (2007) 1064–1074.
[24] J. Iqbal, Y. Onuma, J. Ormiston, A. Abizaid, R. Waksman, P. Serruys, Bioresorbable scaffolds: rationale, current status, challenges, and future, Eur. Heart J. 35 (2014) 765–776.
[25] Y. Onuma, P.W. Serruys, Bioresorbable scaffold: the advent of a new era in percutaneous coronary and peripheral revascularization? Circulation 123 (2011) 779–797.
[26] H.M. Garcia-Garcia, P.W. Serruys, C.M. Campos, T. Muramatsu, S. Nakatani, Y.J. Zhang, et al., Assessing bioresorbable coronary devices: methods and parameters, JACC Cardiovasc. Imaging 7 (2014) 1130–1148.
[27] J. Kohn, J. Zeltinger, Degradable, drug-eluting stents: a new frontier for the treatment of coronary artery disease, Expert Rev. Med. Devices 2 (2005) 667–671.
[28] J. Berglund, Y. Guo, J.N. Wilcox, Challenges related to development of bioabsorbable vascular stents, EuroIntervention 5 (Suppl. F) (2009) F72–F79.
[29] N. Foin, R.D. Lee, R. Torii, J.L. Guitierrez-Chico, A. Mattesini, S. Nijjer, et al., Impact of stent strut design in metallic stents and biodegradable scaffolds, Int. J. Cardiol. 177 (2014) 800–808.
[30] R. Waksman, R. Pakala, Biodegradable and bioabsorbable stents, Curr. Pharm. Des. 16 (2010) 4041–4051.

[31] J.A. Ormiston, F. De Vroey, P.W. Serruys, M.W. Webster, Bioresorbable polymeric vascular scaffolds: a cautionary tale, Circ. Cardiovasc. Interv. 4 (2011) 535–538.
[32] M. Bartkowiak-Jowsa, R. Będziński, A. Kozłowska, J. Filipiak, C. Pezowicz, Mechanical, rheological, fatigue, and degradation behavior of PLLA, PGLA and PDGLA as materials for vascular implants, Meccanica 48 (2013) 721–731.
[33] D.Y. Kwon, J.I. Kim, D.Y. Kim, H.J. Kang, B. Lee, K.W. Lee, et al., Biodegradable stent, J. Biomed. Sci. Eng. 05 (04) (2012) 9.
[34] J. Oberhauser, S. Hossainy, R. Rapoza, Design principles and performance of bioresorbable polymeric vascular scaffolds, EuroIntervention 5 (2009) F15–F22.
[35] C.R. Gajjar, M.W. King, Degradation process, Resorbable Fiber-Forming Polymers for Biotextile Applications, Springer International Publishing, Cham, 20147–10.
[36] G.W. Stone, R. Gao, T. Kimura, D.J. Kereiakes, S.G. Ellis, Y. Onuma, et al., 1-Year outcomes with the Absorb bioresorbable scaffold in patients with coronary artery disease: a patient-level, pooled meta-analysis, Lancet 387 (2016) 1277–1289.
[37] S. Verheye, J.A. Ormiston, J. Stewart, M. Webster, E. Sanidas, R. Costa, et al., A next-generation bioresorbable coronary scaffold system: from bench to first clinical evaluation: 6- and 12-month clinical and multimodality imaging results, JACC Cardiovasc. Interv. 7 (2014) 89–99.
[38] P. Suwannasom, Y. Sotomi, H. Tateishi, E. Tenekecioglu, Y. Zeng, R.P. Kraak, et al., Bioresorbable drug-eluting scaffolds for treatment of vascular disease, Expert Opin. Drug Deliv. 13 (5) (2016) 1–15.
[39] J.F. Granada, Amaranth Fortitude BRS: differentiating features and clinical update, in: TCT, San Francisco, CA, USA, 2015.
[40] P. Serruys, Manli Mirage BRS: differentiating features and clinical update, in: TCT 2015, San Francisco, CA, USA, 2015.
[41] R.A. Costa, H.-B. Liew, A. Abizaid, J.J. de Ribamar Costa, D. Chamié, A. Abizaid, et al., TCT-546 6-month angiographic results of the novel MIRAGE microfiber Sirolimus-eluting bioresorbable vascular scaffold—a quantitative coronary angiography analysis from the prospective, randomized MIRAGE clinical trial, J. Am. Coll. Cardiol. 66 (2015) B223.
[42] E. Durand, M. Lemitre, L. Couty, T. Sharkawi, C. Brasselet, M. Vert, et al., Adjusting a polymer formulation for an optimal bioresorbable stent: a 6-month follow-up study, EuroIntervention 8 (2012) 242–249.
[43] A. Lafont, Bench testing and pre-clinical evaluation of a bioresorbable scaffold with a racemic mixture of L and D-lactides, in: EuroPCR, Paris, France, 2015.
[44] Y. Wu, L. Shen, L. Ge, Q. Wang, J. Qian, F. Zhang, et al., Six-month outcomes of the XINSORB bioresorbable sirolimus-eluting scaffold in treating single de novo lesions in human coronary artery, Catheter. Cardiovasc. Interv. 87 (Suppl. 1) (2016) 630–637.
[45] Y. Wu, L. Shen, Q. Wang, L. Ge, J. Xie, X. Hu, et al., Comparison of acute recoil between bioabsorbable poly-L-lactic acid XINSORB stent and metallic stent in porcine model, J. Biomed. Biotechnol. 2012 (2012) 8.
[46] M.B. Leon, Meril Life Science MeRes: differentiating features and clinical updates, in: TCT, San Francisco, CA, USA, 2015.
[47] V. Tangpasuthadol, S.M. Pendharkar, J. Kohn, Hydrolytic degradation of tyrosine-derived polycarbonates, a class of new biomaterials. Part I: study of model compounds, Biomaterials 21 (2000) 2371–2378.
[48] A. Abizaid, Fantom—Sirolimus-eluting bioresorbable scaffold, in: EuroPCR, Paris, France, 2015.
[49] N. Kumar, R.S. Langer, A.J. Domb, Polyanhydrides: an overview, Adv. Drug Deliv. Rev. 54 (2002) 889–910.

[50] C.V. Bourantas, Y. Onuma, V. Farooq, Y. Zhang, H.M. Garcia-Garcia, P.W. Serruys, Bioresorbable scaffolds: current knowledge, potentialities and limitations experienced during their first clinical applications, Int. J. Cardiol. 167 (2013) 11–21.

[51] S.K. Varshney, O. Hnojewyi, J. Zhang, P. Rivelli, Polyanhydride polymers and their uses in biomedical devices, US Patent 7674285B2, 2010.

[52] C.M. Campos, T. Muramatsu, J. Iqbal, Y.J. Zhang, Y. Onuma, H.M. Garcia-Garcia, et al., Bioresorbable drug-eluting magnesium-alloy scaffold for treatment of coronary artery disease, Int. J. Mol. Sci. 14 (2013) 24492–24500.

[53] R.D. Alexy, D.S. Levi, Materials and manufacturing technologies available for production of a pediatric bioabsorbable stent, Biomed. Res. Int. 2013 (2013) 11.

[54] Y. Huang, H.C. Ng, X.W. Ng, V. Subbu, Drug-eluting biostable and erodible stents, J. Control. Release 193 (2014) 188–201.

[55] S. Cassese, R.A. Byrne, G. Ndrepepa, S. Kufner, J. Wiebe, J. Repp, et al., Everolimus-eluting bioresorbable vascular scaffolds versus everolimus-eluting metallic stents: a meta-analysis of randomised controlled trials, The Lancet, 387 (2016) 537–544.

[56] M. Doshi, D. Sherdiwala, P. Sojitra, Drug-eluting insert able medical device for treating acute myocardial infarction, thrombus containing lesions and saphenous-vein graft lesions, US Patent WO2011089618 A3, 2011.

[57] D. Shah, BIOLUTE-next, fully resorbable scaffold program and update, in: TCT, San Francisco, CA, USA, 2015.

[58] M. Moravej, D. Mantovani, Biodegradable metals for cardiovascular stent application: interests and new opportunities, Int. J. Mol. Sci. 12 (2011) 4250–4270.

[59] W. Lin, G. Zhang, P. Cao, D. Zhang, Y. Zheng, R. Wu, et al., Cytotoxicity and its test methodology for a bioabsorbable nitrided iron stent, J. Biomed. Mater. Res. B Appl. Biomater. 103 (2015) 764–776.

[60] D. Zhang, Lifetech iron-based BRS: differential feature and challenges, in: TCT, San Francisco, CA, USA, 2015.

[61] R. Gao, Y. Yang, Y. Han, Y. Huo, J. Chen, B. Yu, et al., Bioresorbable vascular scaffolds versus metallic stents in patients with coronary artery disease: ABSORB China trial, J. Am. Coll. Cardiol. 66 (2015) 2298–2309.

[62] S.G. Ellis, D.J. Kereiakes, D.C. Metzger, R.P. Caputo, D.G. Rizik, P.S. Teirstein, et al., Everolimus-eluting bioresorbable scaffolds for coronary artery disease, N. Engl. J. Med. 373 (2015) 1905–1915.

[63] P.W. Serruys, B. Chevalier, D. Dudek, A. Cequier, D. Carrié, A. Iniguez, et al., A bioresorbable everolimus-eluting scaffold versus a metallic everolimus-eluting stent for ischaemic heart disease caused by de-novo native coronary artery lesions (ABSORB II): an interim 1-year analysis of clinical and procedural secondary outcomes from a randomised controlled trial, Lancet 385 (2015) 43–54.

[64] G.W. Stone, R. Gao, T. Kimura, D.J. Kereiakes, S.G. Ellis, Y. Onuma, et al., 1-Year outcomes with the Absorb bioresorbable scaffold in patients with coronary artery disease: a patient-level, pooled meta-analysis, The Lancet, 387 (10025) (2016) 1277–1289.

[65] S. Cassese, R.A. Byrne, G. Ndrepepa, S. Kufner, J. Wiebe, J. Repp, et al., Everolimus-eluting bioresorbable vascular scaffolds versus everolimus-eluting metallic stents: a meta-analysis of randomised controlled trials, Lancet 387 (2016) 537–544.

[66] P.W. Serruys, B. Chevalier, Y. Sotomi, A. Cequier, D. Carrié, J.J. Piek, et al., Comparison of an everolimus-eluting bioresorbable scaffold with an everolimus-eluting metallic stent for the treatment of coronary artery stenosis (ABSORB II): a 3 year, randomised, controlled, single-blind, multicentre clinical trial, The Lancet, 388 (10059) (2016) 2479–2491.

[67] S.G. Ellis, G.W. Stone, Everolimus-eluting bioresorbable vascular scaffolds in patients with coronary artery disease: ABSORB III Trial 2-year results, in: ACC, Florida, United States, 2017.

[68] R.A. Abellas-Sequeiros, R. Ocaranza-Sanchez, C. Galvao Braga, S. Raposeiras-Roubin, D. Lopez-Otero, B. Cid-Alvarez, et al., Assessment of effectiveness and security in high pressure postdilatation of bioresorbable vascular scaffolds during percutaneous coronary intervention. Study in a contemporary, non-selected cohort of Spanish patients, Int. J. Cardiol. 219 (2016) 264–270.

[69] D.G. Rizik, J.B. Hermiller, D.J. Kereiakes, The ABSORB bioresorbable vascular scaffold: a novel, fully resorbable drug-eluting stent: current concepts and overview of clinical evidence, Catheter. Cardiovasc. Interv. 86 (2015) 664–677.

[70] P.W. Serruys, J. Ormiston, R.-J. van Geuns, B. de Bruyne, D. Dudek, E. Christiansen, et al., A polylactide bioresorbable scaffold eluting everolimus for treatment of coronary stenosis 5-year follow-up, J. Am. Coll. Cardiol. 67 (2016) 766–776.

Further Reading

[1] R.M. AL-Mangour Bandar, S. Yue, Coronary stents fracture: an engineering approach (review), Mater. Sci. Appl. 4 (2013) 606–621.

Bioabsorbable metallic stents

Y.F. Zheng
Peking University, Beijing, China

6

6.1 Introduction

Currently used stents are mostly made of 316 SS, Co-Cr alloys, and Ni-Ti alloys in clinical trials [1]. They have favorable corrosion resistance and mechanical strength, fulfilling the mechanical requisite of vascular support. However, serious concerns still remain on account of their permanent nature. Persistent presence of stent is claimed to be superfluous supposing that the remodeling of artery wall reaches a new equilibrium under the circumstance of mechanical stress issuing from the deployed stent [2]. The implantation of permanent stents will ensnare the vessel in a metallic cage perpetually. This interferes the restoration/persistence of vessel wall and perhaps renders a second atherosclerosis [3]. Firstly, they correlated with impairment of vessel geometry as it alters arteries curvature at the entrance and exit of the stent where regions with increased and decreased shear stress emerge. These changes in shear stress are associated with asymmetric patterns of in-stent restenosis [4]. In addition, a permanent metallic cage hampers the response of vessel to the vibration of tensile stress due to pulsatile nature of blood flow. However, under normal circumstance, such mechanical stimulation of arterial vascular plays an important role in maintenance of contractile phenotype of smooth muscle cell [5]. On the other side, the potential risks of impaired surgical revascularization, the vascular inflammation, and neoatherosclerosis are attributed to the presence of a stable foreign body [6]. Thus, the motivation for the development of bioabsorbable stents (BAS) as a potential alternative was driven for the sake of overcoming the drawbacks of permanent stents.

Theoretically, only the healed, natural vessel is left behind once the BAS degrades. Hence, long-term clinical problems such as stent thrombosis are avoided compared with permanent metallic stent since no potential initiator for thrombosis exists like uncovered stent struts. The demand of prolonged antiplatelet therapy is also decreased with potential decrease in related bleeding complications. Meanwhile, future treatment options such as PCI, coronary artery bypass grafting, or pharmacologically induced plaque regression are still available for stented segment with a transient implantation.

There are two classes of materials proposed for BAS: (1) polymer-based and (2) metal-based. Up to now, several polymeric stents based on PLLA, PDLA, and PLGA had been developed [7]. The Igaki-Tamai stent (Kyoto Medical Planning Co, Ltd, Japan) is the first BAS implanted in humans, which is entirely composed of PLLA monofilament with a strut thickness of 170 μm, and the resorption of stent strut occurs within third year of follow-up. Preliminary results with accepted major adverse cardiac events similar to those of bare metal stents have raised the interest of future development of polymeric stent [8]. However, bioabsorbable polymeric stent technology

Functionalized Cardiovascular Stents. https://doi.org/10.1016/B978-0-08-100496-8.00006-8

has its limitation. In consideration of relatively poor mechanical properties of polymers, a thicker stent strut is required in contrast with many metallic stents, which will in turn result in smaller residual luminal area [9]. Another limitation is the inability of expansion and optimal scaffold apposition by its nature. It's difficult to fully expand stent with balloon dilatation and an additional heat is required, potentially harmful to the vessel wall [10]. In addition, polymeric stent struts are prone to rupture when overexpanded [11]. Recently, greater interest has been focused on the bioabsorbable metallic stents (BMSs) because of their superior mechanical properties in respect to their polymeric competitors. Biodegradable/bioabsorbable metals (BMs) are mainly composed of essential metal elements since they are expected to corrode and be metabolized by human body gradually without any deleterious side effect to the host. Meanwhile, complete degradation will be achieved after accomplishing the mission of assisting injured tissue healing [12]. So far, two kinds of alloy systems, namely Mg and Fe, have been extensively studied for the stent application.

6.2 General design criterions of bioabsorbable metallic stents

6.2.1 Healing procedure of blood vessels

With reference to cardiovascular application, stent deployment generally conduces to vascular injury subsequent to balloon angioplasty, thus the denuded intimal surface, cracked atheroma, and the stretched normal segment of the vessel circumference [13]. A complex healing process is induced in response to the injured tissue which is similar to the generalized wound healing, including three overlapping phase: inflammation, granulation, and remodeling, as shown in Fig. 6.1 [12]. As a consequence of tissue injury, platelets aggregation and an acute inflammation response come up, along with cellular inhibition by monocytes/macrophages filtration, the migration and proliferation of local tissue cells such as endothelial and smooth muscle cells as well as extracellular matrix deposition and remodeling [13,14].

6.2.2 Desired performance of bioabsorbable metallic stents

In contrast to the bioinert metal corresponding to the traditional metallic stent, BMSs are bioactive and only serve as a temporary supporting role until the vascular healing. Hence, there are some updated requisites proposed for the application of BMSs to ensure their safety and efficacy. BMSs are characterized by their degradation property, which in turn account for new concerns such as corrosion. An inappropriate corrosion rate would have a deteriorate effect on the mechanical properties and biocompatibility of BMSs, resulting in potential toxicity of releasing degradation products and the premature loss of mechanical strength followed by the BMSs fracture. Meanwhile, vessels recoil and negative remodeling may contribute to a narrowing luminal area to bring about ischemia regardless of the restenosis caused by neointimal hyperplasia due to lesion. Therefore, an adequate radial strength should be considered to resist

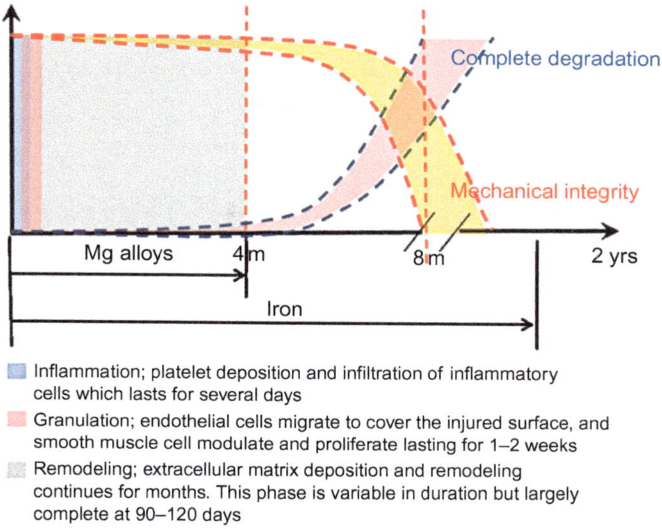

Fig. 6.1 The schematic diagram of degradation behavior and the change of mechanical integrity of magnesium stent during the vascular healing process [12].
Reprinted from Y.F. Zheng, X.N. Gu, F. Witte, Biodegradable metals, Mater. Sci. Eng. R: Rep. 77 (2014) 1–34, Copyright (2014), with permission from Elsevier

vessel recoil and prevent negative remodeling [15]. From this perspective, an ideal BMS is expected to compromise both sufficient mechanical properties and degradation to allow vessel support healing. As demonstrated in Fig. 6.1, at the initial stage the degradation of BMS should occur at a very slow rate to keep the favorable mechanical integrity during vascular remodeling process. Subsequently, a complete degradation is demanded and the degradation products should be absorbed by the human body or easily excreted from the human body. Those two significant variables strongly reply on the selected stent material and implantation site. Hence, the comprehension of interaction between BMSs and implantation environments is necessary to define the optimal materials for BMSs. Another crucial influencing factor is the thickness of stent strut. Generally, a rapid endothelial cell coverage of stent strut to fight against thrombosis with minimal neointimal hyperplasia and no in-stent restenosis is expected after stent deployment. It's claimed that the strut thickness is inverse to the strut coverage related to the late stent thrombosis [16]. In addition, increase in strut thickness also means higher presence of foreign material and worse flow disturbances [7]. Nevertheless, a thicker strut always stands for a better ability of radial support. In this way, BMSs should be designed with a minimal strut thickness under the circumstance of efficient mechanical properties.

Broadly speaking, there should be a balance among diverse parameters in the design of a selected BMS to achieve a desired performance. Some specific design considerations from the current research are summarized by Bowen et al. as listed in Table 6.1 [17]. It is significant to note that a concordant standard of BMSs is still under discussion in research community. Controversies exist on some criteria demonstrated

Table 6.1 General design constraints and criteria for a bioabsorbable metal stent [17]

Criterion	Constraints	Ref.
Bioabsorption	Mechanical integrity for 3–6 months	[18]
	Mechanical integrity for 4 months	[19]
	Mechanical integrity for 6–12 months	[20]
	Full absorption in 12–24 months	[18,21]
Biocompatibility	Nontoxic, non-inflammatory, hypoallergenic	[18]
	No harmful release or retention of particles	[18]
	No aluminum or zirconium content	[22]
Mechanical properties	Yield strength > 200 MPa	[18]
	Tensile strength > 300 MPa	
	Elongation to failure > 15%–18%	
	Elastic recoil on expansion < 4%	
Microstructure	Maximum grain size of ≈ 30 μm	[18]
	Maximum grain size of 10–12.5 μm	[20]
Hydrogen evolution	Evolution < $10\,\mu L\,H^2\,cm^{-2}\,day^{-1}$	[22]
	Penetration rate < $20\,\mu m\,year^{-1}$	[23]

in this table as well. For example, concerns about aluminum-containing magnesium alloys have been raised recently because of their potential toxicity. It is reported that aluminum has an adverse influence on neurons, bone, and osteoblasts, associated with various neurological disorders [24,25]. The opponents, yet, hold the point that relatively low doses of aluminum that contribute little to the serum concentrations can be metabolized by renal action [26,27]. Therefore, the establishment of a rational and unified stent specification is expected for a further clinical application.

6.3 Development of Mg-based bioabsorbable metallic stents

6.3.1 *The physiological function of Mg*

Mg is an essential nutrition element to human body and plays a significant role in supporting and maintaining health and life. There is generally 35 g Mg per 70 kg body weight in human body and the dietary reference ranges from 80 mg/day for a children of 1–3 years old to 420 mg/day for an adult male of 31–70 years old [28,29]. In human body, the majority of the Mg^{2+} is stored in bone, muscle, and soft tissue. Specifically, 50%–60% of Mg exists in bone, binding with calcium and phosphorous; 25%–30% is distributed in muscles and serum Mg^{2+} only represents 1% of body's Mg^{2+} content [30]. Generally, small intestine is the primary absorption site of Mg, covering approximately 30%–50% of Mg intake, which varies in respect to the changes in the amount of Mg intake, senescence, and chronic renal disease [31,32]. Under basal condition, a daily amount of about 2400 mg of Mg^{2+} is filtered by glomeruli along with 95%–99% of resorption by nephron. Supposing that a daily acquisition of 370 mg Mg^{2+}, a net absorption of 100 mg is obtained corresponding to a 100 mg net excretion via urine [30].

As the fourth most abundant mineral and the second most abundant intracellular cation after potassium, Mg holds a post in over 600 enzymatic reactions involved in, but not limited to, energy metabolism, protein synthesis, DNA and RNA synthesis, and stabilizing mitochondrial membranes [29,30]. In addition, the maintenance of some vital physiological functions is closely related to Mg such as cardiac excitability, muscular contraction, vasomotor tone, normal blood pressure, bone integrity, etc. [29]. As for the cardiovascular diseases, in particular, it is claimed that Mg may be beneficial to the improvement of several diseases, such as treatment of cardiac arrhythmia, protection from atherogenesis, and so on [33]. Low serum Mg^{2+} level and low Mg^{2+} intake are related to an increased risk of coronary artery disease, acute myocardial infraction, hypertension, and vascular calcification [30].

6.3.2 The mechanical properties of Mg and its alloys

Without the concerns of releasing alloying element, pure Mg itself is regarded available on account of its existence as an essential nutrient of human body. However, the poor mechanical strength limits its further application. The yield strength (YS), ultimate tensile strength (UTS), and elongation of as-cast pure Mg are around 20 MPa, 84 MPa, and 13%, respectively [34]. Hence, a further improvement is required for the clinical implement. Promotion strategies have been intensively researched such as alloying, heat treatment, and so on. Table 6.2 gives a brief comparison of the mechanical properties of diverse Mg-based alloys with different compositions or working process. Some nontoxic or low-toxic alloying elements are selected to develop new Mg-based alloying system with reliable mechanical properties, and obvious improvement of the mechanical performance can be observed after alloying.

Mg-based alloys generally undergo heat treatments in order to improve the mechanical properties or meet the demand of specific fabrication operations. The most frequent applied heat treatment techniques for Mg alloys are T4 and T6 treatment, thus solution treatment and aging treatment [42]. Jia et al. [43] studied the effect of solution treatment on as-cast Mg-4Zn alloy. It was observed that UTS and elongation of the as-cast Mg-4Zn alloy were obviously improved at the same time because of the solution strength effects of the alloying elements and the decrease of stacking fault energy of the matrix with Zn dissolving into Mg matrix. Feng et al. [44] investigated the effects of solution treatment and aging on microstructure and mechanical properties of as-cast Mg-3Zn-0.9Y-0.6Nd-0.6Zr (ZW30N). Fig. 6.2 shows the tensile properties of ZW30N with different working condition. It can be seen that solution treatment increased the UTS and elongation by about 18.3% and 34.4%, respectively, compared with that of as-cast sample, but the YTS remained relatively stable increasing slightly from 119 to 124 MPa. The main reason was the dissolution of net-like interdendritic W-phase into matrix and the precipitation of small Zr-containing particles at grain interiors. Following by aging, obvious increment of UTS and YS were observed, increasing by almost 23% and 29.8%, respectively, due to the precipitation of fine W, β_1' and β_2' phases.

Table 6.2 Summary of the mechanical properties of different Mg-based alloys

Materials	State	Yield strength (MPa)	Ultimate tensile strength (MPa)	Elongation (%)	Reference
Pure Mg	Cast	~20	~84	~13	[34]
Pure Mg	Rolled	~112	~170	~12.5	[34]
Pure Mg	Extruded	55	167	18	[35]
WE43	Extruded	198	277	17	[36]
Mg-Nd-Zn-Zr	Extruded (extrusion ratio 8) + aging	333 ± 4	334 ± 4	7.9 ± 0.2	[37]
Mg-3Sn-0.5Mn	Extruded	150	240	23	[38]
Mg-4Zn-0.5Zr	Cast	96	176	4	[39]
Mg-4Zn-0.5Zr	Solution-treated	92	134	3	[39]
Mg-3Sr-0.6Y	extruded	~160	~450	~28	[40]
Mg-4Zn-1Sr	Aged		270	12.8	[41]

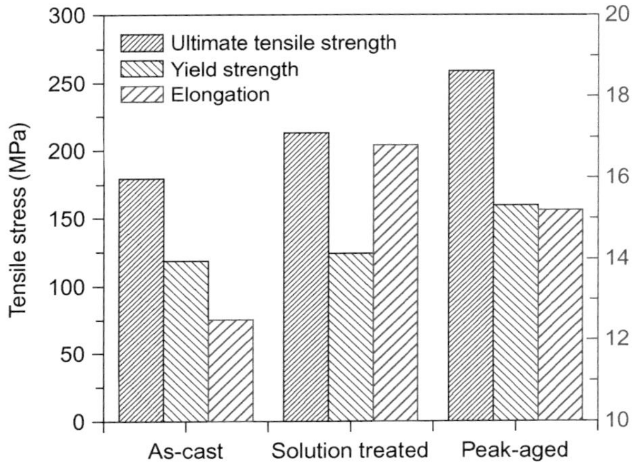

Fig. 6.2 Tensile properties of ZW30N alloy at room temperature in different conditions: as-cast; solution-treated; peak-aged [44].
Reprinted from S. Feng, W. Zhang, Y. Zhang, J. Guan, Y. Xu, Microstructure, mechanical properties and damping capacity of heat-treated Mg-Zn-Y-Nd-Zr alloy, Mater. Sci. Eng. A 609 (2014) 283–292, Copyright (2014), with permission from Elsevier.

6.3.3 *In vitro testing of Mg-based bioabsorbable metals in cardiovascular applications*

Mg-based alloys are promising candidate for bioabsorbable cardiovascular stent application. However, their low corrosion resistance brings a series of concerns. Along with the progression of corrosion, degradation products would accumulate and have a significant effect on the adjacent vascular cells. For instance, the rapid degradation rates of Mg-based implants may contribute to a relatively high extracellular Mg ions concentration in the local microenvironment [45]. Therefore, researches on the complex degradation behavior and interactions between releasing products and vascular cells response at the cellular and molecular are of great interest for developing desired BMS.

Human vascular smooth muscle cells (SMCs) are closely bound up with the pathogenesis of restenosis associated with the stent application and wound healing [46,47]. As a consequence of response to stent deployment-induced vascular injury, the migration of SMCs from media to intima and proliferation contribute to the neointimal thickening and restenosis [48]. Ma et al. [49] designed a cell culture model to investigate the impact of Mg ions concentration on SMCs. The results indicated that Mg ions had biphasic effects on SMCs. Low concentration of Mg^{2+} (10–40 mM) could improve the viability of SMCs, while a higher concentration (50–60 mM) inhibited the cell viability (as shown in Fig. 6.3A). Meanwhile, cell proliferation rate was enhanced with increasing Mg ions concentration when cultured within 10–20 mM concentration range, whereas an opposite trend was observed within 40–60 mM (as exhibited in Fig. 6.3B). The similar phenomenon was also demonstrated in cell adhesion, spreading, and migration, thus low concentration of Mg ions had a positive effect, while higher concentration had an adverse effect. Besides, gene expression assay displayed that Mg^{2+} primarily influences the expression of genes related to coagulation, inflammation, and cell proliferation.

Normal functioning endothelial cells play an important role in promoting vasodilatation and suppressing intimal hyperplasia [46]. Meanwhile, healthy endothelial response is critical for re-endothelialization after stent deployment, reducing the occurrence of restenosis. Zhao et al. [50] reported the human coronary aorta endothelial cells' (HCAECs) response to Mg and some common alloying elements (namely, Ca, Zn, Al, Y, Dy, Nd, and Gd) in Mg-based stent materials. The overall cell viability of HCAECs decreased with the ion concentrations increasing apart from $CaCl_2$-treated

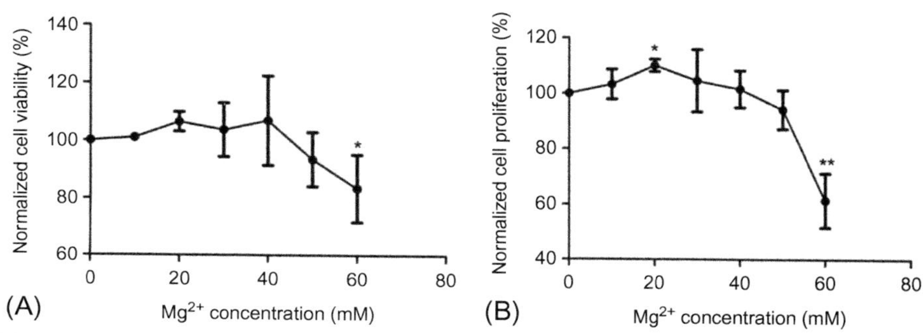

Fig. 6.3 (A) SMC viability analysis. (B) SMC proliferation analysis [49].
Reprinted from J. Ma, N. Zhao, D. Zhu, Biphasic responses of human vascular smooth muscle cells to magnesium ion, J. Biomed. Mater. Res. A 104 (2) (2016) 347–356., Copyright (2016), with permission from John Wiley and Sons.

group, and similar tendency was also observed in cell proliferation for all groups. Low concentration of Mg ions showed positive effect on the cell proliferation and migration, potentially beneficial to re-endothelialization.

The effects of various fluid flow conditions mimicking the blood vessel environment in vivo on the corrosion properties of MgZnCa plates and AZ31 stents were investigated by Wang et al. [51]. The overall corrosions (including localized, uniform, pitting, and erosion corrosions) were speeded up by flow-induced shear stress (FISS) because of the increased mass transfer and mechanical force. The corrosion morphologies of AZ31 stents under the FISS of 0 and 0.056 Pa for 7 days are illustrated in Fig. 6.4. Stents under the static condition exhibited a few of localized corrosion. In contrast, a uniform corrosion mode was observed under dynamic condition. In addition, the strut region facing the flow direction had higher shear stress, leading to the detachment of corrosion products and strut fracture.

6.3.4 In vivo testing of Mg-based bioabsorbable metallic stents within blood vessel

6.3.4.1 Animal testing of Mg-based bioabsorbable metallic stents

As early as in 2003, Heublein et al. [52] conducted the first animal testing of Mg-based cardiovascular stent. They implanted 20 AE21 alloy stents to the coronary artery of 11 domestic pigs with an expected 50% loss of mass within 6 months after implantation. It was reported that the period for a complete breakdown was estimated to be in the order of 3 months. Forty percent of perfused lumen diameter was lost between days 10 and 35 due to the neointima formation and a 25% re-enlargement was observed

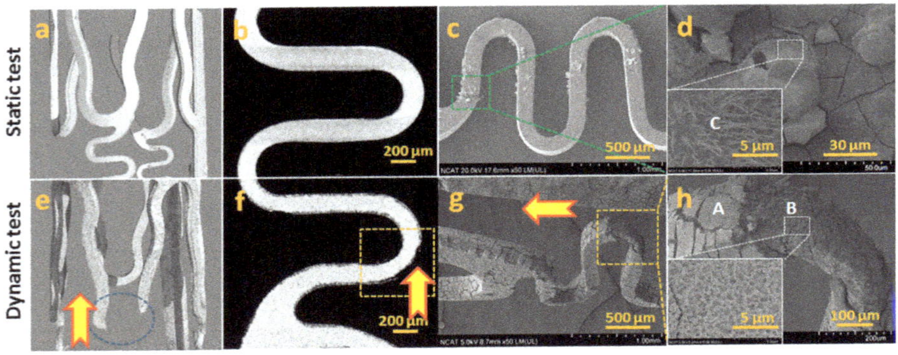

Fig. 6.4 Reconstructions of X-ray micro-CT 3-D (A, E) with representative 2-D slices (B, F) and SEM images (C, D, G and H) of stents under the static (A–D) and dynamic (E–H) conditions for 7 days. *Blue circle* shows a fracture region of the strut (E). *Yellow frames* show a detached region (H). *Green frame* shows localized corrosion (D). *The arrows* indicate the flow direction of corrosion medium [51].
Reprinted from J. Wang, V. Giridharan, V. Shanov, Z. Xu, B. Collins, L. White, et al., Flow-induced corrosion behavior of absorbable magnesium-based stents, Acta Biomater. 10(12) (2014) 5213–5123, Copyright (2014), with permission from Elsevier.

between days 35 and 56 because of vascular remodeling associated with the loss of stent integrity. Nevertheless, stents implantation did not cause any major problems or show signs of initial breakage and thromboembolic events. This work revealed that Mg-based biodegradable stents might be promising alternative to the traditional permanent stent, but a further improvement was required in consideration of prolongation of the degradation and mechanical stability.

In 2006, Di Mario et al. [10] implanted Lekton Magic stents (Biotronik, Bulach, Switzerland) made from WE43 alloy into 33 minipigs. The stent possessed a novel design with circumferential noose-shaped elements connected by unbowed cross-links along its longitudinal axis as shown in Fig. 6.5. The minimal luminal diameter of Mg stent group was higher than that of 316L stainless steel stents (Lekton Motion®; Biotronik) groups at weeks 4 and 12 and showed positive vascular remodeling during these 2 months. Although an inhibitory effect on the smooth muscle cell growth was observed, homogeneous and rapid endothelialization of the Mg stent was found. The stent struts were almost covered by the neointima after 6 days.

Further studies have been conducted to further develop the Lekton Magic stent, which is denominated as absorbable metallic stents (AMS) later. In 2006, Waksman et al. [53] reported a series of animal study. Mg alloy (AMS-1) stents or stain steel (Lekton Motion) stents were randomly implanted into coronary arteries of domestic or minipigs. Domestic pigs were sacrificed at 3 days or 28 days and minipigs at 3 months. As shown in Fig. 6.6, the AMS group remained intact at 3 days and began to show signs of degradation by 28 days. No phenomenon of stent particle embolization, thrombosis, excess inflammation, or fibrin deposition was observed. In addition, the

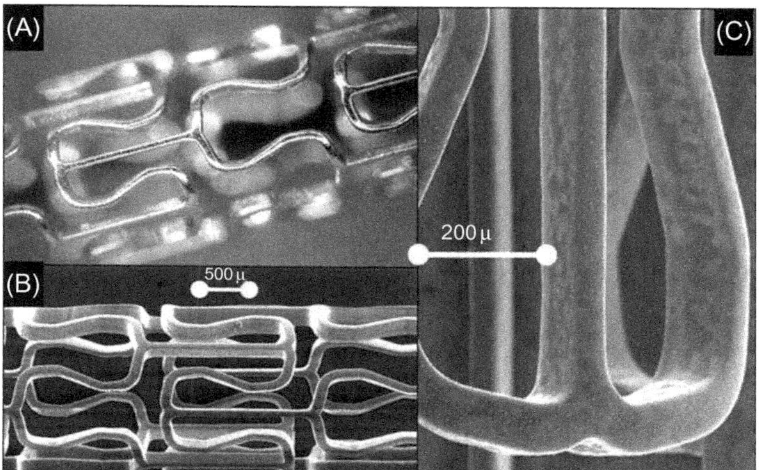

Fig. 6.5 (A) Photograph of the tubular slot balloon expandable magnesium alloy stent after electropolishing. Electron microscopy at low (B) and high (C) magnification [10]. Reprinted from C.D. Mario, H. Griffiths, O. Goktekin, N. Peeters, J. Verbist, M. Bosiers, et al., Drug-eluting bioabsorbable magnesium stent, J. Interv. Cardiol. 17(6) (2004) 391–395, Copyright (2004), with permission from John Wiley and Sons.

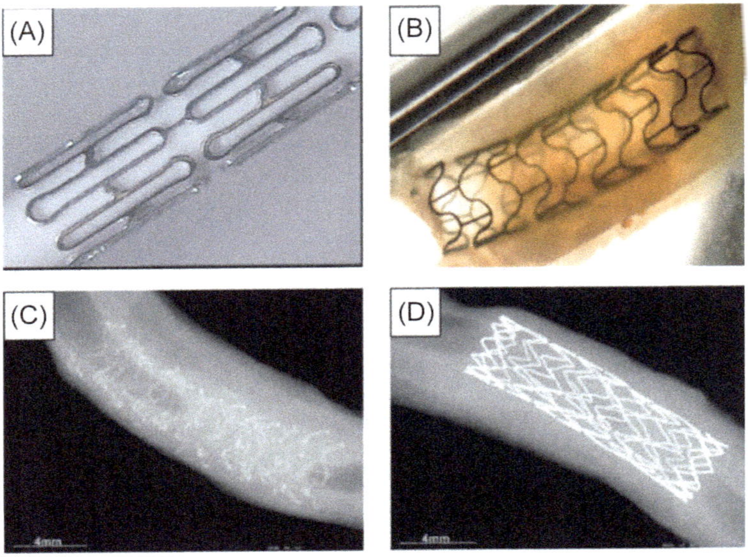

Fig. 6.6 (A) Representative photomicrograph of unexpanded magnesium alloy stent; (B) representative photomicrograph of dehydrated and defatified porcine coronary arteries showing the magnesium alloy stent 3 days after implantation; (C and D) representative X-ray photographs of magnesium alloy stent and stainless steel stent 28 days after implantation in porcine coronary arteries [53].
Reprinted from R. Waksman, R. Pakala, P.K. Kuchulakanti, R. Baffour, D. Hellinga, R. Seabron, et al., Safety and efficacy of bioabsorbable magnesium alloy stents in porcine coronary arteries, Catheter. Cardiovasc. Interv. 68(4) (2006) 607–617, Copyright (2006), with permission from John Wiley and Sons.

neointima thickness and neointimal area were obviously less in AMS segment in contrast to the stainless steel stent segment as exhibited in Fig. 6.7. However, the reduced neointima did not contribute to larger lumen.

In order to prolong vessel scaffolding of AMS-1, Biotronik made advancement by adding some new design considerations as shown in Fig. 6.8 [54]. DREAMS stent (Biotronik AG, Bülach, Switzerland) had a 6-crown 3-link design based on a refined, slower-resorbable WE43 alloy. In addition, the collapse pressure of DREAMS was higher than that of AMS-1 (1.5 vs. 0.8 bar). Compared with the rectangular shape in AMS-1, the cross-sectional profile of the scaffold struts in DREAMs was changed into square-shaped to preserve the radial strength during anisotropic scaffold degradation (Fig. 6.8A and B). The decrease in strut thickness should also facilitate endothelialization and reduced restenosis. Meanwhile, a 1 μm bioresorbable poly(-lactide-*co*-glycolide) polymer matrix (PLGA) containing the antiproliferative drug paclitaxel (0.07 μg/mm^2) was coated on DREAMS in order to reduce neointimal growth.

Wittchow et al. [55] reported the first in vivo evaluation of DREAMS stent with different PLGA composition and subsequently compared the best performing one with two established, paclitaxel-eluting, permanent stents (TAXUS® Liberté®; Boston

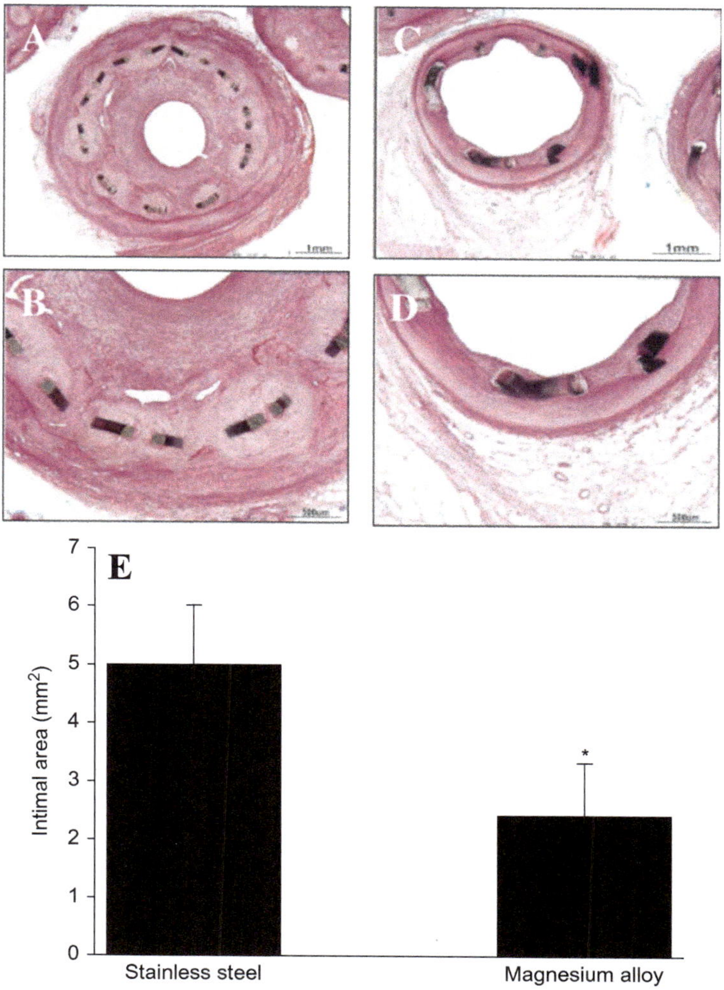

Fig. 6.7 Representative photomicrographs of hematoxylin-eosin stained sections of porcine coronary arteries 28 days after stainless steel stent (A: 403 and B: 1003) and magnesium alloy stent (C: 403 and D: 1003) implantation. (E) Bar graph showing the intimal area [53]. Reprinted from R. Waksman, R. Pakala, P.K. Kuchulakanti, R. Baffour, D. Hellinga, R. Seabron, et al., Safety and efficacy of bioabsorbable magnesium alloy stents in porcine coronary arteries, Catheter. Cardiovasc. Interv. 68(4) (2006) 607–617, Copyright (2006), with permission from John Wiley and Sons.

Scientific, Natick, MA, USA and eucaTAX; eucatech AG, Rheinfelden, Germany). The best-performing DREAM stent-85/15H (ratio of lactide to glycolide of 85/15 and high molecular weight polymer) was equivalent to TAXUS Liberté and superior to eucaTAX regarding late luminal loss, intimal area, fibrin score, and endothelialization. Despite a higher intimal inflammation score observed in 85/15H group compared with

Fig. 6.8 Schematic cross-sectional profile of magnesium scaffolds struts of (A) uncoated, non-eluting, AMS-1 with 80×165 μ; (B) DREAMS 1st generation (DREAMS 1G) with 130×120 μm struts, and (C) DREAMS 2nd generation (2G) with 150×140 μm struts. The poly(lactide-co-glycolide)-coating with paclitaxel elution of the DREAMS 1G scaffold is indicated by the thin light orange layer. The PLA-coating with sirolimus elution of the DREAMS 2G scaffold is indicated by the thin dark orange layer [54].
Reprinted from C.M. Campos, T. Muramatsu, J. Iqbal, Y.J. Zhang, Y. Onuma, H.M. Garcia-Garcia, et al., Bioresorbable drug-eluting magnesium-alloy scaffold for treatment of coronary artery disease, Int. J. Mol. Sci. 14 (12) (2013) 24492–24500, Copyright (2013). Used under the Creative Commons Attribution License.

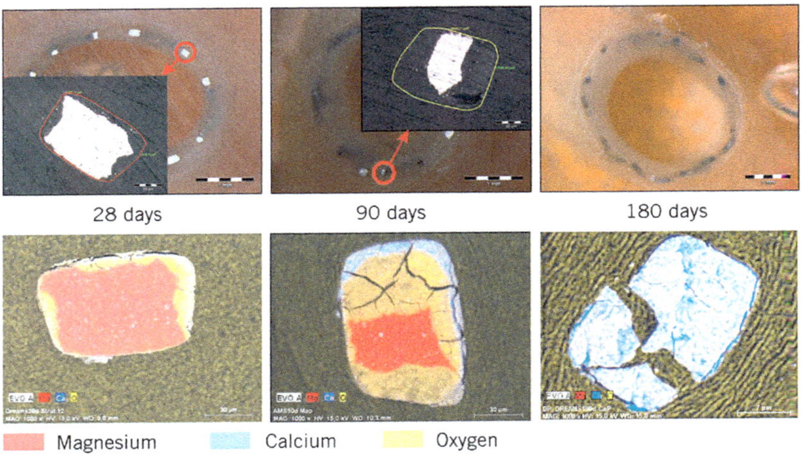

Fig. 6.9 Scanning electron microscope images (upper panels) and energy-dispersive X-ray mapping of DREAMS degradation products (lower panels, unrelated to upper panels) [55].
Reprinted from E. Wittchow, N. Adden, J. Riedmuller, C. Savard, R. Waksman, M. Braune, Bioresorbable drug-eluting magnesium-alloy scaffold: design and feasibility in a porcine coronary model, EuroIntervention 8 (12) (2013) 1441–1450, Copyright (2013), with permission from Europa Digital & Publishing.

the control groups at 28 days, it disappeared at later time points. Fig. 6.9 described the resorption process of DREAMS stent. Instead of complete degradation, Mg alloy was converted into a soft and gel-like amorphous calcium phosphate. During the first 28 days, the average in vivo degradation rates for the DREAMS versions ranged from 0.036 to 0.072 mg/(cm^2 day).

DREAMS 2nd was created based on a further development of DREAMS [54]. As shown in Fig. 6.8C, DREAMS 2nd is made of WE43 alloy with 6 crown 2-link structure and a strut thickness of 150 μm. Compared with DREAMS, two radiopaque markers (made from tantalum) at both ends were added into DREAMS 2nd to make stent deployment and possible post-dilation more precise. Besides, DREAMS 2nd has a superior dismantling and resorption rate. In order to further relieve the neointima formation, the DREAMS 2G incorporated sirolimus drug instead of paclitaxel.

6.3.4.2 Clinical testing of Mg-based bioabsorbable metallic stents

The implement of traditional permanent stents is limited in growing children with small blood vessels because it is generally difficult for the stents to cover full vessel growth, resulting in fixed vessel obstruction. The appearance of bioabsorbable metallic stents may figure out this limitation. In 2005, Zartner et al. [56] reported the first successful implantation of a biodegradable metal stent (AMS; Biotronik, Bülach, Switzerland) with a diameter of 3 and 10 mm length into the completely occluded left pulmonary artery of a preterm baby. Reperfusion of the left lung was recovered and persisted throughout 4 months during which the complete degradation of stent was also achieved. Despite the small size (1.7 kg) of the baby, the degradation process was clinically well-tolerated. In addition, adequate mechanical and degradation characteristics of implanted stent were demonstrated. Unfortunately, the patient died from multiple organ failure caused by severe pneumonia after 5-months implantation [57]. The autopsy results revealed minimal changes of the vessel wall due to the complete degradation of Mg stent. An amorphous to jelly-like substitute of the stent struts mainly composed of calcium phosphate covered by fibrotic tissue was observed, which contributed to a slight increase of intraluminal diameter compared with the original stent diameter. In addition, a mild neointima proliferation and no obvious inflammatory reaction to the stent were reflected.

Also in 2005, Peeters et al. [58] reported the first clinical trial of AMS (Biotronik, Berlin, Germany) implanted in 20 patients with symptomatic critical limb ischemia (CLI). After 3 months, a primary clinical patency of 89.5% and limb salvage rate of 100% were observed, suggesting that AMS devices were potentially promising alternative in the treatment of below-knee lesions in CLI patients. Morphological analysis indicated almost complete degradation of the stent 6 weeks after implantation.

The Progress-AMS clinical trial was the first-in-man study of AMS (Biotronik, Berlin, Germany) in coronary artery, which was a non-randomized, consecutive, multicenter trial with the purpose of assessing the efficacy and safety of AMS [19]. In short, 71 stents were successfully implanted in 63 patients after pre-dilatation and could be safely degraded after 4 months. As shown in Fig. 6.10, a good scaffolding of the vessel was demonstrated by angiography. Diameter stenosis was decreased from $61.5 \pm 13.1\%$ to $12.6 \pm 5.6\%$ with an acute gain of 1.41 ± 0.46 mm. No myocardial infarction, subacute or late thrombosis, or death was recorded. Nevertheless, diameter stenosis increased to $48.4 \pm 17\%$ at 4 months. The main causes of restenosis in this study were neointimal growth and negative remodeling. Therefore, a prolonged degradation and antiproliferative drug elution was suggested for further development.

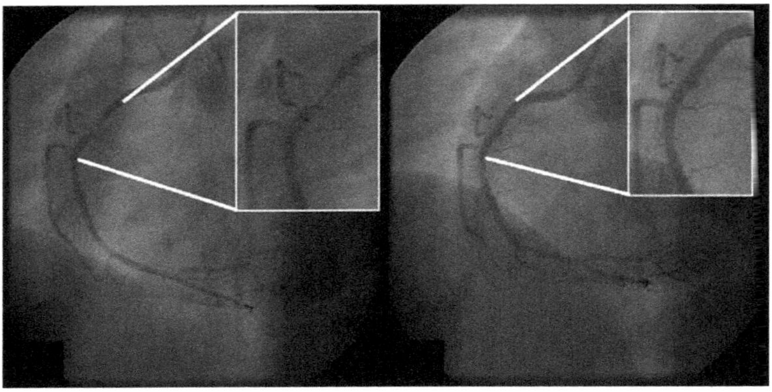

Fig. 6.10 Proximal right coronary artery high-grade stenosis before and immediately after implantation of a 15 mm long and 3.0 mm large bioabsorbable magnesium stent [19]. Reprinted from the R. Erbel, C. Di Mario, J. Bartunek, J. Bonnier, B. de Bruyne, F.R. Eberli, et al., Temporary scaffolding of coronary arteries with bioabsorbable magnesium stents: a prospective, non-randomised multicentre trial, Lancet 369 (9576) (2007) 1869–1875. Copyright (2007), with permission from Elsevier.

In 2013, Haude et al. [59] reported the first-in-man trial (BIOSOLVE-1) of the drug-eluting absorbable metal stent (DREAMS) in 46 patients with 47 lesions at five European centers. All stents were successfully delivered in this procedure and no cardiac death or scaffold thrombosis occurred. A better 12-months rate of clinically driven target lesion revascularization was observed in DREAMS compared with the bare precursor absorbable metal stents (4.7% vs. 26.7%). In addition, the in-stent late lumen loss of DREAMS was 0.65 ± 0.5 mm at 6 months and 0.52 ± 0.39 mm at 12 months, which is much less than that (1.08 ± 0.49 mm) of prior generation bare AMS-1 stent at 4 months reported in PROGRESS study. Besides, a similar rate of target lesion failure with DREAMS at 6 months and 12 months was observed compared with the contemporary drug-eluting stents and the bioabsorbable everolimus-eluting coronary scaffold systems. However, the late lumen loss of DREAMS did not match the excellent performance of currently available drug-eluting stents, showing a need for further modification.

In 2016, the first-in-man trial (BIOSOLVE-II) of the second-generation drug-eluting absorbable metal scaffold (DREAMS 2G) in 123 patients with 123 coronary target lesions was reported by Haude et al. [60]. No malapposed struts were observed after 6 months because the stent struts were embedded into the vessel wall (Fig. 6.11). Compared with DREAMS in BIOSOLVE-I, DREAMS 2G showed a more balanced degradation that the mean stent and mean lumen areas decreased slower than those of DREAMS (−0.5% and −2.4% in DREAMS 2G and −11.1% and −15.3% in DREAMS). In addition, the neointimal hyperplasia area decreased from 0.3 mm^2 in BIOSOLVE to 0.08 mm^2 in BIOSOLVE-II. Mean in-segment late lumen loss of DREAMS 2G also decreased (0.27 ± 0.37 mm at 6 months) in contrast to DREAMS and AMS-1, corresponding to the decreased rate of clinically driven target lesion revascularization.

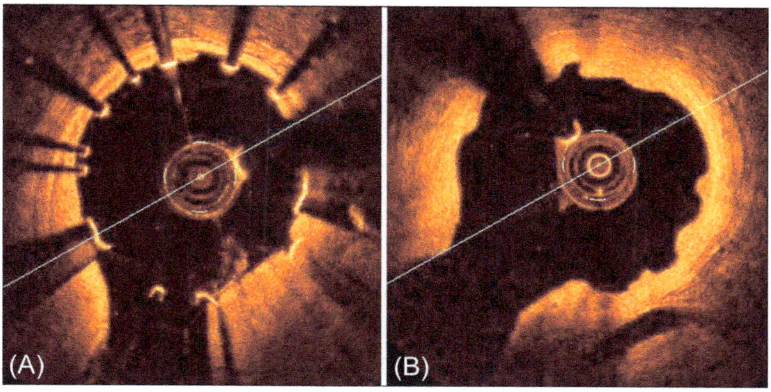

Fig. 6.11 Optical coherence tomography after implantation of the second-generation drug-eluting absorbable metal scaffold (A) and at 6 months (B) [60].
Reprinted from the M. Haude, H. Ince, A. Abizaid, R. Toelg, P.A. Lemos, C. von Birgelen, et al., Safety and performance of the second-generation drug-eluting absorbable metal scaffold in patients with de-novo coronary artery lesions (BIOSOLVE-II): 6 month results of a prospective, multicentre, non-randomised, first-in-man trial, Lancet 387 (10013) (2016) 31–39, Copyright (2016), with permission from Elsevier.

6.4 Development of Fe-based bioabsorbable metallic stents

6.4.1 The physiological function of Fe

Fe is the most abundant transition metal in human body. The amount of Fe in adult women is around 35 mg/kg of body weight and 45 mg/kg in adult men [61]. Approximately 60%–70% of the total body Fe exists in hemoglobin in circulating erythrocytes. Ten percent is present in the forms of myoglobins, cytochromes, and iron-containing enzymes. The residual 20%–30% of surplus Fe is distributed in the hepatocytes and reticuloendothelial macrophages as ferritins and hemosiderins [61]. Fe participates in the formation of several metallic proteins and involves in many vital biochemical activities such as oxygen sensing and transport, electron transfer and catalysis, and DNA synthesis [62]. The biological functions of Fe rely on its chemical properties. Iron can react with organic ligand to form a variety of coordination complexes in a dynamic and flexible mode. In addition, it can switch between the ferrous, Fe(II), and ferric, Fe(III), states under redox potential (+772 mV at neutral pH) [62]. Generally, the bioavailability of Fe is limited because soluble Fe(II) is readily oxidized in aqueous solution into Fe(III) which is almost insoluble in physiological pH [62]. The absorption of Fe takes place by the enterocytes by divalent metal transporter 1, which mainly occurs in the duodenum and upper jejunum. Subsequently, it is transferred through the duodenal mucosa into the blood, then transported to the cells or the bone marrow for erythropoiesis by transferrin [63]. In addition, as a result of formation of free radicals, the concentration of Fe in body tissues should be carefully regulated because excessive Fe leads to tissue damage [63]. The disorder of Fe metabolism is associated with plenty of diseases, ranging from anemia to neurodegenerative diseases [61].

6.4.2 The mechanical properties of Fe and its alloys

In contrast to Mg-based BMs, Fe-based BMs possess superior mechanical properties, similar to stainless steel (Table 6.3). As stated before, good ductility and radical support to the blood vessels are required for the successful implantation of cardiovascular stents. Therefore, Fe-based BMs are attractive from a structural point of view.

Mn is claimed to be an appropriate alloying element in Fe-based BMs. Mn is a trace element and plays an important role in many enzymatic reactions. In addition the addition of Mn can speed up the degradation of Fe [68] and displays anti-ferromagnetic with more than 29 wt.% addition [72]. As seen in Table 6.4, alloying with Mn also improves the mechanical strength of pure Fe (such as Fe-30Mn). Liu et al. added 6% Si into Fe30Mn and found an improvement on the mechanical properties with shape memory effect. A recovery ratio about 53.7% could be reached when the sample was deformed to the total strain of 3% and the exact pre-strain of 2.73%.

Xu et al. prepared a novel Fe-30Mn-1C alloy by vacuum induction melting. It was found that the addition of carbon improved the UTS and elongation to 1010 MPa and 88%, respectively [69]. Besides, Fe-30Mn-1C exhibited lower magnetic susceptibility in contrast to Fe-30Mn, which is beneficial to magnetic resonance imaging.

6.4.3 In vitro testing of Fe-based bioabsorbable metals for cardiovascular application

Mueller et al. designed a cell culture model to investigate the interactions between human umbilical venous smooth muscle cells (SMCs) and ferrous iron (Fe(II)). It was found that high concentration of Fe(II) led to inhabitation of SMCs proliferation and the down-regulation of gene expression required for cell proliferation, cell cycle progression, or DNA replication was observed in the presence of Fe, suggesting that ions released from Fe-based stents might had a suppressive effect on restenosis by reducing the proliferation of SMCs. Zhu et al. [78] studied the effect of ferrous ions on human umbilical vein endothelial cells (HUVECs) and found an interesting biphasic effect. As shown in Fig. 6.12, Fe (II) had a promotion effect on the metabolic activity of HUVECs when ion concentration was less than 10 μg/mL. However, higher ion concentration (>50 μg/mL) caused severe cytotoxicity on cells.

Schaffer et al. [79] conducted a systematic research to quantify the impact of bioabsorbable metals on primary human aortic endothelial cells (EC) and primary human aortic smooth muscle cells (SMC) in four tissue interface investigations (Fig. 6.13). In Fig. 6.13A, specific metal ion gradients across a porous polycarbonate track etch (PTFE) membrane barrier between a cell suspension and cell-free medium were created. It was found Fe^{2+} and Fe^{3+} repressed the migration of SMCs across a porous PTFE membrane at 1000 and 37 μM, respectively. LD_{50} (50% cell death) concentrations of Fe^{2+} and Fe^{3+} were greater than 1 mM. In addition, all tested materials showed excellent EC coverage and proliferation up to 120 h.

Table 6.3 Summary of the properties of reported Fe-based biodegradable metals [12]

Materials	State	Tensile yield strength (MPa)	Ultimate tensile strength (MPa)	Elongation (%)	Ref.
Pure Fe	Cast	220	225	12%	[64]
	Annealed (550°C)	140 ± 10	205 ± 6	25.5 ± 3	[65]
	Electroformed	360 ± 9	423 ± 12	8.3 ± 2	[65]
	ECAPed (8 passes)		470 ± 29		[66]
Nitride Fe		561.4	614.4		[67]
Fe-10Mn[a]	Forged + ht2	650	1300	14	[68]
Fe-10Mn-1Pd[a]	Forged + ht2	850	1450	11	[68]
Fe-30Mn	Cast	124.5	366.7	55.7	[64]
Fe-30Mn-6Si	Cast	177.8	433.3	16.6	[64]
Fe-30Mn	Forged	169	569	60	[69]
Fe-30Mn-1C	Forged	373	1010	88	[69]
Fe-3Co[a]	Rolled	460	648	5.5	[70]
Fe-3W[a]	Rolled	465	712	6.2	[70]
Fe-3C[a]	Rolled	440	600	7.4	[70]
Fe-3S[a]	Rolled	440	810	8.3	[70]
Fe-20Mn	PM	420	700	8	[71]
Fe-25Mn	PM	360	720	5	[71]
Fe-30Mn	PM	240	520	20	[71]
Fe-35Mn	PM	230	430	30	[71]
316 SS		190	490	40	[71]

[a] The chemical composition was in atom percentage, while the others were in weight percentage.

Reprinted from Y.F. Zheng, X.N. Gu, F. Witte, Biodegradable metals, Mater. Sci. Eng. R: Rep. 77 (2014) 1–34, with permission from Elsevier.

Table 6.4 **Comparison of the mechanical properties of different Zn-based materials**

Material	Ultimate tensile strength (MPa)	Tensile yield strength (MPa)	Elongation (%)	Ultimate compressive strength (MPa)	Compressive yield strength (MPa)	Compressive strain (%)	Hardness (Hv)	Ref.
As-cast pure Zn	18	10	0.3				38	[73]
As-rolled pure Zn	48	30	5.8				39	[73]
As-extruded pure Zn	60	33	3.6		103			[73]
As-cast Zn-1Mg	185	128	1.8				78	[73]
As-rolled Zn-1Mg	253	190	12				75	[73]
As-extruded Zn-1Mg	265	205	8.5		285	Superplastic		[73]
As-cast Zn-1Ca	165	119	2.1				73	[73]
As-rolled Zn-1Ca	230	200	12.6				62	[73]
As-extruded Zn-1Ca	240	197	7.7		281	Superplastic		[73]
As-cast Zn-1Sr	171	120	2				62	[73]
As-rolled Zn-1Sr	220	186	19.7				62	[73]
As-extruded Zn-1Sr	260	215	10.5		341	Superplastic		[73]

As-extruded Zn-0.5Al	203	119	33				59	[74]
As-extruded Zn-1Al	223	134	24				73	[74]
As-cast Zn-1Mg-1Ca	120	80	1				92	[75]
As-rolled Zn-1Mg-1Ca	198	138	8.5				107	[75]
As-extruded Zn-1Mg-1Ca	205	255	5.4		300	Superplastic		[75]
As-cast Zn-1Mg-1Sr	135	85	1.3				86	[75]
As-rolled Zn-1Mg-1Sr	202	140	8.6				92	[75]
As-extruded Zn-1Mg-1Sr	252	202	7.5		375	Superplastic		[75]
As-cast Zn-1Ca-1Sr	140	83	1.2				90	[75]
As-rolled Zn-1Ca-1Sr	203	144	8.8				86	[75]
As-extruded Zn-1Ca-1Sr	260	213	6.8		340	Superplastic		[75]
As-cast Zn-1.5Mg-0.1Ca	241	174	1.7				150	[76]
As-cast Zn-1.5Mg-0.1Sr	209	130	2.0				145	[76]
As-cast Zn-1Mg-0.1Mn	132	114	1				98	[77]
As-rolled Zn-1Mg-0.1Mn	299	195	26.1				108	[77]
As-cast Zn-1.5Mg-0.1Mn	122	115	0.8				149	[77]

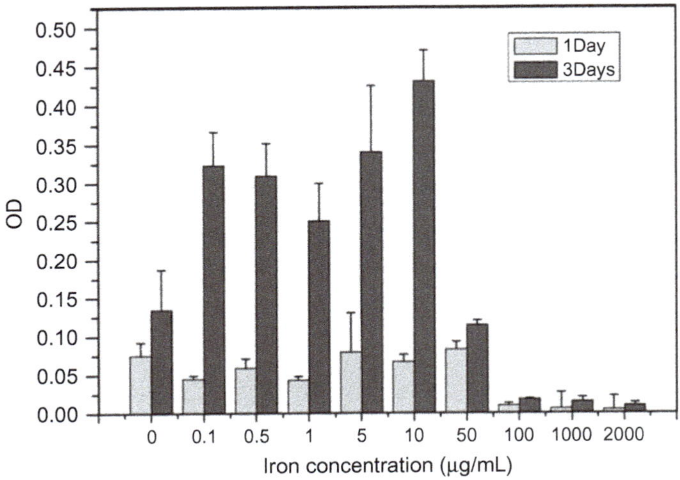

Fig. 6.12 Metabolic activity of HUVECs after cultured in different iron concentrations for 1 day and 3 days [78].
Reprinted from S. Zhu, N. Huang, L. Xu, Y. Zhang, H. Liu, H. Sun, et al., Biocompatibility of pure iron: in vitro assessment of degradation kinetics and cytotoxicity on endothelial cells, Mater. Sci. Eng. C 29 (5) (2009) 1589–92, Copyright (2009), with permission from Elsevier.

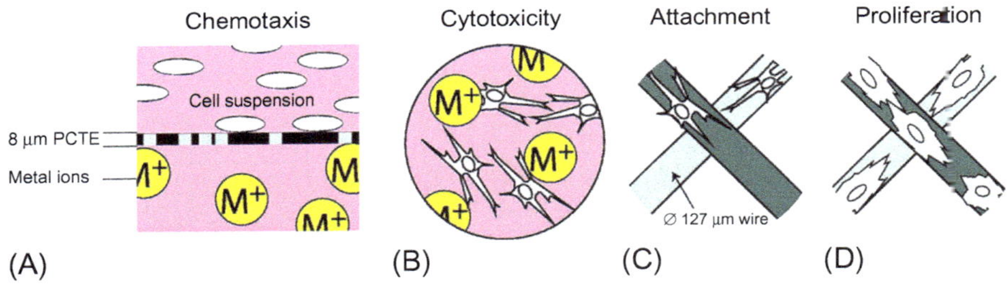

Fig. 6.13 The four biomaterial-tissue interface investigations performed in this study. (A) The chemotaxis of cells to metal ions. (B) The cytotoxicity of metallic ions. (C, D) Schematics of the thin wire metallic substrates used for the cell attachment and proliferation studies, respectively [79].
Reprinted from J.E. Schaffer, E.A. Nauman, L.A. Stanciu, Cold drawn bioabsorbable ferrous and ferrous composite wires: an evaluation of in vitro vascular cytocompatibility, Acta Biomater. 9(10) (2013) 8574–8584, Copyright (2013), with permission from Elsevier.

6.4.4 Animal testing of Fe-based bioabsorbable metals for cardiovascular application

The earliest attempt of Fe-based material as biodegradable stent can ascend to 2001. Peuster et al. [80] implanted 16 pure iron stents into the native descending aorta of 16 New Zealand white rabbits up to 18 months. All stents were implanted successfully without periprocedural complications. There was complete patency of the

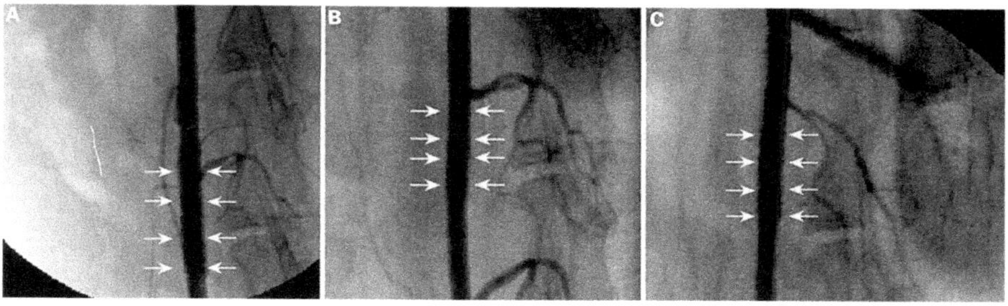

Fig. 6.14 Lateral angiography of the stented descending aorta (A) 6 months, (B) 12 months, and (C) 18 months after implantation. *Arrows* indicate stent implantation site [80]. Reprinted from M. Peuster, P. Wohlsein, M. Brügmann, M. Ehlerding, K. Seidler, C. Fink, et al., A novel approach to temporary stenting: degradable cardiovascular stents produced from corrodible metal—results 6–18 months after implantation into New Zealand white rabbits, Heart 86 (5) (2001) 563–9, Copyright (2001), with permission from BMJ Publishing Group Ltd.

descending aorta in all rabbits with no obstruction or thrombosis at the implantation site 6, 12, and 18 months after the implantation (Fig. 6.14). In addition, no significant neointimal proliferation, pronounced inflammatory response, and systemic toxicity was observed. This work demonstrated the safety of pure Fe as biodegradable stent. Nonetheless, the corrosion rate of pure Fe was too slow to meet the clinical requirement

Subsequently, Peuster et al. [81] implanted pure iron stents into the descending aorta of 29 minipigs with an overstretch injury and the commercially available 316-L stents served as control (Fig. 6.15). There was no difference between 316-L and pure Fe stents with respect to the amount of neointimal proliferation. No signs of iron overload or iron-related organ toxicity was observed in histopathological examination. The corrosion products of iron stent caused no local toxicity. The majority of the iron stent kept intact, even though disintegration of the stent struts was observed after 1 year. Therefore, an accelerated corrosion rate should be pursued in the following work.

In 2008, Waksman et al. [82] implanted pure iron stents into the coronary arteries of juvenile domestic pigs with cobalt chromium stents as control for 28 days. At 28 days, degradation of iron stent was detected. Compared with cobalt chromium stents, a reduction of neointimal formation and inflammation was observed. No significantly statistical difference between iron and control group was found in any of the tested parameters.

In 2012, eight iron stents were randomly implanted into left anterior descending and right circumflex arteries of eight healthy miniature pigs with eight Vision stents (Abbott Vascular, CA, USA) as control [83]. There were no iron overload and abnormal histopathologic changes in heart, lung, liver, kidney, and brain. The similar neointimal proliferation was observed in both groups. Meanwhile, no thrombosis, inflammation, and necrosis existed in both groups.

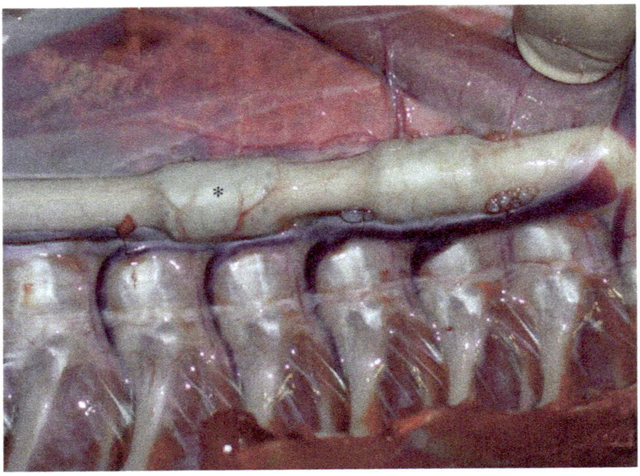

Fig. 6.15 Macroscopical aspect of the stented descending aorta after expiration of the animal. The iron stent is marked with an *asterisk* [81].
Reprinted from M. Peuster, C. Hesse, T. Schloo, C. Fink, P. Beerbaum, C. von Schnakenburg, Long-term biocompatibility of a corrodible peripheral iron stent in the porcine descending aorta, Biomaterials 27 (28) (2006) 4955–4962, Copyright (2006), with permission from Elsevier.

A novel experimental model was developed to clarify the in vivo corrosion mechanism of pure iron by Pierson et al., where an iron wire was implanted into a rat artery lumen or artery wall to simulate bioabsorbable stent blood contact and matrix contact, respectively [84]. At 22 days, a brown-colored product in the tissue surrounding the wire was observed in artery wall wire implant group, but no visible corrosion product was presented in the luminal wire group. The blood-contacting luminal wire showed minimal corrosion at 9 months (Fig. 6.16A). In contrast, iron wires implanted into the arterial wall exhibited severe degradation (Fig. 6.16B). Similarly, wires in lumen for 9 months remained intact, whereas arterial wall penetration site experienced significant degradation (Fig. 6.16C). The results revealed the importance of arterial environment to the corrosion behavior of iron. In addition, the retention of the degradation products of the iron may impair the long-term integrity of the artery.

In 2016, Lin et al. [85] implanted a novel iron-based drug-eluting coronary scaffold (IBS scaffold) made of nitride iron materials into the abdominal aorta of adult New Zealand white rabbits. A Zn barrier (~600 nm) was electroplated on the surface of stents followed by a sirolimus-carrying Poly (D,L-lactide) (PDLLA, amorphous; Evonik Industries, Germany) coating (~12 μm) as shown in Fig. 6.17. IBS scaffold showed no signs of corrosion in the first 3 months and corroded partially 6 months after implantation and nearly completely corroded 13 months after implantation. Complete endothelialization was observed 3 months after implantation. In spite of an incomplete bioresorption of the corrosion products, no identified biological issues were observed after implantation up to 13 months.

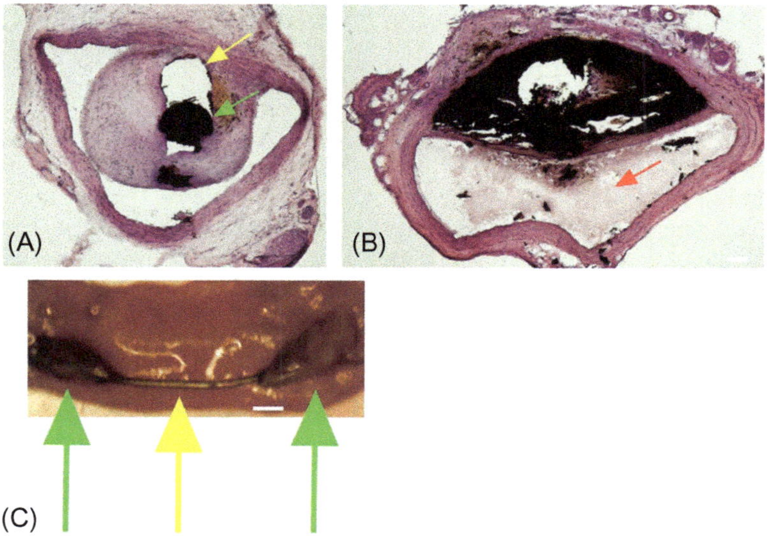

Fig. 6.16 Histological analysis of the 9-month wire implants. Hematoxylin/eosin-stained tissue sections depict the location of the iron wire and corrosion product within the arterial lumen (A) and wall (B). The black circular structure in (A) is the iron wire cross section (identified by the *green arrow*). When the luminal wire was bowed out away from the vessel wall, the central portion of the wire (*yellow arrow* in C) remained free of encapsulating tissue and experienced minimal biodegradation. In contrast, the wall puncture sites (*green arrows* in C) experienced substantial degradation. *Red layer* in (C) background is the luminal surface of the exposed artery [84].
Reprinted from D. Pierson, J. Edick, A. Tauscher, E. Pokorney, P. Bowen, J. Gelbaugh, et al., A simplified in vivo approach for evaluating the bioabsorbable behavior of candidate stent materials, J. Biomed. Mater. Res. B: Appl. Biomater. 100 (1) (2012) 58–67, Copyright (2012), with permission from John Wiley and Sons.

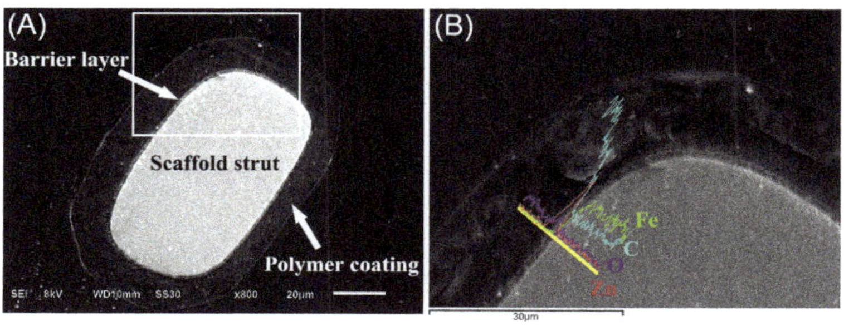

Fig. 6.17 The strut crossing-section (A) structure and (B) composition of the IBS scaffold. The area of (B) is an enlargement of the area in the *white rectangle* of (A) [85].
Reprinted from W.J. Lin, D.Y. Zhang, G. Zhang, H.T. Sun, H.P. Qi, L.P. Chen, et al., Design and characterization of a novel biocorrodible iron-based drug-eluting coronary scaffold, Mater. Des. 91 (2016) 72–79, Copyright (2016), with permission from Elsevier.

6.5 Development of Zn-based bioabsorbable metallic stents

6.5.1 The physiological function of Zn

In human body, 85% of Zn is distributed in the muscle and bone, 11% in the skin and the liver, and the residue in all the other tissues. The daily recommended intake for Zn ranges from 11 mg for adult males to 2 mg for infants [86]. The absorption of Zn occurs throughout the small intestine, but the most rapid absorption exists in the duodenum and proximal jejunum [87]. Intestine is also the major excretion routine for Zn [88]. It is well-acknowledged that Zn plays a significant role in human nutrition. Zn is considered the second most abundant transition metal in human body and associates with many significant biological functions such as nucleic acid metabolism, signal transduction, gene expression, apoptosis regulation, endocrine regulation, etc. [87,89]. More than 300 enzymes have been identified related to Zn in their structure, catalytic, and regulatory action [90]. In addition, the involvement of Zn has been confirmed in bone formation and it can stimulate bone growth and bone mineralization apart from the preservation of bone mass [73].

6.5.2 The mechanical properties of Zn and its alloys

Much attention has been focused on Zn as an alternative in a new generation of BMs due to their promising biocompatibility and corrosion properties. However, the poor mechanical properties of pure Zn limited its further application. It was reported that the tensile strength, elongation, and the hardness of the as-cast pure Zn were approximately 20 MPa, 0.3%, and 25 Hv, respectively, far too low for clinical requisites [91,92]. Therefore, the development of new Zn-based materials with reliable mechanical support is urgent for opening a new avenue in its application. Improvement Methods such as alloying, thermomechanical treatment, and composite technology have been utilized to achieve this goal. The mechanical properties of different Zn-based materials are listed in Table 6.4.

Alloying is one of the most effective ways to improve the mechanical properties of metals. In the past decades, commercial Zn alloys have been applied in the industrial and automotive areas [93,94]. Nonetheless, there are some alloying elements claimed to be potentially harmful in the commercial Zn alloys such as Al (e.g., ZA22 [95], ZA27 [96], and ZA40 [97]). It is reported that Al is detrimental to the neurons, bones, and osteoblasts and related to a series of neurological disorders such as Alzheimer's disease [24,25]. Therefore, both safety and mechanical properties should be taken into consideration in the selection of proper alloying elements.

Li et al. [73] added the nutrition elements Mg, Ca, and Sr into Zn to prepare Zn-1X (X = Mg, Ca, and Sr) binary and investigated the effect of hot rolling and hot extrusion on the mechanical properties as well. The results exhibited that the microhardness, YS, UTS, and elongation were obviously improved after alloying with Mg, Ca, and Sr elements, and hot rolling and hot extrusion process further improved the mechanical properties (as shown in Table 6.4). In addition, the as-extruded Zn-1X alloys exhibited unique superplastic characteristic under compression due to the formation of compressive twin crystal. Besides, a trace of Ca or Sr was added into Zn-1.5Mg alloy to develop Zn-1.5Mg-0.1Ca and Zn-1.5Mg-0.1Sr alloy by Liu et al. [76]. It was reported that the yield strength, ultimate tensile strength, and elongation of Zn-1.5Mg-0.1Ca and Zn-1.5Mg-0.1Sr were significantly improved with respect to Zn-1.5Mg alloy because of the formation of more homogeneous and smaller grains apart from the precipitation strengthening of the new phase $CaZn_{13}$ and $SrZn_{13}$ after alloying with Ca and Sr.

6.5.3 In vitro testing of Zn-based bioabsorbable metals for cardiovascular application

Ma et al. [98,99] reported that Zn^{2+} had an interesting biphasic effect on the cell viability, adhesion, proliferation, migration of human aorta smooth muscle cells (HASMCs), and primary human coronary artery endothelial cells (HCECs). No significant effect on viability but a promotion of cell proliferation was observed in HASMC with the Zn^{2+} concentration increasing while below 80 μM. Yet, the cell viability and proliferation declined with a further increase of Zn^{2+} concentration. Meanwhile, Zn^{2+} also altered the adhesion, spreading, and migration in a dose-manner. Zn^{2+} improved HASMCs adhesion density when below 40 μM, while inhibited cell adhesion when above 40 μM. Similarly, at 40 μM, Zn^{2+} promoted HASMCs spreading with a larger cell area and perimeter, but an opposite result was observed at 120 μM. Zn^{2+} increased HASMCs migration rate when below 80 μM, while decreased cell migration rate significantly when above 100 μM. In terms of HCECs, generally a promotion effect was demonstrated on the cell viability, proliferation, adhesion, migration, and F-actin and vinculin expression except for the cell adhesion strength at low Zn^{2+} concentrations. On the contrary, high concentrations had opposite effects. Therefore, Zn^{2+} concentration can be controlled to achieve specific endothelial and smooth muscle cell behaviors according to the demand of vascular healing.

Li et al. [73] investigated the in vitro biological compatibility of the as-cast and as-rolled pure Zn and Zn-1X (X=Mg, Ca and Sr) alloy. It was found that the hemolysis rates of all samples were quite low (<0.2%), far below the safe requirement of 5%, suggesting they would not cause severe hemolysis reaction according to ISO 10993-4:2002 (Fig. 6.18). In addition, the platelets adhered on the surface of samples displayed normal morphologies, thus round shape with no pseudopodia spreading. Hence, pure Zn and Zn-1X alloys possess excellent anti-platelets adhesion and antithrombotic properties in vivo. The addition of alloying elements into Zn decreased its adhesive platelets number. Fig. 6.19 showed the viability of human umbilical vein endothelial cells (ECV304 cells), rodent vascular smooth muscle cells (VSMC) cultured in pure Zn, and Zn-1X alloy extraction medium for 1, 3, and 5 days. For ECV304 cells, the addition of alloying elements significantly improved the cell viability, while alloying elements didn't have obvious promotion effect on VSMC.

Gong et al. [100] evaluated the cytotoxicity of as-extruded Zn-1Mg alloy on fibroblast cells (L-929). As shown in Fig. 6.20, there was no obvious difference of cell viability between the control group and Zn-1Mg group at 24 and 72 h. Besides, cells in all groups exhibited normal and healthy cell morphologies at 72 h (Fig. 6.21), indicating that Zn-1Mg is of good biocompatibility.

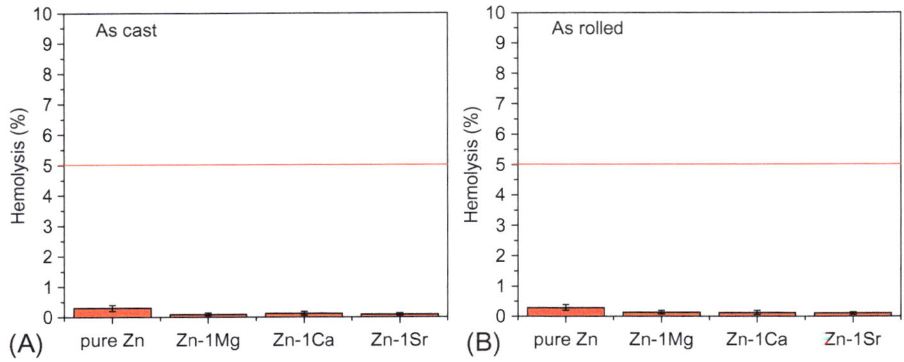

Fig. 6.18 Hemolysis of pure Zn and Zn-based alloys: (A) as-cast, (B) as-rolled [73].
Reprinted from H.F. Li, X.H. Xie, Y.F. Zheng, Y. Cong, F.Y. Zhou, K.J. Qiu, et al., Development of biodegradable Zn-1X binary alloys with nutrient alloying elements Mg, Ca and Sr, Sci. Rep. 5 (2015), Copyright (2015). Used under the Creative Commons Attribution 4.0 International License.

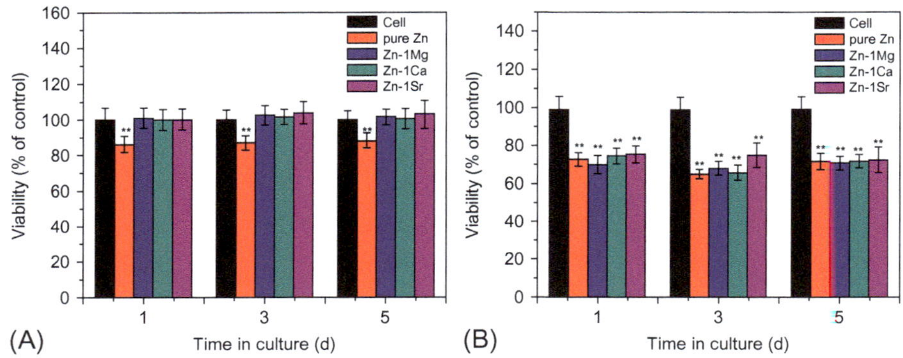

Fig. 6.19 (A) ECV304 cells, (B) VSMC viability cultured in cell culture medium, pure Zn, and Zn-1X alloy extracts (* means $P<.05$, ** means $P<.01$ compared with pure Zn group) [73].
Reprinted from H.F. Li, X.H. Xie, Y.F. Zheng, Y. Cong, F.Y. Zhou, K.J. Qiu, et al., Development of biodegradable Zn-1X binary alloys with nutrient alloying elements Mg, Ca and Sr, Sci. Rep. 5 (2015), Copyright (2015). Used under the Creative Commons Attribution 4.0 International License.

6.5.4 Animal testing of Zn-based bioabsorbable metals within blood vessel

Bowen et al. [17] implanted pure Zn wire into the abdominal aorta of adult male Sprague-Dawley rats for 1.5, 3, 4.5, or 6 months. The results showed that Zn could maintain its mechanical stability for a 4-month period and thereafter an accelerated degradation was observed, thus ensuring timely degradation of the implant. As can be seen in Fig. 6.22, the Zn wires showed relatively shallow and uniform corrosion in the

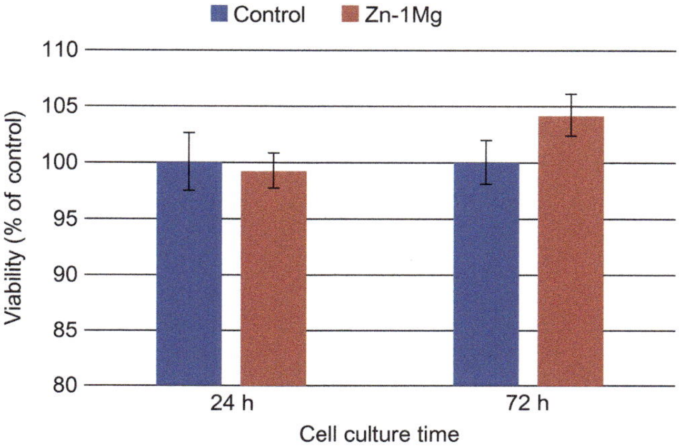

Fig. 6.20 Cell viability of fibroblast cells (L-929) cultured in regular media (control) and extraction media of extruded Zn-1Mg [100].
Reprinted from H. Gong, K. Wang, R. Strich, J.G. Zhou, In vitro biodegradation behavior, mechanical properties, and cytotoxicity of biodegradable Zn-Mg alloy, J. Biomed. Mater. Res. B: Appl. Biomater. 103 (8) (2015) 1632–1640, Copyright (2015), with permission from John Wiley and Sons.

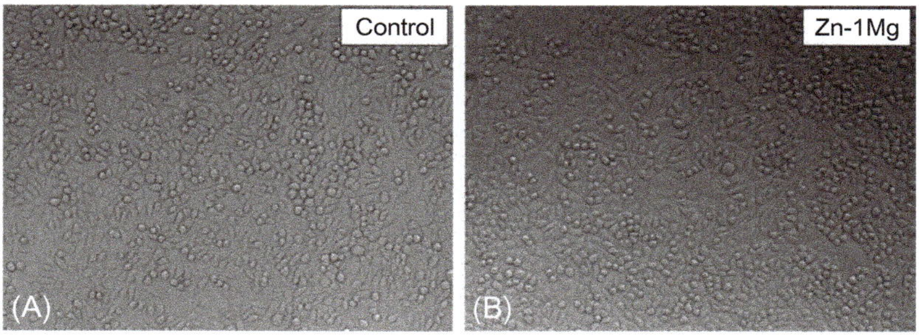

Fig. 6.21 Optical images of fibroblast (L-929) cells after 72h of culture in regular media (A, control) and extraction media of extruded Zn-1Mg (B) [100].
Reprinted from H. Gong, K. Wang, R. Strich, J.G. Zhou, In vitro biodegradation behavior, mechanical properties, and cytotoxicity of biodegradable Zn-Mg alloy, J. Biomed. Mater. Res. B: Appl. Biomater. 103 (8) (2015) 1632–1640, Copyright (2015), with permission from John Wiley and Sons.

first 3 months. The ragged edges of remaining metallic cores in 1.5 and 3 months were hypothesized to be caused by the removal of material due to semi-localized dissolution. After 4.5 and 6 months in vivo, more severe, localized corrosion was observed, but the cases of local corrosion were more numerous in 6 months. After 4.5 months, the corrosion products were mainly composed of zinc oxide and zinc carbonate. This study revealed that the degradation of Zn combined the advantages of Mg and Fe, namely the harmless degradation and in vivo longevity.

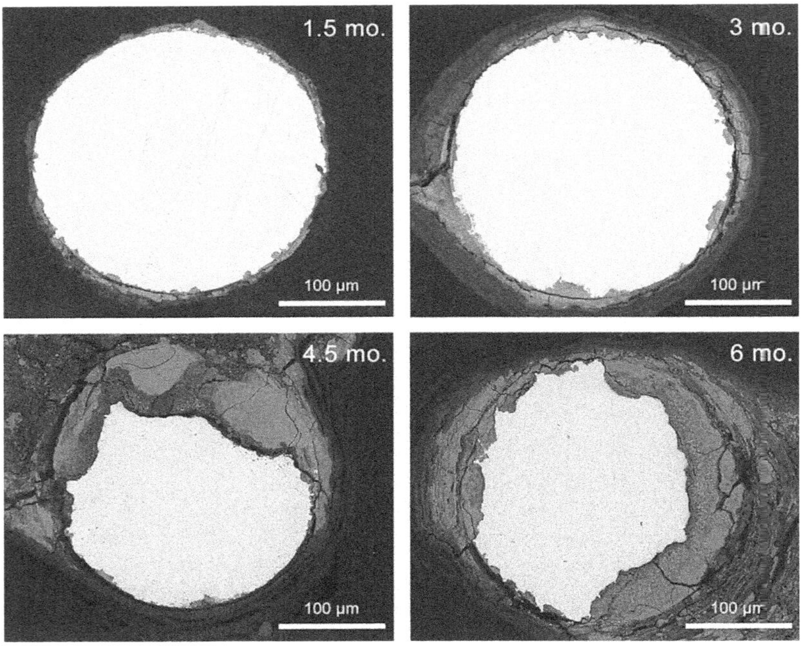

Fig. 6.22 Representative backscattered electron images from zinc explant cross sections after 1.5, 3, 4.5, and 6 months' time in vivo [17].
Reprinted from P.K. Bowen, J. Drelich, J. Goldman, Zinc exhibits ideal physiological corrosion behavior for bioabsorbable stents, Adv. Mater. 25 (18) (2013) 2577–2582, Copyright (2013), with permission from John Wiley and Sons.

Pure Zn wires were punctured and advanced into the lumen of a rat abdominal aorta for up to 6.5 months to further assess the biocompatibility of Zn by Bowen et al. [101]. The Zn wire remains in contact with flowing blood with some area of the wires in direct contact with the arterial wall. No significant chronic inflammatory response, localized necrosis, or progressive intimal hyperplasia were observed. A limited smooth muscle cell (SMC) proliferation and the persistence of a stable endothelial cell (EC) layer were demonstrated (Fig. 6.23). Low cell density and near total lack of SMCs near the implant implied that Zn or its degradation products may have a suppressive effect on SMCs and inflammatory cells. Tissue regeneration was observed within the footprint of the biocorrosion area of implant.

Guillory et al. [102] investigated the chronic inflammatory response to Zn at high purity (4 N) [99.99%], special high grade (SHG) [~99.7%] as well as all alloyed with 1 wt% (Zn-1Al), 3% (Zn-3Al), and 5.5% (Zn-5Al) aluminum. Two wires of the same material composition were implanted into the abdominal aorta wall of an adult Sprague-Dawley. It was found that the inflammatory response between five Zn-dominant materials is dramatically different. Such difference in biocompatibility was attributed to their different compositions which caused various implant corrosion

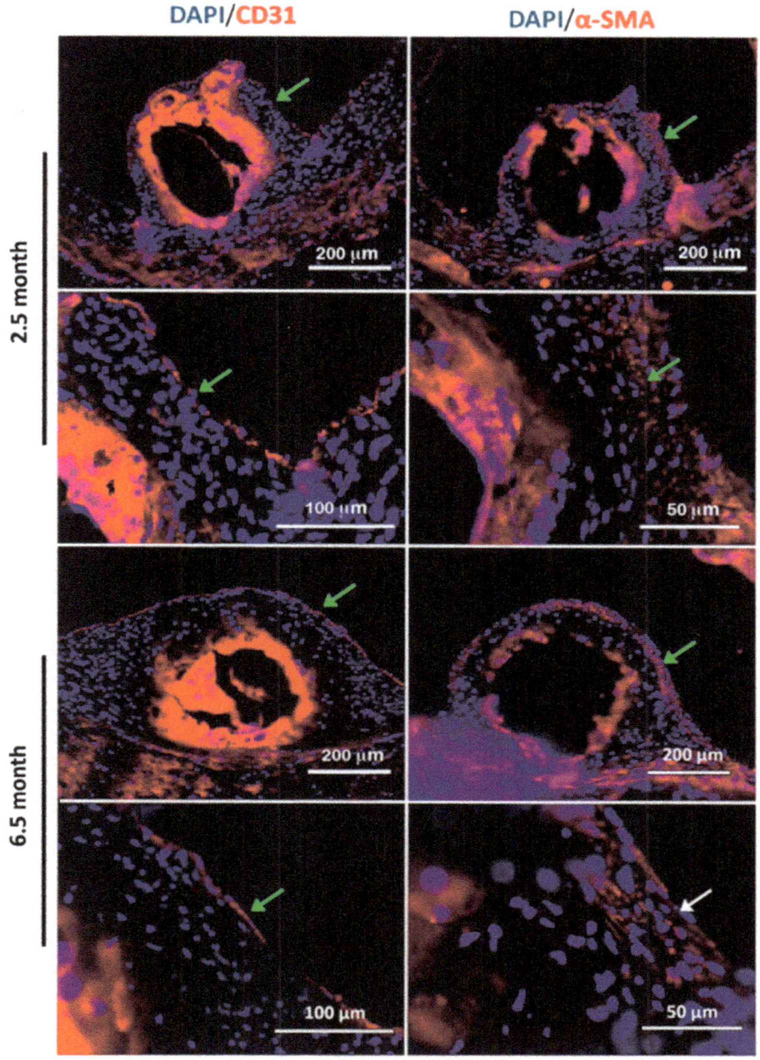

Fig. 6.23 Cross sections were stained for endothelial cells (*red*, CD31, left panels), smooth muscle cells (*red*, α-actin, right panels), and cell nuclei (DAPI counterstained blue, both panels) at 2.5 and 6.5 months. The *green arrow* in each panel identifies a characteristic region of positive staining within the neointimal tissue. Note that the corrosion layer impregnating the center of the neointimal tissue is excited and fluorescence red in these images [101]. Reprinted from P.K. Bowen, R.J. Guillory II, E.R. Shearier, J.-M. Seitz, J. Drelich, M. Bocks, et al., Metallic zinc exhibits optimal biocompatibility for bioabsorbable endovascular stents, Mater. Sci. Eng. C 56 (2015) 467–472, Copyright (2015), with permission from Elsevier.

activities and Zn ion fluxes. Influence factors such as corrosion behavior, corrosion products, and dynamic tissue–material interface are involved in the regulation of inflammatory progression and remodeling of the surrounding and penetrating tissue. In addition, the ability of macrophages to penetrate and remain viable within the corrosion layer played a significant role in eliciting biocompatible inflammatory response near the corrodible metal.

6.6 Challenges and opportunities for bioabsorbable metallic stents

Over the last decades, intensive researches have been conducted in the area of bioabsorbable metallic stent. In contrast to the traditional permanent stents, the superior conformability and flexibility of bioabsorbable stents remit altered distribution of tissue biomechanics and maintained vessel geometry. Meanwhile, the emancipation of the stented vessel from a metallic cage is beneficial to the restoration of physiological vasomotion, mechanotransduction, adaptive shear stress, late luminal enlargement, and late expansive remodeling. The unique degradation behavior also reduces the risk of vascular inflammation, thrombosis, neoatherosclerosis as well as inhibition of surgical revascularization. Among the currently developed biodegradable stents, bioabsorbable metallic stent possesses superior mechanical and surface properties compared with their polymeric counterparts. In addition, the degradation product of BMs exhibited weak base in contrast to the acid products of polymer, which would cause inflammatory response. Therefore, bioabsorbable metallic stent is a promising candidate in the cardiovascular applications.

Moving ahead, high-performance bioabsorbable metallic stents are still on the way to exploring. As mentioned above, all the implants require sufficient mechanical properties and appropriate corrosion rates to match the evolution of tissue healing apart from excellent biocompatibility. In terms of Mg-based stents, rapid corrosion rates make them fail to provide sufficient radical support before completing the mission of assisting vessel healing. In addition, the following accumulation of degradation products such as hydrogen and hydroxide ions may cause damage to the host. In contrary, superior corrosion resistance of Fe-based stents leads to a long-term retention in the artery vessel. Despite numerous measures that have been taken to accelerate the corrosion rate, there is still much endeavor required for clinical application. Zn-based stents have appeared as a new alternative on account of its intermediate corrosion resistance and excellent biocompatibility between Mg and Fe. Nonetheless, long-term in vivo study is absent to evaluate the loss of function and elimination. Besides, the metabolism of degradation products also should be further clear-up.

Meanwhile, a unified testing standard should be put forward. The majority of the testing methods utilized in bioabsorbable metallic stents are proposed for the permanent stents previously, and obviously there are plenty of differences between them. Therefore, specific standards are required in the design and evaluation of bioabsorbable metallic stents for clinical applications, which also make it simple to compare the results from different research groups. Another challenge is the mismatch between the in vitro and in vivo results. The key to breakthrough is to ascertain the influencing variables in vivo and present them in vitro.

References

[1] G. Mani, M.D. Feldman, D. Patel, C.M. Agrawal, Coronary stents: a materials perspective, Biomaterials 28 (9) (2007) 1689–1710.

[2] H. Hermawan, D. Dube, D. Mantovani, Developments in metallic biodegradable stents, Acta Biomater. 6 (5) (2010) 1693–1697.

[3] P.W. Serruys, H.M. Garcia-Garcia, Y. Onuma, From metallic cages to transient bioresorbable scaffolds: change in paradigm of coronary revascularization in the upcoming decade? Eur. Heart J. 33 (1) (2012) 16–25.

[4] J.J. Wentzel, D.M. Whelan, W.J. van der Giessen, H.M.M. van Beusekom, I. Andhyiswara, P.W. Serruys, et al., Coronary stent implantation changes 3-D vessel geometry and 3-D shear stress distribution, J. Biomech. 33 (10) (2000) 1287–1295.

[5] K.G. Birukov, V.P. Shirinsky, O.V. Stepanova, V.A. Tkachuk, A.W.A. Hahn, T.J. Resink, V.N. Smirnov, Stretch affects phenotype and proliferation of vascular smooth muscle cells, Mol. Cell. Biochem. 144 (2) (1995) 131–139.

[6] C.V. Bourantas, Y. Onuma, V. Farooq, Y. Zhang, H.M. Garcia-Garcia, P.W. Serruys, Bioresorbable scaffolds: current knowledge, potentialities and limitations experienced during their first clinical applications, Int. J. Cardiol. 167 (1) (2013) 11–21.

[7] N. Foin, R.D. Lee, R. Torii, J.L. Guitierrez-Chico, A. Mattesini, S. Nijjer, et al., Impact of stent strut design in metallic stents and biodegradable scaffolds, Int. J. Cardiol. 177 (3) (2014) 800–808.

[8] S. Nishio, K. Kosuga, K. Igaki, M. Okada, E. Kyo, T. Tsuji, et al., Long-term (>10 years) clinical outcomes of first-in-human biodegradable poly-L-lactic acid coronary stents: Igaki-Tamai stents, Circulation 125 (19) (2012) 2343–2353.

[9] C.M. Bünger, N. Grabow, K. Sternberg, M. Goosmann, K.P. Schmitz, H.J. Kreutzer, H. Ince, S. Kische, C.A. Nienaber, D.P. Martin, S.F. Williams, E. Klar, W. Schareck, A biodegradable stent based on poly(L-lactide) and poly(4-hydroxybutyrate) for peripheral vascular application: preliminary experience in the pig, J. Endovasc. Ther. 14 (5) (2007) 725–733.

[10] C.D. Mario, H. Griffiths, O. Goktekin, N. Peeters, J. Verbist, M. Bosiers, et al., Drug-eluting bioabsorbable magnesium stent, J. Interv. Cardiol. 17 (6) (2004) 391–395.

[11] J.A. Ormiston, F. De Vroey, P.W. Serruys, M.W. Webster, Bioresorbable polymeric vascular scaffolds: a cautionary tale, Circ. Cardiovasc. Interv. 4 (5) (2011) 535–538.

[12] Y.F. Zheng, X.N. Gu, F. Witte, Biodegradable metals, Mater. Sci. Eng. R. Rep. 77 (2014) 1–34.

[13] J.S. Forrester, M. Fishbein, R. Helfant, J. Fagin, A paradigm for restenosis based on cell biology: clues for the development of new preventive therapies, J. Am. Coll. Cardiol. 17 (3) (1991) 758–769.

[14] D.P. Faxon, W. Coats, J. Currier, Remodeling of the coronary artery after vascular injury, Prog. Cardiovasc. Dis. 40 (2) (1997) 129–140.

[15] M.J. Lipinski, R.O. Escarcega, T. Lhermusier, R. Waksman, The effects of novel, bioresorbable scaffolds on coronary vascular pathophysiology, J. Cardiovasc. Transl. Res. 7 (4) (2014) 413–425.

[16] N.V. Soucy, J.M. Feygin, R. Tunstall, M.A. Casey, D.E. Pennington, B.A. Huibregtse, et al., Strut tissue coverage and endothelial cell coverage: a comparison between bare

metal stent platforms and platinum chromium stents with and without everolimus-eluting coating, EuroIntervention 6 (5) (2010) 630–637.
[17] P.K. Bowen, J. Drelich, J. Goldman, Zinc exhibits ideal physiological corrosion behavior for bioabsorbable stents, Adv. Mater. 25 (18) (2013) 2577–2582.
[18] R. Werkhoven, W. Sillekens, J. Van Lieshout, Processing aspects of magnesium alloy stent tube, Magnesium Technology, Wiley, London, 2011, pp. 419–424.
[19] R. Erbel, C. Di Mario, J. Bartunek, J. Bonnier, B. de Bruyne, F.R. Eberli, et al., Temporary scaffolding of coronary arteries with bioabsorbable magnesium stents: a prospective, non-randomised multicentre trial, Lancet 369 (9576) (2007) 1869–1875.
[20] M. Moravej, D. Mantovani, Biodegradable metals for cardiovascular stent application: interests and new opportunities, Int. J. Mol. Sci. 12 (7) (2011) 4250–4270.
[21] Y. Onuma, J. Ormiston, P.W. Serruys, Bioresorbable scaffold technologies, Circ. J. 75 (3) (2011) 509–520.
[22] G. Song, Control of biodegradation of biocompatable magnesium alloys, Corros. Sci. 49 (4) (2007) 1696–1701.
[23] F. Witte, N. Hort, F. Feyerabend, C. Vogt, Magnesium (Mg) corrosion: a challenging concept for degradable implants A2, in: G.-l. Song (Ed.), Corrosion of Magnesium Alloys, Woodhead Publishing, Sawston, CA, 2011, pp. 403–425.
[24] G. Hercz, D.L. Andress, H.G. Nebeker, J.H. Shinaberger, D.J. Sherrard, J.W. Coburn, Reversal of aluminum-related bone disease after substituting calcium carbonate for aluminum hydroxide, Am. J. Kidney Dis. 11 (1) (1988) 70–75.
[25] S.S.A. El-Rahman, Neuropathology of aluminum toxicity in rats (glutamate and GABA impairment), Pharmacol. Res. 47 (3) (2003) 189–194.
[26] B. Sjögren, V. Lidums, M. Håkansson, L. Hedström, Exposure and urinary excretion of aluminum during welding, Scand. J. Work Environ. Health 11 (1985) 39–43.
[27] H.J. Gitelman, Aluminum exposure and excretion, Sci. Total Environ. 163 (1) (1995) 129–135.
[28] G. Song, S. Song, A possible biodegradable magnesium implant material, Adv. Eng. Mater. 9 (4) (2007) 298–302.
[29] S.L. Volpe, Magnesium in disease prevention and overall health, Adv. Nutr. 4 (3) (2013) 378S–383S.
[30] J.H. de Baaij, J.G. Hoenderop, R.J. Bindels, Magnesium in man: implications for health and disease, Physiol. Rev. 95 (1) (2015) 1–46.
[31] C.G. Musso, Magnesium metabolism in health and disease, Int. Urol. Nephrol. 41 (2) (2009) 357–362.
[32] R.J. Elin, Magnesium metabolism in health and disease, Dis. Mon. 34 (4) (1988) 162–218.
[33] I.E. Dreosti, Magnesium status and health, Nutr. Rev. 53 (9) (1995) S23.
[34] X. Gu, Y. Zheng, Y. Cheng, S. Zhong, T. Xi, In vitro corrosion and biocompatibility of binary magnesium alloys, Biomaterials 30 (4) (2009) 484–498.
[35] M. Diez, H.-E. Kim, V. Serebryany, S. Dobatkin, Y. Estrin, Improving the mechanical properties of pure magnesium by three-roll planetary milling, Mater. Sci. Eng. A 612 (2014) 287–292.
[36] F. Witte, N. Hort, C. Vogt, S. Cohen, K.U. Kainer, R. Willumeit, et al., Degradable biomaterials based on magnesium corrosion, Curr. Opinion Solid State Mater. Sci. 12 (5–6) (2008) 63–72.
[37] X. Zhang, G. Yuan, J. Niu, P. Fu, W. Ding, Microstructure, mechanical properties, bio-corrosion behavior, and cytotoxicity of as-extruded Mg-Nd-Zn-Zr alloy with different extrusion ratios, J. Mech. Behav. Biomed. Mater. 9 (2012) 153–162.

[38] Z. Zhen, T. Xi, Y. Zheng, L. Li, L. Li, In vitro study on Mg–Sn–Mn alloy as biodegradable metals, J. Mater. Sci. Technol. 30 (7) (2014) 675–685.
[39] D. Hong, P. Saha, D.-T. Chou, B. Lee, B.E. Collins, Z. Tan, et al., In vitro degradation and cytotoxicity response of Mg–4% Zn–0.5% Zr (ZK40) alloy as a potential biodegradable material, Acta Biomater. 9 (10) (2013) 8534–8547.
[40] A study on microstructures and creep behaviors in the Mg-3Sr-xY alloys, in: M. Hu, H. Fei, J. Gao, F.F. Zhao (Eds.), Advanced Materials Research, Trans Tech Publ, Guilin, 2012.
[41] R. Guan, A.F. Cipriano, Z.-y. Zhao, J. Lock, D. Tie, T. Zhao, et al., Development and evaluation of a magnesium–zinc–strontium alloy for biomedical applications—alloy processing, microstructure, mechanical properties, and biodegradation, Mater. Sci. Eng. C 33 (7) (2013) 3661–3669.
[42] Z. Zhang, X. Zeng, W. Ding, The influence of heat treatment on damping response of AZ91D magnesium alloy, Mater. Sci. Eng. A 392 (1–2) (2005) 150–155.
[43] H. Jia, X. Feng, Y. Yang, Influence of solution treatment on microstructure, mechanical and corrosion properties of Mg-4Zn alloy, J. Magnes. Alloys 3 (3) (2015) 247–252.
[44] S. Feng, W. Zhang, Y. Zhang, J. Guan, Y. Xu, Microstructure, mechanical properties and damping capacity of heat-treated Mg–Zn–Y–Nd–Zr alloy, Mater. Sci. Eng. A 609 (2014) 283–292.
[45] L. Wu, B.J.C. Luthringer, F. Feyerabend, A.F. Schilling, R. Willumeit, Effects of extracellular magnesium on the differentiation and function of human osteoclasts, Acta Biomater. 10 (6) (2014) 2843–2854.
[46] S.O. Marx, H. Totary-Jain, A.R. Marks, Vascular smooth muscle cell proliferation in restenosis, Circ. Cardiovasc. Interv. 4 (1) (2011) 104–111.
[47] H. Hao, G. Gabbiani, M.L. Bochaton-Piallat, Arterial smooth muscle cell heterogeneity: implications for atherosclerosis and restenosis development, Arterioscler. Thromb. Vasc. Biol. 23 (9) (2003) 1510–1520.
[48] T. Inoue, K. Node, Molecular basis of restenosis and novel issues of drug-eluting stents, Circ. J. 73 (4) (2009) 615–621.
[49] J. Ma, N. Zhao, D. Zhu, Biphasic responses of human vascular smooth muscle cells to magnesium ion, J. Biomed. Mater. Res. A 104 (2) (2016) 347–356.
[50] N. Zhao, D. Zhu, Endothelial responses of magnesium and other alloying elements in magnesium-based stent materials, Metallomics 7 (1) (2015) 118–128.
[51] J. Wang, V. Giridharan, V. Shanov, Z. Xu, B. Collins, L. White, et al., Flow-induced corrosion behavior of absorbable magnesium-based stents, Acta Biomater. 10 (12) (2014) 5213–5223.
[52] B. Heublein, R. Rohde, V. Kaese, M. Niemeyer, W. Hartung, A. Haverich, Biocorrosion of magnesium alloys: a new principle in cardiovascular implant technology? Heart 89 (6) (2003) 651–656.
[53] R. Waksman, R. Pakala, P.K. Kuchulakanti, R. Baffour, D. Hellinga, R. Seabron, et al., Safety and efficacy of bioabsorbable magnesium alloy stents in porcine coronary arteries, Catheter. Cardiovasc. Interv. 68 (4) (2006) 607–617. discussion 18–19.
[54] C.M. Campos, T. Muramatsu, J. Iqbal, Y.J. Zhang, Y. Onuma, H.M. Garcia-Garcia, et al., Bioresorbable drug-eluting magnesium-alloy scaffold for treatment of coronary artery disease, Int. J. Mol. Sci. 14 (12) (2013) 24492–24500.
[55] E. Wittchow, N. Adden, J. Riedmuller, C. Savard, R. Waksman, M. Braune, Bioresorbable drug-eluting magnesium-alloy scaffold: design and feasibility in a porcine coronary model, EuroIntervention 8 (12) (2013) 1441–1450.
[56] P. Zartner, R. Cesnjevar, H. Singer, M. Weyand, First successful implantation of a biodegradable metal stent into the left pulmonary artery of a preterm baby, Catheter. Cardiovasc. Interv. 66 (4) (2005) 590–594.

[57] P. Zartner, M. Buettner, H. Singer, M. Sigler, First biodegradable metal stent in a child with congenital heart disease: evaluation of macro and histopathology, Catheter. Cardiovasc. Interv. 69 (3) (2007) 443–446.
[58] P. Peeters, M. Bosiers, J. Verbist, K. Deloose, B. Heublein, Preliminary results after application of absorbable metal stents in patients with critical limb ischemia, J. Endovasc. Ther. 12 (1) (2005) 1–5.
[59] M. Haude, R. Erbel, P. Erne, S. Verheye, H. Degen, D. Böse, et al., Safety and performance of the drug-eluting absorbable metal scaffold (DREAMS) in patients with de-novo coronary lesions: 12 month results of the prospective, multicentre, first-in-man BIOSOLVE-I trial, Lancet 381 (9869) (2013) 836–844.
[60] M. Haude, H. Ince, A. Abizaid, R. Toelg, P.A. Lemos, C. von Birgelen, et al., Safety and performance of the second-generation drug-eluting absorbable metal scaffold in patients with de-novo coronary artery lesions (BIOSOLVE-II): 6 month results of a prospective, multicentre, non-randomised, first-in-man trial, Lancet 387 (10013) (2016) 31–39.
[61] P.T. Lieu, M. Heiskala, P.A. Peterson, Y. Yang, The roles of iron in health and disease, Mol. Asp. Med. 22 (1–2) (2001) 1–87.
[62] G. Papanikolaou, K. Pantopoulos, Iron metabolism and toxicity, Toxicol. Appl. Pharmacol. 202 (2) (2005) 199–211.
[63] N. Abbaspour, R. Hurrell, R. Kelishad, Review on iron and its importance for human health, J. Res. Med. Sci. 18 (2) (2014) 164–174.
[64] B. Liu, Y.F. Zheng, L. Ruan, In vitro investigation of $Fe_{30}Mn_6Si$ shape memory alloy as potential biodegradable metallic material, Mater. Lett. (Netherlands) 65 (3) (2011) 540–543.
[65] M. Moravej, F. Prima, M. Fiset, D. Mantovani, Electroformed iron as new biomaterial for degradable stents: development process and structure-properties relationship, Acta Biomater. 6 (5) (2010) 1726–1735.
[66] F.L. Nie, Y.F. Zheng, S.C. Wei, C. Hu, G. Yang, In vitro corrosion, cytotoxicity and hemocompatibility of bulk nanocrystalline pure iron, Biomed. Mater. 5 (6) (2010) 065015.
[67] Q. Feng, D. Zhang, C. Xin, X. Liu, W. Lin, W. Zhang, et al., Characterization and in vivo evaluation of a bio-corrodible nitrided iron stent, J. Mater. Sci. Mater. Med. 24 (3) (2013) 713–724.
[68] M. Schinhammer, A.C. Hänzi, J.F. Löffler, P.J. Uggowitzer, Design strategy for biodegradable Fe-based alloys for medical applications, Acta Biomater. 6 (5) (2010) 1705–1713.
[69] W. Xu, X. Lu, L. Tan, K. Yang, Study on properties of a novel biodegradable Fe-30Mn-1C alloy, Acta Metall. Sin. 47 (10) (2011) 1342–1347.
[70] B. Liu, Y.F. Zheng, Effects of alloying elements (Mn, Co, Al, W, Sn, B, C and S) on biodegradability and in vitro biocompatibility of pure iron, Acta Biomater. 7 (3) (2011) 1407–1420.
[71] H. Hermawan, A. Purnama, D. Dube, J. Couet, D. Mantovani, Fe-Mn alloys for metallic biodegradable stents: degradation and cell viability studies, Acta Biomater. 6 (5) (2010) 1852–1860.
[72] H. Hermawan, H. Alamdari, D. Mantovani, D. Dubé, Iron–manganese: new class of metallic degradable biomaterials prepared by powder metallurgy, Powder Metall. 51 (1) (2013) 38–45.
[73] H.F. Li, X.H. Xie, Y.F. Zheng, Y. Cong, F.Y. Zhou, K.J. Qiu, et al., Development of biodegradable Zn-1X binary alloys with nutrient alloying elements Mg, Ca and Sr, Sci. Rep. 5 (2015) Article No. 10719.

[74] E. Mostaed, M. Sikora-Jasinska, A. Mostaed, S. Loffredo, A.G. Demir, B. Previtali, et al., Novel Zn-based alloys for biodegradable stent applications: design, development and in vitro degradation, J. Mech. Behav. Biomed. Mater. 60 (2016) 581–602.
[75] H. Li, H. Yang, Y. Zheng, F. Zhou, K. Qiu, X. Wang, Design and characterizations of novel biodegradable ternary Zn-based alloys with IIA nutrient alloying elements Mg, Ca and Sr, Mater. Des. 83 (2015) 95–102.
[76] X. Liu, J. Sun, K. Qiu, Y. Yang, Z. Pu, L. Li, et al., Effects of alloying elements (Ca and Sr) on microstructure, mechanical property and in vitro corrosion behavior of biodegradable Zn–1.5Mg alloy, J. Alloys Compd. 664 (2016) 444–452.
[77] X. Liu, J. Sun, F. Zhou, Y. Yang, R. Chang, K. Qiu, et al., Micro-alloying with Mn in Zn–Mg alloy for future biodegradable metals application, Mater. Des. 94 (2016) 95–104.
[78] S. Zhu, N. Huang, L. Xu, Y. Zhang, H. Liu, H. Sun, et al., Biocompatibility of pure iron: in vitro assessment of degradation kinetics and cytotoxicity on endothelial cells, Mater. Sci. Eng. C 29 (5) (2009) 1589–1592.
[79] J.E. Schaffer, E.A. Nauman, L.A. Stanciu, Cold drawn bioabsorbable ferrous and ferrous composite wires: an evaluation of in vitro vascular cytocompatibility, Acta Biomater. 9 (10) (2013) 8574–8584.
[80] M. Peuster, P. Wohlsein, M. Brügmann, M. Ehlerding, K. Seidler, C. Fink, et al., A novel approach to temporary stenting: degradable cardiovascular stents produced from corrodible metal—results 6–18 months after implantation into New Zealand white rabbits, Heart 86 (5) (2001) 563–569.
[81] M. Peuster, C. Hesse, T. Schloo, C. Fink, P. Beerbaum, C. von Schnakenburg, Long-term biocompatibility of a corrodible peripheral iron stent in the porcine descending aorta, Biomaterials 27 (28) (2006) 4955–4962.
[82] R. Waksman, R. Pakala, R. Baffour, R. Seabron, D. Hellinga, F.O. Tio, Short-term effects of biocorrodible iron stents in porcine coronary arteries, J. Interv. Cardiol. 21 (1) (2008) 15–20.
[83] C. Wu, X. Hu, H. Qiu, Y. Ruan, Y. Tang, A. Wu, et al., TCT-571 a preliminary study of biodegradable iron stent in mini-swine coronary artery, J. Am. Coll. Cardiol. 60 (17 Suppl.) (2012) B166.
[84] D. Pierson, J. Edick, A. Tauscher, E. Pokorney, P. Bowen, J. Gelbaugh, et al., A simplified in vivo approach for evaluating the bioabsorbable behavior of candidate stent materials, J. Biomed. Mater. Res. B Appl. Biomater. 100 (1) (2012) 58–67.
[85] W.J. Lin, D.Y. Zhang, G. Zhang, H.T. Sun, H.P. Qi, L.P. Chen, et al., Design and characterization of a novel biocorrodible iron-based drug-eluting coronary scaffold, Mater. Des. 91 (2016) 72–79.
[86] H. Tapiero, K.D. Tew, Trace elements in human physiology and pathology: zinc and metallothioneins, Biomed. Pharmacother. 57 (9) (2003) 399–411.
[87] P.J. Aggett, J.T. Harries, Current status of zinc in health and disease states, Arch. Dis. Child. 54 (1979) 909–917.
[88] M.F. Robinson, J.M. McKenzie, C.D. Thomson, A.L. van Rij, Metabolic balance of zinc, copper, cadmium, iron, molybdenum and selenium in young New Zealand women, Br. J. Nutr. 30 (02) (1973) 195–205.
[89] K.M. Hambidge, N.F. Krebs, Zinc deficiency: a special challenge, J. Nutr. 137 (4) (2007) 1101–1105.
[90] J. Brandão-Neto, V. Stefan, B.B. Mendonca, W. Bloise, A.V.B. Castro, The essential role of zinc in growth, Nutr. Res. 15 (3) (1995) 335–358.

[91] D. Vojtěch, J. Kubasek, J. Šerák, P. Novák, Mechanical and corrosion properties of newly developed biodegradable Zn-based alloys for bone fixation, Acta Biomater. 7 (9) (2011) 3515–3522.
[92] D.V. Jiří Kubásek, Zn-based alloys as an alternative biodegradable materials, Metal 5 (2012) 23–25.
[93] N. Kang, H.S. Na, S.J. Kim, C.Y. Kang, Alloy design of Zn–Al–Cu solder for ultra high temperatures, J. Alloys Compd. 467 (1) (2009) 246–250.
[94] T. Tanaka, K. Makii, H. Ueda, A. Kushibe, M. Kohzu, K. Higashi, Study on practical application of a new seismic damper using a Zn-Al alloy with a nanocrystalline microstructure, Int. J. Mech. Sci. 45 (10) (2003) 1599–1612.
[95] M. Furukawa, Y. Ma, Z. Horita, M. Nemoto, R.Z. Valiev, T.G. Langdon, Microstructural characteristics and superplastic ductility in a Zn-22% Al alloy with submicrometer grain size, Mater. Sci. Eng. A 241 (1998) 122–128.
[96] M. Babic, S. Mitrovic, B. Jeremic, The influence of heat treatment on the sliding wear behavior of a ZA-27 alloy, Tribol. Int. 43 (1) (2010) 16–21.
[97] G. Purcek, O. Saray, I. Karaman, T. Kucukomeroglu, Effect of severe plastic deformation on tensile properties and impact toughness of two-phase Zn–40Al alloy, Mater. Sci. Eng. A 490 (1) (2008) 403–410.
[98] J. Ma, N. Zhao, D. Zhu, Endothelial cellular responses to biodegradable metal zinc, ACS Biomater. Sci. Eng. 1 (11) (2015) 1174–1182.
[99] J. Ma, N. Zhao, D. Zhu, Bioabsorbable zinc ion induced biphasic cellular responses in vascular smooth muscle cells, Sci. Rep. 6 (2016) Article No. 26661.
[100] H. Gong, K. Wang, R. Strich, J.G. Zhou, In vitro biodegradation behavior, mechanical properties, and cytotoxicity of biodegradable Zn-Mg alloy, J. Biomed. Mater. Res. B Appl. Biomater. 103 (8) (2015) 1632–1640.
[101] P.K. Bowen, R.J. Guillory II, E.R. Shearier, J.-M. Seitz, J. Drelich, M. Bocks, et al., Metallic zinc exhibits optimal biocompatibility for bioabsorbable endovascular stents, Mater. Sci. Eng. C 56 (2015) 467–472.
[102] R.J. Guillory, P.K. Bowen, S.P. Hopkins, E.R. Shearier, E.J. Earley, A.A. Gillette, et al., Corrosion characteristics dictate the long-term inflammatory profile of degradable zinc arterial implants, ACS Biomater. Sci. Eng. 2 (12) (2016) 2355–2364.

Part Two

Coatings and surface modification of cardiovascular stents

Physico-chemical stent surface modifications

7

A. Foerster, M. Duda*, H. Kraśkiewicz*, M. Wawrzyńska†, H. Podbielska*, M. Kopaczyńska**
*Wrocław University of Technology, Wrocław, Poland, †Wrocław Medical University, Wrocław, Poland

7.1 Introduction

Surface modification gives great possibility for improving properties of materials to be used in biological environment. Materials used for stent fabrication have ideal mechanical bulk properties allowing the stents to be used in vivo; however, the surface properties lack the control over the interaction between the stent surface and blood. Lacking biocompatibility of stents surface can cause adverse biological effects. Desired properties of stent surface should aim at reducing thrombogenicity, preventing denaturation of adsorbed proteins, and ideally accelerating re-endothelialization [1].

Surface modification of stents for improved biocompatibility can be done by numbers of means. Some of them are designed to introduce a layer of material on the stent surface in order to prohibit or reduce undesirable protein adhesion. Examples of these materials include organic and inorganic materials such as poly (methyl methacrylate) [2], heparin, collagen [3], titanium oxides [4], or nanostructured carbon [5]. Another approach involves direct attachment of endothelial progenitor cells (EPCs) to the stent surface. Presence of these cells on the surface gives a great potential to increase biocompatibility of the stents. Since EPCs do not bond on 316L stainless steel stents directly, different approaches are developed for this purpose. These methods utilize precoating techniques with polymers or proteins to support the adhesion of cells to the stent surface [6,7].

Recently, a very promising approach has been proposed in stent technology, which involves substances that accelerate endothelialization [8]. In this technique, antibodies directed toward proteins located on the cell surface of EPCs are immobilized on the stent surface. Thanks to this approach, EPCs migrate from the blood stream to the stent surface. Recent studies show benefits of capturing EPCs by functionalizing stents with immobilized anti-CD34 antibodies [9,10]. The suitable sites for covalent binding of proteins to the stent surface are provided by functional groups such as hydroxyl, amino, thiol, carboxyl, aldehyde, or isocyanate. Further modification of functional groups before immobilization leads to even more effective reactions [11].

For metallic samples, plasma polymerization technique has been applied for effective covalent protein attachment [12,13]. Surfaces modified with plasma and coated with proteins have presented improved blood biocompatibility, considerably reduced blood clogging, and enhanced endothelialization [1].

Functionalized Cardiovascular Stents. https://doi.org/10.1016/B978-0-08-100496-8.00007-X

This chapter provides an overview on different strategies for chemico-physical stent functionalization aiming at reducing the occurrence of thrombosis, restenosis, and accelerating rapid endothelialization after the stent implantation. The functionalization techniques discussed in more details in this chapter concern stents made of stainless steel. Table 7.1 provides an overview of recent functionalization methods of stents made of different materials than stainless steel.

7.2 Stent surface functionalization

7.2.1 Polymer functionalized surface

The use of polymers in stent technology can be explained by their good compatibility related to wettability, electrostatic surface charge, and surface conductivity of these materials. Their main role is to form a neutral barrier between blood and stent surface. Various polymers have been examined, among them polyurethane [20], polyethylene terephthalate [21], polysiloxane [22]. Although in case of these polymers a decrease of blood coagulation on the polymers-coated stent surface was stated, the reduction of cell proliferation was not observed comparing to the stent without a polymer coating.

Natural polymers first introduced by Holmes and Baker [23,24] have been applied until now in stent technology [25,26]. Recently, Harvey et al. [27] investigated fibronectin as a stent-coating material for improving endothelial cell (EC) attachment. The 316L stainless steel was first electrochemically coated with COOH-terminated surface layer and then extracellular matrix (ECM) protein fibronectin was immobilized on the stent surface. The results suggested that fibrinogen-modified stent surface was more biocompatible in regard to attachment of ECs when compared to the unmodified stent surface.

Heparin is a widely used material in stent-coating technology [28–30] mainly due to the presence of ionic group in the structure. Meng et al. [31] employed layer-by-layer process to coat a stent surface by chitosan/heparin (CS/HEP LBL). The in vitro hemocompatibility test results of functionalized and bare 316L stainless demonstrated safety and efficiency in attachment of ECs, as well as good biocompatibility.

Various polymers [17,32,33] have been applied for drug-eluting stents (DES). The antiproliferative drugs can be trapped or bonded into polymer-coated stents and delivered to arterial wall. The role of DES is to reduce the risk of stent restenosis by the drug elution. Functionalization of DES stents aims at improving their long-term activity. As promising alternative for polymers in DES technology, the inorganic porous materials have been proposed [34].

7.2.2 Metal oxides functionalized surface

Once the stent is brought to contact with blood, protein adsorption takes place. Fibrinogen that is one of adsorbed proteins desaturates, leading to activation of platelets. This is followed by cascade reaction and eventual coagulation of blood and

Table 7.1 **Overview of various approaches for stent functionalization**

Stent	Surface functionalization (first layer)	Surface functionalization (second layer)	Results	Reference
NiTi	3-Amininopropylthretoxysilane (APTES)	Heparin	Good biocompatibility	[14]
Co/Cr	Poly-L-lactic acid (PLLA)	CD34-Ab	Improved functionality	[15]
Co/Cr	APTES	Oligonucleotide	Good biocompatibility Increase of EPCs adhesion	[16]
Co/Cr	Polydopamine (PDA)	Cyclodextrins-based polymers	Good biocompatibility	[17]
Pt/Cr	Polyhedral oligomeric silsesquioxane poly(carbonate-urea) urethane (POSS-PCU) nanocomposite polymer with covalently attached anti-CD34 antibodies	None	Good biocompatibility	[18]
Ti	Heparin/fibronectin	VEGF	Good biocompatibility, promotion of endothelialization	[19]

formation of thrombosis [35]. Huang et al. [36] examined biocompatibility of amorphous titanium oxide films for increasing biocompatibility of stents. The stent surface was coated with titanium oxide using ion beam-enhanced deposition technique. The semiconducting nature of the titanium oxide films was demonstrated to play an important role in the reduction of platelets adhesion. Thanks to this property, deposition of fibrinogen could be inhibited, leading to improved compatibility of stents. Their further work [37] indicated surface energy as another meaningful factor in the interaction between blood and stent surface. The results demonstrated that lower interface tension between titanium oxide films and proteins, as well as semiconducting nature of the films, contributed greatly to an increased biocompatibility

7.2.3 Metal functionalized surface

Gold was investigated by a number of researches [38,39] as a potential biocompatible coating material. Work of Hehrlein [40] and Schwartz [41] suggested effectivity and safety of the gold-modified surfaces. However, research of Kastrati [42] demonstrated an increased risk of restenosis after stent implantation. These contradictory results could be explained by the different techniques used for gold coating [43,44].

7.2.4 Endothelial cells functionalized surface

Despite great progress in stent functionalization with organic and inorganic molecules, the ideal surface with blood contact is a native endothelium [45]. The ECs are responsible for keeping the blood vessels resistant to thrombosis by controlling the migration and proliferation of smooth muscles [46], while the negative charge on the cells surface helps to reduce platelet adhesion.

ECM functionalized stent surface increased the adhesion and proliferation of ECs on the surface [47,48]. However, the long time needed for cell culture and financial aspect are limiting factors in using this method as common practise.

Since it has been showed [48] that accelerated re-endothelialization after stent implantation can increase its biocompatibility and lowers the risk of restenosis, a very interesting alternative for coating a stent surface with ECs has been examined. In this case, the emphasis has been placed on the stent surface functionalization with molecules that possess capability to capture EPCs in vivo, which then differentiate into mature ECs [49].

An anti-CD34 antibody functionalized multilayer heparin/collagen was developed via layer-by-layer approach by Lin et al. [50]. The results indicated good hemocompatibility and enhancement of the cell attachment and growth in both, heparin/collagen multilayers with and without anti-CD34 on the surfaces. However, only heparin/collagen surface functionalized with anti-CD34 demonstrated selectivity in promotion of ECs cells. The surface modified that way with attached ECs showed better viability and indicated the specific function characteristic for natural vascular ECs. It has been demonstrated that the presence of anti-CD34 greatly contributes to rapid endothelialization.

Chen et al. [51] investigated covalent immobilization of anti-CD34 antibodies onto material surface and its orientation on EPC capturing ability. The significance of antibody orientation has been highlighted in their work. The antibodies immobilized in an oriented way showed higher immunologic binding activity and progenitor cell capturing capability when compared to randomly immobilized ones.

Aoki et al. [52] evaluated murine monoclonal antihuman CD34 antibodies-coated stainless steel stents to reduce stent thrombosis and restenosis. This first human clinical trial showed effectivity in capturing EC and safety of stents.

7.2.5 Antibody fragments functionalized surface

Single chain antibody fragments (scFvs) are fusion protein obtained by bonding antibody heavy and light chain variable domains with polypeptide linker. Employing scFvs rather than whole immunoglobulin molecules leads to considerably increased detection sensitivities. Their small size contributes to enhancing coating densities [53,54].

In our previous work [55], we presented a technique for stent surface functionalization using antibody fragments. The titania-coated stainless steel stent surface was functionalized with amine groups, then glycosylated antibody fragments specific toward vascular endothelial growth factor were immobilized. The cell viability test proved no toxicity of modified surfaces to EPCs.

7.3 Thiol groups functionalized surface

In present study, we describe stent surface functionalization method with thiol groups and the evaluation of biocompatibility of the functionalized surfaces.

7.3.1 Mercaptosilanization procedure

The 316L stainless steel disks (10 mm in diameter) provided by Balton Sp. z o.o. were firstly covered with titania-based coatings (called T15) by sol-gel method [56] in order to introduce the surface hydroxyl moieties that serve as the binders for the thiol substrate. Sols were prepared with the use of titanium ethoxide, ethanol, triton X-100, and hydrochloric acid for acidic hydrolysis at pH = 2. Disks coated with T15 were subsequently dipped in (3-mercaptopropyl) ethoxysilane (MPTS) solution, also prepared with ethanol and triton X-100 and then incubated in the heater for 1 h at 60°C. After the incubation, the substrates were washed with ethanol and dried in air.

7.3.2 Spectroscopic characterization of mercaptosilanized surface

Samples prepared with mercaptosilanization procedure had been characterized thoroughly by Fourier transform infrared spectroscopy in attenuated total reflection mode (FTIR-ATR) (Fig. 7.1) and Raman spectroscopy (Fig. 7.2).

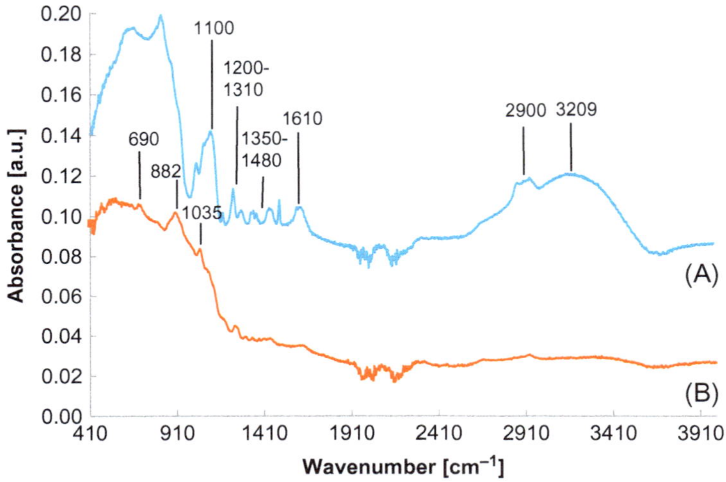

Fig. 7.1 FTIR-ATR spectra of samples coated with T15: (A) before mercaptosilanization and (B) after mercaptosilanization. The mercaptosilanization process smoothened and shifted the spectra slightly. The broad absorption band present at 3209 cm^{-1} in spectrum A represents the stretching vibration of surface hydroxyl groups; on the other hand, such band is almost not visible or significantly decreased on spectrum B, indicating that most of the active hydroxyl moieties partook in the formation of silane bond with MPTS.

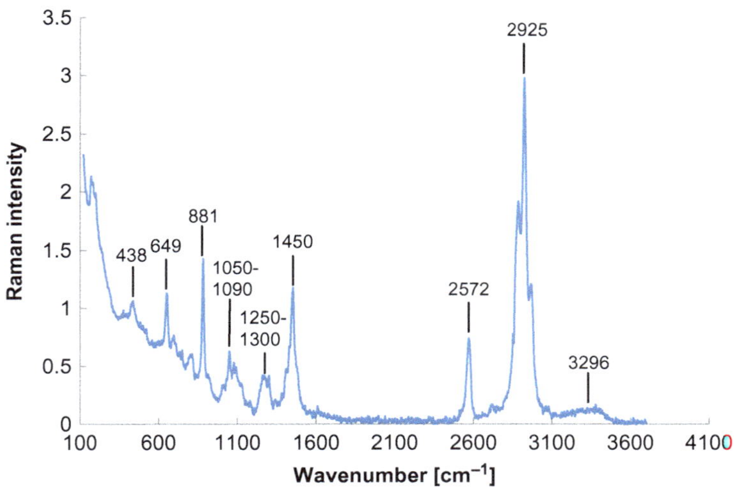

Fig. 7.2 Raman spectroscopy spectra of T15 coating reacted to MPTS compound. With the exception of C–O–C bond (peak at 881 cm^{-1}), all of the other groups detected by FTIR-ATR were also observable in Raman spectra, but in addition some sulfur compounds became visible.

Fig. 7.2 represents the FTIR-ATR spectra obtained from samples coated with T15 before and after the mercaptosilanization process. In spectrum A, the absorption bands at 2900 and 1350–1480 cm^{-1} come from symmetric and asymmetric vibrational stretches of CH_2 groups. Bands visible at 1200–1310 and 1610 cm^{-1} confirm the presence of C–O bonds, whereas the absorption at 1100 cm^{-1} is the evidence of Ti–O–C bridging vibrations.

Spectrum B, on the other hand, is slightly shifted comparing to spectrum A; moreover, the ratio of each band is different. It can be noticed by looking at the absorption bands at 690 and 882 cm^{-1} that are attributed to the vibrations of Ti–O–Ti and Ti–O, respectively. The peak visible at 1035 cm^{-1} also proves that the silane compound is present, because it comes from the vibrations of Si–O–Si bonds. But the most striking difference is the significant decrease of the broad absorption band at 3209 cm^{-1}; it is most likely due to fact that hydroxyl moieties reacted with MPTS and free thiol groups on the surface were formed.

Unfortunately, most of the expected sulfur moieties (S–H, S–S and C–S) appear as very weak in FTIR-ATR spectra and usually cannot be observed due to overwhelming signal of other highly absorbing compounds. Therefore, Raman spectroscopy was employed and the obtained spectrum of T15 coating with MPTS is presented in Fig. 7.2.

Raman spectroscopy revealed that MPTS reacts with T15 by forming various sulfur moieties that are visible in the spectrum as peaks at 438, 649, and 2572 $cm-^{1}$, which confirm the presence of S–S, C–S, and S–H bonds, respectively.

FTIR-ATR and Raman spectroscopy combined indicate that the process of mercaptosilanization of 316L stainless steel is successful and the examined surface is covered with free thiol groups.

7.3.3 In vitro studies

In order to determine the toxicity of the functionalized surfaces, an in vitro study of cell viability was carried out with the use of human EPC line 55.1 EPCs. EPCs were cultured in direct contact with four different surfaces: T15, aminosilanized T15 ($-NH_2$) (as reported in our earlier work in Ref. [42]), mercaptosilanized T15 (–SH), and mercaptosilanized T15 after treatment with reducing agent—dithiothreitol (DTT), which breaks the spontaneously formed disulfide bonds into free thiols on functionalized surface (Fig. 7.3). Cells were marked and counted after 24 and 48 h.

Obtained results show that EPCs growth varies depending on the kind of tested surface. After 24 h of cell cultivating, the most EPCs friendly coatings are the both options of mercaptosilanized T15; on the other hand, the least favorable surface seems to be the aminosilanized T15. This difference might be explained by the closer examination of surface charge of each coating, which is positive for $-NH_2$ and negative for –SH. The general tendency in cell growth after 48 h is maintained with slight changes of bar ratios. In each case, mercaptosilanized T15 exhibits over 20% higher cell growth than the T15 coating itself, and over 40% higher cell growth when compared to aminosilanized surface. These results suggest that the mercaptosilanized surfaces are the most EPCs friendly and the least toxic of all examined coatings.

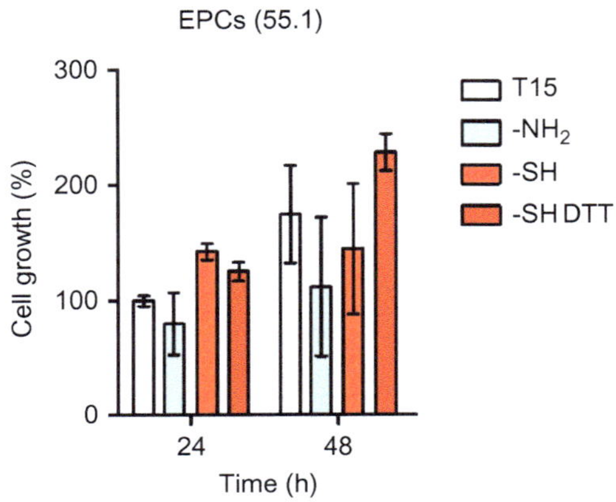

Fig. 7.3 In vitro cell viability study on four different surfaces: T15, aminosilanized T15 ($-NH_2$), mercaptosilanized T15 (−SH), and mercaptosilanized T15 after the treatment with DTT (−SH DTT). Each bar represents the mean of six cell counts.

7.4 Conclusion

Despite a number of studies on stent surface functionalization for its improved biocompatibility, the ultimate design remains a challenge. There is a great potential in future research for developing a stent material with increased detection sensitivities for EC cells, and hence, acceleration of the healing process after stent implantation. The development of functionalized stents allows to minimize stent thrombosis and restenosis. The approach of stent surface functionalization for immobilization of scFvs was demonstrated. Further understanding and actualization of this method will bring thorough perspective in the field of stent technology.

References

[1] T.M. Jeewandara, S.G. Wise, M.K.C. Ng, Biocompatibility of coronary stents, Materials 7 (2014) 769–786.

[2] Y. Shaulov, R. Okner, Y. Levi, N. Tal, V. Gutkin, D. Mandler, et al., Poly(methyl methacrylate) grafting onto stainless steel surfaces: application to drug-eluting stents, ACS Appl. Mater. Interfaces 1 (2009) 519–2528.

[3] K. Zhang, J.Y. Chen, W. Qin, J.A. Li, F.X. Guan, et al., Constructing bio-layer of heparin and type IV collagen on titanium surface for improving its endothelialization and blood compatibility, J. Mater. Sci. Mater. Med. 27 (2016) 81.

[4] S.J. Song, K.W. Jung, Y.J. Park, J. Park, M.D. Cho, M.H. Jeong, et al., Nitrogen-doped TiO_2 films as drug-binding matrices for the preparation of drug-eluting stents, J. Mater. Chem. 21 (2011) 8169–8177.

[5] I.B. Kovalenko, Z.Y. Chefranova, M.I. Boyarintsev, M.V. Filatov, Y.A. Lykov, Bioinert properties of a coronary stent with nanostructural carbon coating, Res. J. Pharm. Biol. Chem. Sci. 5 (2014) 1352–1356.

[6] T. Shirota, H. Yasui, H. Shimokawa, T. Matsuda, Fabrication of endothelial progenitor cell (EPC)-seeded intravascular stent devices and in vitro endothelialization on hybrid vascular tissue, Biomaterials 24 (2003) 2295–2302.
[7] Z. Zhou, S. Shi, M. Song, H. Huang, K. Chen, J. Mi, et al., Development of transgenic endothelial progenitor cell-seeded stents, J. Biomed. Mater. Res. A 91 (2) (2009) 623–628.
[8] M. Avci-Adali, G. Ziemer, H.P. Wendel, Induction of EPC homing on biofunctionalized vascular grafts for rapid in vivo self-endothelialization—a review of current strategies, Biotechnol. Adv. 28 (1) (2010) 119–129.
[9] T. Inoue, M. Sata, Y. Hikichi, R. Sohma, D. Fukuda, T. Uchida, et al., Mobilization of CD34-positive bone marrow-derived cells after coronary stent implantation—impact on restenosis, Circulation 115 (2007) 553–561.
[10] M.A. Beijk, M. Klomp, N.J. Verouden, N. van Geloven, K.T. Koch, J.P. Henriques, et al., Genous endothelial progenitor cell capturing stent vs. the Taxus Liberte stent in patients with de novo coronary lesions with a high-risk of coronary restenosis: a randomized, single-centre, pilot study, Eur. Heart J. 31 (2010) 1055–1064.
[11] R. Williams (Ed.), Surface Modification of Biomaterials: Methods Analysis and Applications, Woodhead Publishing Limited, Oxford, Cambridge, Philadelphia, New Delhi, 2010.
[12] A. Waterhouse, Y.B. Yin, S.G. Wise, D.V. Bax, D.R. McKenzie, M.M.M. Bilek, et al., The immobilization of recombinant human tropoelastin on metals using a plasma-activated coating to improve the biocompatibility of coronary stents, Biomaterials 31 (2010) 8332–8340.
[13] A. Waterhouse, S.G. Wise, Y.B. Yin, B.C. Wu, B. James, H. Zreiqat, et al., In vivo biocompatibility of a plasma-activated, coronary stent coating, Biomaterials 33 (2012) 7984–7992.
[14] F. Wang, Y. Zhang, X. Chen, B. Leng, X. Guo, T. Zhang, ALD mediated heparin grafting on nitinol for self-expanded carotid stents, Colloids Surf. B: Biointerfaces 143 (2016) 390–398.
[15] R. Rebelo, N. Vila, S. Rana, R. Fangueiro, Poly lactic acid fibre based biodegradable stents and their functionalization techniques, in: RILEM. Proceedings of the 2nd International Conference on Natural Fibers; 2015 April 27–29; Azores (Portugal), 2015, pp. 331–342.
[16] M.C. Barsotti, T. Al Kayal, L. Tedeschi, D. Dinucci, P. Losi, S. Sbrana, Oligonucleotide biofunctionalization enhances endothelial progenitor cell adhesion on cobalt/chromium stents, J. Biomed. Mater. Res. A 103 (10) (2015) 3284–3292.
[17] J. Sobocinski, W. Laure, M. Taha, E. Courcot, F. Chai, N. Simon, et al., Mussel inspired coating of a biocompatible cyclodextrin based polymer onto CoCr vascular stents, ACS Appl. Mater. Interfaces 6 (5) (2014) 3575–3586.
[18] A. Tan, Y. Farhatnia, D.G.N. Goh, A. de Mel, J. Lim, S.H. Teoh, et al., Surface modification of a polyhedral oligomeric silsesquioxane poly(carbonate-urea) urethane (POSS-PCU) nanocomposite polymer as a stent coating for enhanced capture of endothelial progenitor cells, Biointerphases 8 (1) (2013) 23.
[19] X. Wang, T. Liu, Y. Chen, K. Zhang, M.F. Maitz, C. Pan, et al., Extracellular matrix inspired surface functionalization with heparin, fibronectin and VEGF provides an anticoagulant and endothelialization supporting microenvironment, Appl. Surf. Sci. 320 (2014) 871–882.
[20] E. Rechavia, F. Litvack, M.C. Fishbien, M. Nakamura, Biocompatibility of polyutherane-coated stents: tissue and vascular aspects, Catheter. Cardiovasc. Diagn. 45 (1998) 202–207.

[21] W.J. Van der Giessen, A.M. Lincoff, R.S. Schwartz, H.M. van Beusekom, P.W. Serruys, D.R. Holmes, et al., Development of a polymer endovascular prosthesis and its implantation in porcine arteries, J. Interv. Cardiol. 5 (1992) 175–185.
[22] J.E. Sousa, M.C. Morice, P.W. Serruys, J. Fajadet, M. Perin, E.B. Hayashi, et al., The RAVEL study: a randomized study with the sirolimus coated BX velocity balloon expandable stent in the treatment of patients with the novo native coronary artery lesions, Circulation (2001). II463.
[23] D. Holmes, A. Camrud, M. Jorgenson, W. Edwards, R. Schwartz, Polymeric stenting in the porcine coronary artery model: differential outcome of exogenous fibrin sleeves versus polyurethane-coated stents, J. Am. Coll. Cardiol. 24 (1994) 525–531.
[24] J. Baker, J. Horn, V. Nikolaychik, N. Kipshidze, Fibrin stent coatings, in: U. Sigwart (Ed.), Endoluminal Stenting, W. B. Saunders Company Ltd, London, Philadelphia, Toronto, Sydney, Tokyo, 1996.
[25] H. Ceylan, A.B. Tekinay, M.O. Guler, Selective adhesion and growth of vascular endothelial cells on bioactive peptide nanofiber functionalized stainless steel surface, Biomaterials 32 (2011) 8797–8805.
[26] E. Michel, P. Chevallier, A. Barrère, D. Letourneur, D. Mantovani, THERMEC'2011, in: Proceeding of the 7th International Conference; 2011 August 1–5; Quebec City (Canada), 2012, pp. 164–169.
[27] J. Harvey, A. Bergdahl, H. Dadafarin, L. Ling, E.C. Davis, S. Omanovic, An electrochemical method for functionalization of a 316L stainless steel surface being used as a stent in coronary surgery: irreversible immobilization of fibronectin for the enhancement of endothelial cell attachment, Biotechnol. Lett. 34 (2012) 1159–1165.
[28] V. Chiono, I. Carmagnola, F. Boccafooschi, P. Gentile, C. Tonda-Turo, M. Del Pilar Camacho Leal, Nanoscale tailoring of the surface properties of biomedical devices by layer by layer technique Document, in: Technical Proceedings of the 2011 NSTI Nanotechnology Conference and Expo, NSTI-Nanotech, 2011, pp. 441–444.
[29] K. Christensen, R. Larsson, H. Emanuelsson, G. Elgue, A. Larsson, Heparin coating of the stent graft—effects on platelets, coagulation and complement activation, Biomaterials 22 (2001) 349–355.
[30] N. Weber, H. Wendel, G. Ziemer, Hemocompatibility of heparin-coated surfaces and the role of selective plasma protein adsorption, Biomaterials 23 (2002) 429–439.
[31] S. Meng, Z. Liu, L. Shen, Z. Guo, L.L. Chou, W. Zhong, et al., The effect of a layer-by-layer chitosan-heparin coating on the endothelialisation and coagulation properties of a coronary stent system, Biomaterials 30 (2009) 2276–2283.
[32] W.J. Van Der Giessen, H.M. Van Beusekom, New drug-eluting stents with biodegradable polymers, Minerva Cardioangiol. 59 (2011) 31–37.
[33] F. Muñoz-Muñoz, J.-C. Ruiz, C. Alvarez-Lorenzo, A. Concheiro, E. Bucio, Temperature- and pH-sensitive interpenetrating polymer networks grafted on PP: Cross-linking irradiation dose as a critical variable for the performance as vancomycin-eluting systems, Radiat. Phys. Chem. 81 (5) (2012) 531–540.
[34] M. Arruebo, Drug delivery from structured porous inorganic materials, Nanomed. Nanobiotechnol. 4 (2012) 16–30.
[35] R.E. Baier, R.C. Dutton, Initial events in interaction of blood with foreign surfaces, J. Biomed. Mater. Res. 3 (1969) 191–206.
[36] N. Huang, Y. Ping, C. Xuan, L. Yongxang, Z. Xiaolan, C. Guangjun, Blood compatibility of amorphous titanium oxide synthesized by ion beam enhanced deposition, Biomaterials 19 (1998) 771–776.

[37] N. Huanga, P. Yanga, Y.X. Lenga, J.Y. Chena, H. Suna, Wanga, et al., Hemocompatibility of titanium oxide films, Biomaterials 24 (2003) 2177–2187.

[38] A. Fisher, H. Wieneke, H. Brauer, R. Erbel, Metallic bioamterials for coronary stents, Z. Kardiol. 4 (2001) 251–262.

[39] S. Windecker, I. Mayer, G. De Pasquale, W. Maier, O. Dirsh, P. De Groot, et al., Stent coating with titanium-nitride-oxide for reduction of neointimal hyperplasia, Circulation 104 (8) (2001) 928–933.

[40] C. Hehrlein, M. Zimmermann, J. Metz, W. Ensinger, W. Kuebler, Influence of surface texture and charge on the biocompatibility of endovascular stents, Coron. Artery Dis. 6 (1995) 581–586.

[41] R.S. Schwartz, K.C. Huber, J.G. Murphy, W.D. Edwards, A.R. Camrud, R.E. Vlietstra, et al., Restenosis and proportional neointimal response to coronary artery injury: results in porcine model, J. Am. Coll. Cardiol. 19 (1991) 267–274.

[42] A. Kastrati, A. Schoeming, J. Dirschinger, J. Mehilli, N. von Welser, J. Pache, et al., Increased risk of restenosis after placement of gold-coated stents, Circulation 101 (2000) 2478–2483.

[43] E.R. Edelman, P. Seifert, A. Groothius, A. Morss, D. Bornstein, C. Rogers, Gold coated NIR stents in porcine coronary arteries, Circulation 193 (2001) 429–434.

[44] I.A. Karoussos, H. Wieneke, T. Sawitowski, S. Wnendt, A. Fischer, O. Dirsh, et al., Inorganic materials as drug delivery systems in coronary artery stenting, Mat-wiss Werkstofftech 33 (2002) 738–746.

[45] M. Avci-Adali, G. Ziemer, H.P. Wendel, Induction od EPC homing on biofunctionalised vascular grafts for rapid in vivo self-endothelialization—a review of current strategies, Biotechnol. Adv. 28 (2010) 119–129.

[46] M. Deutsch, J. Meinhart, T. Fischlein, P. Preiss, P. Zilla, Clinical autologous in vitro endothelialization of infrainguinal ePTFE grafts in 100 patients: a 9-years experience, Surgery 126 (1999) 847–855.

[47] J.G. Meinhart, M. Deutsch, T. Fischlein, N. Howanietz, A. Froschl, P. Zilla, Clinical autologous in vitro endothelialization of 153 infrainguinal ePTFE grfts, Ann. Thorac. Surg. 71 (2001) S327–S331.

[48] N. Kipshidze, G. Dangas, M. Tsapenko, J. Moses, M.B. Leon, M. Kutryk, et al., Role of endothelium in modulating neointimal formation: vasculoprotective approaches to attenuate restenosis after percutaneous coronary interventions, J. Am. Coll. Cardiol. 44 (2004) 733–739.

[49] T. Asahara, T. Murohara, M. Sullivan, R. van der Zee, T. Li, et al., Isolation of putative progenitor endothelial cells for angiogenesis, Science 44 (1997) 733–739.

[50] Q. Lin, X. Ding, F. Qiu, X. Song, G. Fu, J. Ji, In situ endothelialization of intravascular stents coated with an anti-CD34 antibody functionalised heparin-collagen multilayer, Biomaterials 31 (2010) 4017–4025.

[51] J. Chen, Q. Li, J. Li, M.F. Maitz, The effect of anti-CD34 antibody orientation control on endothelial progenitor cell capturing cardiovascular devices, J. Bioact. Compat. Polym. **31** (6) (2016) 583–599.

[52] J. Aoki, P.W. Serruys, H. van Beusekom, A.T.L. Ong, E.P. McFadden, G. Sianos, W.J. van der Giessen, Endothelial progenitor cell capture by stents coated with antibody against CD34, J. Am. Coll. Cardiol. 45 (2005) 10.

[53] V. Afanassiev, V. Hanemann, S. Wolfl, Preparation of DNA and protein micro arrays on glass slides coated with an agarose film, Nucleic Acids Res. 28 (2000) 66.

[54] P.T. Charles, E.R. Goldman, J.G. Rangasammy, C.L. Schauer, M.S. Chen, C.R. Taitt, Fabrication characterization of 3D hydrogel microarrays to measure antigenicity and antibody functionality for biosensor application, Biosens. Bioelectron. 20 (2004) 753–764.

[55] A. Foerster, I. Hołowacz, G.B. Sunil Kumar, S. Anandakumar, J.G. Wall, M. Wawrzynska, et al., Stainless steel surface functionalization for immobilization of antibody fragments for cardiovascular applications, J. Biomed. Mater. Res. A 104 (4) (2016) 821–832.
[56] M. Kopaczyńska, B. Sobieszczańska, A. Ulatowska-Jarża, I. Hołowacz, I. Buzalewicz, Ł. Wasyluk, et al., Photoactivated titania-based nanomaterials for potential application as cardiovascular stent coatings, Biocybern. Biomed. Eng. 34 (2014) 189–197.

Chemical vapor deposition of cardiac stents

8

P. Sojitra
Concept Medical Research Private Limited, Surat, India

8.1 Introduction

Cardiovascular diseases death tolls are on the rise; an estimated 17.3 million people died from cardiovascular diseases in 2008, accounting for 48% of all global deaths [1]. The number of people who died from cardiovascular diseases, mainly from heart disease and stroke, is estimated to reach 23.3 million by 2030 [2].

Stents are made of various metals such as SS316L, L605 cobalt chromium alloy, and platinum chromium alloy, and can be used with or without drug combinations [3–7]. SS316L alloy stents were the first coronary artery stents and contained iron (60%–65%), nickel (12%–14%), and chromium (17%–18%). The manufacturing process of stents includes various mechanical and chemical processes such as laser cutting, descaling, annealing, descaling, electrochemical polishing, acid passivation, and coating. While such processes are occurring, heat stress or chemical stress is generated on the surface. This stress can lead to improper electropolishing, followed by an impact on stent surface. While the electropolishing process provides a good resistance layer to work against corrosion, a confirmative process called passivation is performed [8].

The outer surface of a stent can promote vessel wall responses such as nickel allergy, platelet aggregation, inflammation, and formation of hyperplastic neointima after completion of drug release. In addition, sometimes the polymeric coating on stent can be ablated or delaminated during sterilization steps or in blood vessels with high flow pressures.

Electropolishing is a useful tool not only for smooth surfaces; it also creates an inert layer on a stent surface. Following electropolishing, a passivation process occurs, which creates an active surface to passive (inert) surface. In all bare metal stents (BMS), passivation is an important process which provides thromboresistivity, an unresponsive and electrochemically balanced surface which protects an implant from corrosion in blood.

A very common method which can be performed in laboratories is chemical passivation. In this process, stents are dipped in nonreactive oxidative acid (such as nitric acid or citric acid) at a specific temperature; inert thin-film metal oxides are then generated as passivation surface on the metal stent surface. The stents are then washed with a mild basic material to neutralize the acid traces.

Functionalized Cardiovascular Stents. https://doi.org/10.1016/B978-0-08-100496-8.00008-1

8.2 Chemical vapor deposition

One more technique is the coating technique, in which organic or inorganic coating is performed using various coating compounds. Organic coating includes nondegradable, erodible, and bio-degradable polymers. Nondegradable and erodible polymers are permanent in nature and result in late and very late stent thrombosis. The running example of current organic coating is the BioMatrix stent, which has a parylene coating and has proved to be safe and effective with the drug biolimus.

Inorganic coating includes silicon carbide (SiC), and gold and titanium carbide/nitrides have been tested. SiC is a semiconductor material and believed to be less thermogenic; it has resulted in inhibition of IgG, fibrinogen, and fibronectin adsorption [9]. The semiconductor properties of SiC allow it to adhere to any crystalline surfaces and give better chemical resistance. Stents coated with SiC can produce a smooth and thromboresistant surface, and have not shown acute or subacute stent thrombosis in a large-scale randomized clinical trial [10].

Chemical vapor deposition (CVD) has become the preferred technique of passivation. An example of applying SiC on a stent surface by CVD is as follows. Growth of SiC thin films via homoepitaxy involves the use of an Si (e.g., silane (SiH_4), trichlorosilane (TCS), and dichlorosilane) and a C (e.g., propane (C_3H_8), methane (CH_4), and acetylene (C_2H_2)) containing a precursor transported to the growth surface via carrier gas, which is typically hydrogen or an inert gas such as helium [11]. A hot-wall CVD reactor was adopted [12] and this advancement introduced removal of the hydrogen etching step and the inclusion of a small amount of silane between the carbonization and growth steps resulted in very high-quality films.

Various approaches for stent passivation by CVD include silicon carbide (SiC) coating and TiO_2 coating. Stent passivation with SiC is a potential alternative in which stents are passivated by amorphous silicon carbide (aSiC), reducing the thrombogenecity and possibly improving the biocompatibility of the stent surfaces in preclinical studies. Clinical trials have also shown low rates of stent thrombosis, major adverse cardiac events (MACE), and target lesion revascularization (TLR) in patients with stenotic lesions, with these rates also being low in high-risk patients.

It is now well studied through randomized controlled trials that drug-eluting stents (DES) have shown stent thrombosis (ST) as late ST (later than 1 year) and very late ST. This may be because of nondegradable polymers, which do not allow healing of the vessel in time. Therefore, with such incomplete coverage of the stent struts, a prolonged time required for the formation of a functional vessel endothelial layer [13–15], erosion of the stent polymer coating occurs over time as it releases the drug. This erosion creates irregular surfaces that attract platelets and inflammatory cells.

Various efforts have been made including new coating methods, use of biodegradable polymers, and abluminal coatings. Another development was focused on passivation of stent surfaces by thin inert metallic oxide. Passivation of stents can improve the biocompatibility of the material; this reduces thrombogenecity and finally leads to a decreased restenosis rate and prevents unwanted delay in endothelialization [16–18].

Surface passivation aims to achieve a reduction in the interaction of blood constituents with the stent surface by using semiconducting materials to inhibit electron transfer between the blood constituents and metallic surfaces [19–21].

8.3 CVD passivation process evaluation

8.3.1 Passivation by SiC—in-vitro evaluation

Hydrogen-enriched amorphous SiC (aSiC) stents were coated using the CVD process. The biocompatibility and thrombogenecity of aSiC stents have been evaluated, and indicated reduced thrombogenecity and high biocompatibility of aSiC stents. In addition, there was a significantly lower rate of platelet adhesion and activation on aSiC stents on stainless steel stents compared to on uncoated 316L stainless steel or L605 cobalt-chromium (CoCr) stent surfaces [22].

8.3.2 Passivation by SiC—clinical evaluation

In a preliminary clinical study, aSiC stents used in 44 patients with abrupt vessel closure showed a lower rate (9%) of combined in-hospital complications (including death, emergency revascularization, stent-related MI, and ST) and major bleeding after PCI [23]. The study also showed a restenosis advantage over stainless steel BMS, as the observed restenosis rate of 21% at 6-month follow-up is lower than the restenosis rates observed in studies using cobalt-chromium stents.

Carrie et al. also showed 241 patients at a moderate risk of stent thrombosis who underwent PCI using aSiC stainless steel stents [24]. The patients had either unstable angina and/or recent MI, with most lesions having complex characteristics. The results showed high rates of successful stent deployment (97.4%) and clinical success (95.4%), defined as successful deployment without procedural or clinical event. The restenosis and adverse event rates were quite low at 1-year follow-up, as indicated by a major adverse cardiac event (MACE) rate of 15.8% and target lesion revascularization (TLR) rate of 7.1%.

A recent single-arm stent registry investigated the use of aSiC passivation layer coated on cobalt-chromium stents for PCI for the treatment of 161 lesions in 145 consecutive patients with extended coronary artery disease [25]. The procedural success rate was quite high (97.2%), late lumen loss (0.75 ± 0.71 mm [in-stent]) was similar to that noticed after deployment of DES. The study showed low rate of TLR (4.9%) and MACE (5.6%) with no acute, subacute, or late stent thrombosis [26]. Clinical data shows that the significance of passivation of stents by aSiC layer on various stents is more effective than nonpassivated stents.

Similarly, Song et al. [27] investigated a thin TiO_2 thin film deposition onto a bare metal stent by the plasma-enhanced chemical vapor deposition (PCVD) process and its potential as a drug-combining matrix. During the development of DES, the selection of a material for the drug-combining matrix, i.e., drug polymer matrix, type of polymer, and its coating process, is key. Mechanical stability, adhesion properties, and biocompatibility are equally important factors in the final results. The first DES approved by the FDA [28–31] was the sirolimus-eluting stent (Cypher) with a polymer matrix (nondegradable polymer) of poly(*n*-butyl methacrylate) and poly(ethylene vinyl acetate) copolymer and the paclitaxel-eluting stents (Texus) with a polymer matrix of poly(hydroxystyrene-b-isobutylene-b-hydroxystyrene) or poly(styrene-b-isobutylenestyrene) triblock copolymer. However, such polymeric coating showed a lack of good adhesion properties and mechanical strength of coating, as observed by many studies. Further properties of polymer can also depend on the outcome of DESs. In contrast, CVD TiO_2 stents showed superior biocompatibility [31].

8.4 Discussion

Passivation creates an active medical implant surface which is highly sensitive to corrosion in the human body to passive implants, which can resist for a longer time without metal leaching. Silicon can be considered as a choice of material due to its properties, which can shield implants from resistance. Various bare metal stents (BMS), when passed through much heat and electrochemical treatment, can create a smooth surface, but handling and limitations of stent preparation methods leaves open the risk of pitting corrosion or crevice corrosion. Such BMS are now used very little, and have largely been replaced by drug-eluting stents. Using various passivation techniques, CVD with silicon and titanium has been studied for these elements' effects on stents; this has resulted in positive outcomes when compared to nonpassivation stent. Other passivation techniques have also shown resistance to pitting corrosion as well as stable corrosion potential stabilization curve by bulk and the uncontrolled passivation process of SS316L stents and L605Co-Cr stents [8]. CVD is a controlled passivation technique which has achieved very good results of the passivation layer. Preclinical and clinical trials also favor the passivation process to improve thromboresistance and on-time endothelization.

8.5 Conclusion

Various passivation films such as SiC and TiO_2, by creation of a passivation layer, can improve clinical results as passivation improves biocompatibility and adhesion properties as well as reducing metal leaching from implants. The drug-polymer matrix in DES also needs surface modification by passivation to improve its coating adhesion properties. The passivation process has been favored in various clinical studies.

References

[1] World Health Organization, Global Status Report on Non-Communicable Diseases 2010, World Health Organization, Geneva, 2011.

[2] C.D. Mathers, D. Loncar, Projections of global mortality and burden of disease from 2002 to 2030, PLoS Med. 3 (11) (2006) e442.
[3] M.C. Morice, P.W. Serruys, J.E. Sousa, J. Fajadet, E. Ban Hayashi, M. Perin, et al., A randomized comparison of a sirolimus-eluting stent with a standard stent for coronary revascularization, N. Engl. J. Med. 346 (23) (2002) 1773–1780.
[4] G.W. Stone, S.G. Ellis, D.A. Cox, J. Hermiller, C. O'Shaughnessy, J.T. Mann, et al., One-year clinical results with the slow-release, polymer-based, paclitaxel-eluting taxus stent: the TAXUX-IV trial, Circulation 109 (16) (2004) 1942–1947.
[5] M. Pfisterer, H.P. Brunner-La Rocca, P.T. Buser, P. Rickenbacher, P. Hunziker, C. Mueller, et al., Late clinical events after clopidogrel discontinuation may limit the benefit of drug-eluting stents: an observational study of drug-eluting versus bare-metal stents, J. Am. Coll. Cardiol. 48 (12) (2006) 2584–2591.
[6] G.W. Stone, J.W. Moses, S.G. Ellis, J. Schofer, K.D. Dawkins, M.C. Morice, et al., Safety and efficacy of sirolimus- and paclitaxel-eluting coronary stents, N. Engl. J. Med. 356 (10) (2007) 998–1008.
[7] I. Iakovou, T. Schmidt, E. Bonizzoni, L. Ge, G.M. Sangiorgi, G. Stankovic, et al., Incidence, predictors, and outcome of thrombosis after successful implantation of drug-eluting stents, JAMA 293 (17) (2005) 2126–2130.
[8] P. Sojitra, C. Engineer, A. Raval, A. Jariwala, H. Kotadia, D. Kothwala, G. Mehta, Surface enhancement by novel electrochemical treatment and analysis of L-605 cobalt alloy cardiovascular stent, Trends Biomater. Artif. Organs 23 (2) (2009) 55–64.
[9] Y. Takami, S. Yamane, K. Makinouchi, G. Otsuka, Y. Nose, Protein adsorption onto ceramic surfaces, J. Biomed. Mater. Res. 40 (1998) 24–30.
[10] U. Kalnins, A. Erglis, I. Dinne, I. Kumsars, S. Jegere, Clinical outcomes of silicon carbide coated stents in patients with CAD, Med. Sci. Monit. 8 (2002) P119–P120.
[11] D.J. Larkin, P.G. Neudeck, J.A. Powell, Site-competition epitaxy for superior silicon carbide electronics, Appl. Phys. Lett. 65 (13) (1994) 1659–1661.
[12] O. Kordina, C. Hallin, A. Henry, J.P. Bergman, A. Ivanov Ellision, Growth of SiC by hot wall CVD and HTCVD, Phys. Status Solidi B 202 (1) (1997) 321–324.
[13] H. Duckers, S. Silber, R. de Winter, Circulating endothelial progenitor cells predict angiographic and intravascularultrasound outcome following percutaneous coronary interventions in the HEALING-II trial: evaluation of an endothelial progenitor cell capturing stent, EuroIntervention 3 (2007) 67–75.
[14] F. Liistro, A. Colombo, Late acute thrombosis after paclitaxel eluting stent implantation, Heart 86 (2001) 262–264.
[15] G. Sangiorgi, G. Biondi-Zoccai, P. Agostoni, et al., Appraising the effectiveness and safety of paclitaxel-eluting stents in over 1,000 very high-risk patients: overall results of the Taxus in Real-life Usage Evaluation (TRUE) registry, EuroIntervention 3 (2007) 333–339.
[16] E. Atalar, I. Haznedaroglu, K. Aytemir, et al., Effects of stent coating on platelets and endothelial cells after intracoronary stent implantation, Clin. Cardiol. 24 (2001) 159–164.
[17] M. Co, E. Tay, C.H. Lee, et al., Use of endothelial progenitor cell capture stent (Genous Bio-Engineered R Stent) during primary percutaneous coronary intervention in acute myocardial infarction: intermediate- to long-term clinical follow-up, Am. Heart J. 155 (2008) 128–132.
[18] A.M. Noordeloos, T. Soullie, H.J. Duckers, et al., Promoting vascular regeneration as an alternative to conventional angioplasty-based intervention, Endothelium 13 (2006) 431–439.
[19] P. Bauerschmidt, M. Schaldach, The electrochemical aspects of the thrombogenicity of a material, J. Bioeng. 1 (1977) 261–278.

[20] C. Hehrlein, M. Zimmermann, J. Metz, et al., Influence of surface texture and charge on the biocompatibility of endovascular stents, Coron. Artery Dis. 6 (1995) 581–586.
[21] A. Rzany, M. Schaldach, Smart material silicon carbide: reduced activation of cells and proteins on a sic:h-coated stainless steel, Prog. Biomed. Res. 4 (2001) 182–194.
[22] C. Hansi, A. Arab, A. Rzany, et al., Differences of platelet adhesion and thrombus activation on amorphous silicon carbide, magnesium alloy, stainless steel, and cobalt chromium stent surfaces, Catheter. Cardiovasc. Interv. 73 (2009) 488–496.
[23] C. Ozbek, A. Heisel, B. Gross, et al., Coronary implantation of silicone-carbide-coated Palmaz-Schatz stents in patients with high risk of stent thrombosis without oral anticoagulation, Cathet. Cardiovasc. Diagn. 41 (1997) 71–78.
[24] D. Carrie, K. Khalife, M. Hamon, et al., Initial and follow-up results of the Tenax coronary stent, J. Interv. Cardiol. 14 (2001) 1–5.
[25] J.B. Dahm, T. Willems, H.G. Wolpers, et al., Clinical investigation into the observation that silicon carbide coating on cobalt chromium stents leads to early differentiating functional endothelial layer, increased safety and DES-like recurrent stenosis rates: results of the PRO-Heal Registry (PRO-Kinetic enhancing rapid in-stent endothelialisation), EuroIntervention 4 (2009) 502–508.
[26] J. Fajadet, W. Wijns, G.J. Laarman, et al., Randomized, doubleblind, multicenter study of the Endeavor zotarolimus-eluting phosphorylcholine-encapsulated stent for treatment of native coronary artery lesions: clinical and angiographic results of the ENDEAVOR II trial, Circulation 114 (2006) 798–806.
[27] S.-J. Song, Y.J. Park, J. Park, M. Duck Cho, J.-H. Kim, M. Ho Jeong, Y. Sook Kim, D. Lyun Cho, Preparation of a drug-eluting stent using a TiO_2 film deposited by plasma enhanced chemical vapour deposition as a drug-combining matrix, J. Mater. Chem. 20 (2010) 4792–4801.
[28] J.G. Joseph, Polymers for targeted and/or sustained drug delivery, Polym. Adv. Technol. 20 (2009) 595–606.
[29] H. Tamai, K. Igaki, E. Kyo, K. Kosuga, A. Kawashima, S. Matsui, H. Komori, T. Tsuji, S. Motohara, H. Uehata, Initial and 6-month results of biodegradable poly-*l*-lactic acid coronary stents in humans, Circulation 102 (2000) 399–404.
[30] F. Vogt, A. Stein, G. Rettemeier, N. Krott, R. Hoffmann, J.V. Dahl, A.K. Bosserhoffd, W. Michaelic, P. Hanratha, C. Webere, R. Blindt, Long-term assessment of a novel biodegradable paclitaxel-eluting coronary polylactide stent, Eur. Heart J. 25 (2004) 1330–1340.
[31] R.E. Richard, K. Matyjaszewski, M. Schwarz, S. Ranade, A.K. Chan, Evaluation of acrylate-based block copolymers prepared by atom transfer radical polymerization as matrices for paclitaxel delivery from coronary stents, Biomacromolecules 6 (2005) 3410–3418.

Further Reading

[1] K. Okamoto, T. Nakatani, S. Yamashita, S. Takabayashi, T. Takahagi, Development of surface-functionalized drug-eluting stent with diamond-like carbon nanocoated by using PECVD method, Surf. Coat. Technol. 202 (2008) 5750–5752.

Polymer coatings for biocompatibility and reduced nonspecific adsorption

9

M.C. Ramkumar, P. Cools†, A. Arunkumar*, N. De Geyter†, R. Morent†, V. Kumar‡, S. Udaykumar‡, P. Gopinath‡, S.K. Jaganathan§,¶, K.N. Pandiyaraj**
*Sri Shakthi Institute of Engineering and Technology, Coimbatore, India, †Ghent University, Gent, Belgium, ‡Indian Institute of Technology Roorkee, Roorkee, India, §Ton Duc Thang University, Ho Chi Minh City, Vietnam, ¶Universiti Teknologi Malaysia, Johor, Malaysia

Abbreviations

AAc	acrylic acid
AC	alternate current
Ag	silver
APGD	atmospheric pressure glow discharge
APPJ	atmospheric pressure plasma jet
APTT	activated partial thromboplastic time
Ar	argon
BMS	bare-metal stents
BSA	bovine serum albumin
CC	cyanuric chloride
CF_4	tetrafluoromethane
DBD	dielectric-barrier discharge
DC	direct current
DEGDME	diethylene glycol dimethyl ether
DES	drug-eluting stents
DLC	diamond-like carbon
FEP	poly(tetrafluoroethylene-co-hexafluoropropylene)
FTIR	Fourier transform infrared
HCEC	human corneal epithelial cell
HFF	human foreskin fibroblast
hFOB	human osteoblast
hGF	human fibroblast
HMDSO	hexamethyldisiloxane
hTM	human thrombomodulin
HUVEC	human umbilical vein endothelial cells
IOLs	intraocular lens
LDPE	low-density polyethylene
LST	late stent thrombosis
LTE	local thermodynamic equilibrium
MW	microwave
O_2	oxygen

Functionalized Cardiovascular Stents. https://doi.org/10.1016/B978-0-08-100496-8.00009-3

OPN	osteopontin
PCL	poly(ε-caprolactone)
PDMS	poly(dimethyl siloxane)
PECVD	plasma-enhanced chemical vapor deposition
PEEK	polyether ether ketone
PEG	poly(ethylene glycol)
PEGA	poly(ethylene glycol) acrylate
PEGMA	poly(ethylene glycol) methacrylate
PEO	polyethylene oxide
PET	polyethylene terephthalate
PET/PP	polyethylene terephthalate/polypropylene
PIII	plasma-immersion ion implantation
PLLA	poly(L-lactic acid)
PMMA	polymethyl methacrylate
PMMA IOLs	polymethyl methacrylate intraocular lenses
PP	polypropylene
PP-PEG	plasma-polymerized poly(ethylene glycol)
PPrPE	polyphenol-rich plant extract
PRP	platelet-rich plasma
PT	prothrombin time
PU	polyurethane
RF	radio frequency
RFGD	radio-frequency glow discharge
SBF	stimulated body fluid
SEM	scanning electron microscopy
SS	stainless steel
TCP	tissue culture plate
T_e	temperature of electron
TEGDME	tetraethylene glycol dimethyl ether
T_h	temperature of heavy particle
TiO_2	titanium dioxide
TMS	tetramethylsilane
TT	thromboplastin time
WBCT	whole-blood clotting time
XPS	X-ray photoelectron spectroscopy

9.1 Introduction

Cardiovascular stents are small, expandable metal mesh tubes typically placed inside a coronary artery to open and support narrowed or weakened arteries. This loss of arterial function is most often caused by atherosclerosis and its subsequent residue [1–3]. Atherosclerosis, or narrowing of the arteries, is often the result of plaque depositions on the arterial walls, which are resultant of inflammatory signs produced by local cells and succeeding inflammatory response. Plaque is often composed of cholesterol, fat, and blood components [4]. The accumulation of cholesterol plaque on arteries walls will eventually lead to obstruction of the blood flow, causing acute ischemia, resulting in strokes. Cardiovascular stents offer a slightly invasive means to mechanically back the damaged vessel that reinstates oxygenated blood flow to the connected tissues. Bare-metal stents (BMS) were the first kind of stents, which consisted of different metals and/or alloys (stainless steel, cobalt chromium, tantalum, or nitinol) [5]. This type of stents offered the required mechanical support to the damaged vessel, but formation of thrombosis or restenosis onto these devices would inevitably lead to a second intervention shortly after stent implantation [6]. Furthermore, during the restoration of the endothelium across the stent support, the smooth muscle proliferation persuaded by injury would lead to neointimal growth and restenosis. The rate of restenosis formation was found to be high for bare-metal stents, making them substantially limited in their clinical application. In addition to the abovementioned problem, a second problem linked with BMS is thrombogenicity, for example, the tendency of the device to form blood clots on the surface of the material. This may arise due to the net electric charge differences among the surface of the stent and the blood component, in addition to the surface mismatch between the metal and the contacting blood [7,8]. To overcome these disadvantages, a new type of stent was developed, the drug-eluting stent (DES). A typical DES is composed of three major components: (1) a metallic stent platform, (2) a polymer-based drug delivery coating [9–12], and (3) a pharmacological agent, usually an immunosuppressant and/or an antiproliferative compound [13]. DES offers the benefit and control of delivering the pharmacological agents, intended to prevent restenosis, straight to the damaged vessel, evading the systemic side effects. The polymer coating acts as a drug reservoir and permits the elution of the drugs overtime [14–16]. In advanced DES, the nonabsorbable polymer coating is replaced with a nonthrombogenic absorbable one, which supports the reendothelialization via direct drug release and results in a decreased inflammatory response during polymer degradation. In addition, the existence of restenosis decreased due to the release of antiproliferative agents. However, stent-induced thrombosis still occurred due to different issues, resulting in fibrin deposition and delayed healing [17,18]. In stable single-vessel disease patients, late stent thrombosis (LST) occurs at a constant rate (0.6% per year) [19], with even higher rates recorded (0.9%–3% per year) in clinical studies [20]. Inadequate performance of both BMS and DES has led to further research in stent development, aiming for enhancement of stent biocompatibility. A new type of stent, consisting of a polymer backbone, has accomplished enhanced hemocompatibility by selecting suitable combination of polymer components and processing techniques [7]. Polymer stents exhibit a prolonged reendothelialization, thus stimulating a thrombotic environment compared with BMS. Polymers possess various advantages such as outstanding mechanical properties, easy processing, and low processing cost [21–23]. In general, polymers can be a suitable alternative for metals as they play a substantial part in biomedical industry. Since they are less dense and highly flexible compared with most other materials, it makes them an excellent candidate for blood-contacting purposes, such as heart valves, artificial blood vessels, and stents [24–26]. However, polymeric materials lack as an implant material due to their substandard biocompatibility, caused by their inherent surface properties, for instance, poor wettability and low surface energy [27,28]. This can be avoided by suitability modifying the surface of the polymeric material, thus improving the surface energy, wettability, adhesion by incorporating specific functional moieties, resulting in enhanced biocompatibility. Numerous modification techniques have been employed to alter the surface properties of polymers, for instance, wet-chemical modification, redox reactions, UV irradiation, ozone treatments, and plasma-based technologies [29–35]. Among those techniques, nonthermal plasma surface modification was found to be a promising surface modification tool, because it is reproducible, inexpensive, fast, and solvent-free and can modify the surface of various materials [36–39]. In the following paragraphs, a more in-depth description can be found on what plasma exactly is, how it is usually classified, and how it has been successfully applied to polymeric materials suitable for stent applications.

9.2 Classification of plasma

Plasma is a gaseous mixture of radicals, ions, electrons, and neutrals that is also referred to as the fourth state of matter [40]. In modern literature, plasmas are classified as either thermal or nonthermal, based on their pressure, degree of ionization, and temperature conditions. Thermal plasma, also known as hot plasma, exists in a state of local thermodynamic equilibrium (LTE), that is, electron temperature (T_e) lies within the same range as the heavy particles or sensible temperature (T_h). The core gas temperature usually lies above 10,000 K. Thermal plasmas have a wide range of industrial applications such as plasma spraying, destruction and treatment of waste materials, wire arc spraying, and surface modification [41]. Nonthermal plasmas or cold plasma possesses a lower degree of ionization and is characterized by lower energy densities and huge differences between the temperatures of electrons and the heavier particles. The collision of electrons of higher energies with the background gas results in low levels of dissociation, excitation, and ionization without a significant rise in the enthalpy of the gas. As a result, the temperature of the electrons surpasses the temperature of heavy particles ($T_e \ggg T_h$). Hence, the nonthermal discharge can be maintained at temperatures as low as room temperature. In the field of biomaterials, nonthermal plasma is the most abundantly studied and will therefore be discussed in more detail below.

9.2.1 *Nonthermal plasma*

In general, nonthermal plasmas have gained a broad interest due to advantages such as low cost, low operating temperature, smaller volume, and lack of solvents. Nonthermal plasma is used extensively for surface activation or modification, since the ions, atoms, and molecules are comparatively cold (near room temperature) and do not cause any thermal damage to the surfaces of heat-sensitive materials, such as polymers and biological tissues [42]. This gives it the potential to efficiently amend the surface properties without distressing the bulk properties of the material [43–45]. Various mechanisms occur in parallel when a polymer surface is exposed to plasma environment: surface etching, cross-linking, and chemical modifications. Nonthermal plasma can also be used as an initiation medium for the deposition of polymer-like thin films on most substrates, a process that is known as plasma polymerization. Plasma-based coatings or thin films are found to be homogeneous and pinhole-free with exceptional adhesion to various substrates [46,47], while offering the desired physical, morphological, and chemical properties to the polymers. In the field of surface engineering, nonthermal plasma technology is a well-established technique known to enhance the functionality and tribological properties of artificial implant devices. Within the field of biomedical engineering, plasma processing has been successfully employed for the fabrication of antimicrobial coatings [48–52], functionalization of drug for targeted delivery and controlled release [53–55], and development of functional polymeric substrates for tissue engineering [56,57]. Based on the applied pressure in the plasma system, literature makes a further distinction, namely, (i) low-pressure plasma and (ii) cold atmospheric pressure plasma.

9.2.2 Low pressure plasmas

In most practical circumstances, glow discharge plasma is produced by an electric discharge under low pressures. There are a number of different discharge techniques including the following:

(a) Direct-current (DC) discharge
(b) Radio-frequency (RF) discharge
(c) Microwave discharge

9.2.2.1 DC glow discharge

DC glow discharges were one of the first methods for producing low-temperature plasmas. The discharge is produced between two bare electrodes. The working pressure of conventional DC glow discharge reactor is typically between 10^{-1} and 10 Pa. One reason for the attractiveness of DC glow discharges is the relatively low voltage and current needed to maintain the discharge. One can generate different discharges by suitably amplifying the current. As the pressure is lower than 10 mbar, the reduced field can reach quite high values. Since the generated electrons possess a high energy, they easily excite the neutral atoms and molecules, leading to the typical glow discharge. Subsequent to the formation of glow discharge, the current starts passing through the ionized gas, leading to a drop in voltage across the plasma regime. Conversely, low-pressure plasma is an appropriate method for governing the chemistry of the material surfaces, and it is a single step to develop functional coatings with homogeneous surface chemistry, pinhole-free morphology, and good adherence with substrates.

It is still a relevant technique today, as it gives good control over various plasma operating parameters. In some reactors, water cooling is applied to the electrodes to evade thermal damage of polymers when subjected to longer plasma processing time. However, DC glow discharge plasmas have certain limitations for using them in industrial applications: the need for expensive vacuum systems, the low deposition rates, and the lack of continuous processing. The schematic diagram of DC glow discharge plasma reactor designed by Pandiyaraj et al. [58] is shown in the Fig. 9.1.

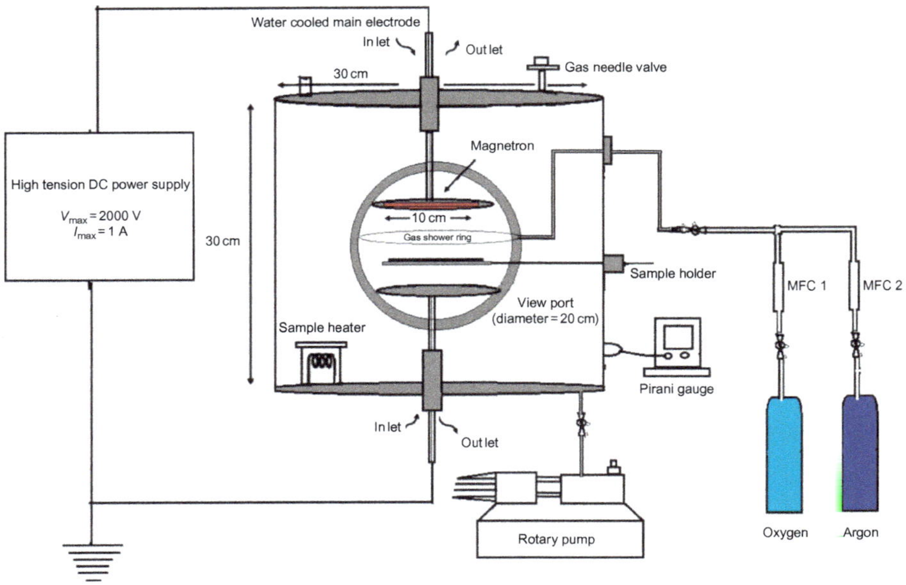

Fig. 9.1 Schematic diagram of DC glow discharge plasma reactor.

9.2.2.2 *Radio frequency discharge*

Plasma can be generated in RF discharge by either capacitively or inductively coupling energy from an AC power supply at radio frequency range (1 kHz to 10^3 MHz, most common value is 13.56 MHz). At distinctive RFs, the electrons and ions have unlike behavior because of their masses. Electron densities obtained in capacitively coupled discharge (CCD) range between 10^9 and 10^{10} cm^{-3}, while for inductively coupled discharge (ICD), the densities are situated around 10^{12} cm^{-3}. RF discharges are usually operated at a low pressure of few Pa.

To date, it is the most successfully commercially available applied discharge system and is extensively used for surface modification of biomaterials. These systems give excellent govern over surface chemistry and have the capability to generate high-energy species during the deposition process. RF discharge plasmas can be used in etching and deposition of biofunctional coatings, fabrication of optical fibers, surface modification of diamond films [59], and also for material processing in various fields that include microelectronics and aerospace [60,61]. The schematic diagram of capacitively and inductively coupled plasma reactor is displayed in Fig. 9.2. RF discharge possesses various advantages over DC discharges. In RF discharge, the electrodes can be placed outside the main discharge chamber, and thus, the contact of the electrodes with plasma can be prevented. This is impossible for DC discharges as they cannot operate in open electronic circuit. For RF discharges, this is not a problem as they are more flexible and scalable than DC discharges and can operate in open electronic circuits. RF discharges can be generated in a variety of different geometries and sizes. The disadvantages of RF discharge

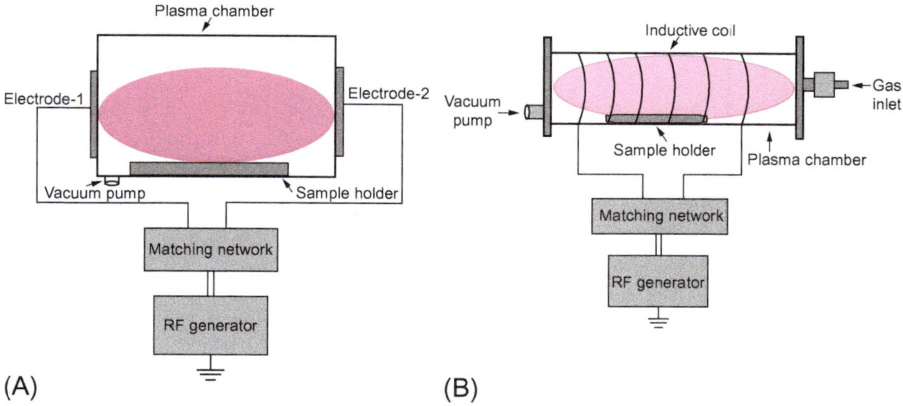

Fig. 9.2 Schematic diagram of (A) capacitively coupled and (B) inductively coupled plasma.

are their low ion flux and high ion energy, as these cannot be varied individually. If damage to the sample is to be expected, processing must be carried out slowly. Heat generation is an often recurring problem for RF systems and can damage more thermosensitive substrates.

9.2.2.3 Microwave discharge

Microwave-powered plasma discharges combine a high energy density with a large area and good uniformity, and this can be used for a wide range of gas pressures. In the microwave region, the wavelength of the electromagnetic field becomes equivalent to the dimensions of the discharge vessel, which requires other coupling mechanisms. When the frequency increases, the dimension of the cavity decreases; hence, the maximum frequency used for discharge applications lies normally below 3 GHz. Most of the microwave discharges are operated at frequency range of 2.45 GHz. At these frequencies, only the light electrons can follow the oscillations of the electric field; for this reason, microwave-driven plasma is typically far from local thermodynamic equilibrium. Microwave discharges can be operated in a wide pressure range from 1 mbar to about atmospheric pressure, and they can operate over a large frequency and pressure range, meaning they can generate large-volume nonequilibrium plasmas of reasonable homogeneity. It is not the most suitable setup for modifying the surface of thermosensitive materials such as polymers, because the higher pressures lead to an increase in electrode temperature, resulting in considerable heat transfer toward the substrates, causing sample damage. Due to the large pressure range under which these discharges can be operated, electron densities ranging between 10^8 and $10^{15}\,cm^{-3}$ have been reported. There exist various types of microwave plasma reactors such as the multislotted planner antenna [62,63], microwave-induced plasma discharge [62], surface-wave discharge [63], and microwave electrode discharge [64]. The mode of operation can be influenced by the plasma parameters such as gas flow, applied potential; electrode separation makes the discharge quite appealing and also suitable for plasma chemical investigations. Microwave-generated plasma normally possesses a high electron kinetic temperature compared with DC or low-pressure RF plasmas, ranging from 5 to 15 eV. Owing to their higher electron kinetic temperature and low pressure, microwave discharges are proficient in offering a higher fraction of ionization and dissociation than DC and RF discharges, a significant benefit in various plasma chemical applications. It can also remain stable over a wide range of background gas pressures, relative to DC or low-frequency RF discharges. Apart from the abovementioned low-pressure plasma types, there is an additional type of low-pressure plasma technique used for surface modification, plasma-immersion ion implantation technique.

9.2.2.4 Plasma immersion ion implantation (PIII)

Plasma-immersion ion implantation (PIII) is a surface modification technique in which an ion beam is extracted from plasma source by applying a high voltage (DC), accelerated to the desired energy and then targeting them into the suitable substrate. Plasma-immersion ion implantation (PIII) has gained interest in the field of material processing, due to its high implantation dose rate, nonflight-of-sight characteristics, and instrument simplicity that have potential for biomedical implants with complex shape. A typical PIII system consists of a power supply, sample stage enclosed in a vacuum chamber and high-voltage pulse modulator. In this system, the substrate is placed in the plasma region, and negative high-voltage pulses ranging from few kV up to about 100 kV are applied to it. When the sample undergoes negative bias, the electrons are repelled from the substrates, and a sheath of positive ions is formed around the sample. As a result, positive ions are accelerated due to induced bias voltage (range 20–200 kV) leading to the implantation of said ions into the surface of sample from all directions. To attain the full ion energy at the surface of sample, the chamber pressure must be kept appropriately low (>0.5 Pa) to evade the ion neutral collisions in the sheath. In order to apply PIII technique effectively to complex-shaped samples, it is essential to have a profound knowledge on plasma sheath dynamics, since it plays a vital role in the energy distribution of the implanted ions. PIII only alters the surface properties of the exposed substrate since the ions have inadequate penetration power (~1 μm). Fig. 9.3 shows the schematic diagram of plasma ion implantation system.

By employing various plasma sources, different elements and chemical groups can be incorporated on to the surface of samples. Different modes such as RF and DC (pulsed or continuous) have been used for substrate biasing. The working pressure and varying biasing influence the structure of thin films and the deposition rate. The implantation effect in PIII system varies for different materials. The variations in physical properties in metal and ceramic biomaterials result from atomic and nuclear collisions; these result in the development of amorphous structures at the surface. The combination of physicochemical variations occurred due to the PIII process makes the surface harder and highly resistant to chemical reactions without modifying the material bulk properties.

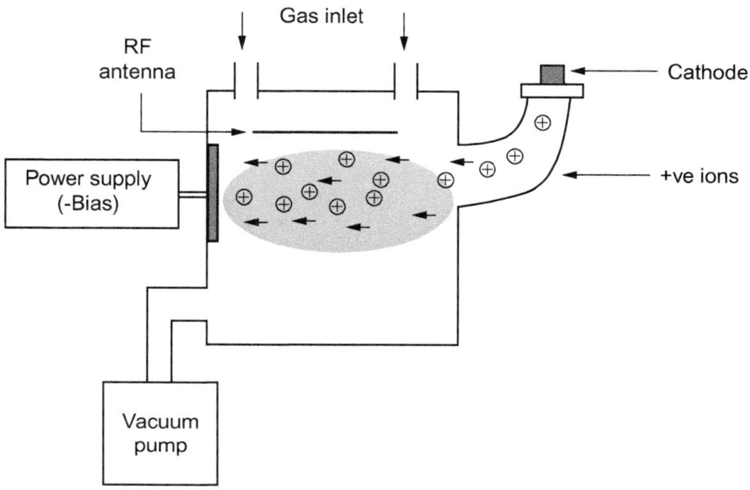

Fig. 9.3 Schematic diagram of plasma-immersion ion implantation system.

9.2.3 Cold atmospheric pressure plasma

In recent years, nonthermal atmospheric pressure plasmas have gained interest over low-pressure plasmas as it possesses various advantages. For instance, it can be operated without expensive vacuum pumping system; it allows for the modification of heat- or vacuum-sensitive materials, gives high deposition rates, and can be implemented onto continuous in-line processing. Atmospheric pressure plasmas are typically classified as follows:

(a) Corona discharge
(b) Dielectric-barrier discharge (DBD)
(c) Atmospheric pressure glow discharge (APGD)
(d) Atmospheric pressure plasma jet (APPJ)

9.2.3.1 Corona discharge

Corona discharges are moderately low-power electric discharges that occur at or near atmospheric pressure. The corona is invariably produced by strong electric fields associated with small diameter wires, needles, or sharp edges on an electrode [65]. The free electron density in corona discharges is approximately 10^8 electrons/cm^3. Plasmas produced using corona discharges are inherently inhomogeneous and require a narrow space between the electrodes. It has widespread industrial applications including electrophotography, printers, textile processing, and in-powder coating. Due to their inherent nonuniformity, it has limited applications for homogeneous treatments and coating depositions. Photograph of corona discharge is displayed in Fig. 9.4. As it can operate in atmospheric pressure, air is generally used as a reagent gas. Air plasma treatment can tailor the physical and chemical properties of the surface of materials

Fig. 9.4 Photograph of (A) multipin-to-plate glow discharge and (B) pin-to-mesh corona discharge.

resulting in enhanced hydrophilicity of the material by incorporating a variety of polar functional groups, such as oxygen- and nitrogen-containing groups. This can act as anchors for a wide range of biological molecules and so render, for example, a textile more biocompatible. For instance, polypropylene textiles used as sutures and hernia patches possess hydrophobic nature. Treatment of these textiles with air or oxygen plasma makes the surface hydrophilic leading to various biological applications.

9.2.3.2 Dielectric barrier discharge (DBD)

Dielectric-barrier discharge (DBD), also known as silent discharge, is a particular type of AC discharge that delivers strong thermodynamic, nonequilibrium plasma at atmospheric pressure and at moderate gas temperatures [66–69]. DBD can be but is not limited to operation at elevated pressures. Dielectric-barrier discharge plasmas have an approximate electron density of 10^{10} electrons/cm^3 and a power density of about 0.1 W/cm^3. In DBD, plasmas are generated when an alternating (1–100 kHz) high voltage (0.5–20 kV) is applied between two electrodes of which at least one is covered with a dielectric material (glass, quartz, and ceramics). Upon discharge, excited, ionized, metastable compounds and radicals are generated via collision of energetic electrons and reactant particles. The presence of a dielectric barrier is one of the easiest ways to produce nonequilibrium atmospheric pressure discharge as it prevents short circuiting and arcing between the two electrodes [70]. Furthermore, the dielectric layer prevents electrode corrosion and etching. DBD discharges at atmospheric pressure are characterized by the formation of large number of short-lived independent current filaments or microdischarges [71,72]. For certain conditions, a homogeneous treatment can be achieved equivalent to the distinctive glow discharge obtained at lower pressures by evading the streamers [73]. Compared with corona dischargers, DBD systems are capable of more homogeneous treatment of large areas. It is geometrically more versatile than corona discharge, where the distance between the two electrodes is usually much smaller, confining the volume. Another advantage over other plasma systems is its capacity to operate without the use of extensive vacuum systems, reducing the acquisition and maintenance cost. Finally, it is characterized by relatively high deposition rates and allows for a straightforward implementation onto in-line processing [74–77]. DBDs have a wide range of applications including sterilization, surface activation, chemical vapor deposition, biotreatment of microorganisms, and material processing [78–83]. Research has proved DBD processes to be suitable for biomedical research [84]. The experimental setup of a typical dielectric-barrier discharge developed by Pandiyaraj et al. [85] is shown in Fig. 9.5.

9.2.3.3 Atmospheric pressure glow discharge (APGD)

Atmospheric pressure glow discharge (APGD) generates plasmas by applying low voltage (~200 V) between two flat metal electrodes at high frequency (MHz). They are characterized by an electron density of 10^{12} electron/cm^{-3} and a power density of more than 10 W/cm^{-3}. Both are the highest of all described atmospheric pressure plasma systems. Helium is the only discharge gas suitable for this configuration [86], making APGD a relatively expensive plasma technique. In some cases, a dielectric barrier is used to cover one of the electrodes to further avoid arcing. Typically, an RF power supply is applied between the two flat electrodes separated by few millimeters. Plasmas formed in APGD were found to be homogeneous and stable. As the electrode distance is usually limited to a few mm, its use in the treatment of biomaterials is rather limited.

9.2.3.4 Atmospheric pressure plasma jet (APPJ)

APPJ is a special geometric configuration of the previously described systems (RF, DBD, and MW). APPJ is a nonthermal, glow discharge plasma functioning at atmospheric pressure and has found growing use in various applications like nanoscience and biodecontamination. The major distinctive property of these plasmas is that the jet is not limited by electrode geometries. In contrast to corona and dielectric-barrier discharge (DBD), APPJ not only is limited to flat and thin substrates but also can be used to modify large three-dimensional structures, for example, inner walls of wells, trenches, or cavities [86]. Furthermore, the jet geometry allows for a better flexibility in terms of substrate distance and electric field, as the surfaces to be treated are not necessarily placed between the electrodes. In addition, the APPJ is considered to be cost- and energy-effective, because it can be generated using relatively cheap components. The surface properties of the materials are tailored due to the presence of electrons at high temperature, whereas the low gas temperature makes it suitable for heat-sensitive materials. The gas temperature of APPJ lies usually between 20°C and 300°C; thus, thermal damage to treated materials can be easily avoided. APPJ systems have been powered with a variety of sources such as DC [87], AC [88], pulse excitation [89,90], RF [91], and microwave [92]. What further makes these plasma systems different from the abovementioned systems is that samples are not exposed to the plasma directly but rather to the long afterglow propagating in open air. When the discharge is initiated by noble gases, the afterglow can propagate into the surrounding air at distance from some mm up to 10 cm [93–95]. However, even if pure noble gases are used, the presence of small amount of oxygen and nitrogen species was found,

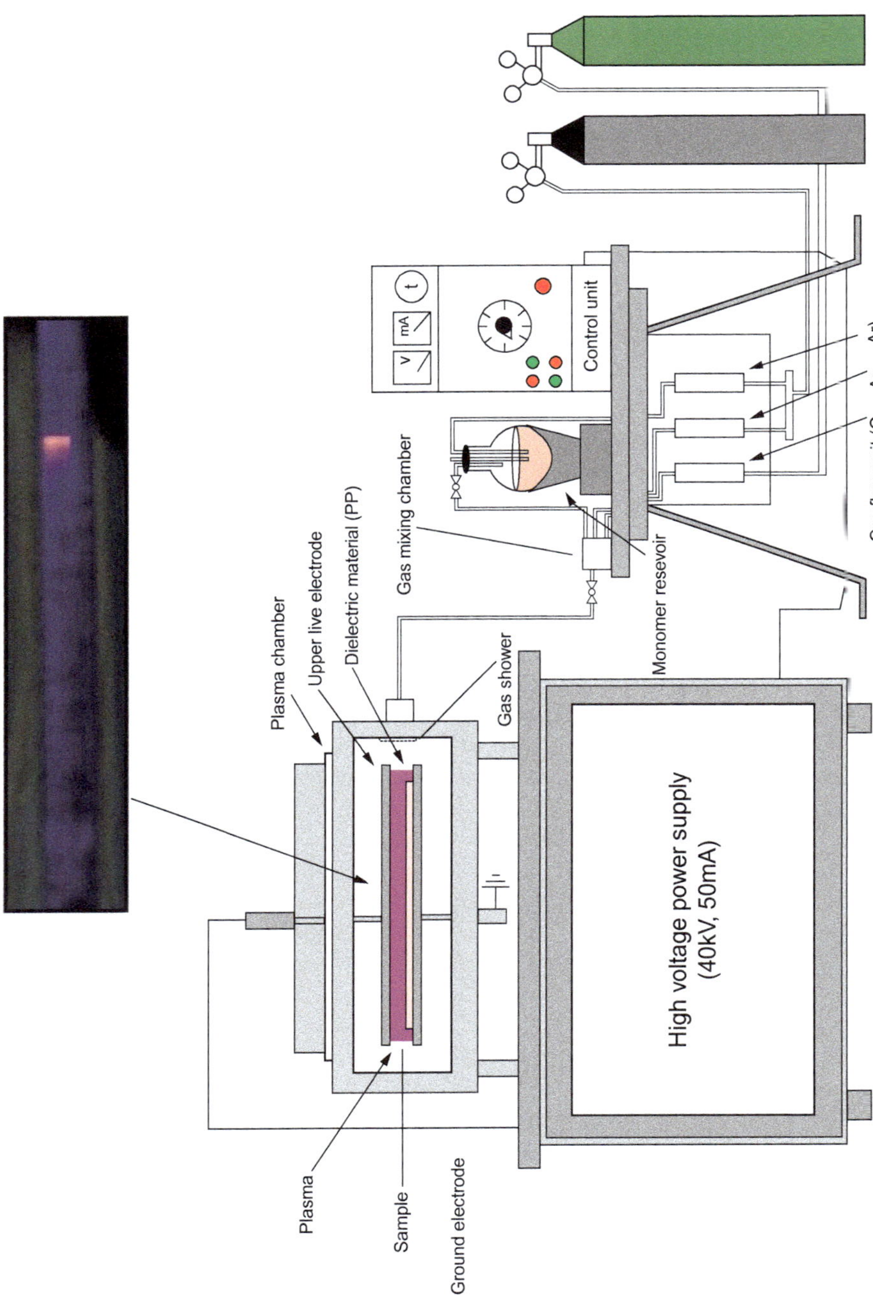

Fig. 9.5 Experimental setup of cold atmospheric pressure plasma-assisted dielectric-barrier discharge (DBD) system.

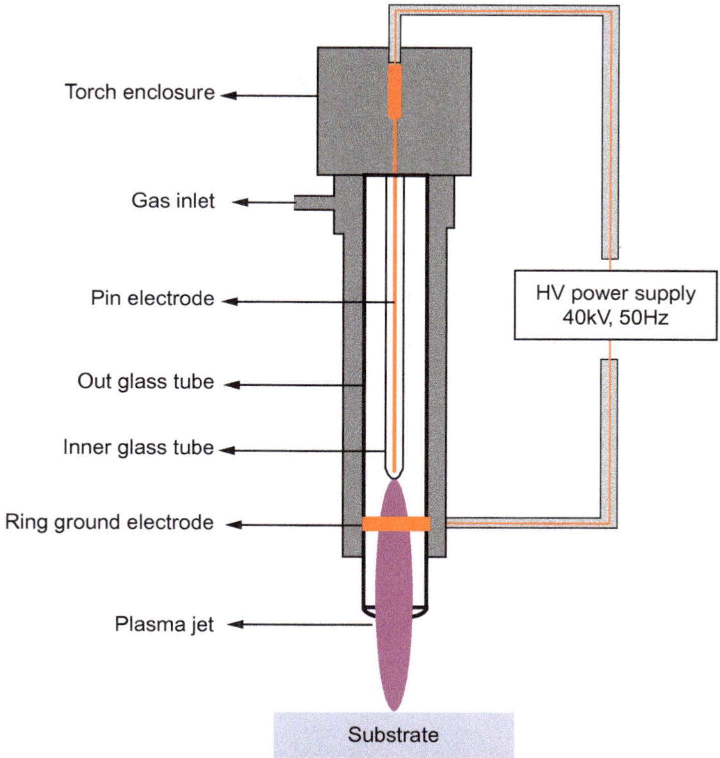

Fig. 9.6 Schematic diagram of atmospheric pressure plasma jet.

since species from the ambient environment will mix and interact with the species present in the afterglow [96–98]. A typical schematic diagram of atmospheric pressure plasma jet (APPJ) is shown in Fig. 9.6.

9.3 Added value of nonthermal plasma for stent applications: Polymer coatings

In general, polymer coatings are very appealing as they offer great versatility in the chemical groups, which can be incorporated at the surface in order to control tissue-biomaterial interactions. Polymer coatings possess mechanical properties comparable with soft biological tissues. Various types of polymer coatings have been studied on biomaterials used for cardiovascular stents in order to reduce platelet adhesion and protein adsorption and prevent from bacterial adhesion and thrombosis formation. In this section, we discuss the polymer coatings produced using nonthermal plasma systems that have proved to enhance stent materials or show potential to do so.

9.3.1 Poly ethylene glycol (PEG): Antifouling coating

Polyethylene glycol (PEG) is a well-known polymer for preparing protein-repellent coatings on various substrates, due to its nontoxic, nonantigenic, and nonimmunogenic characteristics, and can be used for various medical applications. As a result of their low-protein-fouling properties, PEG coatings can also prevent nonspecific protein and bacterial binding on implantable material surfaces. PEG is known to reduce the attractive forces between surfaces and proteins as a result of high mobility in the hydrated state and related steric repulsion. Fig. 9.7 shows the chemical structure of polyethylene glycol (PEG).

PEG is well-known to incorporate hydroxyl functional groups on the surface of the material. The hydroxyl group functionality (OH) represents a neutral, hydrophilic surface; the presence of OH functionality has been found responsible for a decreased plasma protein adsorption and thus greater platelet compatibility [99,100] as shown in Fig. 9.8. Sakthi Kumar et al. used radio-frequency (RF) plasma for polymerization of a PEG-like coating on polyethylene terephthalate (PET) surfaces. The polymerized films exhibited the characteristics of PEG, and hydrophilicity was found to be extremely high (contact angle = 30 degrees). Static platelet adhesion studies using platelet-rich plasma (PRP) revealed significant decrease in adhesion of platelet on the surface of modified films [101]. In a similar study by Fu et al., stainless-steel substrates were modified by coupling agent (SCA), (3-mercaptopropyl) trimethoxysilane. The silanized surfaces were further activated by Ar plasma for different treatment times (5–20 s) and then exposed to UV-induced graft polymerization of poly(ethylene glycol) methacrylate (PEGMA). The substrates that contain high PEG content have great efficiency in averting bovine serum albumin and γ-globulin adsorption [102].

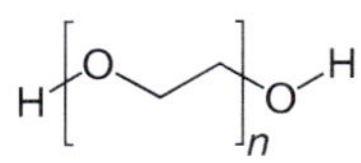

Fig. 9.7 Chemical structure of polyethylene glycol.

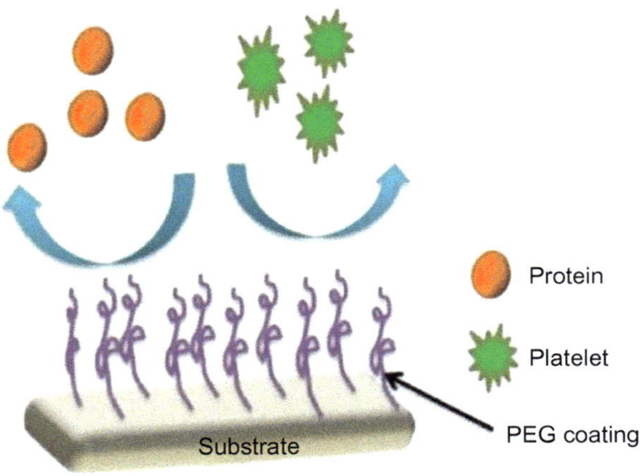

Fig. 9.8 Schematic of PEG coatings for platelet repulsion.

D'Sa et al. functionalized silicon elastomers using atmospheric pressure plasma-induced graft immobilization of PEGMA as a function of different molecular weight (1000 and 2000 Da). The surface was modified by plasma activation of silicon elastomers in order to incorporate reactive oxygen functionalities, followed by in situ grafting of PEGMA of different molecular weights. In vitro cytocompatibility was assessed using lens epithelial cells cultured on the PEGMA-grafted silicon elastomer surfaces. The cells on the unmodified substrates were not proliferated and exhibited shrunken morphology, whereas in the case of pretreated surfaces, the cells were well spread. The films grafted with high molecular weight (2000) exhibited no cell adhesion on their surfaces. In addition, bacterial adhesion to the modified surfaces was performed using *Staphylococcus aureus* NTC8325. Results showed no difference in number of bacteria adhering to any modified surface [103]. Shiheng et al. grafted PEGMA onto chitosan membrane surface via Ar plasma-induced graft polymerization. Improved adhesion of a human corneal epithelial cell (HCEC) on chitosan membrane was observed due to the enhancement in surface free energy and roughness [104]. Pinto et al. modified the surface of poly(dimethyl siloxane) (PDMS) elastomer using Ar plasma treatment subsequent to grafting of pluronic F-68 or PEGMA, and it is evident from the results that the modified surface exhibited a nontoxic nature and was slightly hemolytic [105]. From all the PEGMA studies, it is obvious that the PEGMA with MW = 2000 can be suitable for stent applications as it exhibits no adhesion on its surfaces. Lopez et al. used glow discharge plasma for deposition of tetraethylene glycol dimethyl ether (TEGDME) on a wide variety of substrates (glass, polytetrafluoroethylene, and polyethylene) to produce surfaces that could resist protein adsorption and cellular attachment. Adsorption of proteins (fibrinogen, albumin, and IgG) was found to be lower than all unmodified substrates. They also found that different substrates exhibited different amount of protein adsorption, signifying that the substrates employed may affect the protein adsorption amount. The study of dynamic platelet adhesion, using epifluorescent video microscopy and endothelial cell attachment unveiled the short-term nonadhesiveness of these surfaces [106]. Brétagnol et al. developed fouling and antifouling surfaces using diethylene glycol dimethyl ether (DEGDME) via radiofrequency plasma polymerization as a function of applied power. The films prepared at lower power were found to have a high retention of the monomeric functionalities and exhibited reduced protein adhesion and cell repulsive behavior. The films deposited at higher power showed significantly lower concentration of PEO groups due to high fragmentation, as these films promote more cell adhesion and growth [107]. Wang et al. performed PEG immobilization with different molecular weights (200, 1000, 6000, and 10,000) on polyethylene terephthalate (PET) using plasma grafting. In vitro analysis unveiled that the PEG grafting extended the activated partial thromboplastin time (APTT) and reduced platelet adhesion. The blood compatibility of the modified PET films could be linked to the molecular weight of the PEG, and the same was found to be maximum for PET grafted with molecular weight of 6000. They also stated that the interaction of PEG chains with coagulation factors XI and XII of serine proteinase alters the configuration and structure of the coagulation factors results in reduced activity of coagulation factors. Hence, it is sensible that grafting of PEG chains extended the APTT [108]. Zanini et al. modified polypropylene (PP) membranes by plasma polymerization of acrylic acid followed by covalent attachment of PEG chains of different chain lengths (750 and 2000) on to the modified membranes, and their antifouling properties were evaluated by bovine serum albumin (BSA) and fibrinogen filtration studies. It was confirmed that after grafting of PEG, the antifouling properties were enhanced, and a 95% reduction in protein adsorption was attained by the films with the longer PEG chains [109]. Lin et al. improved the biocompatibility of hydrophobic acrylic intraocular lenses (IOL) by immobilization of PEG by APGD treatment using Ar as the discharge gas. In vitro analysis revealed the PEG-modified IOLs substantially not only reduced the adhesion of platelets, macrophage, and LECs but also managed to inhibit the spreading and growing of cells [110]. Choi et al. prepared plasma-polymerized poly(ethylene glycol) (PP-PEG) films on a wide variety of substrates using PECVD and low-molecular-weight PEG precursor, as this exhibited a high vapor pressure. When subjected to circulating whole blood, the PEG films significantly reduced thrombus formation and were highly resistant to fouling by platelet aggregation and serum protein adsorption. In addition, when implanted in mouse subcutaneous tissue, the deposited film exhibited excellent histocompatibility (i.e., insignificant tissue adhesion, inflammation, and fibrous proliferation) [111]. Zhang et al. pretreated polymethyl methacrylate intraocular lenses (PMMA IOLs) using Ar plasma and modified by immersing in heparin and PEG solutions followed by plasma irradiation. Platelet adhesion tests confirmed that the antithrombogenicity of the modified PMMA samples was enhanced substantially when compared with pristine PMMA [112]. Zanini et al. used Arg plasma-induced graft polymerization to modify the surfaces of polypropylene (PP) by grafting polyethylene glycol acrylate (PEGA) to prepare surfaces with low protein adsorption. The graft polymerization consists of four steps: (1) plasma pretreatment of PP films, (2) immersion in a PEGA solution, (3) argon plasma-induced graft polymerization, and (4) washing and drying the samples. For short plasma treatments, a reduction in adsorbed proteins of 30%–40% was observed compared with the pristine PP. When the plasma treatment was extended, an increased protein adsorption was observed, which was attributed to possible damage of grafted polypropylene as extending the plasma exposure time; etching predominates over graft polymerization [113]. Yang et al. deposited PEO-like films on a stainless-steel substrate via RF discharge plasma using a mixture of tetraglyme vapor with O_2 as carrier gas and studied their interaction with platelets, fibrinogen, and endothelial cells. A high content of C–O–C (70%) groups was achieved at higher ratios of O_2/tetraglyme, and the film showed excellent stability in phosphate buffer solution. In vitro hemocompatibility and endothelial cell adhesion unveiled low platelet adhesion, platelet activation, fibrinogen adhesion, cell adhesion, and proliferation [114].

9.3.2 Heparin: Anticoagulation coatings

Heparin is a highly sulfated, linear polysaccharide having the highest charge density when compared with other biological molecules. Because of its vital role in controlling cell growth and differentiation, immune defense, and blood coagulation, heparin has been used clinically as an anticoagulant drug for over 80 years. Its heterogeneous structure is made up of α-L-iduronic acid, β-D-glucuronic acid, and α-D-glucosamine repeat units, which can bind to the blood protein antithrombin via ionic interactions. This results in a significant acceleration of the rate at which the clotting factors like thrombin can be deactivated by heparin molecules as it acts as an accelerator for antithrombin activity, since the enzyme thrombin plays a vital role in the coagulation cascade by cleaving fibrinogen to generate fibrin monomers. Thrombin is also responsible for an increase in platelet-platelet adhesion and promotes platelet activation and degranulation. Therefore, the accelerated deactivation of thrombin by heparin molecules helps preventing blood coagulation more efficiently. It was found that the amine group present in heparin is one of the main responsible for generating the positive charge on the surface [115]. The behavior observed by heparin was also seen for other $-NH_2$-containing coatings. Studies using fibrinogen and osteopontin (OPN) showed favorable protein conformations after adsorption to the positively charge $-NH_2$ surface [115,116]; the coatings containing amine ($-NH_2$) functional groups were able to form hydrogen bonds with fibrinogen, tethering it to the amine-functionalized surface. Particularly, NH_2 surfaces stimulate the exposure of high-density bound receptors and focal adhesion components by adsorbed fibronectin [115]. This protein adsorption profile frequently leads to an enhancement in endothelial cell growth [116]. Chemical structure of heparin is depicted in Fig. 9.9.

Antithrombogenic coatings can be deposited either by immobilizing heparin on the surface through strong ionic binding or via chemical grafting. Today, heparin is frequently used on commercially available blood-contacting devices such as hemodialysis catheters. Pandiyaraj et al. developed biofunctional coatings on the surface of polypropylene films using acrylic acid and poly(ethylene glycol) in vapor phase followed by immobilization of chitosan, heparin, and insulin on the surface-modified PP films. The existence of heparin and insulin on the surface of the films was validated by XPS analysis as shown in Fig. 9.10. In vitro analysis clearly showed that the adhesion and activation of platelets decreased drastically after grafting of AAc and PEG; moreover, there was no indication of adhesion and activation of platelets on the surface of biomolecules immobilized PP films as shown in the Fig. 9.11. The modified films exhibited a decrease in adhesion and activation of platelets (90%) compared with pristine PP films, and the same was found to be maximum for heparin-immobilized films [117].

Degoutin et al. modified nonwoven polypropylene textiles to enhance its anticoagulation and antibacterial properties. They optimized the grafting of acrylic acid on

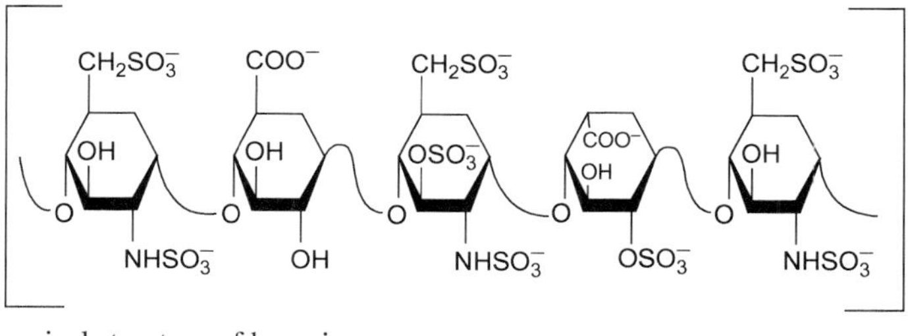

Fig. 9.9 Chemical structure of heparin.

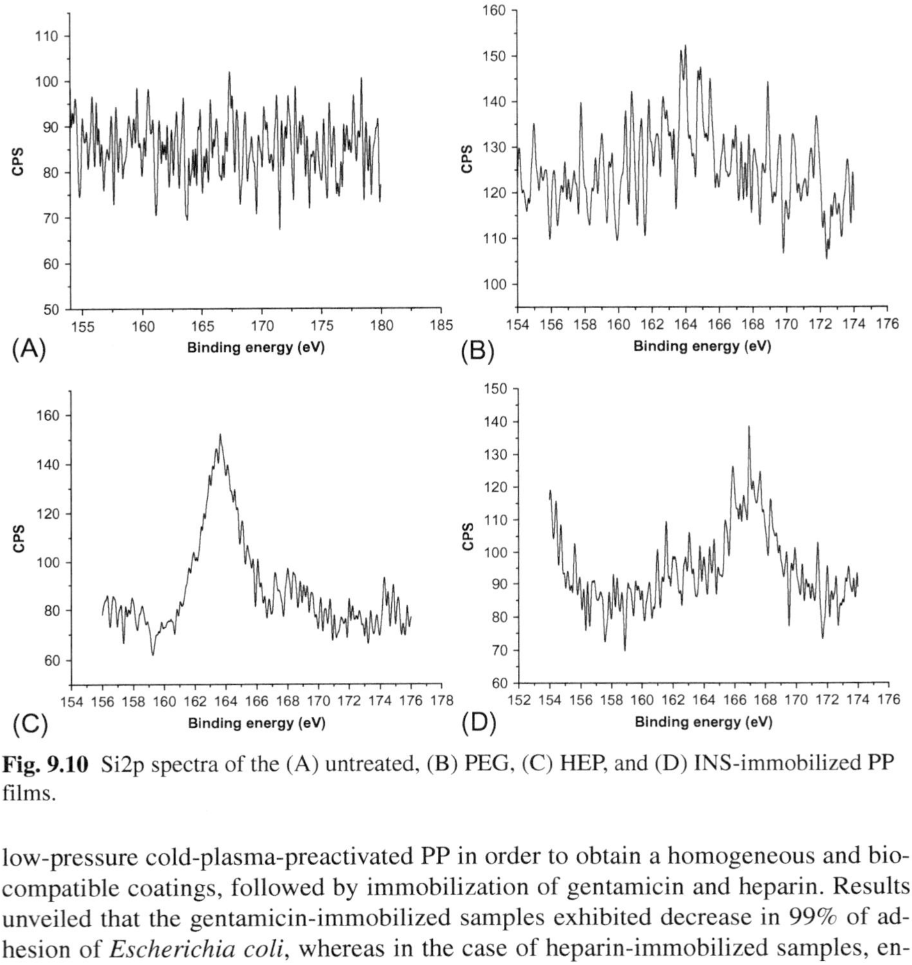

Fig. 9.10 Si2p spectra of the (A) untreated, (B) PEG, (C) HEP, and (D) INS-immobilized PP films.

low-pressure cold-plasma-preactivated PP in order to obtain a homogeneous and biocompatible coatings, followed by immobilization of gentamicin and heparin. Results unveiled that the gentamicin-immobilized samples exhibited decrease in 99% of adhesion of *Escherichia coli*, whereas in the case of heparin-immobilized samples, enhancement in anticoagulant effect up to 35 min was observed [118]. Kim et al. also modified the surface of PET films by oxygen plasma treatment and followed by grafting of acrylic acid and PEO. In a final stage, insulin and/or heparin was immobilized. The acrylic-acid-grafted PET films exhibited a slight decrease in thrombus formation (70%) and decreased significantly by coupling with PEO (55%). However, the heparin-immobilized PET surfaces exhibited the least thrombus formation percentage (28%) compared with other modified films [119]. Gao et al. modified the surface of poly(tetrafluoroethylene-co-hexafluoropropylene) (FEP) films to enhance its hydrophilicity and blood compatibility by allowing acrylic acid on to Ar plasma-pretreated FEP films using UV-induced graft copolymerization. Esterification reaction was employed for covalent immobilization of heparin on the surface of acrylic-acid-grafted FEP films. When compared with other modified samples, the heparin-immobilized acrylic-grafted FEP films exhibited excellent antithrombogenicity [120]. Chandy et al. modified the surface of PTFE and PET films using argon plasma followed by

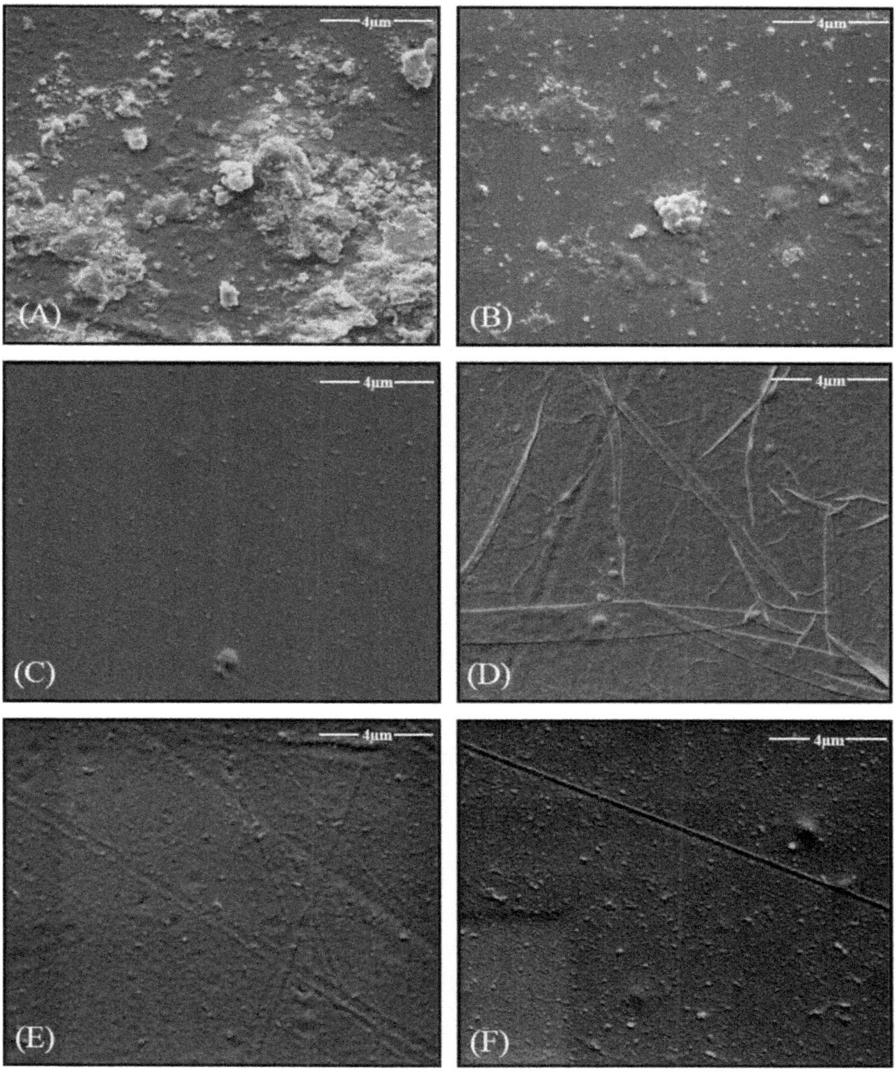

Fig. 9.11 Morphology of the platelet on the surface PP films (A) untreated, (B) Ar plasma treated, (C) PEG, (D) HEP, (E) INS, and (F) CHI immobilized.

series of coatings/grafts such as collagen and laminin. Subsequent to immobilization of biomolecules like PEG, heparin, or phosphatidyl choline via carbodiimide cross-linking, which alters the surface conditions of vascular grafts. Finally, in vitro analysis exhibited substantial reduction in platelet adhesion, and adsorption of fibrinogen on the surface of immobilized films leads to enhancement in biocompatibility [3]. Sask et al. performed immobilization of antithrombin-heparin complex on gold surface with the help of a spacer molecule like PEG that showed a reduced platelet adhesion and prolonged clotting times [121]. Yang et al. studied the improvement in hemocompatibility of 316L stainless-steel (SS) substrates by forming a pulsed plasma polymeric allylamine film on it that exhibited a high degree of cross-linking and possessed a high density of amine groups that could be employed for bonding with heparin. The modified film as a stent coating exhibited good resistance to the mechanical deformation of the stent. In vitro hemocompatibility analysis showed a reduced platelet adhesion and activation and fibrinogen activation, in addition to the activated partial thromboplastin time, which was extended for about 15 s compared with 316L SS. The adhesion and proliferation of endothelial cells were promoted on allylamine films, whereas in the case of heparin-immobilized films, the cell adhesion and proliferation were suppressed slightly after incubation for 3 days. The in vivo analysis specified that the thrombus formation on the heparin surface was restrained by the formation of a homogeneous and intact layer of endothelium on its surfaces [122]. Jin et al. used a two-step process to prepare heparinized polypropylene films using cyanuric chloride (CC) as a trifunctional reagent with the help of spacer poly(ethylene glycol) methacrylate (PEGMA). Initially, the films were activated using oxygen plasma subsequent to grafting of PEGMA and heparin. They also found that the adsorbed proteins were dependent on various factors, for instance, graft density, molecular weight of the monomer, and type of protein. The heparinized polypropylene films exhibited superior anticoagulation compared with PEGMA films as confirmed by hemocompatibility assay. However, a small amount of platelets did adhere on the heparinized films, and the clotting time was not perceived within 60 min, and the films showed a good bioactivity verified by AT III assay, which was not the goal of the study [123]. Hence, heparinized films exhibited better anticoagulation properties making it a suitable candidate for stent applications.

9.3.3 Chitosan: Antimicrobial and antithrombogenic coatings

Biofilms can be formed due to the formation and adhesion of bacteria on to the material surface. Furthermore, the reduced antibiotic susceptibility of a bacterial biofilm is a major concern to remove or replace the infected devices. Hence, it is important for a biomaterial to undergo a suitable preclinical process, in order to protect them from microorganisms. The mechanism of bacterial adhesion and biofilm formation can be evaded by coating the surface with an appropriate antibacterial agent [124,125]. Antimicrobial surface coatings can be formed using various antibiotics that have the potential to kill bacteria attempting to colonize them. However, the major problem is the lack of selectivity, since most of the antibacterial compounds are cytotoxic. For nonbiomedical-related applications, this is of less significance, but in the case of implants, a surface coating must prevent bacterial colonization while assimilating well with human tissue without damaging the surrounding tissue. This need of particular bacterial toxicity with no cytotoxicity is a major task in the development of antimicrobial coatings. Immobilization of functional polymers has a great potential in creating antimicrobial surfaces. Such polymer coating technologies could be economical viable and can be easily incorporated to medical implants, thus leading to an effective method to inhibit infections. Chitosan is a linear polysaccharide polymer, composed of 1,4-linked D-glucosamine and *N*-acetyl-D-glucosamine and is only soluble in aqueous solutions with a $pH < 6.5$. Chitosan is a cationic polysaccharide obtained by alkaline deacetylation of chitin found in the shell of crustaceans, making it the second most abundant natural polysaccharide in the ecosphere after cellulose. The chemical structure of chitosan is shown in Fig. 9.12.

It possess interesting biological properties such as nontoxicity, biocompatibility, low immunogenicity, biodegradability, and acceleration of wound healing and offers various applications in medicine, agriculture, biomaterials, and drug-controlled release systems. Moreover, chitosan is a good candidate for antimicrobial films due to its better film-forming property and antibacterial coating [126]. The schematic diagram of chitosan coating avoiding bacterial adhesion is shown in Fig. 9.13.

Combination of chitosan and heparin has shown to be able to decrease platelet adhesion, results in substantial thromboresistivity and lowers the hemolysis rate [127]. Thanks to its outstanding mucoadhesive nature to a wide variety of hard and soft tissues, chitosan-based hybrid materials may serve as a temporary skeleton in bone tissue engineering [128]. Chitosan has ability to enhance drug absorption and stabilization of drug components to enhance drug targeting [129]. Ziani et al. examined chitosan-based films and solutions against *Aspergillus niger*, *Alternaria alternata*, and *Rhizopus oryzae*; the antifungal activity were tested for the samples prepared at

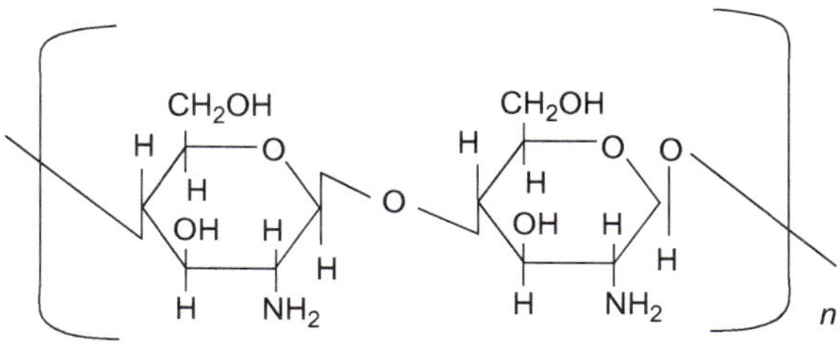

Fig. 9.12 Chemical structure of chitosan.

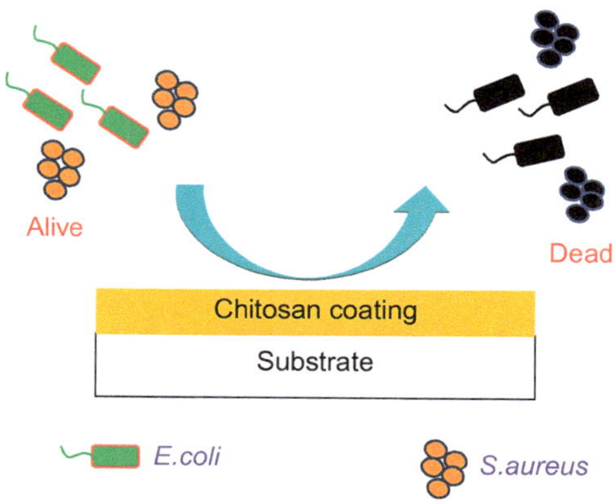

Fig. 9.13 Schematic of chitosan coating preventing bacterial adhesion.

different molecular weights (400 and 149 kDa). The films prepared using chitosan showed an increase in antifungal activity [130]. Ding et al. modified PLLA surfaces by immobilizing chitosan on its surface using plasma-coupling reaction. The results revealed that the morphology of hepatocyte cells tended to become round. The proliferation rate of the cells was found to be similar to that of cell culture plates [131]. Tyan et al. modified the surface of PP nonwoven fabric by activating it using microwave plasma subsequent to acrylic acid grafting and immobilization of chitosan molecules. Bioactivity of the modified samples was examined using activated partial thromboplastin time (APTT), thrombin time (TT), and fibrinogen concentration. Results unveiled the enhancement in antithrombogenic property of surface-modified samples [132]. Meng et al. coated chitosan and heparin onto a coronary stent for the acceleration of the reendothelialization and healing process after coronary stent deployment and tested both in vitro and in vivo. The chitosan and heparin combination resulted in an increase in both hemocompatibility and cell compatibility, which was confirmed by the outcome of the in vitro culturing of porcine iliac artery endothelial cells. In addition, the values of the activated partial thromboplastin time (APPT), prothrombin time (PT), and thromboplastin time (TT) were remarkably extended for the coated samples. Models of porcine coronary injury and arteriovenous shunt were employed for further estimate of the efficiency of this kind of surface-modified stents in vivo, and results verified that the modification process could substantially enhance reendothelialization compared with unmodified stents for its enriched anticoagulation property [133]. Pandiyaraj et al. functionalized low-density polyethylene (LDPE) films using low-pressure DC plasma reactor via in situ grafting of acrylic acid followed by immobilization of polyethylene glycol (PEG) and chitosan molecules to enhance its blood compatibility. FTIR and XPS analysis clearly confirmed the presence of chitosan such as O–C–O, C–O, and O–C=O, as well as C–N, on the surface of LDPE films as shown in Figs. 9.14 and 9.15. In vitro analysis unveiled the increase in blood compatibility by reduction in the adhesion of platelets, adsorption of proteins, and formation of thrombus. Conclusively, the chitosan-immobilized LDPE films exhibited enhanced biocompatibility studied in terms of WBCT (600 s), antithrombogenesis, low protein adsorption, and reduced platelet adhesion [134].

Kara et al. altered the surface of polyurethane films by activating it using oxygen plasma followed by covalent immobilization of chitosan in order to enhance its antibacterial activity. The modified polyurethane films exhibited substantial antibacterial activity against Gram-positive (*S. aureus*) and Gram-negative (*Pseudomonas aeruginosa*) bacteria. It should be highlighted that the number of bacteria colonies were less about 102–105 CFU/mL, and decrease in number of attached viable bacteria was observed due to chitosan modification of polyurethane films [135]. Chang et al. tailored the surface of chitosan films using RF discharge plasma. The modified film exhibited a similar spectrum of unmodified chitosan film but unveiled the presence of substantial hydroxyl and carboxylic groups. It was observed that the surface-modified chitosan films showed less bovine serum albumin adsorption in bicinchoninic acid protein assay due to the presence of carboxylic acid groups on the surface [136].

Douglas et al. coated titanium (Ti) with chitosan and one of three polyphenol-rich plant extracts (PPrPE) for bone-contact applications. They studied the adhesion of

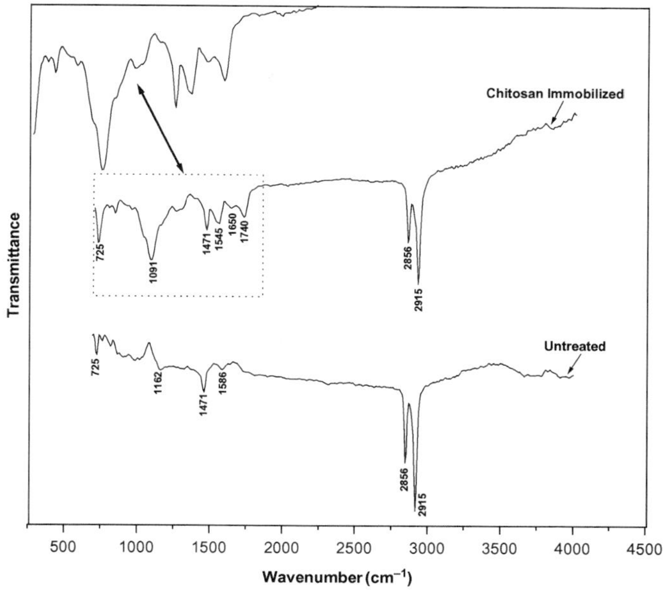

Fig. 9.14 FTIR spectrum of untreated and chitosan-immobilized LDPE films.

HeLa cells and growth of methicillin-resistant *S. aureus* (MRSA). Results showed that the cell adhesion was enhanced for chitosan-coated Ti substrates. In addition, the functionalized PPrPE did not exhibit in any amendment in adhesion of HeLa cell number [137].

9.3.4 Acrylic acid: Cytocompatible coatings

Acrylic acid is the simplest unsaturated carboxylic acid composed of a vinyl group connected to a carboxylic terminus. Acrylic-based coatings on various substrates have received remarkable attention in biomedical applications for the reason that the coatings containing high dense of carboxylic groups can enhance the hydrophilicity of the sample, since the hydrophilic nature of the material is an important characteristic to reduce the adsorption of plasma protein that leads to avoiding the development of thrombus on the material surface [138,139]. In addition, hydrophilic surface can assist in cell adhesion and proliferation on the surface of films through the charge-induced adhesion of cell-adhesion-stimulating proteins. It has a wide range of applications in the field of biomedicine, such as cell culture substrates, and acts as an active site for covalent immobilization of bioactive molecules [140–142]. The chemical structure of acrylic acid is shown in Fig. 9.16.

Carboxylic acid groups exhibit a negatively charged functionality on the material surfaces when stored in pH-neutral solutions. Protein adsorption analysis revealed that fibronectin and albumin are easily prevented from surfaces containing –COOH [143]. In addition, the COOH coatings can also increase the cell growth. It should be noted

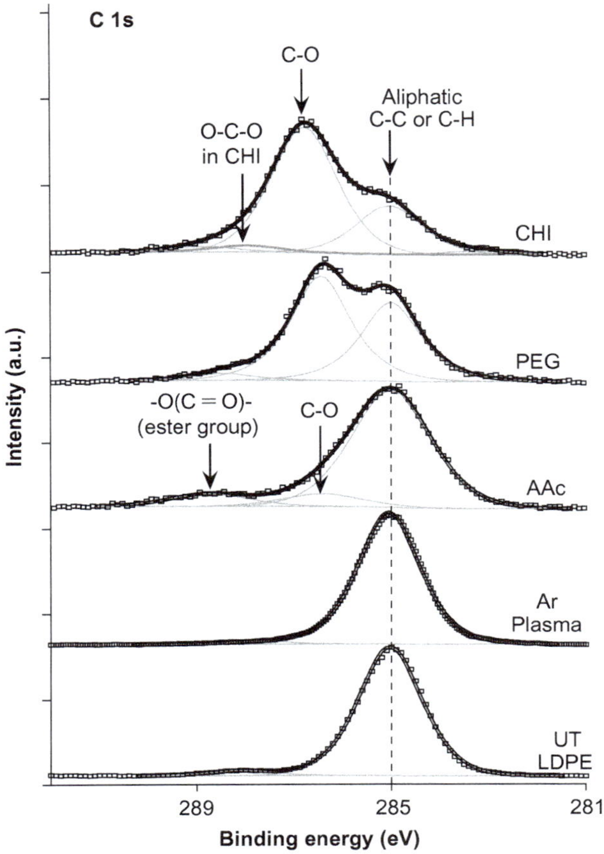

Fig. 9.15 C1s XPS spectra of untreated, Ar plasma-treated, AAc-grafted, PEG-immobilized, and CHI-immobilized LDPE films.

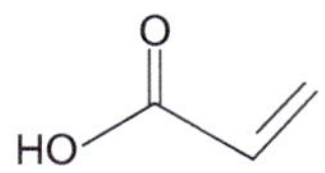

Fig. 9.16 Chemical structure of acrylic acid.

that this phenomenon is highly reliant on the concentration of carboxylic group on the material surface, as escalation in functional group density leads to higher negative charge on the surface, which was shown to prevent cell growth [144]. Acrylic acid can be used as a spacer to bind proteins and substrate. The presence of carbon-carbon double bond in acrylic acid binds easily with polymers. However, the base material can retain their own physicochemical properties subsequent to graft polymerization [145]. Cheng et al. used acrylic acid as a spacer molecule to immobilize collagen on the surface of poly(ε-caprolactone) (PCL) using Ar plasma [146]. Kumar et al. performed plasma polymerization of acrylic acid on the surface of PET mesh followed by immobilization of silver nanoparticles to increase the antibacterial property. Results revealed that the modified sample exhibited reduction in bacterial concentration >99.7% compared with pristine sample [147]. Kang et al. deposited acrylic acid and allylamine on microporous polypropylene membranes. Results showed that the modified films were found to decrease the fouling with bovine serum albumin (BSA) to less than 50%. The enhancement in hydrophilicity and surface changes induced by plasma polymerization influenced the adsorption and removal of BSA [148]. In another study by Dhayal et al., acrylic acid was deposited on glass substrates by plasma polymerization, and their interaction with leukemia cells was examined and compared with polystyrene cell culture. Cell growth was reduced to 60% on the surface-modified films due to the existence of hydroxyl and carbonyl groups [149]. Carton et al. deposited polyacrylic acid thin films using atmospheric pressure nitrogen plasma jet as a function of different operating parameters. Cell adhesion test was performed on the acrylic films using human ovarian carcinoma cells (NIH/OVCAR-3). The films exhibited different stabilities when soaked in water for 24 h. The films with less percent of COOH/COOR showed a comparable cell adhesion and proliferation with positive controls, whereas films with higher concentration of COOH/COOR were found be less stable and exhibited decrease in cell adhesion [150]. Cools et al. developed a different range of stable and unstable acrylic acid coatings as a function of discharge powers and monomer flow rates and also examined their effect of coatings (in) stability on human foreskin fibroblast (HFF) viability and morphology. The stable coatings possess a small concentration of carboxylic acid group (0.7%); these films showed ample stability for 72 h of incubation. Initially, the cells exhibited lower affinity as the films contain lower amounts of carboxylic groups. Conversely, after 72 h, the cells adhered displayed spindle-shaped morphology with noticeable extended filopodia [151]. Vasilets et al. functionalized polytetrafluoroethylene (PTFE) for human thrombomodulin (hTM) binding by plasma activation of CO_2 followed by vapor-phase graft polymerization of acrylic acid. The immobilized hTM activity was assessed by protein C activation test, in which the modified surfaces converted thrombin from a procoagulant protease to an anticoagulant form [152].

9.3.5 Diamond like carbon (DLC): Biocompatible coating

In recent years, diamond-like carbon (DLC) has gained interest in biomedical industry, and it is one of the most attractive blood-contacting biomaterials, for instance, in artificial heart valves, stents, rotary blood pumps, and orthopedic or dentistry because of its inertness, hardness, low frictional coefficient, high wear and corrosion resistance, and excellent smoothness. It consists of a mixture of sp^2 and sp^3 carbon bonds produced using high-energy carbon species. DLC can be produced using various techniques [153–157]. It possesses an amorphous structure; hence, it can be easily doped and alloyed with different elements. In addition to the prevention of reduction of corrosion wear, the coatings also offer a controlled interaction with the ambient; for example, in the case of coronary stents, the coating avoids the interaction with the flowing blood to prevent the occurrence of thrombosis. Here, we discuss the status of the DLC coatings for potential stent applications. Swiatek et al. prepared

Si-Ag-incorporated DLC layers on titanium alloy (Ti6Al7Nb) using modified RF-PACVD technique. This technique has the advantage of doping carbon coatings with silicon and silver in a single-step process. The cell viability of the modified films was assessed using LIVE/DEAD test and endothelial cells, whereas the bactericidal activity was evaluated using *E. coli*. The incorporation of high Ag content led to a reduction in concentration of Si resulting in decreased hardness, higher bactericidal activity, and lower biocompatibility [158]. Avi Bendavid et al. developed DLC containing up to 22% of silicon on to silicon substrates via low-frequency DC plasma-activated CVD. A mixture of methane and tetramethylsilane (TMS) was employed for the fabrication of DLC-Si. An MG-63 osteoblast-like cell culture was used to study the biocompatibility of the modified films that were allowed to grow for 3 days to study, and their proliferation were examined by scanning electron microscopy. The in vitro analysis clearly unveiled that the DLC-Si exhibited good cell adhesion and proliferation [159]. Steffen et al. prepared DLC coatings via plasma polymerization of acetylene subsequent to ammonia plasma treatment, and biomolecules of heparin were immobilized. The effect of heparin was to be expected as discussed earlier and also focuses on the importance of immobilization of heparin on DLC results in extending the blood coagulation time by a factor of 10 [160]. Pandiyaraj et al. developed diamond-like carbon (DLC) films on polyethylene terephthalate (PET) films using PECVD as a function of biasing voltage. The blood compatibility of the DLC film was examined by in vitro analysis. The maximum sp^3 fraction sp^3/sp^2 ratio was attained at 300V of biasing potential confirmed by XPS and Raman analysis. The Raman spectra of DLC prepared at different biasing voltage is shown in the Fig. 9.17. In vitro analysis confirms the enhancement in blood compatibility by reduced platelet adhesion and blood protein adsorption. The platelet adhesion and the quantity of thrombus formed on the DLC films are shown in the Figs. 9.18 and 9.19 [161].

Jones et al. examined the hemocompatibility of hydrogenated amorphous carbon films developed using plasma-enhanced CVD on titanium substrates with interlayers of TiC-TiN. The modified samples did not exhibit any substantial spreading of platelets and did not exhibit any tendency toward thrombus formation [162]. Kwok et al. examined the blood compatibility of P-doped hydrogenated amorphous carbon (a-C/H) films developed using plasma-immersion ion implantation. The coatings exhibited minimal interaction with the plasma proteins, results in the preferential adsorption of albumin that is essential for superior hemocompatibility [163]. Maguire et al. and Okpalugo et al. incorporated silicon on a-C/H films using plasma-enhanced vapor deposition technique; results unveiled decrease in platelet attachment and enhanced the human microvascular endothelial cell attachment [164,165]. Awaja et al. developed amorphous carbon-/diamond-like carbon (a-C/H) coatings to biologically enhance polyether ether ketone (PEEK) for biomedical device integration in human body. Cytotoxicity of the PEEK coatings was investigated using human osteoblast (hFOB) and human fibroblast (hGF). The coated samples exhibited lower toxicity values for hFOB osteoblast and hGF fibroblast cell in comparison with unmodified PEEK. In addition, MTS viability assay results unveiled that the selected coatings showed lower values than the untreated PEEK, owing to their biocompatible property [166].

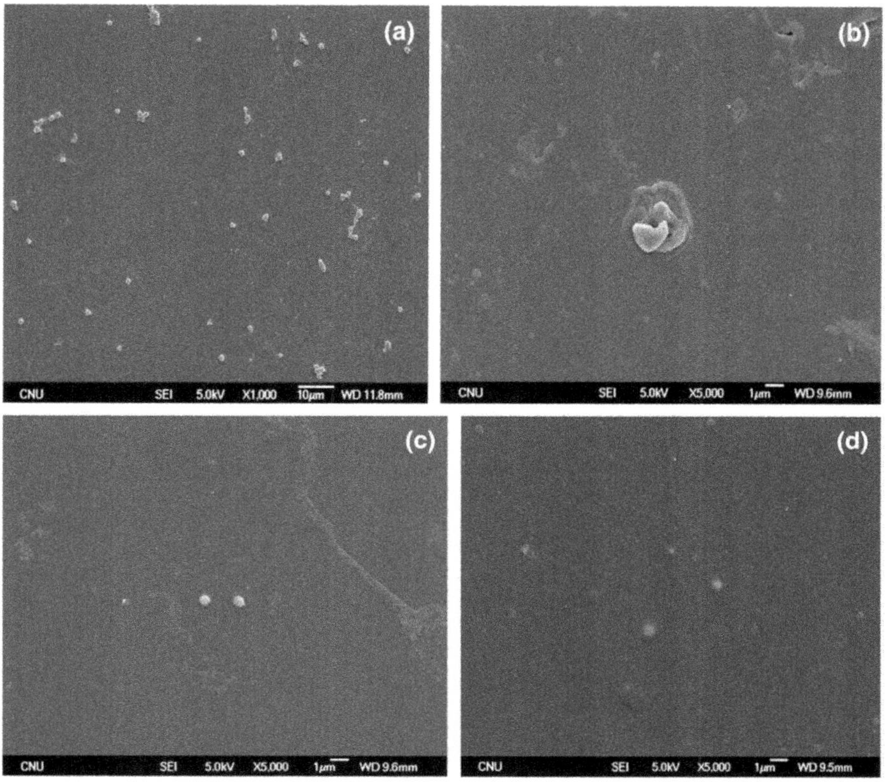

Fig. 9.17 SEM images of platelet adhesion onto a DLC film as a function of biasing potential (A=PET film, B=−150, C=−300, and D=−600V).

9.3.6 Other biocompatible coatings

Pandiyaraj et al. used radio-frequency glow discharge (RFGD) to alter the surface of TiO_2/PET films using oxygen plasma at different treatment times (2–15min). XPS analysis validated the presence of oxygen-containing polar groups on the surface of modified films, in addition to increase in concentration of Ti^{3+} in Ti2p, whereas decrease in Ti^{4+} state was observed. This results in enhancement in surface roughness. The cell compatibility of the modified films was examined using human osteoblast cells. Results unveiled the enhancement in cell compatibility was observed after oxygen plasma treatment when compared with the untreated films. The values of osteoblast cell adhesion are shown in Fig. 9.20. The adhesion of *S. aureus* was highly inhibited when TiO_2/PET was treated using oxygen plasma as shown in the Fig. 9.21 [167]. Su et al. developed transparent TiO_2 films on the surface of plasma-activated polymethyl methacrylate (PMMA) to improve its antibacterial properties. The antibacterial properties were calculated using platelet counting method of *S. aureus* (Gram-positive) and *E. coli* (Gram-negative). Outcome showed that TiO_2-coated PMMA films unveiled

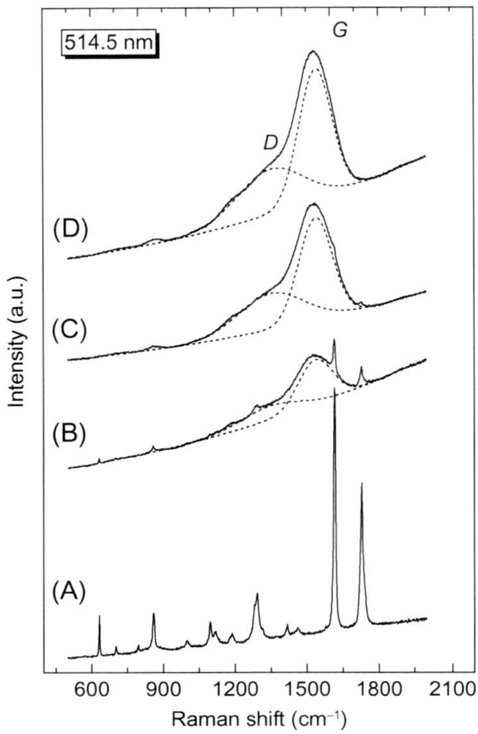

Fig. 9.18 Raman spectra of the DLC films coated at different biasing potential (A = PET film, B = −150, C = −300, and D = −600 V).

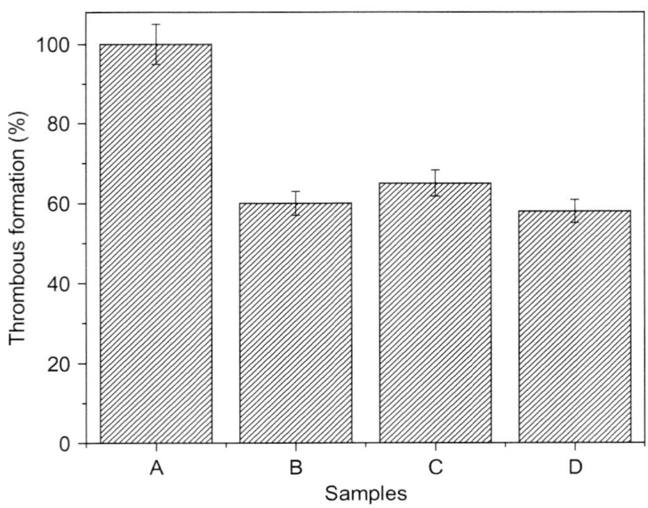

Fig. 9.19 Amount of thrombus formed on DLC film surfaces as a function of biasing potentials (A = PET, B = −150, C = −300, and D = −600 V).

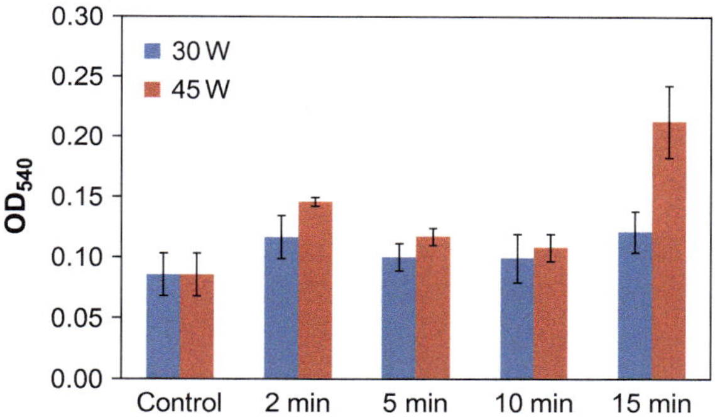

Fig. 9.20 Osteoblast cell adhesion on plasma-treated TiO_2/PET films as a function of exposure time.

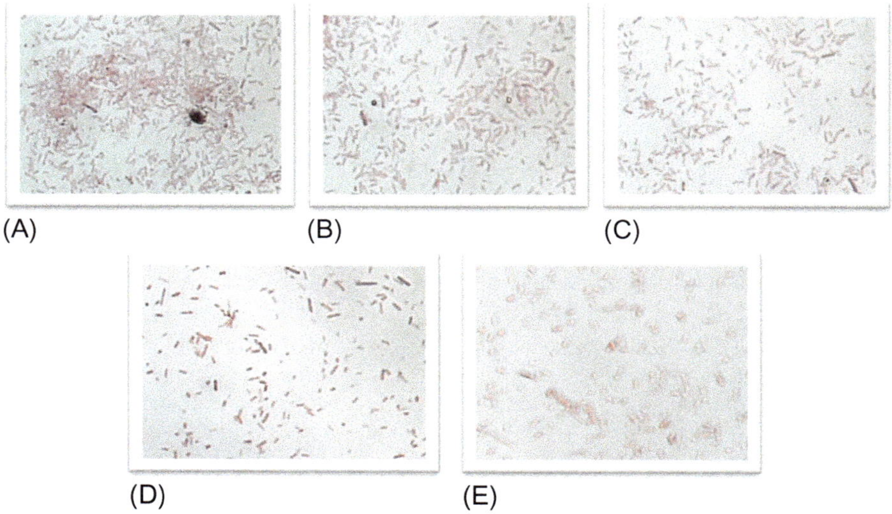

Fig. 9.21 Optical photograph showing adhesion and proliferation of *Staphylococcus* on plasma-treated TiO_2/PET film (A) control PET and (B) untreated TiO_2/PET, (C) 5 min, (D) 10 min, and (E) 15 min at the power level of 45 W.

good performance due to the natural light-induced photocatalytic inactivation and antiadhesion of bacteria [168].

Boudot et al. used vacuum cathodic arc plasma deposition technique to deposit TiO_2 films on the surface of silicone to increase its biological properties that can impair the healing process. The deposited coatings showed good adhesion to the substrates and stable in aqueous environment. In addition, cell biological studies revealed that the in vitro cytocompatibility of fibroblasts can be significantly enhanced without influencing the silicone's nontoxicity [169]. Szymanowski et al. deposited titanium dioxide on glass and cotton textile substrates using titanium tetrachloride as precursor via radio-frequency plasma-enhanced chemical vapor deposition (RF PECVD) technique as a function of different operating power for bactericidal applications. Bactericidal properties were examined using cultures of K12 strain of *E. coli* and the ultraviolet light C (UV-C) irradiation. The results of bactericidal testing unveil that TiO_2-coated surfaces exhibited a substantial high bacteria death rate that increases with increasing the discharge power. They also found that the films deposited at higher discharge power decreased more than 90% of bacteria [170].

Pandiyaraj et al. developed TiOx-based biocompatible coatings on the surface of polypropylene films using DC excited glow discharge plasma, using $TiCl_4/Ar+O_2$ gas mixture as a precursor and deposited as a function of different discharge power. The $TiCl_4/Ar+O_2$ plasma deposition substantially tailored the surface morphology of the polypropylene films; nanosized spherical particles were found on the modified samples. The film deposited at discharge power of 300W was found to be homogeneous. In vitro analysis was used to examine the cytocompatibility of the TiOx/PP films that include cell viability, adhesion, and cytotoxicity using NIH3T3 (mouse embryonic fibroblast) cells as shown in the Fig. 9.22. Additionally, the antibacterial activities of TiOx/PP films were also assessed against two distinct bacterial models, namely, Gram-positive *S. aureus* and Gram-negative *E. coli* DH5α bacteria as shown in the Fig. 9.23. The in vitro study clearly exposed that the tailored films exhibits cell viability similar to tissue culture plate (TCP) and also has impact control on bacterial activity [171].

Liguori et al. developed nanocomposite coating using silver nanoparticles embedded in a plasma-polymerized acrylic acid matrix that was achieved via nonequilibrium atmospheric pressure plasma jet allowing distinct and instantaneous introduction of acrylic acid and Ag nanoparticles dispersed in ethanol in the plasma regime. The antibacterial efficiency of the codeposited films was initially examined with agar disk diffusion tests (using *E. coli*). After incubation for 24h, the growth that was inhibited around the samples may be due to the release of Ag ions from the matrix [172].

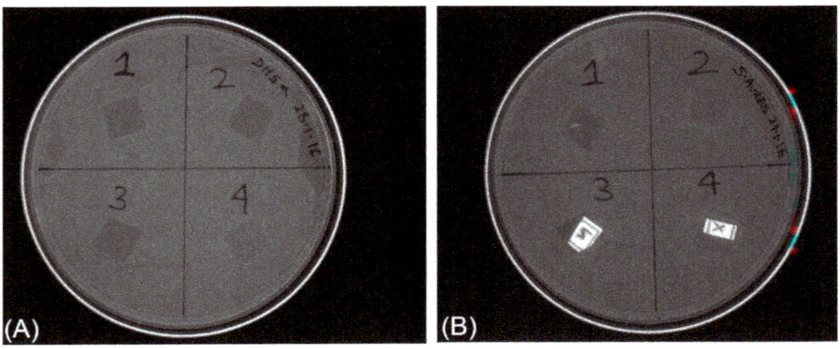

Fig. 9.22 Images of petri plates for evaluation of antibacterial effect of films coated at various discharge power against (A) *E. coli* DH5α and (B) *S. aureus* strains. (1) Uncoated, (2) 100W, (3) 200W, and (4) 300W.

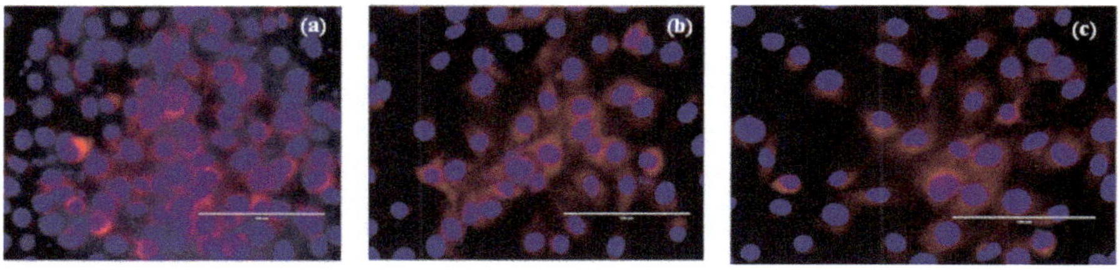

Fig. 9.23 Fluorescence microscopic image of NIH3T3 cells on $TiCl_4$/Ar + O_2 plasma-treated PP films surfaces: (A) TCP, (B) 200 W, and (C) 300 W.

Deng et al. prepared antimicrobial nanosilver nonwoven polyethylene terephthalate (PET) fabric using a three-step process: (1) initially, with the help of atmospheric pressure plasma system, the fabrics were activated by depositing a layer of organosilicon thin film, (2) followed by the incorporation of silver nanoparticles on the surface of fabrics using dipping-dry process and (3) finally deposition of second layer of organosilicon layer (10–50 nm) on the surface of nanoparticles. The deposited films showed good antimicrobial activity against *P. aeruginosa*, *S. aureus*, and *Candida albicans*. They also found that the thickness of the second layer of organosilicon influences the bacterial reduction of the fabrics [173]. Pandiyaraj et al. fabricated a thin layer of TiO_2 on the surface of air plasma-activated PET substrate, and the TiO_2/PET films were further tailored via DC glow discharge air plasma at different discharge potentials. The biocompatibility of the TiO_2/PET films was examined by determining the nucleation and growth of calcium and phosphorous on the surface-modified films subsequent to immersion in simulated body fluids (SBF) as a function of different storage days and studied by scanning electron microscopy (SEM). The changes in physicochemical properties induced by air plasma treatment promote the growth of bone-line apatite layer on the surface-modified TiO_2/PET films as shown in the Fig. 9.24 [174]. Hayakawa et al. performed plasma polymerization of HMDSO onto titanium substrates using radio-frequency technique; results show that the modified HMDSO-coated titanium has ability for applications as a dental implant material [175]. Chen et al. accomplished PEGylation of poly(dimethyl siloxane) (PDMS) surfaces exhibited resistance to nonspecific protein adsorption. Furthermore, by the usage of PEG-lysine conjugates, the incorporation of free ε-amino groups on the surface reduces the surface capable of dissolving fibrin clots due to the adsorption of the fibrinolytic protein, plasminogen from blood plasma [176]. Saulou et al. developed composite thin films composed of silver nanoclusters incorporated in an organosilicon matrix deposited by PECVD. The film formed using two-step process, silver sputtering and immediate plasma polymerization in argon-hexamethyldisiloxane (HMDSO) plasma using RF glow discharge. The organosilicon matrix was effective in preventing the adhesion of eukaryotic microorganism, *S. cerevisiae*. Additionally, the films with high silver content exhibited their antifungal activity against sessile cells [177]. Hsiao et al. altered the surface of polyurethane (PU) by using hexamethyldisiloxane (HMDSO) and tetrafluoromethane (CF_4) as precursors. The modified film exhibits both micro- and nanoscale morphology with superhydrophobic property by augmenting the monomer flow rate. The surface-modified films show promising myoblast cell proliferation that is revealed by

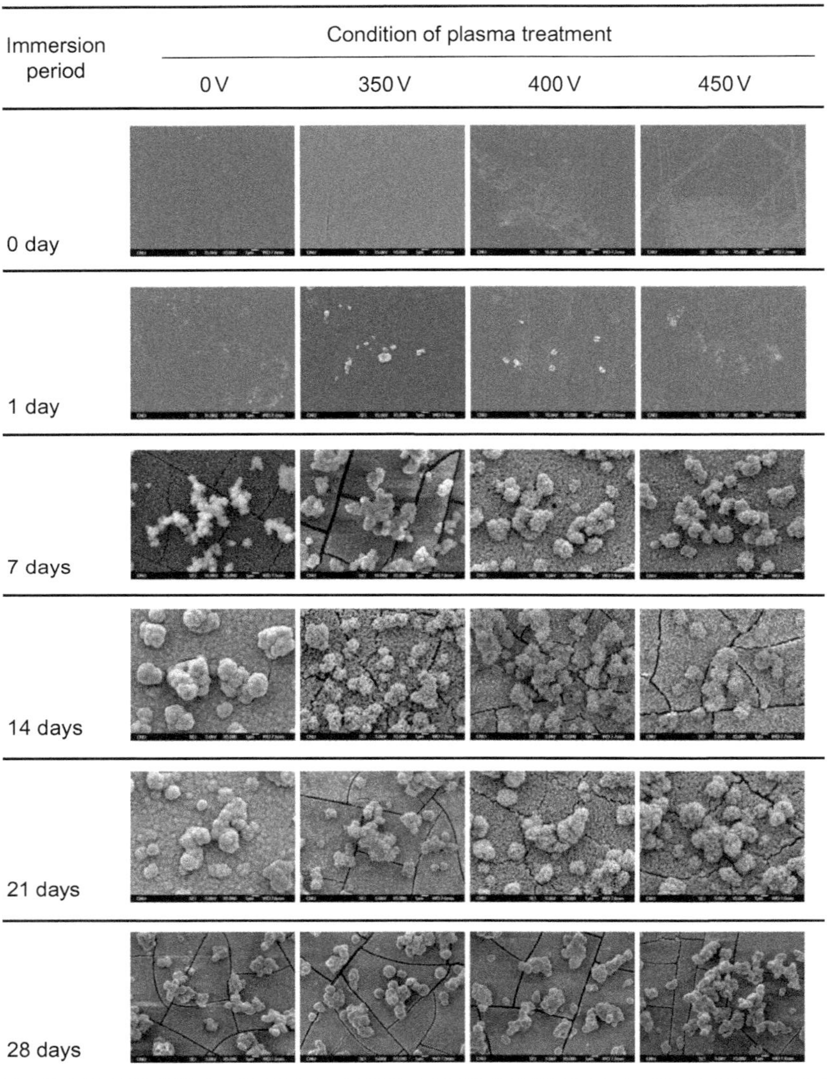

Fig. 9.24 SEM photographs of the surfaces of PET/TiO_2 film surfaces were immersed in SBF for 0, 1, 7, 14, 21, and 28 days as a function of discharge potentials.

in vitro WST-1 cell proliferation test results and also unveiled exceptional low platelet adhesion even after a long incubation time; this may be due to the hierarchical-scale surface roughness and super hydrophobicity. These coatings can inhibit the thrombus formation that can be used for tailoring the surface of blood-contacting devices [178]. Tang et al. used DC plasma reactor to prepare plasma nanocoatings on the surface of two types of intravascular stents made of nitinol (NiTi alloy) and stainless steel by gas mixtures of trimethylsilane and oxygen. The endothelialization effects were examined by seeding the plasma-coated stents with human umbilical vein endothelial cells (HUVECs) for cell culture assay. Results showed that both the nitinol stents and the stainless-steel intravascular stents showed significant enhancement in cell adhesion and proliferation compared with bare stents [179]. Jampala et al. developed quaternary ammonium-based coatings showing the highest antibacterial activity against *S. aureus* in which HMDSO was used as a promoter layer on stainless-steel substrates using low-pressure plasma technique followed by deposition of ethylene diamine polymer to prepare antibacterial coatings [180].

9.4 Conclusion

The fouling of proteins initiated by biological fluids is a major drawback for employing biomedical devices. The performance of these devices can be enhanced by inhibiting or governing nonspecific protein adsorption, resulting in an extended life span. Surface modification using polymer coatings is an effective way of introducing new properties on the surface of material without altering its bulk property of the host material. With cardiovascular stents in mind, this chapter has discussed various types of plasma-based polymeric coatings (PEG, acrylic, heparin, chitosan, DLC coatings, and some others) that are able to reduce platelet adhesion, protein adsorption, and bacterial adhesion while enhancing their antithrombogenic properties and cell compatibility. Plasmas are especially interesting because of their eco-friendly character and allowing for a high degree of tunability of the surface properties. The coating developed using plasmas is highly cross-linked and pinhole-free and is highly adhesive to a wide variety of substrates. In general, coatings based on polyethylene glycol (PEG) showed significant results in resisting protein adsorption. The strong interaction of PEG coatings with water molecules is the main reason for their antibiofouling properties. They also demonstrated outstanding hydrophilic properties, resulting in excellent blood compatibility by inhibiting blood coagulation. Similarly, acrylic acid coatings also enhanced the hydropilic property of the material. This assists the cell to adhere and to proliferate on the surface of materials via cell attachment proteins. It is used as cell culture substrates and as an active layer for covalent immobilization of biomolecules. Heparin coatings unveiled its anticoagulation property; the amine groups present in the heparin molecule generates positive charge on the surface of material prevents blood coagulation. Chitosan-based coatings were found to have good hemocompatibility and exhibited good antimicrobial activity against various microorganisms (*S. aureus*, *Bacillus subtilis*, and *E. coli*). In addition, chitosan-based hybrid materials are used as a temporary skeleton in bone tissue engineering and stabilizer for drug delivery applications. Diamond-like carbon coating showed excellent blood compatible property by preventing the occurrence of thrombosis by inhibiting platelet adhesion and blood protein adsorption. It is evident that the polymer coatings exhibit good biocompatible properties, as it can be used for the modification of stent surface. Since the surface developed and formulated to be either extremely hydrophobic or hydrophilic, to reduce platelet adhesion and accumulation and thus results in thrombosis risks.

Acknowledgments

One of the authors (K.N.) would like to convey his heartfelt thankfulness to the Department of Atomic Energy—Board of Research in Nuclear Sciences (DAE-BRNS) (Ref. 34/14/05/2014-BRNS/2124), Government of India and the Department of Science and Technology—Science and Engineering Research Board (DST-SERB) (SR/FTP/PS-106/2011), Government of India for providing the financial support.

References

[1] A.W. Martinez, E.L. Chaikof, Microfabrication and nanotechnology in stent design, WIREs Nanomed. Nanobiotechnol. 3 (2011) 256–268.
[2] D.C. Miller, A. Thapa, K.M. Haberstroh, T.J. Webster, Endothelial and vascular smooth muscle cell function on poly (lactic-co-glycolic acid) with nano-structured surface features, Biomaterials 25 (2004) 53–61.
[3] T. Chandy, G.S. Das, R.F. Wilson, G.H.R. Rao, Use of plasma glow for surface engineering biomolecules to enhance blood compatibility of Dacron and PTFE vascular prosthesis, Biomaterials 21 (2000) 699–712.
[4] T.J. Reape, P.H.E. Groot, Chemokines and atherosclerosis, Atherosclerosis 147 (1999) 213–225.
[5] N. Grabow, D.P. Martin, K.P. Schmitz, K. Sternberg, Absorbable polymer stent technologies for vascular regeneration, J. Chem. Technol. Biotechnol. 85 (2010) 744–751.
[6] H.S. Tran, M.M. Puc, C.W. Hewitt, D.B. Soll, S.W. Marra, V.A. Simonetti, J.H. Cilley, A.J. DelRossi, Diamond-like carbon coating and plasma or glow discharge treatment of mechanical heart valves, J. Investig. Surg. 12 (1999) 133–140.
[7] T. Peng, P. Gibula, K. Yao, M.F.A. Goosen, Role of polymers in improving the results of stenting in coronary arteries, Biomaterials 17 (1996) 685–694.
[8] M. Gopinath, M.D. Feldman, D. Patel, C.M. Agrawal, Coronary stents: a materials perspective, Biomaterials 28 (2007) 1689–1710.
[9] A. Abizaid, J.R. Costa, New drug-eluting stents: an overview on biodegradable and polymer-free next-generation stent systems, Circ. Cardiovasc. Interv. 3 (2010) 384–393.
[10] L. Pendyala, R. Jabara, K. Robinson, N. Chronos, Passive and active polymer coatings for intracoronary stents: novel devices to promote arterial healing, J. Interv. Cardiol. 22 (2009) 37–48.
[11] T. Parker, V. Davé, R. Falotico, Polymers for drug eluting stents, Curr. Pharm. Des. 16 (2010) 3978–3988.
[12] X. Ma, T. Wu, M.P. Robich, Drug-eluting stent coatings, Interv. Cardiol. 4 (2012) 73–83.
[13] E. Deconinck, J. Sohier, I. De Scheerder, G.V. Den Mooter, Pharmaceutical aspects of drug eluting stents, Pharm. Sci. 97 (2008) 5047–5060.
[14] C. Yang, H.M. Burt, Drug-eluting stents: factors governing local pharmacokinetics, Adv. Drug Deliv. Rev. 58 (2006) 402–411.
[15] B. Tesfamariam, Drug delivery kinetics from stent devise-based delivery systems, J. Cardiovasc. Pharm. 51 (2008) 118–125.
[16] M. Wiemer, A. Seth, P. Chandra, J. Neuzner, G. Richardt, J.J. Piek, M Desaga, C. Macaya, C.J. Bol, K. Miquel-Hebert, K. De Roeck, P.W. Serruys, Systemic exposure of everolimus after stent implantation: a pharmacokinetic study, Am. Heart J. 156 (2008) 751–757.

[17] M. Joner, A.V. Finn, A. Farb, E.K. Mont, F.D. Kolodgie, E. Ladich, R. Kutys, K. Skorija, H.K. Gold, R. Virmani, Pathology of drug-eluting stents in humans: delayed healing and late thrombotic risk, J. Am. Coll. Cardiol. 48 (2006) 193–202.
[18] P. Chu, Enhancement of surface properties of biomaterials using plasma-based technologies, Surf. Coat. Technol. 201 (2007) 8076–8082.
[19] J. Daemen, P. Wenaweser, K. Tsuchida, L. Abrecht, S. Vaina, C. Morger, N. Kukreja, P. Jüni, G. Sianos, G. Hellige, Early and late coronary stent thrombosis of sirolimus-eluting and paclitaxel-eluting stents in routine clinical practice: data from a large two-institutional cohort study, Lancet 369 (2007) 667–678.
[20] E. Kedhi, K.S. Joesoef, E. McFadden, J. Wassing, C. Van Mieghem, D. Goedhart, P.C. Smits, Second-generation everolimus-eluting and paclitaxel-eluting stents in real-life practice (COMPARE): a randomised trial, Lancet 375 (2010) 201–209.
[21] G. Lloyd, G. Friedman, G.S. Jafri, G. Schultz, A. Fridman, K. Harding, Gas plasma: medical uses and developments in wound care, Plasma Process. Polym. 7 (2010) 194–211.
[22] J. Lopez-Garcia, F. Bilek, J. Lehocky, I. Junkar, I. Mozeti, M. Sowe, Enhanced printability of polyethylene through air plasma treatment, Vacuum 95 (2013) 43–49.
[23] S. Guruvenket, G.M. Rao, M. Komath, A.M. Raichur, Plasma surface modification of polystyrene and polyethylene, Appl. Surf. Sci. 236 (2004) 278–284.
[24] K. Ishihara, Y. Iwasaki, S. Ebihara, Y. Shindo, N. Nakabayashi, Photoinduced graft polymerization of 2-methacryloyloxyethyl phosphorylcholine on polyethylene membrane surface for obtaining blood cell adhesion resistance, Colloids Surf. B 18 (2000) 325–335.
[25] J.H. Liu, H.L. Jen, Y.C. Chung, Surface modification of polyethylene membranes using phosphorylcholine derivatives and their platelet compatibility, J. Appl. Polym. Sci. 74 (1999) 2947–2954.
[26] C. Mao, J. Yuan, H. Mei, A. Zhu, J. Shen, S. Lin, Introduction of photocrosslinkable chitosan to polyethylene film by radiation grafting and its blood compatibility, Mater. Sci. Eng. C 24 (2004) 479–485.
[27] R.A. Hoshi, R.V. Lith, M.C. Jen, J.B. Allen, K.A. Lapidos, G. Ameer, The blood and vascular cell compatibility of heparin-modified ePTFE vascular grafts, Biomaterials 34 (2013) 30–41.
[28] Y. Chen, Q. Gao, H. Wan, J. Yi, Y. Wei, P. Liu, Surface modification and biocompatible improvement of polystyrene film by Ar, O_2 and Ar + O_2 plasma, Appl. Surf. Sci. 265 (2013) 452–457.
[29] H. Kaczmarek, J. Kowalonek, A. Szalla, A. Sionkowska, Surface modification of thin polymeric films by air-plasma or UV-irradiation, Surf. Sci. 507–510 (2002) 883–888.
[30] N. Sprang, D. Theirich, J. Engemann, Plasma and ion beam surface treatment of polyethylene, Fourth International Conference on Plasma Surface Engineering Part 2Surf. Coat. Technol. 74–75 (2) (1995) 689–695.
[31] J. Abenojar, R.T. Coque, M.A. Martínez, J.M.M. Martínez, Surface modifications of polycarbonate (PC) and acrylonitrile butadiene styrene (ABS) copolymer by treatment with atmospheric plasma, Surf. Coat. Technol. 203 (16) (2009) 2173–2180.
[32] N. Encinas, B. Díaz-Benito, J. Abenojar, M.A. Martínez, Extreme durability of wettability changes on polyolefin surfaces by atmospheric pressure plasma torch, Surf. Coat. Technol. 205 (2) (2010) 396–402.
[33] Y. Martin, D. Boutin, P. Vermette, Study of the effect of process parameters for n-heptylamine plasma polymerization on final layer properties, Thin Solid Films 515 (17) (2007) 6844–6852.
[34] E. Arenholz, V. Svorcík, T. Kefer, J. Heitz, D. Bauerle, Structure formation in UV-laser ablated poly-ethylene-terephthalate (PET), Appl. Phys. A 53 (4) (1991) 330–331.

[35] K. Walachova, V. Svorcík, L. Bacakova, V. Hnatowicz, Colonization of ion-modified polyethylene with vascular smooth muscle cells in vitro, Biomaterials 23 (14) (2002) 2989–2996.
[36] P. Sioshansi, E.J. Tobin, Surface treatment of biomaterials by ion beam processes, Surf. Coat. Technol. 83 (1996) 175–182.
[37] B.D. Ratner, Plasma deposition for biomedical applications: a brief review, J. Biomater. Sci. Polym. Ed. 4 (1) (1993) 3–11.
[38] P. Mendhe, G.A. Arolkar, S.R. Shukla, R.R. Deshmukh, Low temperature plasma processing for the enhancement of surface properties and dye ability of wool fabric, J Appl. Polym. Sci. 133 (12) (2016). 43097(1–8).
[39] R.R. Deshmukh, N.V. Bhat, Pre-treatments of textiles prior to dyeing: plasma processing, in: P.J. Hauser (Ed.), Textile Dyeing, In Tech Publisher, ISBN: 978-953-307-565-5, 2011. December.
[40] A. Fridman, Plasma Chemistry, Cambridge University Press, New York, 2008.
[41] P.C. Kong, E. Pfender, Plasma processes, in: A.W. Weimer (Ed.), Carbide, Nitride and Boride Materials—Synthesis and Processing, first ed., Chapman & Hall, London, 1997, pp. 359–383.
[42] E. Gomez, D. Amutha Rani, C.R. Cheeseman, D. Deegan, M. Wise, A.R. Boccaccini, Thermal plasma technology for the treatment of wastes: a critical review, J. Hazard. Mater. 161 (2009) 614–626.
[43] Z. Guowei, C. Yashao, D. Tao, W. Xiaoli, Surface modification of polyethylene by heparin for improvement of antithrombogenicity, Plasma Sci. Technol. 9 (2007) 202–205.
[44] R.R. Deshmukh, A.R. Shetty, Surface characterization of polyethylene films modified by gaseous plasma, J. Appl. Polym. Sci. (104) (2007) 449–457.
[45] R. Oosterom, T.J. Ahmed, J.A. Poulis, H.E.N. Bersee, Adhesion performance of UHMWPE after different surface modification techniques, Med. Eng. Phys. 28 (2006) 323–330.
[46] M. Szycher, P. Sioshansi, E.E. Frisch, Biomaterials for the 1990s: polyurethanes, silicones and ion beam modification techniques (Part II), Spire Corporation, Patriots Park, Bedford, 1990.
[47] G.A. Arolkar, S.M. Jacob, K.N. Pandiyaraj, V.R. Kelkar-Mane, R.R. Deshmukh, Effect of TEOS plasma polymerization on corn starch/poly (e-caprolactone) film: characterization, properties and biodegradation, RSC Adv. 6 (2016) 16779–16789.
[48] J.A. Marciano de Paul, J. Realino de Paul, F.C. Pimenta, M.H. Rezende, M.T.F. Bara, Antimicrobial activity of the crude ethanol extract from *Pimenta pseudocaryophyllus*, Pharm. Biol. 47 (10) (2009) 987–993.
[49] R. Matthes, I. Koban, C. Bender, K. Masur, E. Kindel, K.D. Weltmann, T. Kocher, A. Kramer, N.O. Hübner, Antimicrobial efficacy of an atmospheric pressure plasma jet against biofilms of *Pseudomonas aeruginosa* and *Staphylococcus epidermidis*, Plasma Process. Polym. 10 (2) (2013) 161–166.
[50] S.K. Pankaj, C. Bueno-Ferrer, N.N. Misra, L. O'Neill, A. Jiménez, P. Bourke, P.J. Cullen, Surface, thermal and antimicrobial release properties of plasma-treated zein films, J. Renew. Mater. 1 (8) (2014) 77–84.
[51] J. Ehlbeck, U. Schnabel, M. Polak, J. Winter, W. Tvon, R. Brandenburg, H. vondem, K.D. Weltman, Low temperature atmospheric pressure plasma sources for microbial decontamination, J. Phys. D: Appl. Phys. 44 (1) (2011). 18 (013002).
[52] K. Fricke, I. Koban, H. Tresp, L. Jablonowski, K. Schröder, A. Kramer, K.D. Weltmann, T. Woedtke, T. Kocher, Atmospheric pressure plasma: a high-performance tool for the

efficient removal of biofilms. PLoS ONE 7 (8) (2012) e42539. https://doi.org/10.1371/journal.pone.0042539.
[53] C.S. Kwok, T.A. Horbett, B.D. Ratner, Design of infection—resistant antibiotic-releasing polymers II. Controlled release of antibiotics through a plasma deposited thin film barrier, J. Control. Release 62 (1999) 301–311.
[54] S. Yoshida, K. Hagiwara, T. Hasebe, A. Hotta, Surface modification of polymers by plasma treatments for the enhancement of biocompatibility and controlled drug release, Surf. Coat. Technol. 233 (2013) 99–107.
[55] K. Tanaka, M. Kogoma, Y. Ogawa, Fluorinated polymer coatings on PLGA microcapsules for drug delivery system using atmospheric pressure glow plasma, Thin Solid Films 506–507 (2006) 159–162.
[56] I. Djordjevic, L.G. Britcher, S. Kumar, Morphological and surface compositional changes in poly(lactide-co-glycolide) tissue engineering scaffolds upon radio frequency glow discharge plasma treatment, Appl. Surf. Sci. 254 (7) (2008) 1929–1935.
[57] J.P. Chen, C.H. Su, Surface modification of electrospun PLLA nanofibers by plasma treatment and cationized gelatin immobilization for cartilage tissue engineering, Acta Biomater. 7 (1) (2011) 234–243.
[58] K.N. Pandiyaraj, R.R. Deshmukh, I. Ruzybayev, S. Ismat Shah, P.-G. Su, M. Halleluyah Jr., A.S. Halim, Influence of non-thermal plasma forming gases on improvement of surface properties of low density polyethylene (LDPE), Appl. Surf. Sci. 307 (2015) 109–119.
[59] A.Q.D. Faisal, M.O. Dawood, Design and construction of argon dc glow discharge plasma using Al target, J. KUFA Phys. 6 (2) (2014) 1–6.
[60] J. Laimer, H. Stori, Glow discharges observed in capacitive radio-frequency atmospheric-pressure plasma jets, Plasma Process Polym. 10 (2006) 573–586.
[61] B. Bora, H. Bhuyan, M. Favre, E. Wyndham, H. Chuaqui, Diagnostic of capacitively coupled low pressure radio frequency plasma: an approach through electrical discharge characteristic, Int. J. Appl. Phys. Math. 1 (2) (2011) 124–128.
[62] A. Tsuji, Y. Yasaka, S.Y. Kang, T. Morimoto, I. Sawada, A practical simulation scheme for slot-excited microwave plasma reactor equipped with dual shower plate, Thin Solid Films 516 (2008) 4368–4373.
[63] S. Youn, W. Choea, H.S. Uhm, Y.S. Hwang, J.J. Choi, Characteristics of an atmospheric microwave-induced plasma generated in ambient air by an argon discharge excited in an open-ended dielectric discharge tube, Phys. Plasmas 9 (9) (2002) 4045–4051.
[64] I.V. Adamovich, I. Choi, N. Jiang, J.H. Kim, S. Keshav, W.R. Lempert, E. Mintusov, M. Nishihara, M. Samimy, M. Uddi, Plasma assisted ignition and high-speed flow control: non-thermal and thermal effects, Plasma Sources Sci. Technol. 18 (2009) 1–13.
[65] Y.T. Birhane, S.C. Lin, T.Y. Huang, Variation of entrance length effect on EHD gas pump performance, Key Eng. Mater. 649 (2015) 1–8.
[66] P. Attri, B. Arora, E.H. Choi, Utility of plasma: a new road from physics to chemistry, 11 April 2013. http://pubs.rsc.org, doi: 10.1039/C3RA41277F.
[67] V. Nehra, A. Kumar, H.K. Dwivedi, Atmospheric non-thermal plasma sources, Int. J. Eng. (IJE) 2 (1) (2008) 53–68.
[68] V.I. Gibalov, G.J. Pietsch, The development of dielectric barrier discharges in gas gaps and on surfaces, J. Phys. D. Appl. Phys. 33 (20) (2000) 2618–2636.
[69] T. Nozaki, Y. Unno, Y. Miyazaki, K. Okazaki, in: Presented to 15th International Symposium on Plasma Chemistry, Orleans, France, 2001.
[70] X.T. Deng, M.G. Kong, Frequency range of stable dielectric-barrier discharges in atmospheric He and N/sub 2/, IEEE Trans. Plasma Sci. 32 (2004) 1709.

[71] U. Kogelschatz, B. Eliasson, W. Egli, Dielectric-barrier discharges. Principle and applications, J. Phys. IV 7 (1997) 47–66.
[72] H.E. Wagner, R. Brandenburg, K.V. Kozlov, A. Sonnenfeld, P. Michel, J.F. Behnke, The barrier discharge: basic properties and applications to surface treatment, Vacuum 71 (2003) 417.
[73] C. Tendero, C. Tixier, P. Tristant, J. Desmaison, P. Leprince, The atmospheric-pressure plasma jet: a review and comparison to other plasma sources, Spectrochim. Acta B 61 (1) (2006) 2–30.
[74] G. Borcia, N.M.D. Brown, Hydrophobic coatings on selected polymers in an atmospheric pressure dielectric barrier discharge, J. Phys. D. Appl. Phys. 40 (2007) 1927–1936.
[75] S.E. Alexandrov, M.L. Hitchman, Chemical vapor deposition enhanced by atmospheric pressure non-thermal non-equilibrium plasmas, Chem. Vap. Deposition 11 (2005) 457–468.
[76] P. Cools, E. Sainz-García, N. De Geyter, A. Nikiforov, M. Blajan, K. Shimizu, F. Alba-Elías, C. Leys, R. Morent, Influence of DBD inlet geometry on the homogeneity of plasma polymerized acrylic acid films: the use of a microplasma–electrode inlet configuration, Plasma Process. Polym. 12 (2015) 1153–1163.
[77] A. Liguori, A. Pollicino, A. Stancampiano, F. Tarterini, M.L. Focarete, V. Colombo, M. Gherardi, Deposition of plasma-polymerized polyacrylic acid coatings by a non-equilibrium atmospheric pressure nanopulsed plasma jet, Plasma Process. Polym. 13 (2016) 375–386.
[78] K.G. Kostov, R.Y. Honda, L.M.S. Alves, M.E. Kayama, Characteristics of dielectric barrier discharge reactor for material treatment, Braz. J. Phys. 39 (2009) 2322–2325.
[79] M.P. Cal, M. Schluep, Destruction of benzene with non-thermal plasma in dielectric barrier discharge reactors, Environ. Prog. 20 (3) (2014) 151–156.
[80] M. Tanino, W. Xilu, K. Takashima, S. Katsura, A. Mizuno, Sterilization using dielectric barrier discharge at atmospheric pressure, Int. J. Plasma Environ. Sci. Technol. 1 (1) (2007) 102–107.
[81] N. Shainsky, D. Dobrynin, U. Ercan, S.G. Joshi, H. Ji, A. Brooks, G. Fridman, Y. Cho, A. Fridman, G. Friedman, Plasma acid: water treated by dielectric barrier discharge, Plasma Process. Polym. 10 (2012) 1–6.
[82] I. Topala, M. Asandulesa, N. Dumitrascu, G. Popa, J. Durand, Application of dielectric barrier discharge for plasma polymerization processes, J. Optoelectr. Adv. Mater. 10 (8) (2008) 2028–2032.
[83] U. Konelschatz, B. Eliasson, W. Egli, Dielectric-barrier discharges, principle and applications, J. Phys. IV France 7 (1997) 47–66.
[84] C. Sarra-Bournet, S. Turgeon, D. Mantovani, G. Laroche, A study of atmospheric pressure plasma discharges for surface functionalization of PTFE used in biomedical applications, J. Phys. D. Appl. Phys. 39 (16) (2006) 3461–3469.
[85] K.N. Pandiyaraj, M.C. Ram Kumar, A. Arun Kumar, P.V.A. Padmanabhan, R.R. Deshmukh, A. Bendavid, P.-G. Su, A. Sachdev, P. Gopinath, Cold atmospheric pressure (CAP) plasma assisted tailoring of LDPE film surfaces for enhancement of adhesive and cytocompatible properties: influence of operating parameters, Vacuum 130 (2016) 34–47.
[86] R. Foest, T. Bindemann, R. Brandenburg, E. Kindel, H. Lange, M. Stieber, K.D. Weltmann, On the vacuum ultraviolet radiation of a miniaturized non-thermal atmospheric pressure plasma jet, Plasma Process. Polym. 4 (2007) S460–S464.
[87] W. Zhu, J.L. Lopez, A DC non-thermal atmospheric-pressure plasma micro jet, Plasma Sources Sci. Technol. 21 (2012) 034018.
[88] I. Onyshchenko, N. De Geyter, A.Y. Nikiforov, R. Morent, Atmospheric pressure plasma penetration inside flexible polymer tubes. Plasma Process Polym. (2015), https://doi.org/10.1002/ppap.201400190.

[89] J.H. Liu, X.Y. Liu, K. Hu, D.W. Liu, X.P. Lu, F. Iza, M.G. Kong, Plasma plume propagation characteristics of pulsed radio frequency plasma, Appl. Phys. Lett. 98 (15) (2011) 151502.

[90] H.M. Joh, S.J. Kim, T.H. Chung, S.H. Leem, Reactive oxygen species-related plasma effects on the apoptosis of human bladder cancer cells in atmospheric pressure pulsed plasma jets, Appl. Phys. Lett. 101 (5) (2012) 053703.

[91] D.B. Kim, J.K. Rhee, B. Gweon, S.Y. Moon, W. Choe, Comparative study of atmospheric low pressure and radio frequency microjet plasmas produced in a single electrode configuration, Appl. Phys. Lett. 91 (2007) 151502.

[92] G. Arnoult, R.P. Cardoso, T. Belmonte, G. Henrion, Flow transition in a small scale microwave plasma jet at atmospheric pressure, Appl. Phys. Lett. 19 (93) (2008) 191507.

[93] M. Laroussi, W. Hynes, T. Akan, X.P. Lu, C. Tendero, The plasma pencil: a source of hypersonic cold plasma bullets for biomedical applications, IEEE Trans. Plasma Sci. 36 (2008) 1298–1299.

[94] X.C. Li, L.F. Dong, N. Zhao, Z.Q. Yin, T.Z. Fang, L. Wang, A simple device of generating glow discharge plasma in atmospheric pressure argon, Appl. Phys. Lett. 91 (2007) 161507.

[95] J.L. Walsh, J.J. Shi, M.G. Kong, Contrasting characteristics of pulsed and sinusoidal cold atmospheric plasma jets, Appl. Phys. Lett. 88 (2006) 171501.

[96] A. Sarani, A.Y. Nikiforov, C. Leys, Atmospheric pressure plasma jet in Ar and Ar/H_2O mixtures: optical emission spectroscopy and temperature measurements, Phys. Plasmas 17 (2010) 063504–063508.

[97] N. De Geyter, A. Sarani, T. Jacobs, A.Y. Nikiforov, T. Desmet, P. Dubruel, Surface modification of poly-epsilon-caprolactone with an atmospheric pressure plasma jet, Plasma Chem. Plasma Process. 33 (2013) 165–175.

[98] S. Bornholdt, M. Wolter, H. Kersten, Characterization of an atmospheric pressure plasma jet for surface modification and thin film deposition, Eur. Phys. J. D 60 (2010) 653–660.

[99] M. Lestelius, B. Iedberg, P. Tengvall, In vitro plasma protein adsorption on ω functionalized alkanethiolate self-assembled monolayers, Langmuir 13 (22) (1997) 5900–5908.

[100] M. Lindblad, M. Lestelius, A. Johansson, P. Tengvall, P. Tomsen, Cell and soft tissue interactions with methyl and hydroxyl terminated alkane thiols on gold surfaces, Biomaterials 18 (15) (1997) 1059–1068.

[101] D. Sakthi Kumar, M. Fujioka, K. Asano, A. Shoji, A. Jayakrishnan, Y. Yoshida, Surface modification of poly(ethylene terephthalate) by plasma polymerization of poly(ethylene glycol), J. Mater. Sci. Mater. Med. 18 (2007) 1831–1835.

[102] Z. Fu, E.T. Kang, K.G. Neoh, P. Wang, K.L. Tan, Surface modification of stainless steel by grafting of poly(ethylene glycol) for reduction in protein adsorption, Biomaterials 22 (2001) 1541–1548.

[103] R.A. D'Sa, J. Raj, M.A.S. McMahon, D.A. McDowell, G.A. Burke, B.J. Meenan, Atmospheric pressure plasma induced grafting of poly(ethylene glycol) onto silicone elastomers for controlling biological response, J. Colloid Interface Sci. 375 (2012) 193–202.

[104] Y. Shiheng, R. Li, W. Yingjun, Argon plasma-induced graft polymerization of PEGMA on chitosan membrane surface for cell adhesion improvement, Plasma Sci. Technol. 15 (10) (2013) 1041–1046.

[105] S. Pinto, P. Alves, C.M. Matos, A.C. Santos, L.R. Rodrigues, J.A. Teixeira, M.H. Gil, Poly (dimethyl siloxane) surface modification by low pressure plasma to improve its characteristics towards biomedical applications, Colloids Surf. B 8 (1) (2010) 20–26.

[106] G.P. Lopez, B.D. Ratner, C.D. Tidwell, C.L. Haycox, R.J. Rapoza, T.A. Horbett, Glow discharge plasma deposition of tetraethylene glycol dimethyl ether for fouling-resistant biomaterial surfaces, J. Biomed. Mater. Res. 26 (1992) 415–439.

[107] F. Brétagnol, M. Lejeune, A. Papadopoulou-Bouraoui, M. Hasiwa, H. Rauscher, G. Ceccone, P. Colpo, F. Rossi, Fouling and non-fouling surfaces produced by plasma polymerization of ethylene oxide monomer, Acta Biomater. 2 (2006) 165–172.

[108] J. Wang, C.J. Pan, N. Huang, H. Sun, P. Yang, Y.X. Leng, J.Y. Chen, G.J. Wan, P.K. Chu, Surface characterization and blood compatibility of poly(ethylene terephthalate) modified by plasma surface grafting, Surf. Coat. Technol. 196 (2005) 307–311.

[109] S. Zanini, M. Muller, C. Riccardi, M. Orlandi, Polyethylene glycol grafting on polypropylene membranes for anti-fouling properties, Plasma Chem. Plasma Process. 27 (2007) 446–457.

[110] L. Lin, Y. Wang, X.-D. Huang, Z.-K. Xu, K. Yao, Modification of hydrophobic acrylic intraocular lens with poly(ethylene glycol) by atmospheric pressure glow discharge: a facile approach, Appl. Surf. Sci. 256 (2010) 7354–7364.

[111] C. Choi, I. Hwang, Y.L. Cho, S.Y. Han, D.H. Jo, D. Jung, D.W. Moon, E.J. Kim, C.S. Jeon, J.H. Kim, T.D. Chung, T.G. Lee, Fabrication and characterization of plasma-polymerized poly(ethylene glycol) film with superior biocompatibility, ACS Appl. Mater. Interfaces 5 (2013) 697–702.

[112] L. Zhang, Di Wu, Y. Chen, X. Wang, G. Zhao, H. Wan, C. Huang, Surface modification of poly methyl methacrylate intraocular lenses by plasma for improvement of antithrombogenicity and transmittance, Appl. Surf. Sci. 255 (2009) 6840–6845.

[113] S. Zanini, C. Riccardi, E. Grimoldi, C. Colombo, A.M. Villa, A. Natalello, P. Gatti-Lafranconi, M. Lotti, S.M. Doglia, Plasma-induced graft-polymerization of polyethylene glycol acrylate on polypropylene films: chemical characterization and evaluation of the protein adsorption, J. Colloid Interface Sci. 341 (2010) 53–58.

[114] Z. Yang, J. Wang, X. Li, Q. Tu, H. Sun, N. Huang, Interaction of platelets, fibrinogen and endothelial cells with plasma deposited PEO-like films, Appl. Surf. Sci. 258 (2012) 3378–3385.

[115] B.G. Keselowsky, D.M. Collard, A.J. Garcia, Surface chemistry modulates focal adhesion composition and signals through changes in integrin binding, Biomaterials 25 (2004) 5947–5954.

[116] L. Liu, S. Chen, C.M. Giachelli, B.D. Ratner, S. Jiang, Controlling osteopontin orientation on the surfaces to modulate endothelial cell adhesion, J. Biomed. Mater. Res. 74A (2005) 23–31.

[117] K. Navaneetha, M.C. Ram Kumar, A. Arun Kumar, P.V.A. Padmanabhan, R.R. Deshmukh, M. Bah, S. Ismat Shah, P.-G. Su, A.S. Halim, Tailoring the surface properties of polypropylene films through cold atmospheric pressure (CAPP) plasma assisted polymerization and immobilization of biomolecules for enhancement of anti-coagulation activity. Appl. Surf. Sci. (2016), https://doi.org/10.1016/j.apsusc.2016.02.137.

[118] S. Degoutin, M. Jimenez, M. Casetta, S. Bellayer, F. Chai, N. Blanchemain, C. Neut, I. Kacem, M. Traisnel, B. Martel, Anticoagulant and antimicrobial finishing of non-woven polypropylene textiles, Biomed. Mater. 7 (2012). 035001 (13 pp.).

[119] Y.J. Kim, I.K. Kang, M.W. Huh, S.C. Yoon, Surface characterization and in vitro blood compatibility of poly(ethylene terephthalate) immobilized with insulin and/or heparin using plasma glow discharge, Biomaterials 21 (2) (2000) 121–130.

[120] Q. Gao, Y. Chen, Y. Wei, X. Wang, Y. Luo, Heparin-grafted poly(tetrafluoroethylene-co-hexafluoropropylene) film with highly effective blood compatibility via an esterification reaction, Surf. Coat. Technol. 228 (2013) S126–S130.

[121] K.N. Sask, W.G. McClung, L.R. Berry, A.K. Chan, J.L. Brash, Immobilization of an antithrombin-heparin complex on gold: anticoagulant properties and platelet interactions, Acta Biomater. 7 (5) (2011) 2029–2034.
[122] Z. Yang, J. Wang, R. Luo, M.F. Maitz, F. Jing, H. Sun, N. Huang, The covalent immobilization of heparin to pulsed-plasma polymeric allylamine films on 316L stainless steel and the resulting effects on hemocompatibility, Biomaterials 31 (2010) 2072–2083.
[123] J. Jin, W. Jiang, Q. Shi, J. Zhao, J. Yin, P. Stagnaro, Fabrication of PP-g-PEGMA-g-heparin and its hemocompatibility: from protein adsorption to anticoagulant tendency, Appl. Surf. Sci. 258 (2012) 5841–5849.
[124] Y.H. An, R.J. Friedman, Laboratory methods for studies of bacterial adhesion, J. Microbiol. Methods 30 (2) (1997) 141–152.
[125] K. Hori, S. Matsumoto, Bacterial adhesion: from mechanism to control, Biochem. Eng. J. 48 (3) (2010) 424–434.
[126] D. Raafat, H.-G. Sahl, Chitosan and its antimicrobial potential—a critical literature survey, Microb. Biotechnol. 2 (2) (2009) 186–201.
[127] X. Wang, N. Shi, Y. Chen, C. Li, X. Du, W. Jin, P.R. Chang, Improvement in hemocompatibility of chitosan/soy protein composite membranes by heparinization, Biomed. Mater. Eng. 22 (1–3) (2012) 143–150.
[128] A.I. Cañas, J.P. Delgado, C. Gartner, Biocompatible scaffolds composed of chemically 6 crosslinked chitosan and gelatin for tissue engineering, J. Appl. Polym. Sci. 133 (2016) 43814.
[129] T.A. Ahmed, B.M. Aljaeid, Preparation, characterization, and potential application of chitosan, chitosan derivatives, and chitosan metal nanoparticles in pharmaceutical drug delivery, Drug Des. Dev. Ther. 10 (2016) 483–507.
[130] K. Ziani, I. Fernández-Pan, M. Royo, J.I. Maté, Antifungal activity of films and solutions based on chitosan against typical seed fungi, Food Hydrocoll. 23 (8) (2009) 2309–2314.
[131] Z. Ding, J. Chen, S. Gao, J. Chang, J. Zhang, E.T. Kang, Immobilization of chitosan onto poly-l-lactic acid film surface by plasma graft polymerization to control the morphology of fibroblast and liver cells, Biomaterials 25 (6) (2004) 1059–1067.
[132] Y-C. Tyan, J-D. Liao, S-P. Lin, Surface properties and in vitro analyses of immobilized chitosan onto polypropylene non-Woven fabric surface using antenna-coupling microwave plasma, J. Mater. Sci. Mater. Med. 14 (2003) 775–781.
[133] S. Meng, Z. Liu, L. Shen, Z. Guo, L.L. Chou, W. Zhong, Q. Du, J. Ge, The effect of a layer-by-layer chitosan–heparin coating on the endothelialization and coagulation properties of a coronary stent system, Biomaterials 30 (2009) 2276–2283.
[134] K.N. Pandiyaraj, A.M. Ferraria, A.M. Botelho do Rego, R.R. Deshmukh, P.-G. Su, M. Halleluyah Jr., A.S. Halim, Low-pressure plasma enhanced immobilization of chitosan on low-density polyethylene for bio-medical applications, Appl. Surf. Sci. 328 (2015) 1–12.
[135] F. Kara, E.A. Aksoy, Z. Yuksekdag, N. Hasirci, S. Aksoy, Synthesis and surface modification of polyurethanes with chitosan for antibacterial properties, Carbohydr. Polym. 112 (2014) 39–47.
[136] S.H. Chang, C.H. Chian, Plasma surface modification effects on biodegradability and protein adsorption properties of chitosan films, Appl. Surf. Sci. 282 (2013) 735–740.
[137] T.E.L. Douglas, S. Kumari, K. Dziadek, M. Dziadek, A. Abalymov, P. Cools, G. Brackman, T. Coenye, R. Morent, M.K. Mohan, A.G. Skirtach, Titanium surface functionalization with coatings of chitosan and polyphenol-rich plant extracts. Mater. Lett. (2017), https://doi.org/10.1016/j.matlet.2017.03.065.

[138] B. Gupta, C. Plummer, I. Bisson, P. Frey, J. Hilborn, Plasma-induced graft polymerization of acrylic acid onto poly(ethylene terephthalate) films: characterization and human smooth muscle cell growth on grafted films, Biomaterials 23 (2002) 863–871.
[139] L. Detomaso, R. Gristina, G.S. Senesi, R. d'Agostino, P. Favia, Stable plasma-deposited acrylic acid surfaces for cell culture applications, Biomaterials 26 (2005) 3831–3841.
[140] H. Muguruma, I. Karube, Plasma-polymerized films for biosensors, Trends Anal. Chem. 18 (1) (1999) 62–68.
[141] B.R. Pistillo, L. Detomaso, E. Sardella, P. Favia, R. d'Agostino, RF-plasma deposition and surface characterization of stable (COOH)-rich thin films from cyclic L-lactide, Plasma Process. Polym. 4 (2007) S817–S820.
[142] K.S. Siow, L. Britcher, S. Kumar, H.J. Griesser, Plasma methods for the generation of chemically reactive surfaces for biomolecule immobilization and cell colonization—a review, Plasma Process. Polym. 3 (2006) 392–418.
[143] C.D. Tidwell, S.I. Ertel, B.D. Ratner, B.J. Tarasevich, S. Atre, D.L. AllaraL, Endothelial cell growth and protein adsorption on terminally functionalized, selfassembled monolayers of alkanethiolates on gold, Langmuir 13 (1997) 3404–3413.
[144] Y. Ohya, H. Matsunami, T. Ouchi, Cell growth on porous sponges prepared from poly (depsipeptide-co-lactide) having various functional groups, J. Biomater. Sci. Polym. Ed. 215 (1) (2004) 111–123.
[145] A. Sodergard, Preparation of poly (e-caprolactone)-co-poly (acrylic acid) by radiation-induced grafting, J. Polym. Sci. A 36 (1998) 1805–1812.
[146] Z. Cheng, S.H. Teoh, Surface modification of ultra-thin poly (ε-caprolactone) films using acrylic acid and collagen, Biomaterials 25 (11) (2004) 1991–2001.
[147] V. Kumar, C. Jolivalt, J. Pulpytel, R. Jafari, F. Arefi-Khonsari, Development of silver nanoparticle loaded antibacterial polymer mesh using plasma polymerization process, J. Biomed. Mater. Res. A 101A (4) (2013) 1121–1132.
[148] M.S. Kang, B. Chun, S.S. Kim, Surface modification of polypropylene membrane by low-temperature plasma treatment, J. Appl. Polym. Sci. 81 (6) (2001) 1555–1556.
[149] M. Dhayal, S.I. Cho, Leukaemia cells interaction with plasma-polymerized acrylic acid coatings, Vacuum 80 (6) (2006) 636–642.
[150] O. Carton, D.B. Salem, S. Bhatt, J. Pulpytel, F. Arefi-Khonsari, Plasma polymerization of acrylic acid by atmospheric pressure nitrogen plasma jet for biomedical applications. Plasma Process. Polym. (2012), https://doi.org/10.1002/ppap.201200044.
[151] P. Cools, H. Declercq, N. De Geyter, R. Morent, A stability study of plasma polymerized acrylic acid films. Appl. Surf. Sci. (2017), https://doi.org/10.1016/j.apsusc.2017.04.015.
[152] V.N. Vasilets, G. Hermel, U. Konig, C. Werner, M. Miiller, F. Simon, K. Grundke, Y. Ikada, H.J. Jacobasch, Microwave CO_2 plasma-initiated vapour phase graft polymerization of acrylic acid onto polytetrafluoroethylene for immobilization of human thrombomodulin, Biomaterials 18 (1997) 1139–1145.
[153] J.C. Sánchez-López, C. Donnet, J. Fontaine, M. Belin, A. Grill, V. Patel, C. Jahnes, Diamond-like carbon prepared by high density plasma, Diamond Relat. Mater. 9 (2000) 638–642.
[154] C.S. Lee, K.R. Lee, K.Y. Eun, K.H. Yoon, J.H. Han, Structure and properties of Si incorporated tetrahedral amorphous carbon films prepared by hybrid filtered vacuum arc process, Diamond Relat. Mater. 11 (2002) 198–203.
[155] Y.S. Zou, W. Wang, G.H. Song, H. Du, J. Gong, R.F. Huang, L.S. Wen, Influence of the gas atmosphere on the microstructure and mechanical properties of diamond-like carbon films by arc ion plating, Mater. Lett. 58 (2004) 3271–3275.

[156] G. Thorwarth, C. Hammerl, M. Kuhn, W. Assmann, B. Schey, B. Stritzker, Investigation of DLC synthesized by plasma immersion ion implantation and deposition, Surf. Coat. Technol. 193 (2005) 206–212.
[157] N.A. Sánchez, C. Rincón, G. Zambrano, H. Galindo, P. Prieto, Characterization of diamond-like carbon (DLC) thin films prepared by r.f. magnetron sputtering, Thin Solid Films 373 (2000) 247–250.
[158] L. Swiatek, A. Olejnik, J. Grabarczyk, A. Jedrzejczak, A. Sobczyk-Guzenda, M. Kaminska, W. Jakubowski, W. Szymanski, D. Bociaga, Multi-doped diamond like-carbon coatings (DLC-Si/Ag) for biomedical applications fabricated using the modified chemical vapour deposition method. Diamond Relat. Mater. (2016), https://doi.org/10.1016/j.diamond.2016.03.005.
[159] A. Bendavid, P.J. Martin, C. Comte, E.W. Preston, A.J. Haq, F.S. Magdon Ismail, R.K. Singh, The mechanical and biocompatibility properties of DLC-Si films prepared by pulsed DC plasma activated chemical vapor deposition, Diamond Relat. Mater. 16 (2007) 1616–1622.
[160] H.J. Steffen, J. Schmidt, A. Gonzalez-Elipe, Biocompatible surfaces by immobilization of heparin on diamond-like carbon films deposited on various substrates, Surf. Interface Anal. 29 (2000) 386–391.
[161] K.N. Pandiyaraj, V. Selvarajan, J. Heeg, F. Junge, A. Lampka, T. Barfels, M. Wienecke, Y.H. Rhee, H.W. Kim, Influence of bias voltage on diamond like carbon (DLC) film deposited on polyethylene terephthalate (PET) film surfaces using PECVD and its blood compatibility, Diamond Relat. Mater. 19 (2010) 1085–1092.
[162] M.I. Jones, I.R. McColl, D.M. Grant, K.G. Parker, T.L. Parker, Hemocompatibility of DLC and TiC–TiN interlayers on titanium, Diamond Relat. Mater. 8 (1999) 457–462.
[163] S.C.H. Kwok, J. Wang, P.K. Chu, Surface energy, wettability, and blood compatibility phosphorus doped diamond-like carbon films, Diamond Relat. Mater. 14 (2005) 78–85.
[164] P.D. Maguire, J.A. McLaughlin, T.I.T. Okpalugo, P. Lemoine, P. Papakonstantinou, E.T. McAdams, M. Needham, A.A. Ogwu, M. Ball, G.A. Abbas, Mechanical stability, corrosion performance and bioresponse of amorphous diamond-like carbon for medical stents and guidewires, Diamond Relat. Mater. 14 (2005) 1277–1288.
[165] T.I.T. Okpalugo, A.A. Ogwu, P.D. Maguire, J.A.D. McLaughlin, Platelet adhesion on silicon modified hydrogenated amorphous carbon films, Biomaterials 25 (2004) 239–245.
[166] F. Awaja, P. Cools, B. Lohberger, A.Y. Nikiforov, G. Speranza, R. Morent, Functionalized, biocompatible, and impermeable nanoscale coatings for PEEK, doi: 10.1016/j.msec.2017.03.153.
[167] K.N. Pandiyaraj, R.R. Deshmukh, R. Mahendiran, P.-G. Su, E. Yassitepe, I. Shah, S. Perni, P. Prokopovich, M.N. Nadagouda, Influence of operating parameters on surface properties of RF glow discharge oxygen plasma treated TiO_2/PET film for biomedical application, Mater. Sci. Eng. C 36 (2014) 309–319.
[168] W. Su, S. Wang, X. Wang, X. Fu, J. Weng, Plasma pre-treatment and TiO_2 coating of PMMA for the improvement of antibacterial properties, Surf. Coat. Technol. 205 (2010) 465–469.
[169] C. Boudot, M. Kühn, M. Kühn-Kauffeldt, J. Schein, Vacuum arc plasma deposition of thin titanium dioxide films on silicone elastomer as a functional coating for medical applications. Mater. Sci. Eng. C (2016), https://doi.org/10.1016/j.msec.2016.12.045.
[170] H. Szymanowski, A. Sobczyk, M. Gazicki-Lipman, W. Jakubowski, L. Klimek, Plasma enhanced CVD deposition of titanium oxide for biomedical applications, Surf. Coat. Technol. 200 (2005) 1036–1040.

[171] K.N. Pandiyaraj, A. Arun Kumar, M.C. Ramkumar, A. Sachdev, P. Gopinath, P. Cools, N. De Geyter, R. Morent, R.R. Deshmukh, P. Hegde, C. Han, M.N. Nadagouda, Influence of non-thermal TiCl4/Ar + O2 plasma-assisted TiOx based coatings on the surface of polypropylene (PP) films for the tailoring of surface properties and cytocompatibility. Mater. Sci. Eng. C (2016), https://doi.org/10.1016/j.msec.2016.02.042.

[172] A. Liguori, E. Traldi, E. Toccaceli, R. Laurita, A. Pollicino, M.L. Focarete, V. Colombo, M. Gherardi, Co-deposition of plasma-polymerized polyacrylic acid and silver nanoparticles for the production of nanocomposite coatings using a non-equilibrium atmospheric pressure plasma jet, Plasma Process. Polym. 13 (2016) 623–632.

[173] X. Deng, A.Y. Nikiforov, T. Coenye, P. Cools, G. Aziz, R. Morent, N. De Geyter, C. Leys, Antimicrobial nano-silver nonwoven polyethylene terephthalate fabric via an atmospheric pressure plasma deposition process. Sci. Rep. 5 (2015) 10138, https://doi.org/10.1038/srep10138.

[174] K.N. Pandiayaraj, V. Selvarajan, Y.H. Rhee, H.W. Kim, M. Pavesec, Effect of dc glow discharge plasma treatment on PET/ TiO_2 thin film surfaces for enhancement of bioactivity, Colloids Surf. B 79 (2010) 53–60.

[175] T. Hayakawa, M. Yoshinari, K. Nemoto, Characterization and protein-adsorption behavior of deposited organic thin film onto titanium by plasma polymerization with hexamethyldisiloxane, Biomaterials 25 (2004) 119–127.

[176] H. Chen, L. Wang, Y. Zhang, D. Li, W.G. McClung, M.A. Brook, H. Sheardown, J.L. Brash, Fibrinolytic poly(dimethyl siloxane)surfaces, Macromol. Biosci. 8 (2008) 863–870.

[177] C. Saulou, B. Despax, P. Raynaud, S. Zanna, P. Marcus, M. Mercier-Bonin, Plasma deposition of organosilicon polymer thin films with embedded nanosilver for prevention of microbial adhesion, Appl. Surf. Sci. 256S (2009) S35–S39.

[178] C.R. Hsiao, C.W. Lin, C.M. Chou, C.J. Chung, J.L. He, Surface modification of blood-contacting biomaterials by plasma-polymerized superhydrophobic films using hexamethyldisiloxane and tetrafluoromethane as precursors, Appl. Surf. Sci. 346 (2015) 50–56.

[179] C.J. Tang, G.X. Wang, Y. Shen, L.J. Wan, L. Xiao, Q. Zhang, Q.S. Yu, L.S. Liu, G.B. Wen, A study on surface endothelialization of plasma coated intravascular stents, Surf. Coat. Technol. 204 (2010) 1487–1492.

[180] S.N. Jampala, M. Sarmadi, E.B. Somers, A.C.L. Wong, F.S. Denes, Plasma-enhanced synthesis of bactericidal quaternary ammonium thin layers on stainless steel and cellulose surfaces, Langmuir 24 (2008) 8583–8591.

Coating stability for stents

10

V. Montaño-Machado, E.C. Michel, D. Mantovani
Laval University, Quebec, QC, Canada

The stability of a stent coating is critical and needs to be satisfied during processing, shelf life, and the use of the implant. It is, thus, clear that the term "stability" is inclusive of different points of view and several properties and as many characterization modes. From a clinical point of view, a stable stent implies that no unwanted biological responses occur; as for the shelf life, a stable stent means that no deteriorations occur in the coatings, and the stability in processing concerns the chemical structure of a polymer during the coating procedure.

This chapter will focus on the stability from a material point of view, as a key prerequisite for surface-modified materials applied on the implanted device to ensure its security in vivo. In terms of the characterization of the stability, it is important to define first which property of the stent coating system is investigated. Commonly, the most required properties are the chemical composition, the chemical structure, the morphology, the surface texture, the degradation which includes corrosion, dissolution and cohesion, and the mechanical properties which include the adhesion to the substrate. Stability tests generally target the assessment and the evolution of these properties for the whole coating system (which include the substrate and the coating, with all its components, and the interfacial layers between them). Hence, the more complex the coating is, the more the stability of all the coating system is investigated. For example, for drug-eluting stents (DESs), not only the stability of the polymer carrier is important, but also the drug's one.

Stability tests are designed to recreate the chemical, physical, and mechanical environment in which the device will be implanted. For most common tests, samples are immersed in a solution during a certain amount of time at 37°C (body temperature). Depending on which property is investigated, the parameters of the tests will vary such as the type of solution or the presence of mechanical stimuli (agitation, flow, deployment, etc.) to get closer to the physiological environment without hindering the characterization. Not only the coating surface is investigated in a stability test, but also the comprehension of the interaction of the coating and the substrate, and more specifically, the studies of the interface will provide data about the long-term stability of the coating. These investigations will be detailed in the following sections.

10.1 Static tests

A common test found in the literature for biomedical devices and specifically for cardiovascular stents is an immersion of the samples in specific solutions of interest. In these tests, the changes occurring on the surface are evaluated after a specific period of

Functionalized Cardiovascular Stents. https://doi.org/10.1016/B978-0-08-100496-8.00010-X

time during which the sample is in contact with the selected solution. In this context, we will call these tests "static" when no shaking, shear stress, or other dynamic stimuli are applied.

Several parameters are considered when performing static tests, the most common being: The solution to be used as test fluid, the period of time to be investigated, and the temperature. Commonly, for biological application, the temperature during the tests is the body temperature, 37°C. As for the duration of the test and the type of fluid used, they vary depending on the final purpose of the device (see Table 10.1). One of the big challenges is to select a fluid that will provide relevant information in terms of stability of the coating. In general, a compromise must be found between simulating the biological fluids and allowing the surface characterization after the stability test. For instance, a very complex fluid containing biomolecules can interestingly simulate blood; however, this kind of fluid can induce the deposition of its components on the surface, which might be a hindrance for further characterization of the samples.

Several research groups investigated the interaction of samples with distilled water as the simplest fluid. In this way, it is possible to obtain a surface clean of any

Table 10.1 Parameters that need to be carefully selected when studying coatings stability, as a function of the specific site where they are expected to be implanted

Parameter	Values	Justification
Temperature	37, 50, 4, −25°C	A temperature of 37°C is used in order to simulate the biological system Other temperatures than 37°C can be used to simulate extreme conditions and hence extrapolate pertinent results on stability
Solution	Water Buffer (PBS, TRIS, Hanks) Culture media	Simplicity of analysis with no risk of contamination of the samples with deposits of salts or organic compounds To mimic pH and salts present in blood To mimic the presence of proteins and other blood components
Time	Some hours up to 6 months	It depends mainly on the purpose of the coating
Sample shape/ geometry	Flat surfaces Stent with complex geometries	Simplicity of the analysis and characterization. Model surfaces easy to characterize and highly reproducible High understanding of the expected results in vivo in a complex geometry
Dynamic/stimuli	None, static Shaking Shear stress	Simplest, it evaluated the effect of the solution on the stability of the coating Simplest dynamic, to mimic a minimal dynamism of the body Dynamic closest to the body, especially for arteries

deposition from the fluid, and hence, very appropriate for characterization. This is the case of Touzin et al. [1] who investigated the contact of fluorocarbon coatings on stainless steel with water during 1, 2, 3, and 4 weeks at 37°C. Samples were further characterized by atomic force microscopy (AFM), water contact angle (WCA), ellipsometry, Fourier-transform infrared spectroscopy (FTIR), and X-ray photoelectron spectroscopy (XPS). Yang et al. also studied the stability of coatings for cardiovascular application after immersion in water [2]. In their work, they studied the immobilization of heparin on allylamine films on stainless steel up to 30 days, under shaking, at 37°C. The characterization was performed by scanning electron microscopy (SEM) and toluidine blue method was used to measure the concentration of heparin. The great advantage of testing with simple deionized water as a fluid is the convenience of the further characterization of the samples. However, the simplicity of the fluid avoids a proper similarity to complex fluids such as blood.

A second fluid commonly used for static tests is a buffer solution such as: phosphate-buffered saline (PBS) solution, tris(hydroxymethyl)aminomethane solution (TRIS), and Hank's buffered salt solution (HBSS); PBS being the most common buffer found in the literature. Doro et al. studied the stability of hybrid coatings on stainless steel after immersion in PBS at 37°C up to 21 days [3]. The mechanical adhesion and leaching of specific elements were studied. Levy et al. studied the adhesion of polymeric films on metallic surfaces for DES applications [4]. Samples were immersed in PBS-containing SDS (sodium dodecyl sulfate) at 37 and 60°C. Characterizations were performed by means of cyclic voltammetry (CV), alternating current voltammetry (ACV), electrochemical impedance spectroscopy (EIS), light microscopy, SEM, XPS, peeling, and high pressure liquid chromatography (HPLC). As a third example, Kaufmann et al. used TRIS-buffered saline solution to study the stability of methyl and carboxylic acid-terminated phosphonic acid self-assembled monolayers (SAMs) at 37°C for up to 28 days [5].

The main advantage of using buffer solutions over distilled water is the presence of salts found in the biological system as well as the preservation of the physiological pH leading to a better simulation of the living environment. Nevertheless, the deposition of salts or other components (such as glucose in the case of Hank's solution) on the surface represents a hindrance to further characterization. Besides, despite the controlled pH and the presence of salts simulating the biological fluid, the solution remains simple when compared to complex biological fluids such as blood.

Researchers have worked with more complex fluids to simulate the biological environment as are culture media. In these solutions, different biomolecules as carbohydrates and proteins are added, which leads to more similarities with blood. For example, Micksch et al. investigated eight peptide sequences regarding their adsorption on zirconium oxide (ZrO_2). Stability tests after immersion of the samples in PBS (up to 30 days) and with culture media with fetal calf serum (for 15 min) were performed [6]. Probably, the greatest drawback of studying samples after the interaction with culture media is the surface characterization after tests, mainly if the coatings are very thin and/or contain biomolecules. Indeed, in these cases the coating may be completely covered by the deposition of macromolecules from the culture media and impede the topographical analysis of the coating after the test, for example. Moreover, if the coating contains also biomolecules, it could be difficult to distinguish the original biomolecules of the coating from the new deposition of biomolecules of the culture media.

As it can be seen, every fluid to be used to evaluate the stability of coatings may have its advantages and drawbacks regarding a better simulation of biological fluids or a better coating characterization. It is clear that a compromise should be found between the information the authors are willing to obtain and the utility of the test in terms of the final application. Moreover, the techniques involved during the characterization of the samples can vary significantly, which complicate the comparison among the investigations of different research groups.

Regarding the period of time of the tests, it can vary from several hours to several months. This variation can be both related to the nature of the coating as well as to its final purpose. However, it can also depend on the nature of the stability test itself. For example, according to the literature, 7 days is considered a reasonable amount of time to complete endothelialization process after the implantation of a stent. Hence, when bioactive coatings are developed to induce the endothelialization process as a final purpose, the stability tests can last 7 days [7]. For different applications, the coating might require to stay stable for several months. Indeed, in general the periods of time selected are not justified in the literature.

In the case of the temperature, as it was observed in the examples given in this section, most of the time it is 37°C, the body temperature. However, some tests will be performed at room temperature in order to facilitate the experiment. In some other cases, higher temperatures can be also investigated in order to simulate extreme conditions and/or to extrapolate the results. Table 10.1 presents the variation of several parameters of this kind of tests.

10.2 Dynamic tests

In the last decade, several research groups worldwide have been focused on simulating the biological systems when developing tests to evaluate the stability of coatings for medical devices. In the context of cardiovascular stents, simulating the dynamism of the arteries is one of the desired aspects to be accomplished. There are several parameters to consider as the fluid, the period of time, the viscosity of the fluid, the pressure, the shear stress, the flow, the shear rate, etc. A standard method has not been defined yet and this is probably the reason why the tests can differ significantly among them.

The simplest dynamic test is similar to the static tests previously mentioned, with the addition of shaking of the samples during the contact with fluids. Sometimes, the specifications of the shaking process are not provided in the literature; however, most of the time it is rotational shaking. For example, Chen et al. deposited by a one-step dip-coating method a copolymer of gallic acid and hexamethylenediamine that was further immersed during 30 days in PBS with dynamic conditions that are not further explained. Samples were mainly characterized by SEM [8]. In the work of Chen et al. [9], a chemically hydroxylated titanium surface was aminosilanized. First, PEG600 or PEG4000 was covalently bonded, then anti-CD34 antibody was grafted. Samples were immersed in PBS in an orbital shaker incubator at 80 rpm and 37°C for 5, 10, and 15 days. Toluidine Blue O assay and immunofluorescent staining were used to characterize the samples. Joung et al. studied hydrogels composed of metal-adhesive and enzyme-reactive amphiphilic block copolymers (tetronic-tyramine/dopamine) and enzyme-reactive heparin derivatives (heparin-tyramine or heparin-polyethylene glycoltyramine). Samples were immersed in PBS and incubated with shaking at 100 rpm and 37°C. Toluidine blue assay was used to measure the amount of heparin remaining as coating [10].

Compared to static tests, the presence of shaking when a coating is immersed in a fluid adds a constraint that helps to simulate better the dynamic of the biological environment. More complex dynamic tests found in the literature include shear stress, with the aim of simulating the one present in the arteries. In this context, Oh et al. designed a microfluidic perfusion chamber in order to test the stability of hydrogel scaffolds for cardiovascular stent applications. The solution used was PBS with 2% of fetal bovine serum, at 37°C up to 7 days [11]. Some groups have developed complexes systems with different parameters as shear stress, shear rate, flow, different solutions, and temperature. Andukuri et al. studied the deposition of peptide amphiphiles using a rotational coating technique on stainless steel substrate. Samples were expanded to 14 atm in deionized water at 37°C, PBS was perfused at 10 dynes cm^{-2} for 3 days, in order to simulate the expansion the stent undergoes when implanted. Characterizations were performed by SEM, AFM, and FTIR [12]. Touzin et al. developed a “test bench” in order to test the stability under shear stress of biodegradable metals. Even if the application of these tests was to evaluate the degradation of metals and not directly for coatings, it can be perfectly applied to test the stability of different coatings. In their work, they used PBS as solution, 37°C, and tested different shear stresses, simulating low and high pressure in the arteries. Levesque et al. also developed dynamic tests by the use of a test bench [13]. In their work, they went further to simulate the viscosity of blood, by the addition of bermocoll to the PBS buffer solution. However, even if it could be interesting to simulate the viscosity of blood while performing in vitro tests, the subsequent characterization of the surface is not possible by common techniques as XPS, FTIR, AFM, etc., mainly because of the deposition of bermocoll that avoids the detection of the coating. Indeed, PBS without bermocoll was later used in the same test by Montaño-Machado et al. in order to compare the stability of fibronectin adsorption or grafting on fluorocarbon films that covered stainless steel substrates [14]. In this way, the coating stability was to be analyzed by common surface characterization techniques. The test was performed at 37°C during 7 days at a shear stress of 4 Pa. Characterization was performed by AFM, XPS, WCA, immunostaining, and ToF-SIMS analysis. These dynamic tests show great advantages when compared to the static ones since they have a remarkable high similitude to the biological environment.

As it has been observed, both static and dynamic tests could be performed to obtain the same kind of response; nevertheless, a remarkable variation on the different parameters is still present among them. Moreover, as in the case of the static tests, different approaches are used to study the stability, and hence, comparing coatings prepared by different research groups is still not trivial. Table 10.1 shows a brief summary of the different parameters considered when performing static and dynamic tests.

10.3 Adhesion

Another common—and often simple—test performed to assess the stability of a coating deals with its adhesion to the substrate. Indeed, the deformation that the stent undergoes during its deployment might cause failures such as cracks or delamination, mainly in poor adhered coatings. These failures can lead to clinical complications by exposing the substrate. In this context, some standard and generic methods can be applied; however, some research groups have developed more complex tests with the aim of obtaining more specific information in terms of the final application for cardiovascular stents.

Several groups apply different ASTM tests to evaluate the adhesion of coatings for stents. Indolfi et al. applied a standard method of Single-Lap-Joint (S-L-J) and Pull Off (P-O) (ASTM: D 1002-05) using microtest specimens in order to test the stability of a spray-coating of poly(2-hydroxy-ethyl-methacrylate) (pHEMA) on stainless steel [15]. In the works of Doro et al., a steel surface was coated with ormosils prepared from tetraethylsiloxane and 3-glycidoxypropyltrimethoxysilane or polydimethylsiloxane and the mechanical adhesion of the coatings was investigated by a peel strength adhesion test according to ASTM D 100042 [3]. In the works of Levy et al., adhesion tests were performed by applying and removing adhesion tape according to the ASTM D3359-02 standard test and the peeled surface calculated [4].

Other groups have developed expansion-crimping tests in order to simulate the deployment of stents [15]. For example, Kim et al. prepared a coating of diamond-like carbon (DLC) on nitinol vascular stents. They performed a cyclic of contraction and extension applied to the samples using a stent-crimping system [16,17]. In that work, the cracking or delamination of the DLC coating occurred predominantly near the hinge, connecting the V-shape segments of the stent, where the maximum strain is induced. In this last work, the authors were interested in the relevance of modifying the interface between the substrate and the coating. The group of Mantovani et al. has developed a small punch test in order to simulate the 25% of deformation of the metal, which was calculated to be the maximal deformation the stent undergoes during its deployment [1,18]. In their works, flat samples were used as models. Samples were immersed into different fluids, as water and PBS, after expansion in order to further analyze the stability of the coatings. Samples are characterized by XPS, SEM, and AFM and different coatings have been tested until 7 days [19]. There are also several groups studying the adhesion of coatings directly on stents instead of using model samples. Zhang et al. studied the stability of poly(lactic-co-glycolic acid) (PLGA) coatings loaded with dexamethasone. Coatings were created directly on stents and a balloon expansion was performed. They characterized the samples by FTIR and SEM. The drug release behavior in vitro was also investigated, accordingly to the final application searched [20]. Gollwitzer et al. coated implants of stainless steel with poly(D,L-lactic acid) (PDLLA). They studied the mechanical stability of the coatings during dilatation of coronary artery stents. The total coating mass was determined with an electronic micro-balance [21]. A drawback of the characterization by mass is the lack of information of the homogeneity and possible cracks of the coating. This is the reason why it is always better to couple this kind of study to surface characterization. In this case, the ductility of the coating was examined by light microscopy and SEM.

Finally, regarding the adhesion tests, there are several research groups developing computational models in order to simulate the expansion of a stent and study delamination. This helps to determine the optimum treatment to coat (mainly in terms of thickness) [22,23]. In the works of Nayebpashaee et al., failure mode of the representative volume was numerically simulated as thermal stress localization during thermal cycle [24]. Comparisons of computed results with experiments verified that, the computational method can successfully predict residual stress and crack initiation mode of the studied thermal barrier coatings.

In terms of the evaluation of the coating adhesion, a large variety of tests have been observed since ASTM standards until complex procedures developed by different research groups. Unfortunately, this great variety complicates the comparison of the stability of coatings among different laboratories, and hence, this delays the further syntheses of highly stable coatings in terms of their adhesion.

10.4 DES and biodegradable polymers

In the specific case of DESs, the accomplishment of their stability could be very different due to the purpose of these kinds of stents. Indeed, more than stability tests, the rate of delivery of the drugs is studied, which normally implies the evaluation of the degradation of the loaded coatings.

For example, in the work of Bedair et al., cobalt-chromium (Co-Cr) alloy surfaces were modified with hydrophilic poly(2-hydroxyethyl methacrylate) (PHEMA) or hydrophobic poly(2-hydroxyethyl methacrylate)-grafted-poly(caprolactone) (PHEMA-g-PCL) brushes. The modified Co-Cr samples were then ultrasonically spray-coated with sirolimus (SRL)-containing PDLLA and drug free PDLLA. Both types of samples were immersed in 2 mL of PBS solution under the following conditions: 100 rpm, pH 7.4, and 37°C up to 8 weeks [25]. The degradation morphology of the drug-free PDLLA was examined using a field emission-SEM and SRL-loaded samples were examined by drug release tests.

In the work of Strickler et al., paclitaxel-eluting coronary stents, which are coated with poly(styrene-b-isobutylene-b-styrene) were expanded in a water bath at 37°C. Their tests were developed to simulate the condition in a blood vessel during a 2-year implantation cycle. Characterization was performed by SEM. Different compositions of styrene in terms of the mechanical properties, chemical stability, and vascular compatibility were studied. A fatigue test was also performed. Distention pulses were created to simulate blood vessel conditions. The fatigue tester cycles the stent through a specified distention in aqueous conditions at 37°C. Ten million cycles at 60 Hz were used to simulate 3-month implantation in a beating heart. [26]

Stoebner et al. studied the effects of different thermal treatments on the release of paclitaxel. In this work, Co-Cr alloy samples were coated with SAMs, then with paclitaxel before being thermally treated. Specimens were characterized with differential scanning calorimetry (DSC), SEM, AFM, and a drug elution test to evaluate the stability of the drug in terms of chemical structure, morphology, and distribution. Specimens were immersed in a mixture of PBS and Tween 20, at 37°C up to 56 days and drug release profiles were obtained. Paclitaxel showed a better stability after treatments at 100 and 140°C and a lesser percentage of drug release when compared to room temperature and treatment at 70°C [27].

A particular case is presented by Oh et al. where coated surfaces with polymeric nanofibers by electrospinning were further loaded with nanoparticles containing β-estradiol. Indeed, in this case the authors were interested firstly in studying the stability of nanoparticles. To accomplish this, the degradation profiles of nanoparticles in the organic solvent mixture were examined through the ultracentrifuge technique performed at 14,000 rpm. Then, it was spectrophotometrically analyzed using UV-vis at 270 nm. The drug (β-estradiol) concentration in supernatant was used for the assessment of stability of nanoparticles in the organic solvent [28]. Furthermore, the stability of the coating containing the nanoparticles and β-estradiol was evaluated after contact with PBS at 37°C with shaking of 120 rpm. The characterization was performed by UV-vis.

Carlyle et al. developed several computational models in order to determine whether a drug delivery stent product could reduce the peaks and valleys in drug concentration. Indeed, the objective was to distribute the drug evenly in the initial area rather than concentrating it on the struts. The absorbable-coating sirolimus-eluting stent demonstrated enhanced drug stability (crystallinity of the sirolimus) under simulated use conditions and consistent drug delivery balanced with coating erosion in a porcine coronary implant model. The initial drug burst was eliminated and drug release was sustained after implantation [29].

10.5 Stability tests involving endothelial cells

As it has been previously mentioned, one of the aims of a coating for cardiovascular stents is the stimulation of the endothelialization process. In this context, some research groups have been interested in studying the growth of endothelial cells on the surface of the coating and the further stability of this cell layer.

For example, McCracken et al. studied the growth of endothelial cells on a surface with a patterning of RGD-nanoparticle-nanowell arrays. The samples were further exposed to shear stress. The authors were interested in studying the effect of the flow in parallel or perpendicular as well as its effects on the cell adhesion and spreading [30].

Kumar et al. studied the adhesion of endothelial cells on surfaces treated with fibrin glue, growth factor, and gelatin on polytetrafluoroethylene (PTFE) and polyethyleneterephthalate (Dacron) substrates. Shear stress parallel to the samples was imposed on the monolayer of endothelial cells in order to test its stability [31].

These investigations are relevant for two different reasons: (1) to predict the interaction of endothelial cells with coatings and (2) to evaluate the stability of the potential endothelial cell layer that would be created once the stent is implanted and in contact with the endothelium. However, it is important to mention that once the biomaterial is in contact with the biological system, quickly a layer of proteins will be adsorbed on its surface, and hence the endothelium will not interact with exactly the same surface, but with this new layer formed by the interaction of proteins with the biomaterial surface.

10.6 Conclusions and perspectives

From a review of the literature, it is evident that prior to perform any stability evaluation, the property to be investigated has to be clearly determined, which will accommodate the adequate assay. With all the possibilities of materials for stent coatings and properties to consider present in the literature, few standards exist to test stability and many procedures exist and still are created.

Nevertheless, as this chapter dealt with stability evaluations to predict the coating material behavior once the device is implanted, usual experiments involve immersion in aqueous medium at body temperature. It has been observed that immersion testing may evolve from static to dynamic tests with shaking or shear stress stimulation, to get closer to the body environment by mimicking the mechanical stress. Furthermore, the medium used may evolve from simple water to buffer solutions containing proteins or cells. Particular attention has to be drawn here, as a compromise has to be found between the complexity of the solution and the subsequent coating characterization, especially for coatings containing biomolecules. For chemical characterization, water or PBS solution would be preferred compared to solutions containing biomolecules.

The principal properties of the coatings tested were the chemical composition and structure stability as well as the coating morphology in general. Specifically, the coating adhesion to the substrate is usually tested via simple peel tests or dynamic test involving stent expansion. This last one is often followed by coating morphology characterization by topography, to observe eventual coating degradation such as delamination. Degradation of the coating could be wanted as for some DES; thus, the stability of the coating is also evaluated toward the degradation profile of the carrier material and the drug release profile.

It has been observed that for a complex coating system, each part of this system has to be studied towards stability properties, which may imply different experiments for a same property. As for example, in DES stents stability of the drug and the carrier could be tested separately and together. Some very complex coating system has been tested under dynamic conditions with shear stress and flow, containing three parts: the substrate, the coating, and cell layers. These last ones mostly focus on cell adhesion and spreading.

Finally, the plurality of the testing procedure complicates the comparison between different works, but they could be regrouped under the final purpose of the coating. Studying the stability of the coatings remains substantial to ensure the holding of the device and its success after implantation.

References

[1] M. Touzin, et al., Study on the stability of plasma-polymerized fluorocarbon ultra-thin coatings on stainless steel in water, Surf. Coat. Technol. 202 (2008) 4884–4891.

[2] Z. Yang, et al., The covalent immobilization of heparin to pulsed-plasma polymeric allylamine films on 316L stainless steel and the resulting effects on hemocompatibility, Biomaterials 31 (2010) 2072–2083.

[3] F.G. Doro, et al., Deposition of organic-inorganic hybrid coatings over 316L surgical stainless steel and evaluation on vascular cells, Can. J. Chem. 92 (2014) 987–995.
[4] Y. Levy, et al., Drug-eluting stent with improved durability and controllability properties, obtained via electrocoated adhesive promotion layer, J. Biomed. Mater. Res. B Appl. Biomater. 91 (2009) 819–830.
[5] C. Kaufmann, G. Mani, D. Marton, D. Johnson, C.M. Agrawal, Long-term stability of self-assembled monolayers on electropolished L605 cobalt chromium alloy for stent applications, J. Biomed. Mater. Res. B Appl. Biomater. 98 (B) (2011) 280–289.
[6] T. Micksch, N. Liebelt, D. Scharnweber, B. Schwenzer, Investigation of the peptide adsorption on ZrO_2, TiZr, and TiO_2 surfaces as a method for surface modification, ACS Appl. Mater. Interfaces 6 (2014) 7408–7416.
[7] N. Kipshidze, et al., Role of the endothelium in modulating neointimal formation: vasculoprotective approaches to attenuate restenosis after percutaneous coronary interventions, J. Am. Coll. Cardiol. 44 (2004) 733–739.
[8] S. Chen, et al., A simple one-step modification of various materials for introducing effective multi-functional groups, Colloids Surf. B Biointerfaces 113 (2014) 125–133.
[9] J. Chen, et al., Biofunctionalization of titanium with PEG and anti-CD34 for hemocompatibility and stimulated endothelialization, J. Colloid Interface Sci. 368 (2012) 636–647.
[10] Y.K. Joung, S.S. You, K.M. Park, D.H. Go, K.D. Park, In situ forming, metal-adhesive heparin hydrogel surfaces for blood-compatible coating, Colloids Surf. B Biointerfaces 99 (2012) 102–107.
[11] B. Oh, R.B. Melchert, C.H. Lee, Biomimicking robust hydrogel for the mesenchymal stem cell carrier, Pharm. Res. 32 (2015) 3213–3227.
[12] A. Andukuri, et al., Evaluation of the effect of expansion and shear stress on a self-assembled endothelium mimicking nanomatrix coating for drug eluting stents in vitro and in vivo, Biofabrication 6 (2014). 035019-1-8.
[13] J. Lévesque, Conception et validation d'un banc d'essai pour évaluer la dégradaton d'alliages de magnesium dans des conditions quasi-physiologiques, Université Laval, Quebec, 2004.
[14] V. Montaño-Machado, et al., A comparison of adsorbed and grafted fibronectin coatings under static and dynamic conditions, Phys. Chem. Chem. Phys. 18 (2016) 24704–24712.
[15] L. Indolfi, F. Causa, P.A. Netti, Coating process and early stage adhesion evaluation of poly(2-hydroxy-ethyl-methacrylate) hydrogel coating of 316L steel surface for stent applications, J. Mater. Sci. Mater. Med. 20 (2009) 1541–1551.
[16] H.J. Kim, et al., Mechanical stability of the diamond-like carbon film on nitinol vascular stents under cyclic loading, Thin Solid Films 517 (2008) 1146–1150.
[17] J.H. Kim, et al., Comparison of diamond-like carbon-coated nitinol stents with or without polyethylene glycol grafting and uncoated nitinol stents in a canine iliac artery model, Br. J. Radiol. 84 (2011) 210–215.
[18] F. Lewis, D. Mantovani, Methods to investigate the adhesion of soft nano-coatings on metal substrates—application to polymer-coated stents, Macromol. Mater. Eng. 294 (2009) 11–19.
[19] E. Gallino, S. Massey, M. Tatoulian, D. Mantovani, Plasma polymerized allylamine films deposited on 316L stainless steel for cardiovascular stent coatings, Surf. Coat. Technol. 205 (2010) 2461–2468.
[20] J. Zhang, et al., In vitro hemocompatibility and cytocompatibility of dexamethasone-eluting PLGA stent coatings, Appl. Surf. Sci. 328 (2015) 154–162.
[21] H. Gollwitzer, et al., Biomechanical and allergological characteristics of a biodegradable poly(D,L-lactic acid) coating for orthopaedic implants, J. Orthop. Res. 23 (2005) 802–809.

[22] C.G. Hopkins, P.E. McHugh, J.P. McGarry, Computational investigation of the delamination of polymer coatings during stent deployment, Ann. Biomed. Eng. 38 (2010) 2263–2273.
[23] C. Hopkins, P.E. McHugh, N.P. O'Dowd, Y. Rochev, J.P. McGarry, A combined computational and experimental methodology to determine the adhesion properties of stent polymer coatings, Comput. Mater. Sci. 80 (2013) 104–112.
[24] N. Nayebpashaee, S.H. Seyedein, M.R. Aboutalebi, H. Sarpoolaky, S.M.M. Hadavi, Finite element simulation of residual stress and failure mechanism in plasma sprayed thermal barrier coatings using actual microstructure as the representative volume, Surf. Coat. Technol. 291 (2016) 103–114.
[25] T.M. Bedair, et al., Effects of interfacial layer wettability and thickness on the coating morphology and sirolimus release for drug-eluting stent, J. Colloid Interface Sci. 460 (2015) 189–199.
[26] F. Strickler, et al., In vivo and in vitro characterization of poly(styrene-b-isobutylene-b-styrene) copolymer stent coatings for biostability, vascular compatibility and mechanical integrity, J. Biomed. Mater. Res. A 92 (2010) 773–782.
[27] S.E. Stoebner, G. Mani, Effect of processing methods on drug release profiles of anti-restenotic self-assembled monolayers, Appl. Surf. Sci. 258 (2012) 5061–5072.
[28] B. Oh, C.H. Lee, Advanced cardiovascular stent coated with nanofiber, Mol. Pharm. 10 (2013) 4432–4442.
[29] W.C. Carlyle, et al., Enhanced drug delivery capabilities from stents coated with absorbable polymer and crystalline drug, J. Control. Release 162 (2012) 561–567.
[30] K.E. McCracken, P.L. Tran, D.J. You, M.J. Slepian, J.-Y. Yoon, Shear vs. nanotopography-guided control of growth of endothelial cells on RGD-nanoparticle-nanowell arrays, J. Biol. Eng. 7 (2013) 11.
[31] T.R.S. Kumar, L.K. Krishnan, A stable matrix for generation of tissue-engineered non-thrombogenic vascular grafts, Tissue Eng. 8 (2002) 763–770.

Simple one-step covalent immobilization of bioactive agents without use of chemicals on plasma-activated low thrombogenic stent coatings

11

M. Santos[*,†], *A. Waterhouse*[*], *B.S.L. Lee*[*,†], *A.H.P. Chan*[*,†], *R.P. Tan*[*,†], *P.L. Michael*[*,†], *E.C. Filipe*[*,†], *J. Hung*[*,†], *S.G. Wise*[*,†], *M.M.M. Bilek*[†]
[*]The Heart Research Institute, Sydney, NSW, Australia, [†]University of Sydney, Camperdown, NSW, Australia

11.1 Functionalization of stents to improve their clinical performance

Improving the clinical performance of implantable vascular devices such as stents is a complex, multifaceted problem. Endovascular delivery of stents is already widespread in the coronary circulation and becoming increasingly common in the peripheral arteries supplying blood to the legs. Common to all stents is the inherent thrombogenicity of metallic implants in contact with blood and destruction of the protective endothelial cell layer lining arterial walls [1]. Stent implantation is also associated with chronic inflammation, leading to hyperproliferation of smooth muscle cells in the vascular wall and renarrowing (restenosis) of the treated vessel over time [2]. For peripheral stents, the additional mechanical and torsion forces acting on the stent due to leg movement lead to particularly aggressive restenosis that is resistant to traditional drug-elution strategies [3]. Accordingly, improving the clinical performance of stents requires an approach that can simultaneously increase the blood compatibility of the device while differentially regulating the behavior of vascular endothelial and smooth muscles cells.

While no clinical device has yet been able to meet these goals, one approach to achieving multifunctional regulation of local vascular biology is delivery of active biological agents including proteins, enzymes, and growth factors. However, robustly adhering biomolecules to metal interfaces is an inherently challenging task. Metals are hydrophobic, chemically nonuniform, and electropositive, which makes their surface generally unfavorable for biomolecule binding [4]. Biomolecules interact with metal alloys through relatively nonspecific, van der Waals-type interactions, resulting in variable substrate association and susceptibility to exchange with host blood proteins, removal by physical forces (e.g., by washing) and to surface-induced unfolding, and subsequent loss of activity [5,6]. For these reasons, uniform covalent immobilization is preferred. An important consideration for surface modifications that can facilitate covalent binding

Functionalized Cardiovascular Stents. https://doi.org/10.1016/B978-0-08-100496-8.00011-1

and retain biological activity is that they must also be able to withstand stent crimping and expansion as well as exposure to long-term pulsatile blood flow. It is therefore unsurprising that few coatings have adequately met these diverse criteria.

11.1.1 Methods of covalent immobilization of biomolecules on metals

Immobilization of biomolecules, including proteins, enzymes, and antibodies, is used in many applications including the construction of biomedical devices, microarrays, biosensors, and industrial biotechnology [7]. Due to the inert nature of metals, several techniques have been developed to facilitate the robust immobilization of biomolecules. Established methods include the use of chemical linkers [5] and atomic force microscopy nanografting [8]. Techniques that have been used to bind molecules specifically to stents include ionic bonding, protein incorporation using an intermediate polymer overlying the stent metal, and chemical modification of the metal to allow binding of linkers or spacers. For example, heparin has been copolymerized onto metallic surfaces with a variety of polymers including poly(methylmethacrylate) or poly(vinyl alcohol) [4]. This method was also utilized to immobilize phosphorylcholine onto the Endeavor stent by copolymerization with 2-methacryloyloxyethyl [9]. However, polymers sprayed or dip-coated onto metals create a mechanical mismatch between the metal and the polymer, which frequently causes delamination [10]. Heparin has been ionically bonded to metals via polyethyleneimine and dextran sulfate; however, this is not stable upon exposure to citrated plasma [11]. Of the immobilization techniques available, the use of chemical linkers or spacers has received the most attention as it can be tailored to the biomolecule and surface of interest. However, chemical linkers and spacers have important limitations, as described in Section 11.1.2.

11.1.2 Chemical linkers and spacers

Linker chemistry creates reactive groups on the surface of metallic substrates, facilitating covalent binding a range of bioactive molecules. The most common method involves the use of a silane linkage, which reacts with the oxide layer on the metal. This can then be further reacted with various coupling compounds to link to the biomolecule [12–18]. Using this methodology, trypsin was immobilized onto cobalt-chromium using aminopropyltriethoxysilane as the silane linkage with glutaraldehyde to link to the biomolecule [18]. Similarly, alkaline phosphatase was immobilized on titanium using aminoalkylsilane and glutaraldehyde [17]. Likewise, extracellular matrix (ECM) molecules such as hyaluronan [13] and type I collagen [12] were also immobilized onto stainless steel using silane chemistry. Another common chemical linker is p-nitrophenylchloroformate, which has been used to immobilize trypsin to titanium and cobalt-chromium [19] and type I collagen onto titanium [20]. Spacer molecules such as polyethylene glycol have also been used in conjunction with hexamethylene diisocyanate to immobilize heparin onto 316L stainless steel [21]. Although successful in vitro, all these methods of chemical immobilization require complex and time-consuming wet chemistry steps, and in some cases, the linker can impair biomolecule function [22] or stability [23]. Additionally, the chemical modifications can potentially result in unknown pharmacological and toxicological responses in vivo, making these methods undesirable for bioengineering stents [22]. Preferably, linker-free covalent immobilization would occur by directly binding proteins to the surface. This process would require a surface coating to comprise active groups necessary for immobilization. In line with the additional considerations for stent applications, the coating should also be resilient upon stent crimping and expansion, hemocompatible and allow a straightforward immobilization of biomolecules without causing their unfolding.

11.2 Bioengineering of plasma-activated coatings for stents

The deposition of biofunctional thin-film coatings, suitable for coronary stents, has been previously demonstrated using plasma-activated coatings (PAC) on various substrates [24–30]. PAC is a hemocompatible and mechanically robust coating that is capable of one-step, radical-mediated biofunctionalization of stents with bioactive molecules in contact with the modified stent surface. The coatings are synthesized using low-temperature and low-pressure plasma polymerization (PP) with enhanced ion bombardment [31]. This technology allows for the deposition of tailored coatings on a myriad of substrates, without compromising the inherent mechanical properties of the underlying stent material [32]. Molecules that interact with surface radicals on PAC are robustly immobilized and retain their bioactivity, paving the way for the cost-effective design of stents that could proactively interact with the vasculature. In this section, we first review the mechanisms of PP used for the design of PAC and show how relevant process parameters can be optimized to achieve a coating with mechanical properties and surface chemistry optimal for vascular stents.

11.2.1 The plasma deposition process

The plasma state is the most common form of fundamental matter found in the universe. To generate a plasma, energy can be coupled to a gas either through an increase of its temperature or by applying a sufficiently strong electric field. The medium becomes ionized when some of the background electrons, heated to a thermal energy greater than the gas ionization threshold, collide with the gas atoms. Since most of the electric power used to generate a plasma is coupled to the more mobile electrons, plasmas can be generated with low energy inputs and at low temperatures. Plasmas can also be easily sustained in a controlled environment and harvested for specific applications. For instance, surface modification of a wide range of materials can be performed by depositing thin-film coatings using reactive PP [33–41]. The versatility of PP allows the coatings to be synthesized with numerous properties by controlling the conditions at which the plasma is generated. The plasma is usually sustained using a capacitively coupled radio-frequency (CCRF) discharge, but other plasma generation platforms, such as DC [42,43] or microwave discharges [44], can be adopted.

PP of PAC is performed using in a CCRF reactor (Fig. 11.1) with the assistance of a pulsed high-voltage power supply that enhances ion bombardment upon a sample holder (HV electrode). The radio-frequency power, coupled to a top electrode (RF electrode), is partially channeled to fragment and activate a reactive gaseous mixture comprising acetylene, nitrogen, and argon. The breakdown of the reactive mixture is mostly driven by electron impact collisions, resulting in the formation of a myriad of reactive (radicals), charged (ions), and excited species.

Optical emission spectroscopy (OES) is used in the process as a noninvasive diagnostic tool to monitor the plasma chemical fingerprint in real time during PAC deposition [24]. Electronic and vibrational deexcitations result in the emission of radiation, which can be collected and analyzed using OES. This allows collection of information about critical plasma populations relevant in the growth of PAC, including the radicals, which contain unpaired electrons and can easily recombine with other species to form more complex molecules. These molecules, formed in the plasma phase, eventually diffuse out of the active plasma boundaries and condense on surfaces of the reactor (e.g., electrodes, walls, or any sample type), initiating the growth of a solid plasma-polymerized coating. Further coating growth occurs spontaneously via plasma-surface interactions between plasma species and radical sites in the condensed polymerized molecules. Additional unpaired electrons can be formed on the surface of the coating if the energy gained by the ions in the plasma sheath is sufficiently high to break chemical bonds between atoms. Therefore, stable molecules can still be activated upon the impact of energetic ions or UV irradiation on the coating. Coating growth is exacerbated when the newly created radicals at the interface react with other carbon-based radical species from the plasma, a process often known as ion-activated growth of plasma polymers [45].

Conversely, coating ablation can take place if the energetic ions bombard the coating interface with a sufficient energy to remove material from the coating (i.e., physical etching). Therefore, the deposition of plasma-polymerized coatings is a competition between material growth, assisted by radical chain reactions, and ablation caused by the impact of energetic ions on the surface of the coating. The radical density in plasma polymer coatings can also be significantly bolstered if the precursor is fully activated in higher energy plasma processes [31,46].

In summary, the array of different radical species in each process typically relies on (i) the type of precursor, (ii) power coupled to the discharge (electrons), and (iii) the introduction of buffer gases that can in turn be excited and react with the radical species. For this reason, the plasma-phase chemistry during PP is extremely rich and comprises a vast number of the radical-assisted propagation reactions. Consequently, PAC does not usually retain the same homogeneous and repeating structure that is typical of conventional polymers. In contrast, PAC can be denser, more cross-linked, and characterized by a complex, highly branched, heterogeneous, and random network of structural units. The density, the cross-linking degree, and structural complexity of PAC can be tailored by a suitable manipulation of the plasma parameters.

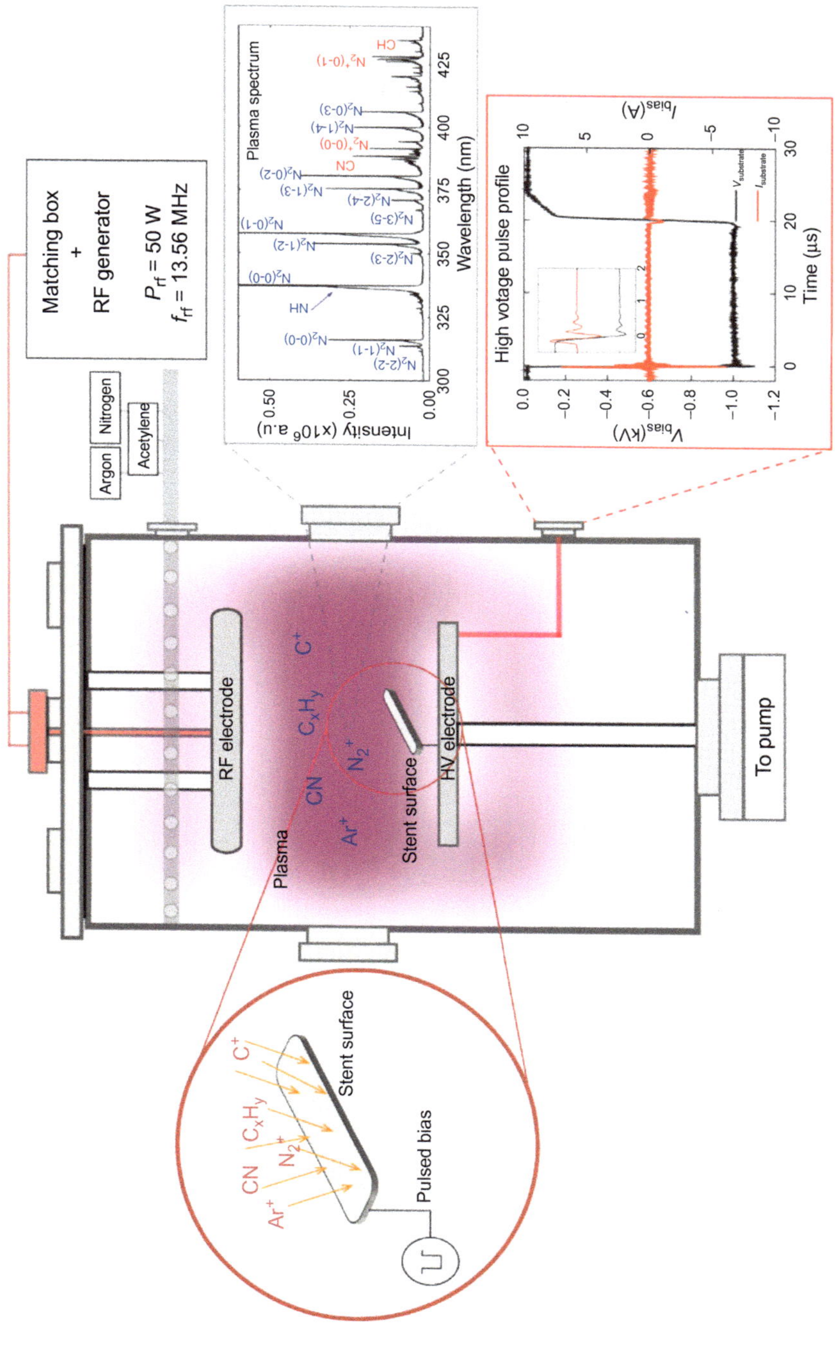

Fig. 11.1 Schematic showing the deposition process of plasma-activated coatings (PAC) on stents using plasma polymerization with ion-enhanced bombardment. The gaseous mixture is activated by applying radio-frequency power, creating radicals and ions that recombine and diffuse toward the stent surface to form a coating. The formation of reactive species is monitored in real time using OES diagnostics. The significant growth mechanisms are identified by cross-checking OES data with analysis of PAC surface chemistry, hence enabling process optimization.

11.2.2 Mechanically resilient and functional PAC for vascular stents

The mechanical properties of PAC are ultimately modulated by the degree of activation of the precursor molecules and hence by the parameters at which the plasma is sustained [24,26]. The deposition of PAC can be empirically described by plotting the deposition rate as a function of the overall energy input per unit mass, the precursor flow rate, and the applied high-voltage pulsed voltage [24]. The overall energy per unit mass is determined by considering the fraction of power delivered by both the radio-frequency supply and the high-voltage pulsed power supply in the active plasma. Overall, results show that the physical properties and the mechanical performance of PAC on stents are ultimately modulated by the energy input per unit mass (see Fig. 11.2) at a constant RF power and pressure.

An optimal parameter window was found for energy inputs between 4.2×10^8 and 6.5×10^8 J/kg (see green region in Fig. 11.2), corresponding to deposition rates

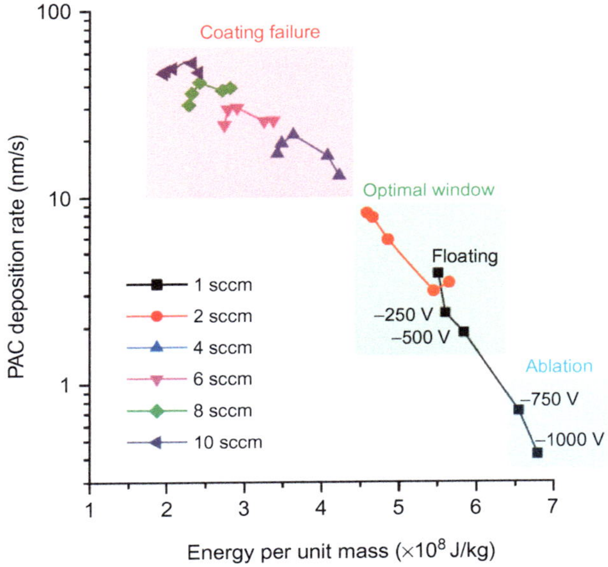

Fig. 11.2 PAC deposition rate as a function of the energy input per unit mass in the active plasma for various acetylene flow rates and applied pulsed bias (from left to right, floating; −250V, −500V, −750V, and −1000V for each acetylene flow rate). The deposition rate is tailored by the acetylene flow rate and significantly suppressed by increasing the applied bias at flow rates lower than 2 sccm. The mechanical resilience of PAC is ultimately modulated by the energy input per unit mass. Coating failure upon stent crimping and balloon expansion is observed at energies below 4.2×10^8 J/kg, that is, at deposition rates higher than 9 nm/s (red window). The optimal deposition window is shown in green, where PAC can be synthesized to resist extreme compressive and tensile stresses while being also hemocompatible. At energies per unit mass higher than 6.5×10^8 J/kg, ablation dominates PAC growth on stents. Experiments were performed at a constant RF power (50 W) and pressure (110 mTorr) [24].

between 8.1 nm/s and 1.9 nm/s, respectively. In this optimal window, PAC is mechanically resilient upon stress and able to resist cracking and delamination after plastic deformation of the stent, even after immersion of the stent in solution [24]. At lower energy inputs ($<4.2\times10^8$ J/kg), PAC delaminates following stent crimping and balloon expansion, while at very high energy inputs ($>6.5\times10^8$ J/kg), PAC ablation dominates over deposition. The acetylene flow rate, together with the pumping speed, defines the density of precursor molecules and their residence time in the discharge. While a sufficiently high RF power promotes a high level of acetylene fragmentation, an increase in acetylene density in these conditions would naturally result in the creation of more radical species, hence a higher rate of polymerization as shown in Fig. 11.2. However, it was shown that the atomic interfacial mixing (AIM) dominates the level of PAC adhesion to the stent surface. A high level of AIM is achieved by a competition between substrate ablation and PAC deposition rates. The significant decrease in the PAC deposition rate at higher energy corresponds also to a higher ablation of the stent surface. Interfacial mixing occurs at the PAC/stent surface interface when ablated metal from the stent surface blends with the deposited coating. Greater material blending can be achieved by increasing the applied pulsed bias. Ion bombardment upon the stent surface is enhanced at higher bias voltages, significantly boosting material ablation as shown in Fig. 11.2 by a significant decrease in PAC deposition at lower acetylene flow rates.

Enhanced ion bombardment also disrupts chemical bonds on PAC, generating a reservoir of stable and long-lived radicals that increases surface reactivity [31,46]. The radical areal density is greater for thicker coatings [47] and can be significantly enhanced with an increased applied pulsed bias [24] as shown in Fig. 11.3. Surface oxidation occurs when radicals from the coating bulk diffuse toward the surface and react with atmospheric oxygen. Functionalization of PAC-coated stents makes use of the radicals formed in the synthesis process by simply incubating the coated stent in a solution containing biomolecules. The biomolecules suspended in solution are covalently and spontaneously immobilized on the PAC surface upon reaction with the radicals. Thus, radical-mediated functionalization is a simple and one-step process that mitigates the need for further functionalization with chemical linkers. Since PAC retains hydrophilic properties, the immobilized molecules also maintain their activity and native conformation.

11.2.3 Deposition of PAC on stents of varied design and composition

Stents of varying size and composition are used throughout the vasculature to open vessels and restore blood flow to ischemic tissue. The dominant use is in the coronary circulation, reopening narrowed arteries supplying the myocardium, but stents are also frequently used in the carotid artery and in the lower limbs. Accordingly, stents are an heterogenous collection of devices with significantly different materials and design characteristics. For coronary applications, stents range in diameter from 2 to 4 mm and length from 8 to 38 mm. They are most commonly made from 316L stainless steel or alloys of cobalt-chromium or platinum-chromium to facilitate crimping and balloon

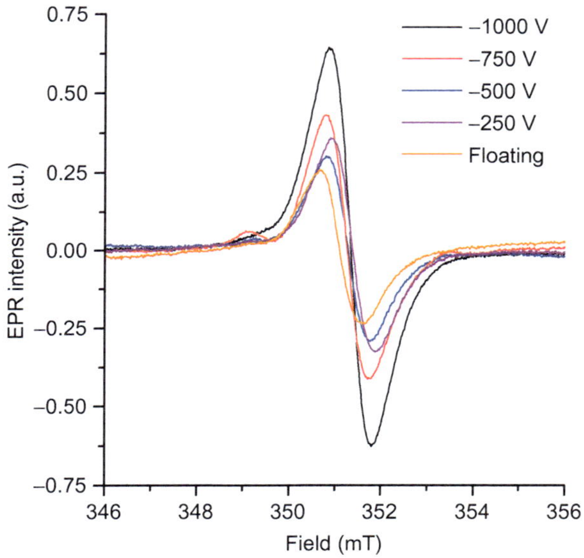

Fig. 11.3 Electron spin resonance (ESR) spectroscopy performed on PAC synthesized at different substrate pulsed bias demonstrates the presence of long-lived radicals, the density of which increases with applied pulsed bias. Radicals diffuse toward the coating surface continuously and over time, enabling covalent immobilization of biomolecules on the stent.

expansion for delivery. Their design can vary widely with respect to the width and number of struts and corresponding open or closed cell design. In contrast, peripheral stents for the femoropopliteal segment are made from shape-memory nickel-titanium (nitinol) and have a larger diameter (4.5–10 mm) and significantly longer length (20–200 mm). Carotid stents are also self-expanding nitinol, with an intermediate length of 20–40 mm and diameter ranging from 6 to 10 mm. Taken together, the diversity of size, shape, and underlying substrate make robustly and uniformly applying PAC challenging.

In addition, comparative studies in both rabbit and porcine models represent a well-established preclinical and frequently used pathway for development of novel stents [48]. The models are complimentary, as each represents the best-validated approach for assessing stent thrombogenicity, endothelialization, and neointimal hyperplasia in vivo, respectively. Stents in these models are similar to those used for the coronaries, with the rabbit iliac model accommodating 3 mm diameter stents, of 10 mm length [49]. However, mechanistic studies of stent performance can now also be performed in mice, using a donor-acceptor model where the stented artery of the donor mouse is grafted in the carotid of the recipient. Stents for this model are significantly smaller (0.6 mm diameter and 2.5 mm length), though also manufactured from stainless steel [50].

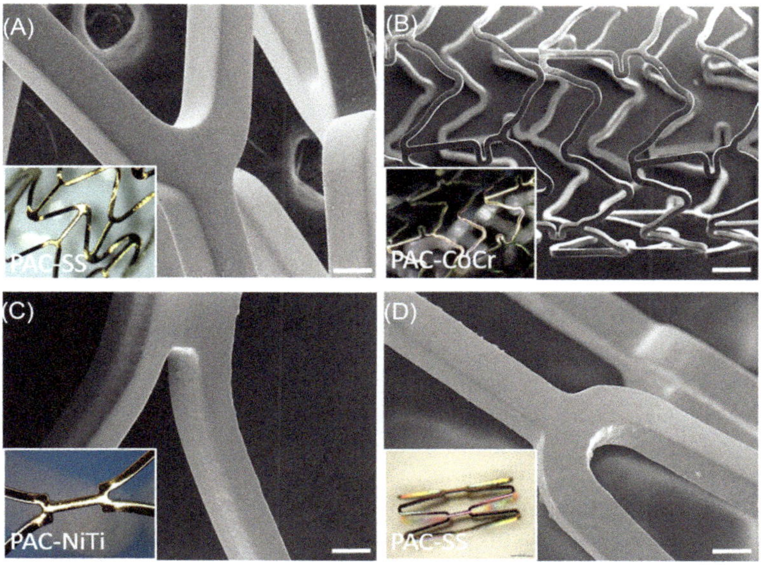

Fig. 11.4 Representative images of PAC deposited on SS (A) and CoCr (B) coronary stents, a NiTi peripheral stent (C) and an SS mouse stent (D). All stents were subjected to plastic deformation carried out by crimping and balloon expansion. Scale bars are 100 μm (A), 600 μm (B), 270 μm (C), and 50 μm (D).

We have successfully applied PAC to stents of varied composition, including stainless steel (SS), cobalt-chromium (CoCr), and nitinol (NiTi) (Fig. 11.4A–C), and sizes appropriate for coronary (Fig. 11.4A and B), peripheral (Fig. 11.4C), and mouse use (Fig. 11.4D). In each case, we have demonstrated the application of a smooth, uninterrupted coating, resistant to delamination following crimping and expansion, in sharp contrast to commercial drug-eluting stents (DES) under the same conditions [26]. Stent size and cell design do not appear to affect the uniform application of the coating, assisted by the biasing of the stents. PAC is extremely smooth (1–2 nm rms roughness) and wear-resistant using a 3-week pulsed flow of 500 mL/min and 100 pulses/min [51]. PAC stainless stents have also been successfully delivered in vivo in the rabbit iliac model, remaining stable following the rigors of surgical implantation and a week subjected to arterial flow [25].

While we have undertaken significant parameter optimization to maximize the adhesion of PAC to stainless steel, we found that adhesion of PAC to carbide-forming substrates such as zirconium and titanium is even more robust [29,52]. Other early transition metals such as chromium and niobium would be expected to similarly form additional bonds and demonstrate even greater adhesion than stainless steel. Across multiple platforms, directly relevant to commercial stents in current clinical use, PAC can be applied to and adhere robustly to stents for all applications.

11.3 Biological properties of PAC coated stents

11.3.1 Blood compatibility

The metals and drug-carrying polymers comprising modern stents all promote thrombosis. Metal (oxides) are hydrophobic, chemically heterogenous, and electropositive, properties that cause denaturation of proteins that in turn leads to coagulation and thrombosis. Polymer-coated stents are also hydrophobic (in order to elute drugs), and they delaminate, exposing the underlying thrombogenic metal [26]. The drugs eluted (taxus and limus family) cause endothelial dysfunction and hypersensitivity, contributing to thrombogenic potential, and newly developed degradable stents have high strut profiles, causing increased blood flow disturbance, also contributing to thrombotic potential of stents. Accordingly, stent thrombosis remains a problematic complication, especially for carotid and coronary applications. In the case of coronary stenting, the introduction of dual antiplatelet therapy (DAPT) has reduced overall stent thrombosis rates to 1%–2%. Nevertheless, stent thrombosis remains a catastrophic complication, being consistently associated with mortality rates of 25%. Moreover, stent implantation in the context of an acute coronary syndrome (ACS) is associated with much higher rates of stent thrombosis than elective stenting [53]. As the prognostic benefits are largely confined to its application in the context of ACS, the issue of stent thrombosis remains very significant [54]. Finally, the current low rates of stent thrombosis require complete adherence to DAPT with aspirin and a thienopyridine. This is problematic for patients requiring surgery or at high bleeding risk, especially considering the lengthy DAPT regimens now proposed [55].

PAC was designed to overcome many of these problematic properties. It is mildly hydrophilic and chemically uniform and can bind proteins covalently in their native conformation, preventing protein denaturation. Consequently, it displays significantly reduced thrombogenicity compared with bare metal. This was demonstrated by reduced platelet binding from platelet-rich plasma and whole blood and reduced thrombus formation in modified Chandler loops with partially heparinized, whole blood [26,56]. However, this reduction was only observed in the presence of plasma proteins. When isolated platelets were exposed to PAC and metal, they adhered to the same extent, suggesting that the type and conformation of plasma proteins that adhere to PAC are responsible for its low thrombogenic properties. Therefore, the two most likely mechanisms by which PAC reduces thrombogenicity are (1) its ability to covalently bind the first protein that arrives at the surface, which would be albumin in blood due to the Vroman effect and (2) its hydrophilic and chemically uniform nature that allows proteins to remain in their native conformation, preventing subsequent coagulation and platelet adhesion and activation. It is likely that both these mechanisms play a role, and further investigation is required to elucidate the mechanism of PACs low thrombogenicity. Importantly, the underlying improvement in blood compatibility of PAC can be further augmented by the covalent retention of a diverse range of biomolecules.

11.3.2 Covalent protein immobilization

In addition to acute failure due to thrombosis, long-term failure from smooth muscle cell-driven restenosis is common in bare metal stents. In the coronaries, DES releasing antiproliferative agents such as sirolimus and paclitaxel not only are highly effective in inhibiting restenosis but also profoundly delay healing and reendothelialization at the stent deployment site. Consequently, the long-term safety of DES has come into question. In contrast, peripheral stents develop restenosis that is resistant to traditional drug-elution strategies, leaving the treatment of femoropoliteal peripheral artery disease as the least effective of all endovascular procedures in terms of long-term patency. As such, biomolecules that can suppress the growth of smooth muscle cells, in a manner distinct from drug elution, while promoting rapid recruitment and growth of endothelial cells, are of significant interest for stent modification.

11.3.2.1 Tropoelastin

A range of biomolecules in the subendothelial space have been shown to play critical roles in local regulation of thrombosis, endothelialization, and smooth muscle cell proliferation, making these attractive candidates for modulation of vascular biocompatibility. Elastin, one of the major structural components of the vasculature, provides critical mechanical and biological properties and is uniquely the major regulator of SMC proliferation in vivo [57]. Recombinant human tropoelastin (TE) is the soluble precursor of elastin. TE is identical to the naturally secreted form of human tropoelastin, is recognized by mammalian cells to form elastin, and can be cross-linked to form a range of elastin-like biomaterials. In vitro, TE has been shown to have favorable interactions with endothelial cells and low thrombogenicity and inhibit the growth of smooth muscle cells.

TE was first successfully immobilized on flat stainless steel sheets. PAC facilitated the binding of TE in a linker-free, one-step process, comprising a simple overnight incubation with the protein solution at 4°C [26,51,58]. Investigation into the nature of the bound TE showed that at a concentration of 0.5 mg/mL, the bound layer was a porous monolayer [51]. Bound TE was resistant to stringent washes in 5% SDS at 90°C, suggesting covalent binding to PAC. TE was detected on the surface using an ELISA assay, demonstrating that the attachment of TE on PAC did not disrupt antibody access to the central region of the molecule and suggesting a free conformation. Most importantly, the functional benefits of applying TE to PAC-functionalized stainless steel were shown in vitro [26]. The attachment and proliferation of endothelial cells were found to be significantly higher in the TE-treated samples relative to PAC-only and stainless steel controls. This demonstrates that TE was able to confer an enhanced biological effect, even when tethered to PAC.

During the optimization of the PAC formulation for cardiovascular applications, the importance of nitrogen incorporation was recognized, with respect to increasing the elasticity of the coating, reducing thrombogenicity, and binding proteins [26]. Nitrogen containing PAC deposited with argon not only was most resistant to delamination but also bound the highest amount of TE relative to other recipes, such as those containing hydrogen or oxygen instead. The increase in binding was accompanied by a reduced contact angle (63° vs. 94° for hydrogen PAC) contributing to conformational stability and hence biological activity of the bound molecules. Consistent with previous results, TE bound to PAC robustly enhanced endothelial cell proliferation, while PAC alone has no cell binding effect above stainless steel controls. Interestingly, the addition of TE did not affect the thrombogenicity of PAC, though the baseline level of clot formation on PAC alone was so low that any further TE-driven improvements were not visible.

A common consideration with full-length protein functionalization is the loss of the protein or its activity due to enzymic degradation once implanted. TE is susceptible to degradation by serine proteases in vivo, particularly kallikrein and thrombin present in human plasma. To reduce this loss of bioactivity, the preferred kallikrein/thrombin cleavage site within TE was removed by mutation of the amino acid at position 515 from arginine to alanine [59]. This change significantly improved the retention of immobilized TE following enzymatic digest. Following from this work, smaller and more specific subsections of TE retain the functional benefits of the full-length protein, while minimizing enzymatic degradation were screened. Two fragments of TE, called N10 and N18, derived from the first 10 and 18 n-terminal domains of TE, respectively, were identified to have similar endothelial cell attachment and proliferation to full-length TE [60]. These constructs also retained the low thrombogenicity of full-length TE, demonstrated in both static and flowing whole blood assays. Enzymatic digest of these constructs immobilized on PAC showed that both truncations of TE were significantly less prone to degradation, with the shorter construct N10 superior to N18 in this respect [60]. Shorter constructs, ultimately leading to the identification of short functional peptides, are not only most resistant to proteolytic degradation but also easiest to sterilize and translate to the clinical.

11.3.2.2 Other ECM proteins

While tropoelastin remains a promising candidate to modulate the biocompatibility of stents, other proteins present in the arterial wall have also been immobilized on PAC to demonstrate its versatility. Two classical matrix proteins, well known to enhance endothelial cell attachment and proliferation, are collagen and fibronectin. Endothelial cells in vivo are in contact with the basement membrane and internal elastic lamina. Together, these structures contain a variety of ECM components that have the potential to enhance endothelialization. Basement membrane ECM proteins include collagen IV and fibronectin. These ECM proteins contain cell adhesive motifs, supporting adhesion, spreading, and cytoskeletal assembly. Additionally, these ECM proteins play important roles in cell signaling, controlling phenotype, and proliferation. However, despite having favorable endothelial cell interactions, both collagen and fibronectin are known to be prothrombogenic. Additionally, collagen (IV and III) and fibronectin promote smooth muscle cell migration. The proclotting behavior of these matrix proteins arises from their interactions with platelets, 3–5 μm anuclear cells that can adhere to foreign surfaces or damaged vasculature and form aggregates through a cohort of platelet-specific adhesion receptors and integrins. For example,

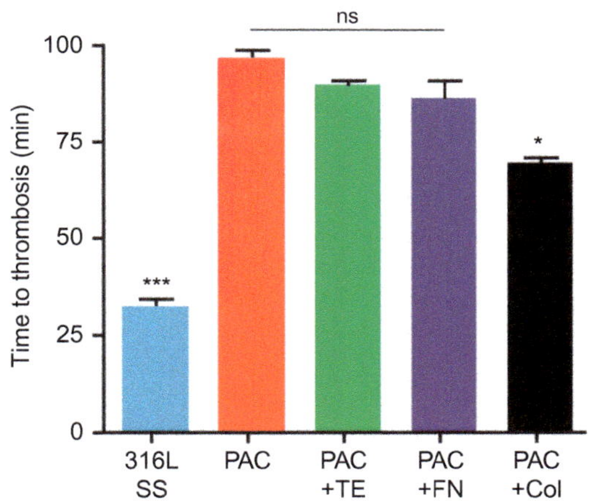

Fig. 11.5 Time to thrombus formation in modified Chandler loops containing 316L SS, PAC, PAC + TE, PAC + FN, or PAC + Col with heparinized whole blood (0.5 U/mL).

fibronectin and collagen specifically bind multiple platelet integrins most notably $\alpha_v\beta_3$ and GPVI. Platelet activation in response to these proteins triggers thrombosis and an inflammatory cascade coinciding with the onset of hyperplasia. These properties have been characterized for fibronectin and collagen passively coated in tissue culture well or on surfaces, but not when covalently immobilized. We compared passive absorption with PAC immobilization to determine if this altered their effect (see Fig. 11.5).

The thrombogenicity of fibronectin (FN) and collagen type 1 fibrils (Col) bound to PAC was assessed under flow conditions in a modified Chandler loop. Measuring time to thrombus formation, 316L SS thrombosed at 32 ± 2 min, is significantly quicker than all PAC samples. Somewhat surprisingly, FN coating did not make PAC more thrombogenic. These findings suggest that the native confirmation of FN was retained when bound to PAC that did not expose platelet binding sites. In contrast, Col-coated PAC resulted in a decrease in the time to thrombus formation by 28%, possibly because all the PAC surface was masked by the collagen fibrillar gel. These results highlight that PAC properties can potentially improve the behavior of some molecules when immobilized on PAC, for example, reducing the thrombogenicity of fibronectin.

11.3.3 Bioactive attachment of enzymes

Orientation, correct folding, and freedom for conformational changes are especially important criteria for the function of immobilized enzymes. For enzymes to perform their task, they need to be correctly orientated on the substrate for the active site to be exposed and folded correctly to preserve activity. Lastly, enzymes require a certain degree of movement within its structure to catalyze its destined reaction. Having already established that PAC can covalently tether ECM proteins important for vascular biocompatibility, we assessed the binding and subsequent functionality of a series of conformationally sensitive enzymes.

Horseradish peroxidase (HRP) is a well-characterized enzyme extensively utilized in biochemical assays ranging from western blot to ELISA. Using hydrogen peroxide as the oxidizing agent, HRP can oxidize organic or inorganic substrates. Depending on the substrate, the oxidation can result in spectrophotometric changes to the substrate or chemiluminescent emission, subsequently detected for its location and intensity. Its high turnover rate and low cost of production make HRP the most common protein in the biochemical field, thus an ideal candidate for testing enzymatic activities on PAC. We first demonstrated retention of HRP activity following PAC immobilization [30,61]. Untreated and PAC-treated silicon wafers were incubated with HRP (50 μg/mL) in 10 mM phosphate buffer (PB) at pH 7.0 for 20 h, followed by five PB washes. Samples were then incubated with tetramethylbenzidine (TMB), a spectrometric substrate to HRP, with absorbance analyzed via a spectrometer at 450 nm. The data revealed enzymatic activity doubled upon PAC immobilization. The same activity assay was repeated ten days after HRP incubation, revealing prolonged HRP activity in treated samples, whereas no activity was observed in untreated silicon wafers. This demonstrates the ability of PAC to immobilize HRP, without compromising its activity and suggesting the immobilization can maintain this activity over time.

Having demonstrated proof of principle for HRP, we sought to demonstrate the activity of fibrinolytic enzymes relevant to improving blood compatibility for stents, beginning with plasmin [62]. Plasmin is a serine protease that specifically degrades fibrin; a key protein that forms the fiber networks that stabilize and adhere blood clots. This breakdown of fibrin is known as fibrinolysis, and it is the best understood mechanism by which blood clots are physiologically degraded and cleared within the body. Plasmin was immobilized on PAC-treated stainless steel samples by incubation with increasing concentrations of plasmin (2, 20, and 200 mU) in Milli-Q water at 37°C overnight [62]. Plasmin enzyme activity was examined using a D-Val-Leu-Lys-p-Nitroanilide (VALY) chromogenic substrate, incubated with PAC treatment only, PAC with plasmin and free plasmin in solution. VALY showed no plasmin activities present in PAC-only samples, whereas a linear increase in absorbance was observed in plasmin-immobilized PAC samples, reaching a value of 1.5 after 290 min, compared with a value of 2.0 in free plasmin solutions [62]. We further confirmed the activity of plasmin using whole blood adhesion assays and thrombus weight as a measurement of clot formation. Stainless steel, PAC only, and PAC with covalently bound plasmin were incubated with heparinized whole blood (0.3 U/mL) for 60 min at 37°C while rocking. Increasing concentrations of plasmin-coated PAC (2, 20, and 200 mU) caused a decrease in clot formation (31.0 ± 2.9, 21.0 ± 3.3, and 3.7 ± 0.7 mg, respectively), significantly lower than stainless steel (63.1 ± 4.7 mg) [62].

Like plasmin, streptokinase has thrombolytic properties often employed for clinical use. Contrary to plasmin, streptokinase acts in a similar manner to tPA and activates unbound plasminogen to produce free plasmin. As above, we tested the activity of streptokinase bound to PAC. Strikingly, PAC + streptokinase had virtually no clots formed on the surface, with only 6.2 ± 1.2 mg of thrombus weight measured after 90 min (Fig. 11.6). In contrast, untreated samples were entirely covered in clots, producing approximately 111.2 ± 2.2 mg of thrombus after 90 min of incubation (Fig. 11.1), and PAC-only samples were only marginally lighter than untreated, showing 86.5 ± 2.6 mg

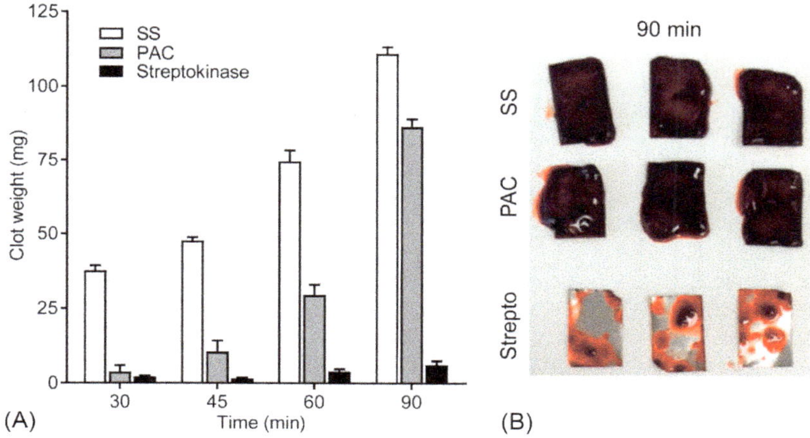

Fig. 11.6 (A) Thrombogenicity of SS, PAC, and PAC + streptokinase for various time points using heparinized whole blood. While PAC alone could reduce thrombogenicity compared with bare SS in the first time points, the immobilization of streptokinase on PAC had a striking effect in mitigating thrombus formation. (B) Representative images of samples following a 90 min incubation with heparinized whole blood.

of thrombus weight at the 90 min mark. Together, all these enzyme studies exemplified the utility of PAC. It can covalently bind enzymes on to any desired surface; furthermore, the functions of these enzymes were retained, with some showing prolonged activity.

11.3.4 Summary

We have shown that PAC can be simply and cost-effectively synthesized for the clinical application on vascular stents using a reactive PP process with ion-enhanced bombardment. Optimization of the plasma characteristics allowed the deposition of a mechanically resilient PAC, capable of withstanding fluid immersion and the extreme plastic deformation that stents undergo during typical deployment conditions. We demonstrated the versatility of our plasma-based synthesis platform by applying PAC on a variety of stent designs with different alloy compositions, geometries, and sizes. The coating resisted crimping and balloon expansion in all tested stent modes, demonstrating the potential use of PAC for rabbit iliac, mouse carotid, and human coronary and peripheral models. PAC is universally reactive, a property inherent to the synthesis in a reactive plasma, and contains radicals that diffuse over time to the surface. A wide range of biomolecules can be easily immobilized through reaction with the radical groups in PAC. Due to PAC's mildly hydrophilic properties, functional conformations and the biological activity of the immobilized molecules are retained. We have successfully demonstrated bioactive immobilization of tropoelastin, collagen, fibronectin, HRP, plasmin, and streptokinase on PAC surfaces. While PAC alone is well-tolerated in vivo, formulations with TE, plasmin, or streptokinase were also able to significantly enhance hemocompatibility and reendothelialization against bare metal substrates.

References

[1] P.H. Grewe, et al., Acute and chronic tissue response to coronary stent implantation: pathologic findings in human specimen, J. Am. Coll. Cardiol. 35 (1) (2000) 157–163.

[2] F.G. Welt, C. Rogers, Inflammation and restenosis in the stent era, Arterioscler. Thromb. Vasc. Biol. 22 (11) (2002) 1769–1776.

[3] J.R. Laird, M. Hong, Drug-eluting stents in the superficial femoral artery: the long and winding road, Circulation 133 (15) (2016) 1435–1437.

[4] G. Mani, et al., Coronary stents: a materials perspective, Biomaterials 28 (9) (2007) 1689–1710.

[5] V.E. Ferrero, et al., Protein and electrode engineering for the covalent immobilization of P450 BMP on gold, Anal. Chem. 80 (22) (2008) 8438–8446.

[6] M.J. Desroches, S. Omanovic, Adsorption of fibrinogen on a biomedical-grade stainless steel 316LVM surface: a PM-IRRAS study of the adsorption thermodynamics, kinetics and secondary structure changes, Phys. Chem. Chem. Phys. 10 (18) (2008) 2502–2512.

[7] L.S. Wong, F. Khan, J. Micklefield, Selective covalent protein immobilization: strategies and applications, Chem. Rev. 109 (9) (2009) 4025–4053.

[8] Y. Hu, et al., Nanografting de novo proteins onto gold surfaces, Langmuir 21 (20) (2005) 9103–9109.

[9] E.J. Lobb, et al., Facile synthesis of well-defined, biocompatible phosphorylcholine-based methacrylate copolymers via atom transfer radical polymerization at 20 degrees C, J. Am. Chem. Soc. 123 (32) (2001) 7913–7914.

[10] M. Wiemer, et al., Scanning electron microscopic analysis of different drug eluting stents after failed implantation: from nearly undamaged to major damaged polymers, Catheter. Cardiovasc. Interv. 75 (6) (2010) 905–911.

[11] O. Larm, R. Larsson, P. Olsson, A new non-thrombogenic surface prepared by selective covalent binding of heparin via a modified reducing terminal residue, Biomater. Med. Devices Artif. Organs 11 (2–3) (1983) 161–173.

[12] R. Muller, et al., Surface engineering of stainless steel materials by covalent collagen immobilization to improve implant biocompatibility, Biomaterials 26 (34) (2005) 6962–6972.

[13] W.G. Pitt, et al., Attachment of hyaluronan to metallic surfaces, J. Biomed. Mater. Res. A 68 (1) (2004) 95–106.

[14] K. Udipi, et al., Modification of inflammatory response to implanted biomedical materials in vivo by surface bound superoxide dismutase mimics, J. Biomed. Mater. Res. 51 (4) (2000) 549–560.

[15] H.P. Jennissen, et al., Biocoating of implants with mediator molecules: surface enhancement of metals by treatment with chromosulfuric acid, Materwiss. Werksttech. 30 (12) (1999) 838–845.

[16] A. Rezania, et al., Bioactivation of metal oxide surfaces. 1. Surface characterization and cell response, Langmuir 15 (20) (1999) 6931–6939.

[17] A. Nanci, et al., Chemical modification of titanium surfaces for covalent attachment of biological molecules, J. Biomed. Mater. Res. 40 (2) (1998) 324–335.

[18] D.A. Puleo, Activity of enzyme immobilized on silanized Co-Cr-Mo, J. Biomed. Mater. Res. 29 (8) (1995) 951–957.

[19] L.J. Mikulec, D.A. Puleo, Use of p-nitrophenyl chloroformate chemistry to immobilize protein on orthopedic biomaterials, J. Biomed. Mater. Res. 32 (2) (1996) 203–208.

[20] J. van den Dolder, J.A. Jansen, The response of osteoblast-like cells towards collagen type I coating immobilized by p-nitrophenylchloroformate to titanium, J. Biomed. Mater. Res. A 83 (3) (2007) 712–719.

[21] T.W. Chuang, D.T. Lin, F.H. Lin, Immobilization of NaIO4-treated heparin on PEG-modified 316L SS surface for high anti-thrombin-III binding, J. Biomed. Mater. Res. A 86 (3) (2008) 648–661.
[22] M.M. Bilek, D.R. McKenzie, Plasma modified surfaces for covalent immobilization of functional biomolecules in the absence of chemical linkers: towards better biosensors and a new generation of medical implants, Biophys. Rev. 2 (2) (2010) 55–65.
[23] Y. Ikada, Surface modification of polymers for medical applications, Biomaterials 15 (10) (1994) 725–736.
[24] M. Santos, et al., Mechanically robust plasma-activated interfaces optimized for vascular stent applications, ACS Appl. Mater. Interfaces 8 (15) (2016) 9635–9650.
[25] A. Waterhouse, et al., In vivo biocompatibility of a plasma-activated, coronary stent coating, Biomaterials 33 (32) (2012) 7984–7992.
[26] A. Waterhouse, et al., The immobilization of recombinant human tropoelastin on metals using a plasma-activated coating to improve the biocompatibility of coronary stents, Biomaterials 31 (32) (2010) 8332–8340.
[27] S.G. Wise, et al., TCT-433 plasmin immobilization for reduced thrombogenicity of metallic implants, J. Am. Coll. Cardiol. 64 (11 Suppl.) (2014) B127.
[28] S.G. Wise, et al., Plasma-based biofunctionalization of vascular implants, Nanomedicine (London, England) 7 (12) (2012) 1907–1916.
[29] G.C. Yeo, et al., Plasma-activated tropoelastin functionalization of zirconium for improved bone cell response, ACS Biomater. Sci. Eng. 2 (4) (2016) 662–676.
[30] Y. Yin, et al., Acetylene plasma polymerized surfaces for covalent immobilization of dense bioactive protein monolayers, Surf. Coat. Technol. 203 (10–11) (2009) 1310–1316.
[31] M.M.M. Bilek, Biofunctionalization of surfaces by energetic ion implantation: review of progress on applications in implantable biomedical devices and antibody microarrays, Appl. Surf. Sci. 310 (2014) 3–10.
[32] M. Santos, M.M.M. Bilek, S.G. Wise, Plasma-synthesised carbon-based coatings for cardiovascular applications, Biosurf. Biotribol. 1 (3) (2015) 146–160.
[33] H. Biederman, Plasma Polymer Films, Imperial College Press, London, 2004.
[34] M.C. Vasudev, et al., Exploration of plasma-enhanced chemical vapor deposition as a method for thin-film fabrication with biological applications, ACS Appl. Mater. Interfaces 5 (10) (2013) 3983–3994.
[35] H.K. Yasuda, et al., Atomic interfacial mixing to create water insensitive adhesion, J. Adhes. 13 (3–4) (1982) 269–283.
[36] H. Yasuda, M. Gazicki, Biomedical applications of plasma polymerization and plasma treatment of polymer surfaces, Biomaterials 3 (2) (1982) 68–77.
[37] H. Yasuda, Glow discharge polymerization, J. Polym. Sci. Macromol. Rev. 16 (1) (1981) 199–293.
[38] H. Yasuda, Plasma Polymerization, Academic Press, Inc., Orlando, FL, 1985.
[39] H. Yasuda, T. Hsu, Plasma polymerization investigated by the comparison of hydrocarbons and perfluorocarbons, Surf. Sci. 76 (1) (1978) 232–241.
[40] H. Yasuda, C.R. Wang, Plasma polymerization investigated by the substrate-temperature dependence, J. Polym. Sci. A Polym. Chem. 23 (1) (1985) 87–106.
[41] H. Yasuda, T. Hsu, Some aspects of plasma polymerization investigated by pulsed R.F. discharge, J. Polym. Sci. A Polym. Chem. Ed. 15 (1) (1977) 81–97.
[42] Q.S. Yu, C. Huang, H.K. Yasuda, Glow characterization in direct current plasma polymerization of trimethylsilane, J. Polym. Sci. A Polym. Chem. 42 (5) (2004) 1042–1052.
[43] T. Suwa, et al., Plasma polymerization in direct current glow: characterization of plasma-polymerized films of benzene and fluorinated derivatives, Microelectron. Technol. 614 (1995) 471–484.

[44] J. Schwarz, M. Schmidt, A. Ohl, Synthesis of plasma-polymerized hexamethyldisiloxane (HMDSO) films by microwave discharge, Surf. Coat. Technol. 98 (1) (1998) 859–864.
[45] R. D'Agostino, Plasma Deposition, Treatment, and Etching of Polymers, Academic Press, Boston, 1990.
[46] M.M. Bilek, et al., Free radical functionalization of surfaces to prevent adverse responses to biomedical devices, Proc. Natl. Acad. Sci. U. S. A. 108 (35) (2011) 14405–14410.
[47] Y. Yin, et al., Covalently bound biomimetic layers on plasma polymers with graded metallic interfaces for in vivo implants, Plasma Process. Polym. 6 (10) (2009) 658–666.
[48] R. Virmani, et al., Drug eluting stents: are human and animal studies comparable? Heart 89 (2) (2003) 133–138.
[49] M. Joner, et al., Endothelial cell recovery between comparator polymer-based drug-eluting stents, J. Am. Coll. Cardiol. 52 (5) (2008) 333–342.
[50] Z.A. Ali, et al., Increased in-stent stenosis in ApoE knockout mice: insights from a novel mouse model of balloon angioplasty and stenting, Arterioscler. Thromb. Vasc. Biol. 27 (4) (2007) 833–840.
[51] Y. Yin, et al., Covalent immobilisation of tropoelastin on a plasma deposited interface for enhancement of endothelialisation on metal surfaces, Biomaterials 30 (9) (2009) 1675–1681.
[52] B. Akhavan, S.G. Wise, M.M. Bilek, Substrate-regulated growth of plasma-polymerized films on carbide-forming metals, Langmuir 32 (42) (2016) 10835–10843.
[53] T.R. Tolleson, et al., Frequency of stent thrombosis after acute coronary syndromes (from the SYMPHONY and 2nd SYMPHONY trials), Am. J. Cardiol. 92 (3) (2003) 330–333.
[54] W.E. Boden, et al., Optimal medical therapy with or without PCI for stable coronary disease, N. Engl. J. Med. 356 (15) (2007) 1503–1516.
[55] L. Mauri, et al., Twelve or 30 months of dual antiplatelet therapy after drug-eluting stents, N. Engl. J. Med. 371 (23) (2014) 2155–2166.
[56] M.M.M. Bilek, et al., Free radical functionalization of surfaces to prevent adverse responses to biomedical devices, Proc. Natl. Acad. Sci. 108 (35) (2011) 14405–14410.
[57] S.G. Wise, et al., Tropoelastin: a versatile, bioactive assembly module, Acta Biomater. 10 (4) (2014) 1532–1541.
[58] D.V. Bax, et al., Linker-free covalent attachment of the extracellular matrix protein tropoelastin to a polymer surface for directed cell spreading, Acta Biomater. 5 (9) (2009) 3371–3381.
[59] A. Waterhouse, et al., Stability of a therapeutic layer of immobilized recombinant human tropoelastin on a plasma-activated coated surface, Pharm. Res. 28 (6) (2011) 1415–1421.
[60] M.A. Hiob, et al., The use of plasma-activated covalent attachment of early domains of tropoelastin to enhance vascular compatibility of surfaces, Biomaterials 34 (31) (2013) 7584–7591.
[61] Y. Yin, et al., Plasma polymer surfaces compatible with a CMOS process for direct covalent enzyme immobilization, Plasma Process. Polym. 6 (1) (2009) 68–75.
[62] S.G. Wise, et al., Immobilization of bioactive plasmin reduces the thrombogenicity of metal surfaces, Colloids Surf. B: Biointerfaces 136 (2015) 944–954.

Part Three

Biofunctionalisation of cardiovascular stent surfaces

Chemistry of targeted immobilization of biomediators

12

A. Srivastava
National Institute of Pharmaceutical Education and Research, Ahmedabad, India

12.1 Introduction

Metal stents are functionally incompatible with the vascular system and promote thrombosis due to their native surface properties. The major cause of failure of metal stents is restenosis (leading to reblocking of arteries) due to neointimal proliferation (migration of smooth muscle cells) within the stent. Moreover, an uncontrolled immunological response, disruption of the native endothelium and damage of vessel wall promote in-stent restenosis. Hence, there is a need to overcome these limitations to prevent unwanted biological reactions on the surface of metal stents. The conjugation of proteins, carbohydrates, drugs, and other biomediators on the surface of biomaterials has strengthened the development of biocompatible medical devices. The main aims of biofunctionalization are to enhance the hemocompatibility and endothelialization of artificial vascular grafts [1]. For example, immobilization of heparin onto biomaterial surfaces prevents the thrombus formation [2], which in turn enhances surface hemocompatibility [3,4]. Similarly, gelatin is also used in surface modification to enhance biocompatibility [5]. Current conjugation approaches are directed towards achieving correctly oriented and functionally active protein molecules on the surfaces of polymeric or nonpolymeric biomaterial devices. A careful selection of immobilization chemistry is required to connect the biomolecules on the biomaterial surface depending upon the availability of chemical groups [6]. Increasing knowledge of the chemistries involving cysteine and lysine residues, in particular, allows general and robust chemical conjugation method to develop "functional biomaterials" for in vivo applications. Surface modifications of vascular grafts are vital to control cellular response, hemocompatibility, and complete success of the graft after implantation [7]. The surface properties of metal, stainless steel, and alloy used in the fabrication of coronary stent allows the immobilization of biomediators such as proteins, heparin, and drug molecules via direct conjugation or via polymeric coating on coronary stent. The immobilization of biomolecules is also helpful in the reduction of adverse events (for example, restenosis) associated with cardiovascular stenting procedures. There are various approaches for the immobilization of peptides/proteins. Protein/peptide conjugation methods can be tailored according to material surface, protein/peptide, and their intended application. These methods are crucial to avoid undesirable effects, e.g., generation of free radicals [8], inflammatory response [9], modification of active sites, etc., which may lead to the failure. The covalent coupling enables homogenous distribution of biomolecules on the material surface. The choice of appropriate covalent modification depends on the functional

Functionalized Cardiovascular Stents. https://doi.org/10.1016/B978-0-08-100496-8.00013-5

group present on material surface and protein/peptides [10]. Uniformly conjugated biomolecules throughout the surface of the material have shown to extend beneficial effects in cardiovascular stents for promoting neovascularization and angiogenesis to enhance endothelium development. Recruitment of endothelial cells (EC) on a vascular graft requires proper angiogenesis to facilitate transinterstitial migration of EC [11]. Adsorption of peptides/proteins is most commonly employed method for surface modification on cardiovascular stents. However, simple adsorption provides limited control over orientation of the attached ligands or biomolecules. There are limited examples available for targeted immobilization of biological molecules on the surface of cardiovascular stents. This chapter is focused on summarizing the reports on targeted chemical conjugation of biomolecules and their mechanism.

Bioconjugated cardiovascular stents have shown numerous examples in enhancing endothelial cell attachment. The covalently modified Arg-Gly-Asp (RGD) and heparin-conjugated poly(carbonate-urea)urethane graft conduit have shown better retention of EC when compared to simple adsorption [12]. In another study, antibodies/peptide motifs were immobilized onto EDC (N-ethyl-N′-(3-dimethylaminopropyl)carbodiimide hydrochloride)—NHS (N-hydroxy-succinimide) activated polyhedral oligomeric silsesquioxanepoly (carbonate-urea) urethane (POSS-PCU) [13]. The conjugated antibody had significantly improved the hemocompatibility of POSS-PCU and promoted in situ endothelialization through the attachment of endothelial progenitor cells (EPCs) [13]. Conjugation of anti-CD133 antibody onto the biodegradable polymeric material was successfully carried out by amide bond formation using EDC/NHS-based covalent conjugation. A large number of metal surfaces form a passivation layer of metal oxide that exposes hydroxyl groups on their surface, allowing the binding of silanes on the metal [14,15]. To ensure optimal silanization, several chemical and physical surface treatments are able to activate the metallic surface by increasing the amount of OH groups on the surface [16,17]. The oxygen plasma is an effective method for the removal of contaminants and provides an excellent tool for surface modification [18,19]. Sodium hydroxide (NaOH) etching can also be used to create free hydroxyls on titanium surfaces for developing bioactive surfaces [20]. In a study, oxygen plasma and NaOH etching method were compared for the activation on the surface of cobalt-chromium (CoCr) alloy. Both activation treatments increase the hydroxyl groups formation, but oxygen plasma treatment was more efficient compared to NaOH etching and polished CoCr surfaces. It was also confirmed by deconvolution of singlet oxygen peak of X-ray photoelectron spectroscopy (XPS) that the ratio of OH^- / O_{2^-} is higher for plasma-treated samples compared to NaOH etching and polished CoCr surfaces. Moreover, oxygen plasma process also decreases the percentage of carbon and nitrogen contaminant and increased the amount of oxygen when compared to NaOH etching activation.

Silanization has been widely used to covalently immobilize functional biomolecules on metallic supports [10,21,22]. These methods of chemical immobilization have been successful in in vitro, but it requires series of complex and time-consuming steps of chemical reactions and, in some cases, it may require the linker, which can compromise bimolecular function or stability [23]. However, linker-free covalent immobilization may be preferable which would allow direct binding of proteins to the surface. The plasma-based treatments have been used to modify metallic and polymeric surface for covalent binding of biomolecules. Plasma treatments in the presence of highly reactive gases, e.g., ammonia, oxygen, nitrogen, hydrogen, and ozone, introduce many functional groups (carboxyl, carbonyl, hydroxyl, amine, etc.) on material surface [24]. Grafting of polymers on a material surface involves bombardment of high energetic plasma gaseous species, which can transfer the energy and dissipate it through the solid by a variety of chemical and physical processes. The functional groups can be utilized in chemical immobilization of biomolecules via grafting [25–27].These modifications are readily applied to Dacron® and expanded polytetrafluoroethylene polymer, which are widely used in vascular grafts including stents. The metal surface modified by using acetylene, nitrogen, and argon gas mix plasma deposition results in a formation of cross-linked carbon layer called plasma-activated coating (PAC) [28]. It is different from conventional plasma polymerization that utilizes energetic ion to create a functional group for protein attachment [29]. Infrared spectra demonstrate a cross-linked layer predominantly composed of carbon and nitrogen with an oxidized outermost layer [30]. The free radicals generated on the surface are also constantly renewed by diffusion from under surface regions [29]. The horseradish peroxidase and catalase are covalently attached on plasma-treated surface in proper orientation while retaining bioactivity [30,31]. Plasma-treated surface conjugate with human tropoelastin has also enhanced endothelial cell attachment and proliferation [32]. Acetylene plasma treatment was applied to the stainless steel surfaces to deposit a hydrophilic interface for covalent attachment of tropoelastin. The tropoelastin bound to metallic surfaces retained its function and facilitated modulation of host responses in endovascular stenting [31]. Poly(L-lactide) (PLLA)-coated stent surface was activated with terminal amino groups by NH_3-plasma, aminolysis with hexamethyldiamine, and O_2-plasma with silanization with 3-aminopropyltriethoxylsilane (APTES) [33]. Biodegradable poly(L-lactide) stents also allow the functionaltionalization via plasma exposure [34]. Naturally present heparin along with other sulphated glycans on the endothelial cells is the potent anticoagulant. The diamond-like carbon (DLC) films is an excellent biomaterial in combination with immobilized heparin to develop hemocompatible stent materials [35]. DLC films were deposited on the surface of the polymer by an acetylene plasma treatment, followed by exposure of ammonia plasma. The functionalized surface was then used to covalently couple heparin by an end-point attachment method. Apart from bare metal stent, bioengineered stents are the next generation stents with polymer and/or protein-coated metal stent [36]. Polymers such as poly(methyl methacrylate) (PMMA), polyethylene (PE), polyurethanes (PUs), polylactide (PLA), and polyglycolide (PGA) have been investigated for cardiovascular implants and other medical devices [1]. Among all these, PU is most commonly employed polymer in developing blood-containing devices [37]. This is due to its better mechanical properties, nontoxic degradation, reduced activation of blood cells, and biocompatibility [38]. The biomaterial substrate and surface modification techniques discussed above have enabled the conjugation of biomolecules using targeted conjugation methods. In the following section, the various chemical conjugation reaction mechanisms which are currently applied are discussed, as well as their potential in targeted immobilization of biomediators on cardiovascular stents.

12.2 Targeted immobilization chemistries

The use of site-specific conjugation reactions mediated by chemical groups present in the biomolecules and the biomaterial improves the success of developed medical devices. Primary amine (–NH_2), carboxyl (–COOH), sulfhydryl (–SH), and carbonyl (RCOR′) groups, widely present in biomolecules as well as in several biomaterials, have been frequently exploited for the site-specific conjugation methods [39]. Site-selective protein modification reactions require precise control of chemical reaction at particular site under physiological conditions [6] and the properties of biomolecules and material influence the choice of conjugation techniques. Covalent tethering of proteins utilizes lysine amino acids, cysteine residues, and carbohydrate moieties [6]. Amine and sulphydryl conjugation methods are most commonly used in immobilization of biomolecules on the stent materials as discussed below. Other methods of surface functionalization of stent material and targeted immobilization of biomolecules are also discussed in the following sections.

12.2.1 Amine conjugation

Amine conjugation involves the conjugation of biomolecules to materials via an amine group present on the surface of either of the conjugating partners [40,41]. These reactions are rapid and obtain high yield to establish stable amide or secondary amine bonds. Carbodiimide crosslinker can react with carboxylic acids groups to form highly reactive, short-lived O-acylisourea derivatives. O-acylisourea derivatives are reactive towards nucleophilic attack, e.g., primary amine to form an amide bond [42]. However, oxygen atoms, those in water molecules, may also act as the nucleophile. The hydrolysis by water containing reaction medium plays the major role in inactivating 1-ethyl-3-(-3-dimethylaminopropyl) carbodiimide (EDC) or N′,N′-dicyclohexyl carbodiimide (DCC) and deactivating the ester intermediate, which led to the formation of an isourea and regeneration of the carboxylate group (Scheme 12.1) [43]. Currently, majority of amine-reactive conjugation methods utilize NHS esters. An NHS ester usually forms by the reaction of a carboxylic acid with N-hydroxysuccinimide (NHS) in the presence of a carbodiimide. NHS esters react immediately with the target molecules in aqueous reaction condition (Scheme 12.1). Water-soluble EDC can be transformed into an active ester in the presence of NHS or sulfo-NHS. Sulfo-NHS esters are the active groups and readily couple with targeted amines with similar specificity and reactivity as shown by NHS esters [44]. Sulfo-NHS esters are relatively water-soluble unlike other NHS esters. Upon amine nucleophilic attack on electron-deficient carbonyl group of the activated ester, the sulfo-NHS group rapidly leaves and creates a stable amide linkage.

EDC/NHS-mediated amine conjugation strategies are widely used for the immobilization of antibodies on polymeric-coated stents to enhance endothelialization [45,46]. Heparin was also covalently immobilized on plasma and UV-treated, polyacrylic acid grafted biomaterial surface using a 4-dimethylaminopyridine as catalyst and dicyclohexylcarbodiimide (DCC) coupling agent [47]. Heparin modified surfaces are achieved by using a condensation reaction between the EDC-NHS activated carboxylic acid groups of heparin and free amines on the surfaces [48,49].

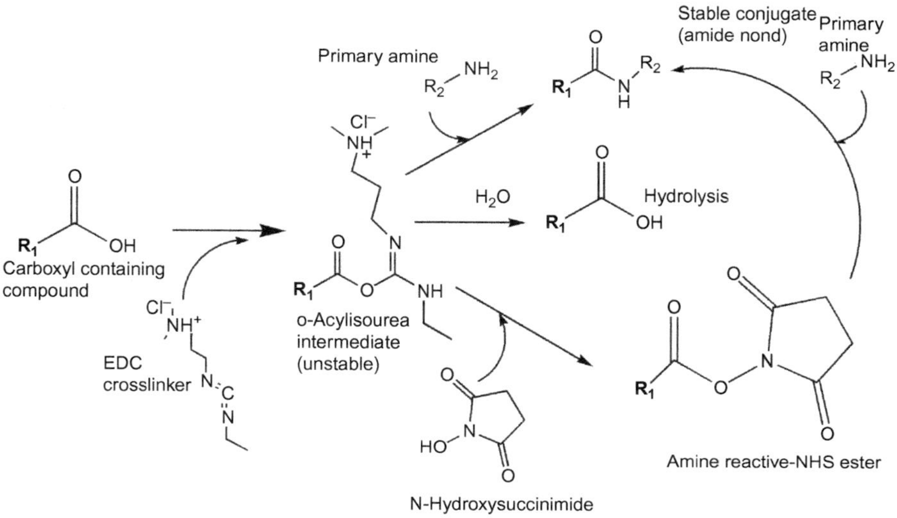

Scheme 12.1 EDC and NHS-based reaction mechanism for coupling amine and carboxyl-containing compounds.

Hydroxyl group containing surface of poly(hydroxyethyl methacrylate) (PHEMA) was also modified with heparin using 1,10-carbonyldiimidazole (CDI) [50]. Yin and co-workers successfully immobilized bovine serum albumin (BSA) on poly(acrylic acid) (poly(AAc)) grafted onto O_2 plasma treated and the UV-irradiated polypropylene membranes using EDC and NHS [51]. Moreover, biomaterial surface with sufficiently high hemocompatibility is achieved by the immobilization of tissue plasminogen activator (t-PA), urokinase, and streptokinase [52–56]. These biomolecules can be conjugated on polymeric surface using EDC and sulfo-NHS [53]. RGD peptide can also be immobilized onto the material surface through an amide bond formation by the reaction of carboxylic acid with the N-terminal amino groups of the RGD, in the presence of carbodiimide and NHS [57]. However, conjugation of biomolecules in multiple orientations due to the potential occurrence of amine moieties in proteins and other biomolecules has been reported. Despite this limitation, amine conjugation method has been successfully applied for linking molecules to biomaterials in a myriad of medical applications. Other chemical reactions used in targeted protein bioconjugation include reductive alkylation of amine groups [6]. Native chemical ligation (NCL) is an advanced chemoselective reaction that has been used to conjugate proteins to materials via the formation of a linking peptide bond [58]. The reaction occurs under physiological conditions via the formation of a thioester bond between cysteine residues in the protein and support material, followed by spontaneous rearrangement of a S→N acyl shift to form the peptide bond at the ligation site. NCL has been used in a variety of conjugation of oligopeptides and recombinant proteins to dendrimers [59]. The covalent bond formation comprising a 1,2 dicarbonyl moiety (RCOCOR′) and atleast one guanidino moiety (R″NHC(NH)NH_2) for immobilizing biomolecule on a biomaterial surface does not require coupling molecules [60]. A gold-plated stent has been coated with various amine-containing chemical for oligonucleotide immobilization [61]. A copolymerized coating of dopamine and hexamethylendiamine (HD) (PDAM/HD) onto a vascular stent yield high density of amine groups (30 nmol cm^{-2}) for biofunctionalization [62]. Vascular stents coated with anti-integrin antibody was prepared by addition of amine group on the surface of a stent by plasma polymerization and the conjugation of antiintegrin antibody with the amine group-terminated stent surface [63]. Selective modification at the N-terminal amino group of proteins in the presence of multiple lysine residues is usually possible due to the fact that the pK value of amino group at the N terminus is slightly less basic than the lysine. Hence, the N-terminal amino group will become more nucleophilic by adjusting the pH and can be selectively acylated for further modification. Derivatives of 2-pyridinecarboxyaldehyde selectively react with N-terminus of proteins to form stable conjugates with substrate [64]. Site-specific conjugation can be achieved through the introduction of an aldehyde moiety on the material surface [65]. The aldehyde activated surface can be coupled to amine groups present on the biomolecules, via the formation of Schiff base intermediate. The Schiff' base can further reduce to form a stable secondary amine (Scheme 12.2) [66–69]. The above method has enabled the site-specific functionalization of the *N*-terminal amine group without modifying the lysine side chain [70–73]. Reductive alkylation has been used to immobilize human serum albumin (HSA) to methacrylate-based polymers [74]. The targeted immobilization of heparin using reductive amination has been shown involving partially controlled depolymerization of heparin [75,76] and the terminal aldehyde moiety formed on the sugar unit of heparin subsequently linked with the primary amine on biomaterial surface. It has shown to prevent platelet activation [49,77] and reduce coagulation [78,79] and decrease inflammation to the heparin-coated surface [80]. The developed immobilization method is more efficient than the bioconjugation method involving carbodiimide [81], epoxy [82], and succinimidyl carbonate [83,84]. The acetal moiety was also introduced in reductive alkylation method to ensure longer shelf life [85,86]. This approach was utilized in surface immobilization of streptavidin on acetal-protected aldehyde-decorated methacrylate surface [87].

Isothiocyanate compounds are highly selective for the modification of ε-amino groups in lysine and N-terminal α-amines in proteins or other primary amines [88]. The attack of the nucleophile on the electrophilic carbon of the isothiocyanate group results in electron shift and proton loss to create a thiourea linkage (Scheme 12.3).

12.2.2 Sulfhydryl-reactive conjugations

Functional groups reacting with sulfhydryl-containing molecules are the most common groups present on biomaterial surface or in crosslinking reagents. The designs

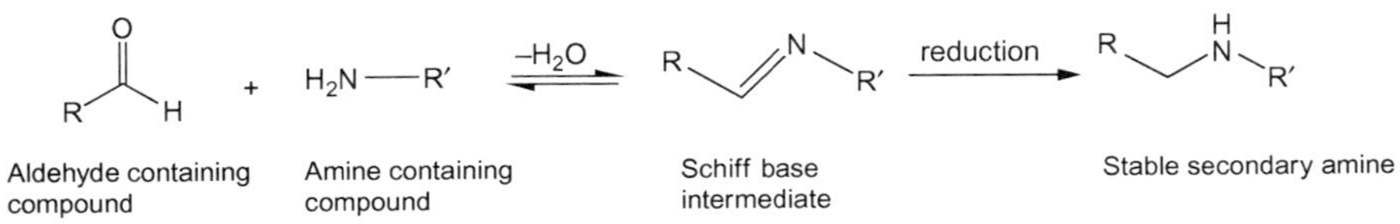

Scheme 12.2 Reaction mechanism of reductive amination.

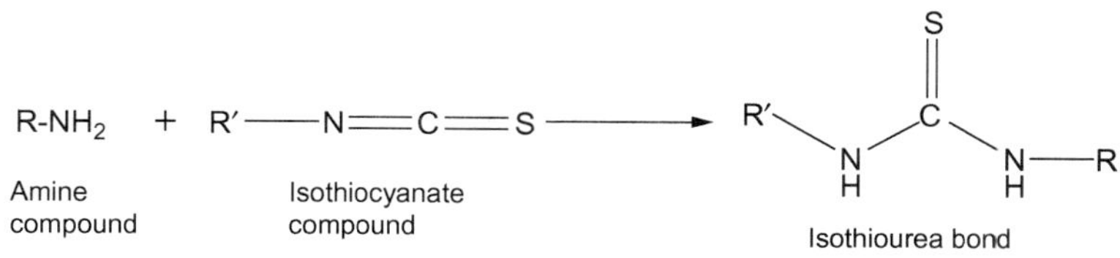

Scheme 12.3 Reaction mechanism of amine conjugation through isothioether bond formation.

of a sulfhydryl-reactive group containing heterobifunctional crosslinkers often use an amine-reactive group. The primary coupling reactions for the modification of sulfhydryls can proceed by either alkylation or disulfide interchange. These sufhydryl-reactive groups are stable in aqueous medium to allow sequential conjugation strategy with stable intermediate. These reactions are rapid and occur at high yield to give thioether or disulfide bonds. Maleimides derivatives are the product of the reaction between maleic anhydride and amine-containing compounds. The double bond of maleimides undergoes an alkylation reaction with sulfhydryl groups to form stable thioether bonds (Scheme 12.4). Maleimide reactions are highly specific for thiol-containing mediators in the pH range of 6.5–7.5 [89–91]. The reaction of the maleimide proceeds at 1000 times greater rate with sulfhydryls groups than with amines at pH 7. One of the carbons adjacent to the maleimide double bond undergoes nucleophilic attack by the thiolate anion to generate the addition product.

Another strategy for the thiol conjugation involves the exchange of disulfide bonds with another thiol group. For example, the thiol group of the cysteine residue present in protein/peptide reacts with a disulfide group on a polymeric surface. A main advantage of this reaction is that the disulfide group reacts specifically with only thiols under a pH range (pH 3–10). The reaction is completely reversible where disulfide bond can be reduced using dithiothreitol (DTT). Most of disulfide crosslinking reactions are done under physiological conditions to maintain stability of the protein or other biomolecule. The most popular activated disulfide is o-pyridyl disulfide in developing crosslinkers. It reacts with the end product; as it is removed from the substrate as a nonreactive compound, it can no longer be reactive in the thiol-disulfide exchange reaction [92]. Recently, an initiator was developed to introduce a maleimide functional group on PCL surface via the ring-opening polymerization ε-caprolactone in the presence of stannous octoate as a catalyst. The maleimide functional group then covalently linked to reduced BSA via the maleimide-sulfhydryl reaction [93].

The thiol-Michael addition reaction has gained significant attention for surface modification of biomaterials [94]. It also offers an excellent spatiotemporal control on

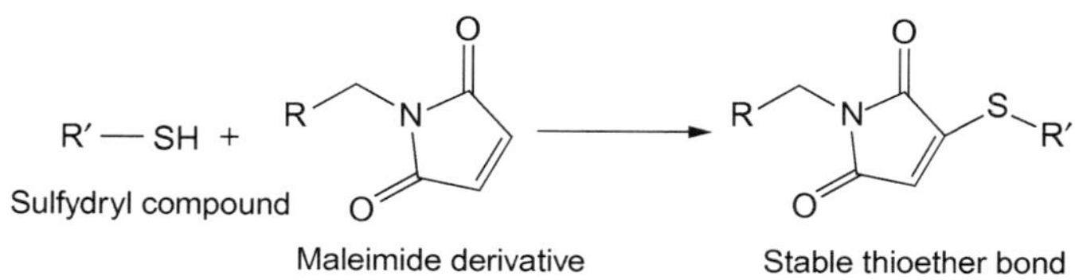

Scheme 12.4 Reaction mechanism of coupling of maleimide derivatives with sulfydryl groups which leads to the formation of stable thioether bond.

modified surface by controlling several reaction parameters [95–98]. These reactions are extensively implemented in the modification of materials. The thiol-maleimide, thiolvinyl sulfone, thiol-yne, and thiol-acrylate are the most commonly employed Michael addition reactions [99,100]. These reactions are classified based on the Michael acceptors in thiol-Michael addition reactions, which are electron-deficient enes, e.g., maleimides, acrylates, vinyl sulfones, and methacrylates (Scheme 12.5). The RGD peptide functionalized with thiol groups was immobilized on acrylic ester or maleinimide functionalized surfaces via the Michael addition reaction [101]. Furthermore, click chemistry [102] and photo-induced reactions have also been employed to immobilize the RGD peptide to material surfaces [103].The multicomponent novel coating can simultaneously perform the azide-alkyne click and thiol-maleimide click reaction. The azide-terminated PEG were first coupled with methyl propionate groups and maleimide groups were utilized for the conjugation of Cys-Arg-Glu-Asp-Val (CREDV) peptides [104]. Michael-type addition reaction can be used to conjugate thiol-containing antibody-fragment (Fab) with the co-polymer of a linear poly(dimethylamino)ethyl methacrylate (pDMAEMA) and poly(ethyleneglycol) methylether acrylate (PEGMEA). The reaction involves base-catalyzed thio-acrylate addition mechanism. Whereas, the thiolate ion and unsaturated bond of acrylate act as Michael donor and Michael acceptor, respectively [105]. In another approach, site-specific

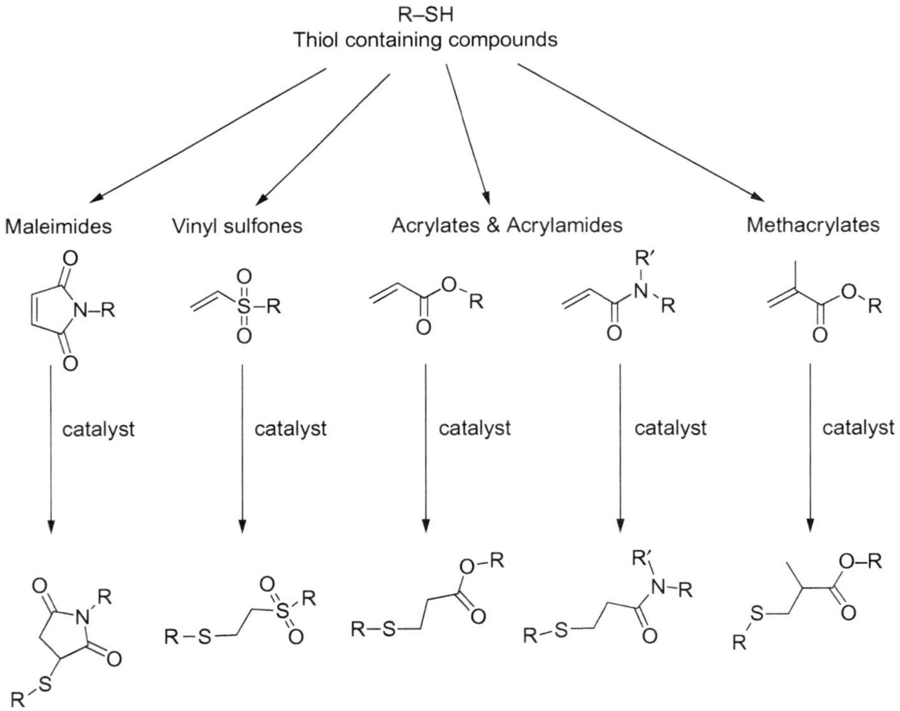

Scheme 12.5 Reaction mechanism of conjugation of thiol groups with vinyl groups containing compounds.

conjugation of PEG-maleimide (PEG-mal) to a small-chain variable fragment (scFv) of antibody was also reported. It was also demonstrated that the monovalent and bivalent PEG-mals can also be conjugated to the scFv [106]. Targeted immobilization of Cys-Ala-Gly (CAG) tripeptide has shown to enhance endothelial cell adhesion on polycarbonate urethane (PCU) surface. The CAG tripeptide was grafted via photo-initiated thiol-ene click chemistry [107]. Steel surfaces coated with polyallylamine bisphosphonate comprising latent thiol groups were generated by the treatment of dithiothreitol (DTTT). Recombinant CD47 or CD47 peptide were attached to the thiol-terminated surface using sulfo-LCSPDP (sulfosuccinimidyl 6-(3'-(2-pyridyldithio) propionamido)hexanoate) crosslinker. Also, the thiol-reactive protein moiety is attached via disulfide bridges to the functionalized surface in the course of reaction between surface thiols and protein-bound pyridyldithio (PDT) groups [108].

12.2.3 Sialanization

The bifunctional silane crosslinkers have shown their application in bioconjugation of biomacromolecules [108a]. Commercially available silane coupling agents contain functional groups to covalently link biomolecules to inorganic substrates [108b]. Inorganic surfaces treated with a suitable silane crosslinker can couple antibodies [108c,108d], oligonucleotides [108e–108g], or other biomolecules [108h] containing appropriate functional groups for reaction. By careful choice of the proper organo-functional component, a silane coupling agent can be used to design virtually any bioconjugation complex. The general structure of silane crosslinker is shown in Scheme 12.6A. Silicon is able to form four covalent bonds with four other carbon atoms. The organic chain attached to Si atom usually terminates in functional groups, which facilitates covalent bonding with other reactive molecules. Another terminus of the silane crosslinker consists of reactive groups directly attached to the silicon atom. Functional groups attached directly to silicon atom may comprise a simple hydrogen atom called silicon hydride or a chlorine atom (chlorosilane), –OH group (silanol), or most importantly groups containing methyl ether called methoxy- or ethoxysilane. These silane derivatives are very reactive to form covalent linkages with other biomolecules or material surfaces. The functional groups attached to organic part of organosilanes are more useful, which allows the conjugation of the silane molecule to other functional organic compounds. The reactive groups attached to silane atom are typically unreactive toward organic molecules. But it can covalently couple with inorganic substrates. The bifunctional silane derivatives promote the bonding of a functional organic molecule to an inorganic surfaces. Alkoxysilanes are widely used for the modification of inorganic surfaces due to the instability of silanol derivatives in aqueous environment by forming hydrogen bonding [108i]. Alkoxysilanes do not react with hydroxyl groups present on substrate under ambient temperature. Similarly, ethoxysilanes are virtually unreactive to –OH groups containing substrates without prior hydrolysis. In contrast, methoxysilanes are very reactive, but it can only react slowly at room temperature. However, under appropriate conditions, both methoxy- and chlorosilane groups are reactive with the functionalities on inorganic substrate without hydrolysis. The addition of a catalyst to the reaction has shown to enhance the

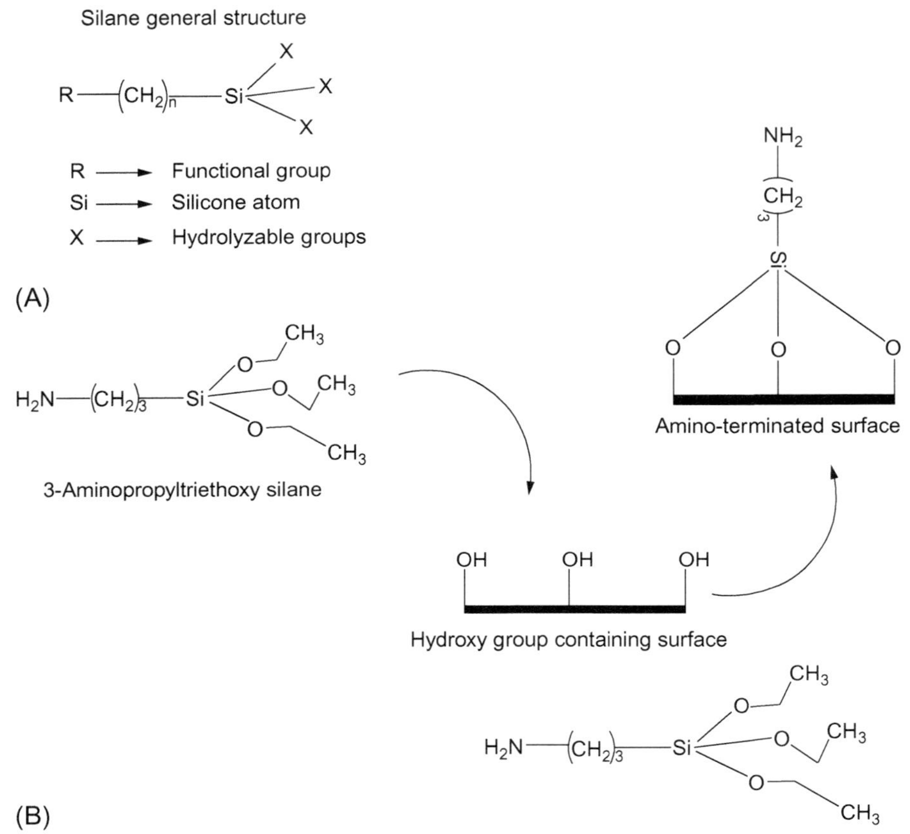

Scheme 12.6 (A) General structure of silane. (B) Amino functionalization of surface (hydroxyl) using 3-aminopropyltriethoxysilane (APTES).

hydrogen bonding capability of the hydroxyl of substrate hydroxyls, which increases the rate of methoxysilane reactions (Scheme 12.6B) [109]. Chlorosilanes and methoxysilanes can also react in an organic solvent (tetrahydrofuran, toluene, or hydrocarbon solvents) under refluxing conditions. In this case, a siloxane polymer network does not form in solution, because no hydrolysis occurs to form silanols on the silane-coupling agents. Silanation has advantage in forming monolayer via the formation of sioxane bond, instead of a thick polymer layer forming on the substrate to create functionality. Silane-conjugated surfaces protrude the organic functional group for conjugation of biological molecules. They have been widely used for the conjugation of protein molecules on the surface of cardiovascular stents [109a,109b].

Periodate-oxidized (PO) site-directed immobilization provides better control on the orientation of the attached antibodies and retaining immunoactivity. The 316L stainless steel (316LSS) surface was initially coated with ethylene (vinylacetate) and followed by oxygen plasma treatment and finally functionalization of surface by (3-aminopropyl) triethoxysilane (APTES). The periodate-oxidized anti-CD34 antibodies were then immobilized on the modified 316LSS surface [110]. Poly(L-lactide)-coated vascular stents with the aim of immobilizing anti-CD34 antibody required surface activation treatments to create the formation of terminal amino groups and covalent attachment antibody through APTES crosslinking agent [33]. The immobilization of recombinant antibody fragments (scFv) on stainless steel 316L was performed to enhance human EPC growth and viability. The stent surface was modified using titania-based coating to increase the hydroxyl groups for silanization. The silanization was performed with APTES to provide amine groups on the surface. The glycosylated scFv were successfully immobilized on silanated surface and most uniform coating with scFv was obtained at 37°C after the oxidation of glycan chain [111]. Biofunctionalization of cobalt chromium (CoCr) alloy surfaces of stents with REDV sequence-modified elastin-like recombinamers (ELR) was done to enhance the endothelialization. CoCr surfaces were activated with plasma treatment and etched with sodium hydroxide before silanization with 3-chloropropyltriethoxysilane. CoCr stent surfaces were successfully functionalized with ELR with an REDV sequence. It has demonstrated higher cell adhesion of HUVEC cells on the biofunctionalized CoCr surface [21].

12.2.4 Reversible addition fragmentation chain transfer

A living radical polymerization that does not end on its own, but continues to synthesize long polymeric chains with controlled molecular weights is known as reversible addition fragmentation chain transfer (RAFT). It directs a controlled reaction mechanism which allows engineering of polymer bioconjugates. The RAFT approach has ability to control the polymerization reaction using wide range of monomers in common and easy reaction conditions involving solvents including water, moderate temperatures, chain transfer agents, and free radical initiators. It has gained much attention in recent years for its potential use in conjugation of biomolecules to polymeric biomaterials [112]. RAFT-mediated bioconjugation allows the preparation of protein-polymer conjugates with well-defined, site-directed, physical, and chemical bioconjugate architectures. The impact of the RAFT technique on bioconjugate preparation should be considerable, particularly in applications such as drug delivery, diagnostics/biosensors, and biopurification in which the physicochemical features of the conjugates need to be precisely tuned for consistent performance. Importantly, the "click" reaction along with RAFT provides a versatile toolbox for highly efficient and highly selective protein conjugation on biomaterials and does not occur naturally and so does not interfere with the functionality of biomolecules [113].

12.2.5 Chemoselective ligation

Chemoselective bioorthogonal reagents contain a reactive group that will only react with another specific group without any potential cross-reactivity with other functionalities present on the biomolecules. The Diels-Alder reaction is known for carbon-carbon bond formation in organic synthesis [114]. Dienophile is the double bond reactant and electron-withdrawing groups (–COOH, –CHO, and –COR groups–) present next to the alkene enhance rate of chemical reaction. The maleimide containing dienophile is present in broad range of commercially available bioconjugation reagents. 3,5-hexadiene phosphoramidite-modified oligonucleotides were prepared by solid phase synthesis using a 3,5-hexadiene phosphoramidite derivative. The functional group was incorporated at the 5′ end of oligonucleotide [115]. Sun et al. [116] have demonstrated sequential Diels-Alder and azide-alkyne (click chemistry) cycloaddition reactions for the immobilization of proteins and carbohydrate ligands on solid surface. The "alkyne-azide" click chemistry was also used for the heparin immobilization with enhanced number of functional groups [117]. The glass slides functionalized with maleimidocaproyl groups (dienophile) react with cyclopentadiene group of heterobifunctional PEG spacer containing an alkyne at another end. The alkyne groups of PEG spacer attach to the azide-containing ligand using click chemistry. The "click" reaction chemistry includes cycloaddition reactions, such as strain-promoted alkyne-azide cycloaddition (SPAAC); 1,3-dipolar family, and hetero Diels-Alder reactions [118]; carbonyl chemistry, such as the oxime ethers ligation, hydrazones, and aromatic heterocycles; in addition to carbon-carbon multiple bonds, such as epoxidation [119] and dihydroxylation [120] and azide-phosphine coupling (Staudinger ligation) [121,122], copper(I)-catalyzed alkyne-azide cycloaddition (CuAAC) [123], thiol-ene [124], and thiol-Michael addition [94]. It allows peptide-functionalization on material's surface in a controlled manner. Alternatively, surface hydroxyl groups of material can also be preactivated by tresyl chloride [125] and *p*-nitrophenyl carbonate [126] for immobilization of carboxy groups containing surfaces. The YIGSR peptide segment of laminin is crucial for binding of integrin and was grafted on polyethylene terephthalate (PET) and polytetrafluoroethylene (PTFE). The hydroxylated surface of PET and PTFE were used to conjugate YIGSR peptide through N-terminal amine using tresyl chloride chemistry [127].

12.3 Future trends

Conjugation of biomediators on biomaterial surface has attracted considerable interest in developing biocompatible cardiovascular grafts and stents. A wide range of chemical compounds are available for targeted immobilization of important biomolecules in correct orientation to further strengthen the clinical success of cardiovascular stents. Moreover, the applications of above-mentioned target conjugation methods are still under investigation to improve site-specificity and functionality when conjugated with the applied biomaterials. There is a need to apply these conjugation methods on cardiovascular stents to further enhance endothelial attachment and decrease toxicity.

References

[1] X. Ren, Y. Feng, J. Guo, H. Wang, Q. Li, J. Yang, X. Hao, J. Lv, N. Ma, W. Li, Surface modification and endothelialization of biomaterials as potential scaffolds for vascular tissue engineering applications, Chem. Soc. Rev. 44 (15) (2015) 5680–5742.

[2] S. Murugesan, J. Xie, R.J. Linhardt, Immobilization of heparin: approaches and applications, Curr. Top. Med. Chem. 8 (2) (2008) 80–100.

[3] P.K. Shireman, H.P. Greisler, Mitogenicity and release of vascular endothelial growth factor with and without heparin from fibrin glue, J. Vasc. Surg. 31 (5) (2000) 936–943.
[4] S. Wang, Y. Zhang, H. Wang, Z. Dong, Preparation, characterization and biocompatibility of electrospinning heparin-modified silk fibroin nanofibers, Int. J. Biol. Macromol. 48 (2) (2011) 345–353.
[5] J.P. Chen, C.H. Su, Surface modification of electrospun PLLA nanofibers by plasma treatment and cationized gelatin immobilization for cartilage tissue engineering, Acta Biomater. 7 (1) (2011) 234–243.
[6] C.D. Spicer, B.G. Davis, Selective chemical protein modification, Nat. Commun. 5 (2014) 4740.
[7] A.J. Melchiorri, N. Hibino, J.P. Fisher, Strategies and techniques to enhance the in situ endothelialization of small-diameter biodegradable polymeric vascular grafts, Tissue Eng. Part B Rev. 19 (4) (2013) 292–307.
[8] C.H. Coyle, S. Mendralla, S. Lanasa, K.N. Kader, Endothelial cell seeding onto various biomaterials causes superoxide-induced cell death, J. Biomater. Appl. 22 (1) (2007) 55–69.
[9] Y. Hirano, D.J. Mooney, Peptide and protein presenting materials for tissue engineering, Adv. Mater. 16 (1) (2004) 17–25.
[10] A. de Mel, G. Jell, M.M. Stevens, A.M. Seifalian, Biofunctionalization of biomaterials for accelerated in situ endothelialization: a review, Biomacromolecules 9 (11) (2008) 2969–2979.
[11] L.P. Brewster, D. Bufallino, A. Ucuzian, H.P. Greisler, Growing a living blood vessel: insights for the second hundred years, Biomaterials 28 (34) (2007) 5028–5032.
[12] B. Krijgsman, A.M. Seifalian, H.J. Salacinski, N.R. Tai, G. Punshon, B.J. Fuller, G. Hamilton, An assessment of covalent grafting of RGD peptides to the surface of a compliant poly(carbonate-urea)urethane vascular conduit versus conventional biological coatings: its role in enhancing cellular retention, Tissue Eng. 8 (4) (2002) 673–680.
[13] A. Tan, D. Goh, Y. Farhatnia, G. Natasha, J. Lim, S.H. Teoh, J. Rajadas, M.S. Alavijeh, A.M. Seifalian, An anti-CD34 antibody-functionalized clinical-grade POSS-PCU nanocomposite polymer for cardiovascular stent coating applications: a preliminary assessment of endothelial progenitor cell capture and hemocompatibility, PLoS ONE 8 (10) (2013) e77112.
[14] Y. Chen, X. Zheng, H. Ji, C. Ding, Effect of Ti-OH formation on bioactivity of vacuum plasma sprayed titanium coating after chemical treatment, Surf. Coat. Technol. 202 (3) (2007) 494–498.
[15] H. Tamura, A. Tanaka, K. Mita, R. Furuichi, Surface hydroxyl site densities on metal oxides as a measure for the ion-exchange capacity, J. Colloid Interface Sci. 209 (1) (1999) 225–231.
[16] X. Chen, P. Sevilla, C. Aparicio, Surface biofunctionalization by covalent co-immobilization of oligopeptides, Colloids Surf. B Biointerfaces 107 (2013) 189–197.
[17] T. Hanawa, A comprehensive review of techniques for biofunctionalization of titanium, J. Periodontal Implant Sci. 41 (6) (2011) 263–272.
[18] X. Liu, P.K. Chu, C. Ding, Surface modification of titanium, titanium alloys, and related materials for biomedical applications, Mater. Sci. Eng. R Rep. 47 (3–4) (2004) 49–121.
[19] V. Paredes, E. Salvagni, E. Rodriguez, F.J. Gil, J.M. Manero, Assessment and comparison of surface chemical composition and oxide layer modification upon two different activation methods on a cocrmo alloy, J. Mater. Sci. Mater. Med. 25 (2) (2014) 311–320.
[20] C. Aparicio, J.M. Manero, F. Conde, M. Pegueroles, J.A. Planell, M. Vallet-Regi, F.J. Gil, Acceleration of apatite nucleation on microrough bioactive titanium for bone-replacing implants, J. Biomed. Mater. Res. A 82 (3) (2007) 521–529.

[21] M.I. Castellanos, A.S. Zenses, A. Grau, J.C. Rodriguez-Cabello, F.J. Gil, J.M. Manero, M. Pegueroles, Biofunctionalization of REDV elastin-like recombinamers improves endothelialization on CoCr alloy surfaces for cardiovascular applications, Colloids Surf. B: Biointerfaces 127 (2015) 22–32.
[22] T. Hanawa, An overview of biofunctionalization of metals in Japan, J. R. Soc. Interface 6 (Suppl 3) (2009) S361–S369.
[23] Y. Ikada, Surface modification of polymers for medical applications, Biomaterials 15 (10) (1994) 725–736.
[24] S.G. Wise, A. Waterhouse, A. Kondyurin, M.M. Bilek, A.S. Weiss, Plasma-based biofunctionalization of vascular implants, Nanomedicine (London) 7 (12) (2012) 1907–1916.
[25] J. Davies, C.S. Nunnerley, A.C. Brisley, R.F. Sunderland, J.C. Edwards, P. Krüger, R. Knes, A.J. Paul, S. Hibbert, Argon plasma treatment of polystyrene microtiter wells. Chemical and physical characterisation by contact angle, ToF-SIMS, XPS and STM, Colloids Surf. A Physicochem. Eng. Asp. 174 (3) (2000) 287–295.
[26] V.B. Ivanov, J. Behnisch, A. Holländer, F. Mehdorn, H. Zimmermann, Determination of functional groups on polymer surfaces using fluorescencelabelling, Surf. Interface Anal. 24 (4) (1996) 257–262.
[27] F. Poncin-Epaillard, G. Legeay, Surface engineering of biomaterials with plasma techniques, J. Biomater. Sci. Polym. Ed. 14 (10) (2003) 1005–1028.
[28] Y. Yin, N.J. Nosworthy, H. Youssef, B. Gong, M.M.M. Bilek, D.R. McKenzie, Acetylene plasma coated surfaces for covalent immobilization of proteins, Thin Solid Films 517 (17) (2009) 5343–5346.
[29] M.M. Bilek, D.R. McKenzie, Plasma modified surfaces for covalent immobilization of functional biomolecules in the absence of chemical linkers: towards better biosensors and a new generation of medical implants, Biophys. Rev. 2 (2) (2010) 55–65.
[30] Y. Yin, M.M.M. Bilek, D.R. McKenzie, N.J. Nosworthy, A. Kondyurin, H. Youssef, M.J. Byrom, W. Yang, Acetylene plasma polymerized surfaces for covalent immobilization of dense bioactive protein monolayers, Surf. Coat. Technol. 203 (10–11) (2009) 1310–1316.
[31] Y. Yin, S.G. Wise, N.J. Nosworthy, A. Waterhouse, D.V. Bax, H. Youssef, M.J. Byrom, M.M. Bilek, D.R. McKenzie, A.S. Weiss, M.K. Ng, Covalent immobilisation of tropoelastin on a plasma deposited interface for enhancement of endothelialisation on metal surfaces, Biomaterials 30 (9) (2009) 1675–1681.
[32] A. Waterhouse, Y. Yin, S.G. Wise, D.V. Bax, D.R. McKenzie, M.M. Bilek, A.S. Weiss, M.K. Ng, The immobilization of recombinant human tropoelastin on metals using a plasma-activated coating to improve the biocompatibility of coronary stents, Biomaterials 31 (32) (2010) 8332–8340.
[33] S. Petersen, A. Strohbach, R. Busch, S.B. Felix, K.P. Schmitz, K. Sternberg, Site-selective immobilization of anti-CD34 antibodies to poly(L-lactide) for endovascular implant surfaces, J. Biomed. Mater. Res. B Appl. Biomater. 102 (2) (2014) 345–355.
[34] E.N. Bolbasov, M. Rybachuk, A.S. Golovkin, L.V. Antonova, E.V. Shesterikov, A.I. Malchikhina, V.A. Novikov, Y.G. Anissimov, S.I. Tverdokhlebov, Surface modification of poly(L-lactide) and polycaprolactone bioresorbable polymers using RF plasma discharge with sputter deposition of a hydroxyapatite target, Mater. Lett. 132 (2014) 281–284.
[35] H.J. Steffen, J. Schmidt, A. Gonzalez-Elipe, Biocompatible surfaces by immobilization of heparin on diamond-like carbon films deposited on various substrates, Surf. Interface Anal. 29 (6) (2000) 386–391.

[36] P.W. Serruys, H.M. Garcia-Garcia, Y. Onuma, From metallic cages to transient bioresorbable scaffolds: change in paradigm of coronary revascularization in the upcoming decade? Eur. Heart J. 33 (1) (2012) 16–25.
[37] A. Strohbach, R. Busch, Polymers for cardiovascular stent coatings, Int. J. Polym. Sci. 2015 (2015) 11.
[38] L. Xue, H.P. Greisler, Biomaterials in the development and future of vascular grafts, J. Vasc. Surg. 37 (2) (2003) 472–480.
[39] A. Srivastava, I.B. O'Connor, A. Pandit, J. Gerard Wall, Polymer-antibody fragment conjugates for biomedical applications, Prog. Polym. Sci. 39 (2) (2014) 308–329.
[40] G.T. Hermanson, Functional Targets for Bioconjugation, Bioconjugate Techniques, third ed., Academic Press, Boston, 2013. pp. 127–228 (Chapter 2).
[41] A. Srivastava, C. Cunningham, A. Pandit, J.G. Wall, Improved gene transfection efficacy and cytocompatibility of multifunctional polyamidoamine-cross-linked hyaluronan particles, Macromol. Biosci. 15 (5) (2015) 682–690.
[42] A. Williams, I.T. Ibrahim, A new mechanism involving cyclic tautomers for the reaction with nucleophiles of the water-soluble peptide coupling reagent 1-ethyl-3-(3'-(dimethylamino)propyl)carbodiimide (EDC), J. Am. Chem. Soc. 103 (24) (1981) 7090–7095.
[43] M.A. Gilles, A.Q. Hudson, C.L. Borders Jr., Stability of water-soluble carbodiimides in aqueous solution, Anal. Biochem. 184 (2) (1990) 244–248.
[44] J.V. Staros, N-hydroxysulfosuccinimide active esters: bis(N-hydroxysulfosuccinimide) esters of two dicarboxylic acids are hydrophilic, membrane-impermeant, protein crosslinkers, Biochemistry 21 (17) (1982) 3950–3955.
[45] J. Li, D. Li, F. Gong, S. Jiang, H. Yu, Y. An, Anti-CD133 antibody immobilized on the surface of stents enhances endothelialization, Biomed. Res. Int. 2014 (2014) 902782.
[46] C.L. Song, Q. Li, Y.P. Yu, G. Wang, J.P. Wang, Y. Lu, J.C. Zhang, H.Y. Diao, J.G. Liu, Y.H. Liu, J. Liu, Y. Li, D. Cai, B. Liu, Study of novel coating strategy for coronary stents: simutaneous coating of VEGF and anti-CD34 antibody, Rev. Bras. Cir. Cardiovasc. 30 (2) (2015) 159–163.
[47] Q. Gao, Y. Chen, Y. Wei, X. Wang, Y. Luo, Heparin-grafted poly(tetrafluoroethylene-co-hexafluoropropylene) film with highly effective blood compatibility via an esterification reaction, Surf. Coat. Technol. 228 (Suppl. 1) (2013) S126–S130.
[48] M.F. Elahi, G. Guan, L. Wang, X. Zhao, F. Wang, M.W. King, Surface modification of silk fibroin fabric using layer-by-layer polyelectrolyte deposition and heparin immobilization for small-diameter vascular prostheses, Langmuir 31 (8) (2015) 2517–2526.
[49] R.A. Hoshi, R. Van Lith, M.C. Jen, J.B. Allen, K.A. Lapidos, G. Ameer, The blood and vascular cell compatibility of heparin-modified ePTFE vascular grafts, Biomaterials 34 (1) (2013) 30–41.
[50] G. Bayramoğlu, M. Yılmaz, E. Batislam, M.Y. Arıca, Heparin-coated poly(hydroxyethyl methacrylate/albumin) hydrogel networks: in vitro hemocompatibility evaluation for vascular biomaterials, J. Appl. Polym. Sci. 109 (2) (2008) 749–757.
[51] C. Zhang, J. Jin, J. Zhao, W. Jiang, J. Yin, Functionalized polypropylene non-woven fabric membrane with bovine serum albumin and its hemocompatibility enhancement, Colloids Surf. B: Biointerfaces 102 (2013) 45–52.
[52] S. Absar, Y.M. Kwon, F. Ahsan, Bio-responsive delivery of tissue plasminogen activator for localized thrombolysis, J. Control. Release 177 (2014) 42–50.
[53] E. Arenas, F.F. Castillón, M.H. Farías, EDC and sulfo-NHS functionalized on PVC-g-PEGMA for streptokinase immobilization, Des. Monomers Polym. 15 (4) (2012) 369–378.

[54] H. Chen, Y. Teramura, H. Iwata, Co-immobilization of urokinase and thrombomodulin on islet surfaces by poly(ethylene glycol)-conjugated phospholipid, J. Control. Release 150 (2) (2011) 229–234.
[55] S. Li, J.J. Henry, Nonthrombogenic approaches to cardiovascular bioengineering, Annu. Rev. Biomed. Eng. 13 (2011) 451–475.
[56] Z. Tang, X. Liu, Y. Luan, W. Liu, Z. Wu, D. Li, H. Chen, Regulation of fibrinolytic protein adsorption on polyurethane surfaces by modification with lysine-containing copolymers, Polym. Chem. 4 (22) (2013) 5597–5602.
[57] S. Jo, P.S. Engel, A.G. Mikos, Synthesis of poly(ethylene glycol)-tethered poly(propylene fumarate) and its modification with GRGD peptide, Polymer 41 (21) (2000) 7595–7604.
[58] J. Xiao, R. Chen, M.A. Pawlicki, T.J. Tolbert, Targeting a homogeneously glycosylated antibody fc to bind cancer cells using a synthetic receptor ligand, J. Am. Chem. Soc. 131 (38) (2009) 13616–13618.
[59] K.L. Wark, P.J. Hudson, Latest technologies for the enhancement of antibody affinity, Adv. Drug Deliv. Rev. 58 (5-6) (2006) 657–670.
[60] J.R. Keogh, Method for covalent attachment of biomolecules to surfaces of medical devices, Google Patents, 2000.
[61] E.J. Jung, J.D. Lee, S.Y. Lee, H.O. Park, Gold-coated stent coated/plated the chemicals and oligonucleotide binding gold-coated stent and process for producing the same, Google Patents, 2009.
[62] Y. Yang, P. Qi, Y. Ding, M.F. Maitz, Z. Yang, Q. Tu, K. Xiong, Y. Leng, N. Huang, A biocompatible and functional adhesive amine-rich coating based on dopamine polymerization, J. Mater. Chem. B 3 (1) (2015) 72–81.
[63] Y.S. Jang, Y.S. Ryu, K.C. Hwang, K.H. Chung, D.L. Cho, M.H. Jeong, A stent coated with anti-integrin antibody and process for preparing the same, Google Patents, 2006.
[64] F.P.J.T. Rutjes, Bioconjugation: how to pick a single amine? Nat. Chem. Biol. 11 (5) (2015) 306–307.
[65] J.M. Harris, E.C. Struck, M.G. Case, M.S. Paley, M. Yalpani, J.M. Van Alstine, D.E. Brooks, Synthesis and characterization of poly(ethylene glycol) derivatives, J. Polym. Sci. Polym. Chem. Ed. 22 (2) (1984) 341–352.
[66] S.M. Chamow, T.P. Kogan, M. Venuti, T. Gadek, R.J. Harris, D.H. Peers, J. Mordenti, S. Shak, A. Ashkenazi, Modification of CD4 immunoadhesin with monomethoxypoly(ethylene glycol) aldehyde via reductive alkylation, Bioconjug. Chem. 5 (2) (1994) 133–140.
[67] J.M. Goddard, J.N. Talbert, J.H. Hotchkiss, Covalent attachment of lactase to low-density polyethylene films, J. Food Sci. 72 (1) (2007) E036–E041.
[68] J. Holland, L. Hersh, M. Bryhan, E. Onyiriuka, L. Ziegler, Culture of human vascular endothelial cells on an RGD-containing synthetic peptide attached to a starch-coated polystyrene surface: comparison with fibronectin-coated tissue grade polystyrene, Biomaterials 17 (22) (1996) 2147–2156.
[69] A.B. Jozwiak, C.M. Kielty, R.A. Black, Surface functionalization of polyurethane for the immobilization of bioactive moieties on tissue scaffolds, J. Mater. Chem. 18 (19) (2008) 2240–2248.
[70] A. Basu, K. Yang, M. Wang, S. Liu, R. Chintala, T. Palm, H. Zhao, P. Peng, D. Wu, Z. Zhang, J. Hua, M.C. Hsieh, J. Zhou, G. Petti, X. Li, A. Janjua, M. Mendez, J. Liu, C. Longley, M. Mehlig, V. Borowski, M. Viswanathan, D. Filpula, Structure-function engineering of interferon-beta-1b for improving stability, solubility, potency, immunogenicity, and pharmacokinetic properties by site-selective mono-PEGylation, Bioconjug. Chem. 17 (3) (2006) 618–630.

[71] O. Kinstler, G. Molineux, M. Treuheit, D. Ladd, C. Gegg, Mono-N-terminal poly(ethylene glycol)-protein conjugates, Adv. Drug Deliv. Rev. 54 (4) (2002) 477–485.
[72] O.B. Kinstler, D.N. Brems, S.L. Lauren, A.G. Paige, J.B. Hamburger, M.J. Treuheit, Characterization and stability of N-terminally PEGylated rhG-CSF, Pharm. Res. 13 (7) (1996) 996–1002.
[73] H. Lee, I.H. Jang, S.H. Ryu, T.G. Park, N-terminal site-specific mono-PEGylation of epidermal growth factor, Pharm. Res. 20 (5) (2003) 818–825.
[74] R. Mallik, T. Jiang, D.S. Hage, High-performance affinity monolith chromatography: development and evaluation of human serum albumin columns, Anal. Chem. 76 (23) (2004) 7013–7022.
[75] S. Beni, J.F. Limtiaco, C.K. Larive, Analysis and characterization of heparin impurities, Anal. Bioanal. Chem. 399 (2) (2011) 527–539.
[76] S. Gore, J. Andersson, R. Biran, C. Underwood, J. Riesenfeld, Heparin surfaces: Impact of immobilization chemistry on hemocompatibility and protein adsorption, J. Biomed. Mater. Res. B Appl. Biomater. 102 (8) (2014) 1817–1824.
[77] P.H. Lin, R.L. Bush, Q. Yao, A.B. Lumsden, C. Chen, Evaluation of platelet deposition and neointimal hyperplasia of heparin-coated small-caliber ePTFE grafts in a canine femoral artery bypass model, J. Surg. Res. 118 (1) (2004) 45–52.
[78] J.-H. Jiang, L.-P. Zhu, X.-L. Li, Y.-Y. Xu, B.-K. Zhu, Surface modification of PE porous membranes based on the strong adhesion of polydopamine and covalent immobilization of heparin, J. Membr. Sci. 364 (1–2) (2010) 194–202.
[79] L. Pol-Fachin, H. Verli, Structural glycobiology of heparin dynamics on the exosite 2 of coagulation cascade proteases: implications for glycosaminoglycans antithrombotic activity, Glycobiology 24 (1) (2014) 97–105.
[80] K.T. Lappegard, G. Bergseth, J. Riesenfeld, A. Pharo, P. Magotti, J.D. Lambris, T.E. Mollnes, The artificial surface-induced whole blood inflammatory reaction revealed by increases in a series of chemokines and growth factors is largely complement dependent, J. Biomed. Mater. Res. A 87 (1) (2008) 129–135.
[81] E. Valeur, M. Bradley, Amide bond formation: beyond the myth of coupling reagents, Chem. Soc. Rev. 38 (2) (2009) 606–631.
[82] K. Bergstrom, K. Holmberg, A. Safranj, A.S. Hoffman, M.J. Edgell, A. Kozlowski, B.A. Hovanes, J.M. Harris, Reduction of fibrinogen adsorption on PEG-coated polystyrene surfaces, J. Biomed. Mater. Res. 26 (6) (1992) 779–790.
[83] T. Miron, M. Wilchek, A simplified method for the preparation of succinimidyl carbonate polyethylene glycol for coupling to proteins, Bioconjug. Chem. 4 (6) (1993) 568–569.
[84] S. Zalipsky, R. Seltzer, S. Menon-Rudolph, Evaluation of a new reagent for covalent attachment of polyethylene glycol to proteins, Biotechnol. Appl. Biochem. 15 (1) (1992) 100–114.
[85] M.D. Bentley, M.J. Roberts, J.M. Harris, Reductive amination using poly(ethylene glycol) acetaldehyde hydrate generated in situ: applications to chitosan and lysozyme, J. Pharm. Sci. 87 (11) (1998) 1446–1449.
[86] L. Tao, G. Mantovani, F. Lecolley, D.M. Haddleton, Alpha-aldehyde terminally functional methacrylic polymers from living radical polymerization: application in protein conjugation "pegylation", J. Am. Chem. Soc. 126 (41) (2004) 13220–13221.
[87] K.L. Christman, M.V. Requa, V.D. Enriquez-Rios, S.C. Ward, K.A. Bradley, K.L. Turner, H.D. Maynard, Submicron streptavidin patterns for protein assembly, Langmuir 22 (17) (2006) 7444–7450.
[88] A. Jobbagy, K. Kiraly, Chemical characterization of fluorescein isothiocyanate-protein conjugates, Biochim. Biophys. Acta 124 (1) (1966) 166–175.

[89] X. Chen, Y.W. Wu, Selective chemical labeling of proteins, Org. Biomol. Chem. 14 (24) (2016) 5417–5439.
[90] J.R. Heitz, C.D. Anderson, B.M. Anderson, Inactivation of yeast alcohol dehydrogenase by N-alkylmaleimides, Arch. Biochem. Biophys. 127 (1) (1968) 627–636.
[91] D.G. Smyth, O.O. Blumenfeld, W. Konigsberg, Reactions of N-ethylmaleimide with peptides and amino acids, Biochem. J. 91 (3) (1964) 589–595.
[92] C.T. Kuan, Q.C. Wang, I. Pastan, Pseudomonas exotoxin A mutants. Replacement of surface exposed residues in domain II with cysteine residues that can be modified with polyethylene glycol in a site-specific manner, J. Biol. Chem. 269 (10) (1994) 7610–7616.
[93] Z. Liu, C. Dong, X. Wang, H. Wang, W. Li, J. Tan, J. Chang, Self-assembled biodegradable protein-polymer vesicle as a tumor-targeted nanocarrier, ACS Appl. Mater. Interfaces 6 (4) (2014) 2393–2400.
[94] D.P. Nair, M. Podgórski, S. Chatani, T. Gong, W. Xi, C.R. Fenoli, C.N. Bowman, The thiol-michael addition click reaction: a powerful and widely used tool in materials chemistry, Chem. Mater. 26 (1) (2014) 724–744.
[95] V.S. Khire, D.S.W. Benoit, K.S. Anseth, C.N. Bowman, Ultrathin gradient films using thiol-ene polymerizations, J. Polym. Sci. A Polym. Chem. 44 (24) (2006) 7027–7039.
[96] V.S. Khire, A.M. Kloxin, C.L. Couch, K.S. Anseth, C.N. Bowman, Synthesis, characterization and cleavage of linear polymers attached to silica nanoparticles formed using thiol-acrylate conjugate addition reactions, J. Polym. Sci. A Polym. Chem. 46 (20) (2008) 6896–6906.
[97] V.S. Khire, T.Y. Lee, C.N. Bowman, Surface modification using thiol-acrylate conjugate addition reactions, Macromolecules 40 (16) (2007) 5669–5677.
[98] H. Seto, M. Takara, C. Yamashita, T. Murakami, T. Hasegawa, Y. Hoshino, Y. Miura, Surface modification of siliceous materials using maleimidation and various functional polymers synthesized by reversible addition-fragmentation chain transfer polymerization, ACS Appl. Mater. Interfaces 4 (10) (2012) 5125–5133.
[99] C.E. Hoyle, A.B. Lowe, C.N. Bowman, Thiol-click chemistry: a multifaceted toolbox for small molecule and polymer synthesis, Chem. Soc. Rev. 39 (4) (2010) 1355–1387.
[100] A.B. Lowe, Thiol-ene "click" reactions and recent applications in polymer and materials synthesis, Polym. Chem. 1 (1) (2010) 17–36.
[101] Y.Y. Wang, L.X. Lu, J.C. Shi, H.F. Wang, Z.D. Xiao, N.P. Huang, Introducing RGD peptides on PHBV films through PEG-containing cross-linkers to improve the biocompatibility, Biomacromolecules 12 (3) (2011) 551–559.
[102] F. Shamsi, H. Coster, K.A. Jolliffe, Characterization of peptide immobilization on an acetylene terminated surface via click chemistry, Surf. Sci. 605 (19–20) (2011) 1763–1770.
[103] M. Mizutani, S.C. Arnold, T. Matsuda, Liquid, phenylazide-end-capped copolymers of epsilon-caprolactone and trimethylene carbonate: preparation, photocuring characteristics, and surface layering, Biomacromolecules 3 (4) (2002) 668–675.
[104] M.-Y. Tsai, Y.-C. Chen, T.-J. Lin, Y.-C. Hsu, C.-Y. Lin, R.-H. Yuan, J. Yu, M.-S. Teng, M. Hirtz, M.H.-C. Chen, C.-H. Chang, H.-Y. Chen, Vapor-based multicomponent coatings for antifouling and biofunctional synergic modifications, Adv. Funct. Mater. 24 (16) (2014) 2281–2287.
[105] M. Monaghan, U. Greiser, H. Cao, W. Wang, A. Pandit, An antibody fragment functionalized dendritic PEGylated poly(2-(dimethylamino)ethyl diacrylate) as a vehicle of exogenous microRNA, Drug Deliv. Transl. Res. 2 (5) (2012) 406–414.
[106] A. Natarajan, C.Y. Xiong, H. Albrecht, G.L. DeNardo, S.J. DeNardo, Characterization of site-specific ScFv PEGylation for tumor-targeting pharmaceuticals, Bioconjug. Chem. 16 (1) (2005) 113–121.

[107] M. Khan, J. Yang, C. Shi, J. Lv, Y. Feng, W. Zhang, Surface tailoring for selective endothelialization and platelet inhibition via a combination of SI-ATRP and click chemistry using Cys-Ala-Gly-peptide, Acta Biomater. 20 (2015) 69–81.

[108] J.B. Slee, I.S. Alferiev, C. Nagaswami, J.W. Weisel, R.J. Levy, I. Fishbein, S.J. Stachelek, Enhanced biocompatibility of CD47-functionalized vascular stents, Biomaterials 87 (2016) 82–92.

[108a] B. Lom, K.E. Healy, P.E. Hockberger, A versatile technique for patterning biomolecules onto glass coverslips, J. Neurosci. Meth. 50 (3) (1993) 385–397.

[108b] G.T. Hermanson, Chapter 13 – Silane Coupling Agents, Bioconjugate Techniques, third ed., Academic Press, Boston, 2013, pp. 535–548.

[108c] H.H. Weetall, Preparation of immobilized proteins covalently coupled through silane coupling agents to inorganic supports, Appl. Biochem. Biotechnol. 41 (3) (1993) 157–188.

[108d] H.H. Weetall, M.J. Lee, Antibodies immobilized on inorganic supports, Appl. Biochem. Biotechnol. 22 (3) (1989) 311–330.

[108e] P.T. Charles, G.J. Vora, J.D. Andreadis, A.J. Fortney, C.E. Meador, C.S. Dulcey, D.A. Stenger, Fabrication and Surface Characterization of DNA Microarrays Using Amine- and Thiol-Terminated Oligonucleotide Probes, Langmuir 19 (5) (2003) 1586–1591.

[108f] E. Mavrogiannopoulou, P.S. Petrou, G. Koukouvinos, D. Yannoukakos, A. Siafaka-Kapadai, K. Fornal, K. Awsiuk, A. Budkowski, S.E. Kakabakos, Improved DNA microarray detection sensitivity through immobilization of preformed in solution streptavidin/biotinylated oligonucleotide conjugates, Colloids Surf. B Biointerfaces 128 (2015) 464–472.

[108g] T. Ratajczak, B. Uszczyńska, E. Frydrych-Tomczak, M.K. Chmielewski, The "clickable" method for oligonucleotide immobilization onto azide-functionalized microarrays, in: P.C.H. Li, A. Sedighi, L. Wang (Eds.), Microarray Technology: Methods and Applications, Springer, New York, 2016, pp. 25–36.

[108h] J.L. Zimmermann, T. Nicolaus, G. Neuert, K. Blank, Thiol-based, site-specific and covalent immobilization of biomolecules for single-molecule experiments, Nat. Protoc. 5 (6) (2010) 975–985.

[108i] J.H. Seo, L.-J. Chen, S.V. Verkhoturov, E.A. Schweikert, A. Revzin, The use of glass substrates with bi-functional silanes for designing micropatterned cell-secreted cytokine immunoassays, Biomaterials 32 (23) (2011) 5478–5488.

[109] S.M. Kanan, W.T.Y. Tze, C.P. Tripp, Method to double the surface concentration and control the orientation of adsorbed (3-aminopropyl)dimethylethoxysilane on silica powders and glass slides, Langmuir 18 (17) (2002) 6623–6627.

[109a] V. Hynninen, L. Vuori, M. Hannula, K. Tapio, K. Lahtonen, T. Isoniemi, E. Lehtonen, M. Hirsimäki, J.J. Toppari, M. Valden, V.P. Hytönen, Improved antifouling properties and selective biofunctionalization of stainless steel by employing heterobifunctional silane-polyethylene glycol overlayers and avidin-biotin technology, Sci. Rep. 6 (2016) 29324.

[109b] O. Majumder, A.K. Singh Bankoti, T. Kaur, A. Thirugnanam, A.K. Mondal, The influence of silane and silane-PMMA coatings on the in vitro biodegradation behavior of AE42 magnesium alloy for cardiovascular stent applications, RSC Adv. 6 (109) (2016) 107344–107354.

[110] Y. Yuan, M. Yin, J. Qian, C. Liu, Site-directed immobilization of antibodies onto blood contacting grafts for enhanced endothelial cell adhesion and proliferation, Soft Matter 7 (16) (2011) 7207–7216.

[111] A. Foerster, I. Hołowacz, G.B. Sunil Kumar, S. Anandakumar, J.G. Wall, M. Wawrzyńska, M. Paprocka, A. Kantor, H. Kraskiewicz, S. Olsztyńska-Janus, S.J. Hinder, D. Bialy, H. Podbielska, M. Kopaczyńska, Stainless steel surface functionalization for immobilization of antibody fragments for cardiovascular applications, J. Biomed. Mater. Res. A 104 (4) (2016) 821–832.

[112] C. Boyer, V. Bulmus, T.P. Davis, V. Ladmiral, J. Liu, S. Perrier, Bioapplications of RAFT polymerization, Chem. Rev. 109 (11) (2009) 5402–5436.

[113] W. Tang, M.L. Becker, "Click" reactions: a versatile toolbox for the synthesis of peptide-conjugates, Chem. Soc. Rev. 43 (20) (2014) 7013–7039.

[114] C.S. McKay, M.G. Finn, Click chemistry in complex mixtures: bioorthogonal bioconjugation, Chem. Biol. 21 (9) (2014) 1075–1101.

[115] K.W. Hill, J. Taunton-Rigby, J.D. Carter, E. Kropp, K. Vagle, W. Pieken, D.P.C. McGee, G.M. Husar, M. Leuck, D.J. Anziano, D.P. Sebesta, Diels-Alder bioconjugation of diene-modified oligonucleotides, J. Org. Chem. 66 (16) (2001) 5352–5358.

[116] X.L. Sun, C.L. Stabler, C.S. Cazalis, E.L. Chaikof, Carbohydrate and protein immobilization onto solid surfaces by sequential Diels-Alder and azide-alkyne cycloadditions, Bioconjug. Chem. 17 (1) (2006) 52–57.

[117] S. Dimitrievska, C. Cai, A. Weyers, J.L. Balestrini, T. Lin, S. Sundaram, G. Hatachi, D.A. Spiegel, T.R. Kyriakides, J. Miao, G. Li, L.E. Niklason, R.J. Linhardt, Click-coated, heparinized, decellularized vascular grafts, Acta Biomater. 13 (2015) 177–187.

[118] K.A. Jorgensen, Catalytic asymmetric hetero-Diels-Alder reactions of carbonyl compounds and imines, Angew. Chem. Int. Ed. Engl. 39 (20) (2000) 3558–3588.

[119] H. Adolfsson, A. Converso, K.B. Sharpless, Comparison of amine additives most effective in the new methyltrioxorhenium-catalyzed epoxidation process, Tetrahedron Lett. 40 (21) (1999) 3991–3994.

[120] H.C. Kolb, M.S. VanNieuwenhze, K.B. Sharpless, Catalytic asymmetric dihydroxylation, Chem. Rev. 94 (8) (1994) 2483–2547.

[121] C.I. Schilling, N. Jung, M. Biskup, U. Schepers, S. Brase, Bioconjugation via azide-staudinger ligation: an overview, Chem. Soc. Rev. 40 (9) (2011) 4840–4871.

[122] S.S. van Berkel, M.B. van Eldijk, J.C. van Hest, Staudinger ligation as a method for bioconjugation, Angew. Chem. Int. Ed. Engl. 50 (38) (2011) 8806–8827.

[123] E. Lallana, R. Riguera, E. Fernandez-Megia, Reliable and efficient procedures for the conjugation of biomolecules through huisgen azide-alkyne cycloadditions, Angew. Chem. Int. Ed. 50 (38) (2011) 8794–8804.

[124] M. Kuhlmann, O. Reimann, C.P.R. Hackenberger, J. Groll, Cysteine-functional polymers via thiol-ene conjugation, Macromol. Rapid Commun. 36 (5) (2015) 472–476.

[125] P. Banerjee, D.J. Irvine, A.M. Mayes, L.G. Griffith, Polymer latexes for cell-resistant and cell-interactive surfaces, J. Biomed. Mater. Res. 50 (3) (2000) 331–339.

[126] J.T. Li, J. Carlsson, J.N. Lin, K.D. Caldwell, Chemical modification of surface active poly(ethylene oxide)-poly (propylene oxide) triblock copolymers, Bioconjug. Chem. 7 (5) (1996) 592–599.

[127] S.P. Massia, J.A. Hubbell, Human endothelial cell interactions with surface-coupled adhesion peptides on a nonadhesive glass substrate and two polymeric biomaterials, J. Biomed. Mater. Res. 25 (2) (1991) 223–242.

Functionalized cardiovascular stents: Cardiovascular stents incorporated with stem cells

13

B. Oh, C.H. Lee
University of Missouri-Kansas City, Kansas City, MO, United States

13.1 Introduction

Tissue engineering integrates advanced technologies from cell and molecular biology, genetics, biochemistry, and biomedical engineering to facilitate the development of a highly beneficial platform. As the ultimate success in the transplantation of engineered organ/device resides in the immunological acceptance of the product by the host, reducing the rejection of the transplanted organ/device by the body immune system remains a major challenge in the tissue engineering field.

Nanofiber scaffolds for artificial system need to simulate extracellular matrix (ECM), especially its porous structure and functions [1]. It was found that three-dimensional ECM structure could provide cells with a suitable environment to be proliferated and properly differentiated into tissues [2]. In contrast to polymeric porous matrix or micro-fiber scaffolds, nanofiber has a much similar structure to biological ECM. The mean diameter of nanofibers ranges from 50 to 500 nm [3], and nanofibers have a higher surface to volume ratio, which provides an ample space for cell loading [4,5]. Owing to these versatile properties, nanofibers have been deemed as one of the most biocompatible formulations, further guaranteeing its application to cardiovascular tissue engineering.

Recently, the application of stem cell for regenerative medicine has received an increased attention from interdisciplinary fields including biotechnology, material science, engineering, and even pharmaceutical science [6–9]. Stem cell therapy is an intervention strategy that introduces new adult stem cells into damaged organ/tissue to treat pathological conditions without immune rejection. However, most stem cell applications need to figure out how to control the specialized property, by which stem cells have pluripotency to differentiate into tissue/organ in response to the specific signal from a microenvironment.

As stem cells hardly survive in the body under the harsh conditions, such as hypoxia or enzymatic degradation, enormous efforts have been placed in development of novel biomaterial, which not only mimics extracellular three-dimensional matrix in cardiovascular tissue regeneration, but also enhances in vivo stability of stem cells. The viable approaches for development of novel biomaterial include hypoxic preconditioning, genetic modification, drug combination, and nanofibers releasing biomolecules [10].

Functionalized Cardiovascular Stents. https://doi.org/10.1016/B978-0-08-100496-8.00014-7

In this paper, we thoroughly reviewed on characterizations of nanofiber loaded with stem cells for the treatment of cardiac diseases including atherosclerosis and myocardial infarction (MI). This review will surely help the readers comprehensively understand nanofiber-mediated stem cell therapy in the cardiovascular diseases.

13.2 Adventitial biology for coronary artery disease (CAD)

13.2.1 Significance of macrophage in atherosclerotic plaque

Macrophages possess unique versatility and plasticity of response to environmental signals with varying forms of polarization [11,12]. Foam cells derived from macrophage necrosis are known to be associated with the initial stage of atherosclerosis [13]. The uptake of modified lipoproteins by oxidized-low density lipoprotein (ox-LDL) is an integral step for foam cell formation.

The recruited-monocyte activated by external stimuli (i.e., intercellular uptake of ox-LDL) caused their polarization to M1 macrophage [14], which was implicated in the early phases of atherosclerosis [15]. M1 macrophages released pro-inflammatory cytokines and reactive oxygen or nitrogen species that would aggravate oxidative stress in plaque lesion [16]. M2 macrophages (also known as alternatively activated macrophages), on the other hand, are immunosuppressive and resolve plaque inflammation [17]. Histological analysis of human plaques demonstrated that M1 macrophages are enriched in lipids and localized to inflammatory areas that are distinct from those in which M2 macrophages are localized [18]. As a novel strategy would be to reduce the inflammatory state of plaque macrophages, re-polarizing macrophages to be transformed into M2 phenotype might be particularly effective [16]. Because M2 macrophages secrete antiinflammatory factors and promote tissue remodeling through the clearance of dying cells and debris, the relevant polarizer inducing M2 phenotypes can be emerged as an alternative route for an atherosclerosis treatment.

13.2.2 Macrophage-autophagy (MA) dysfunction in atherosclerotic plaque

Autophagy plays a critical role in turning over organelles and proteins to maintain the cellular homeostasis [19] at the infection site [20]. As autophagy becomes activated by external stimuli (i.e., reactive oxygen species), it induces smooth muscle cells (SMCs) and endothelial cells (ECs) proliferation by eradicating intracellular debris [21]. However, autophagy under excessive stimulation can cause necrotic death of both SMCs and ECs, resulting in thinning fibrous cap and weakening the stability of plaque [22].

MA has contributed to the disruption of vascular pathology, resulting in chronic atherosclerosis [23]. Macrophage accumulated the cholesterol as a form of lipid droplets by uptake of ox-LDL, leading to the formation of foam cells [24]. With excessive oxidative stress from ox-LDL, macrophage ultimately ruptures and accumulates additional lipids into arterial intima. The high level of atherosclerosis-relevant factors triggered the stress reaction of endoplasmic recticulum (ER) in macrophage, causing the activation of ER apoptotic pathways [25]. The necrotic death of macrophage associated with defective phagocytic clearance of the apoptotic cells destabilized plaque, which orchestrates an acute atherothrombotic event [26,27].

The functional relationships between MA and defective efferocytosis have been studied [26]. As MA was inhibited by silencing autophagy mediators (autophagy-related protein-5, ATG5), the increases in apoptosis and oxidative stress mediated from NADH oxidase were observed. During the observations, the necrotic cells (apoptotic debris) were not recognizable through efferocytes. In vivo data using macrophage ATG5-null mice in fat-fed Ldlr−/− suggest that plaque destabilization associated with the necrotic disruption of foam cells and worse in lesional efferocytosis were highly correlated with an autophagy expression. In addition, ATG5-null mice co-incubated with lipid crystals showed remarkable induction of inflammasomes, resulting in deficient stability of plaque in part of hyperactivation of inflammasomes.

13.3 Role of stem/progenitor cells in atherosclerosis

Cell therapy is believed to be the most influential research due to a variety of stem cell differentiation capacity [28]. The cumulated evidences suggested that stem/progenitor cells (SPCs) contributed to the vascular remodeling process and disease progress [29].

Over the past decade, the identification of SMCs-related SPCs have been extensively conducted [30]. This is mainly because the differentiation of SPCs into SMCs lineage could result in the accumulation of monocytes into the atheroma, where they differentiate into pro-inflammatory stage of macrophages [31,32]. A widely accepted explanation of the origin of SPCs is that SPCs are derived from two identical origins: (1) bone marrow-derived SPCs (BM-SPCs) circulating throughout the blood vessels and (2) locally existed SPCs, which are de-differentiated from mature or contractile SMCs (Table 13.1) [39–41].

13.3.1 Migration of BM-SPCs

Among various source of stem cells, antiinflammatory or immunosuppressive ability of mesenchymal stem cells (MSCs) made these cells potentially useful as a cell therapy source [42]. However, the migratory mechanism of BM-SPCs from their original site needs to be elucidated [34–36].

The *trans*-endothelial migration of murine-MSCs (mMSCs) has been affected by the level of shear stress and chemokines [43]. Under continuous flow (0.1 Pa) in a capillary chamber where murine-ECs seeded, the lack of recruitment of mMSCs was observed. However, when the flow paused sometimes (for 10 min), crawling on endothelial surface was observed, resulting in extending microvillous processes. Interestingly, in the presence of chemokines such as CXCL9, CXCL16, and CCL25, it markedly stimulated trans-endothelial recruitment across the EC layer.

Table 13.1 Role of stem/progenitor cells (SPCs) in atherosclerosis

Cell type	Mediator	Differential lineage	Description	References
SPCs migration via blood circulation				
CD34+ Sca-1+ c-kit Flk-1	MMP-8	Hematopoietic cells Leulocytes Macrophages	In vivo $ApoE^{-/-}/MMP\text{-}8^{-/-}$ mice contained small number of SPCs and small size of atherosclerotic lesions compared to MMP-8 expressed mice model. $MMP\text{-}8^{-/-}$ mice contained small	[33]
CD34+/KDR+/CD45+	–	Neointima development	Randomized-clinical trials (recruiting 155 consecutive stable angina patients) Large population of CD34+/KDR+/CD45+ (1.41 ± 0.64 cells/μL) were found in the restenosis groups as compared to that in the stable and control groups (0.95 ± 0.44 cells/μL)	[34]
BM-MSCs	–	Calcified neointima	The calcified lesion by von Kossa staining was obviously observed in neointima layer MSCs-induced calcification might be associated with increase in BMP-2 expression in medial layers	[35]
BM-MSCs	TGF-β1	Foam cells	α-SMA and SM-MHC-1 were highly expressed in the response to TGF-β1 BM-SMCs were differentiated into foam cells after the treatment with ox-LDL, which upregulates the SR-A and down regulate ABCA1	[36]

De-differentiation of SPCs in the adventitia				
Trans-differentiation of SMCs	Phosphate BMP-2	Osteoblasts Chondrocytes	SMCs phenotypic transition toward osteochondroprogenitors since BMP-2 enhanced excess amount of phosphate Calcified vascular lesion was found as trans-differentiation of SMCs by Erk signaling was dominated	[37]
Sca-1+	PDGF-BB	SMCs	In vitro, Sca-1+ was differentiated into SMCs lineage in response to PPDF-BB In vivo, Sca-1+ carrying with LacZ gene enhanced β-gal+cells, leading to increase in atherosclerotic lesions	[31]
Multivascular Stem Cells (MVSCs)	PDGF-B bFGF TGF-β1	MSC-like cells SMCs	Upon vascular endothelial injury, MVSCs expressing Sox10 were found inside the elastic lamina layers TGF-β1 substantially increased early SMC differentiation markers including SMA and CNN1	[38]

In addition, enzyme-mediated crawling or transmigration across ECs has been reported. As the plaque developed by migration of monocytes, mast cells, and cholesterol across EC layer and differentiation of monocytes into macrophages, the latent stage of various type of matrix metalloproteinases (MMPs) such as MMP-9, MMP-14, and MMP-8 were activated [44].

Among MMPs, MMP-8 played a critical role in proteolysis of fibrillar collagen (collagen I) [45]. It was recently reported that MMP-8 enhanced embryonic Sca-1^{+}SPCs recruitment and accumulation in atheroma in vivo [33]. As compared to ApoE$^{-/-}$/MMP-8$^{-/-}$, ApoE$^{-/-}$/MMP-8$^{+/+}$ mice model showed increase in CD34, Sca-1, c-kit, and Flk-1 positive cells trans-located in atherosclerotic lesion. The migrated-SPCs in ApoE$^{-/-}$/MMP-8$^{+/+}$ mice expressed hematopoietic marker (CD45) and macrophage marker (CD68). It would be plausible that recruited-SPCs will be differentiated into macrophages resulting in developing atherosclerotic lesion with multiple mechanisms.

Taken together, these findings have provided significant insight into the role of fluidal mechanics and biological molecules involved with the SPCs-associated pathogenesis of atherosclerosis.

13.3.2 Significance of adventitia SPCs

Cumulative evidences revealed that SPCs are presented in the media layer and adventitia, which contributed to vascular remodeling under pathological condition [31,37,38,46]. Interestingly, as the expression of smooth muscle myosin heavy chain upon vascular injury decreased, it leaded to proliferative and synthetic SMCs [47]. The contractile or phenotypic plasticity of SMCs would de-differentiate into proliferative SMCs [48]. It meant that under pathological condition, releasing a panel of cytokine would orchestrate SPCs differentiation toward proliferative SMCs. To identify the existence of vascular SPCs, vascular tissues from ApoE$^{-/-}$ mice were rigorously investigated [31]. Immunohistochemical staining of adventitia in aortic roots showed large number of stem cell positively having markers such as Sca-1, c-kit, CD34, and Flk-1. In vitro expansion of isolated Sca-1 positive cells showed the differentiation of the cells into SMCs in response to PDGF-BB (platelet-derived growth factor).

13.3.3 Mesenchymal stem cells (MSCs)

MSCs have been considered the most suitable cell source for tissue engineering/cell therapy due to its effective proliferation and being differentiated into specific lineages, such as chondrocytes, adipocytes, and osteocytes [49,50]. In addition, MSCs expressed endothelial, neural, smooth muscle, and cardiac myocyte differentiations, which is indicative of its potential for the treatment of cardiac diseases. Allogeneic MSCs isolated from human bone marrow generated a little immune rejection by T cells, alleviating immune response and fulfilling patients' compliance [50,51].

However, MSC injection into the body showed a very low survival rate. Recently, allogeneic MSC isolated from male Wister rats, which showed fibroblast-like morphology in TEM (transmission electron microscopy) images, was modified with hepatocyte growth factor (HGF) via transfect retrovirus [52]. Since HGF had versatile capabilities, such as antiapoptotic, antiinflammatory, and angiogenic activity, it can enhance a survival rate of MSC [53]. The survival rate of HGF-MSCs greatly increased even 4 weeks after injection, demonstrating the effectiveness of HGF as the controlling factor for ischemic heart disease.

The unique versatility and plasticity of macrophages can be educated with different forms of polarization (M1 and M2) as interacted with external signals from MSCs [54]. The efforts to reveal the MSCs-guided transition from pro-inflammatory stage of M1 polarization to alternative M2 macrophages have been gradually reported [55]. Special consideration of the use of MSCs in atherosclerosis is warranted as the interaction of MSCs with macrophages was extensively studied [56,57]. It was found that the high level of mannose receptor (MR or CD206) expression, well-known as a surface marker of antiinflammatory macrophages, was activated as the macrophages were co-cultured with bone marrow-derived MSCs (BM-MSCs). Furthermore, macrophages co-cultured with BM-MSCs expressed high level of interleukin-10 and -6 expression (IL-10 and IL-6), whereas the level of IL-12 and tumor necrosis factor-alpha (TNF-α) expression was relatively low [56].

In addition to BM-MSCs, the variety of MSCs (derived from human gingiva (GMSCs) and cardiac adipose tissue (AT-MSCs)) has been gradually investigated [58,59]. GMSCs with similar stem cell-like characteristics including immunosuppressive and antiinflammatory function could educate the M1-polarized macrophages, which were re-polarized into M2 phenotype. The increase in the expression of CD206, IL-10, and IL-6 in M1 macrophages with reduction of TNF-α Th-17 cell expansion supported the repolarization of pro-inflammatory stage into antiinflammatory phenotype [59]. It was reported that human cardiac adipose tissue-derived mesenchymal stem cells (AT-MSCs) could be relevant to polarize M1 phenotype into antiinflammatory (M2) phenotype [58].

A morphological assessment showed that there was a star-shaped morphology, which represents a typical marker of M2 phenotype macrophages, such as CD206 and CD163. It was also reported that there was a bidirectional interaction between AT-MSCs and macrophages, indicating that MSCs can be reprogramed by macrophages and the secretion of M2 inducers including IL-4 and IL-13 can be enhanced by MSCs. It can be concluded that various types of MSCs are capable of eliciting M2 phenotype macrophages and contribute to a marked reduction of atherosclerotic lesion.

Despite the great potential of MSCs, a relatively few or none of coating strategy has been introduced. This is mainly due to the lack of viability and the low growth rate of MSCs. It was demonstrated that M2-associated cytokines, such as IL-10, TGF-β1, and VEGF, promoted the growth and viability of MSCs, whereas M1-associated cytokines, such as IL-1b, IL-6, TNF-α, and IFN-γ, inhibited the growth of MSCs in vitro [60]. The re-polarization of M1 phenotype into M2 was observed after 3 days of co-culture, indicating that the timing of implantation could act as a key role in the ultimate success of MSCs therapy against atherosclerosis. The difficulty in the development of successful cell therapy lies mostly in the loss of MSCs encapsulated into a polymeric scaffold from the blood flow and lack of an appropriate delivery carrier.

Along with rigorous efforts, the recent study on the protective role of MSCs in MI and hind limb ischemia has demonstrated that a systemic infusion of MSCs was effective to alleviate atherosclerosis via improvement of endothelial function [61–63]. In vitro co-culture of ECs with MSCs ameliorated stresses exerted by oxidized low-density lipoprotein (oxLDL) and restored the endothelial nitric oxide synthase (eNOS) level. Due to the restoration of eNOS, nitric oxide (NO) production also increased. In addition, transplantation of MSCs on ApoE(−/−) fed with high-fat diet improved endothelial function, resulting in reduction of plaque formation in ApoE(−/−) model [61]. However, the systemic infusion approach is considered to be restricted due to the rapid loss of transfused cells by washout in bloodstream, immunogenicity, and severe environment where cells cannot survive after the injection [64–66]. Cumulative evidences indicated that the beneficial role of stem cells in the treatment of atherosclerosis is closely correlated with the amount of paracrine factors, which are known to improve intrinsic repair mechanisms in the injury sites (i.e., atherosclerotic lesion) [67–70], released from stem cells.

13.3.4 Endothelial progenitor cells (EPCs)

Among various regenerative medicines, cardiac repair has been recognized as one of the most sophisticated fields. Various cardiac progenitor cell (CPC) populations, such as side population CPCs, c-kit+CPCs, Sca-1+CPCs, Islet-1+CPCs, SSEA-1+CPCs, and cardiosphere PCs, have been considered as a suitable cell source for cardiac regeneration [71]. As compared to other stem cells, progenitor cells have an ability to differentiate into specific organs or tissues through a microenvironment adaptation [38]. It also has an advantage of not being overproliferated, implying that it is free of the risk of becoming a cancer cell [72].

It was reported that stem cell antigen-1 positive (Sca-1+) isolated from murine model enhanced the cardiac functions in rat heart after MI. The recovery of the cardiac functions represents the therapeutically essential process in which Sca-1+ was differentiated into cardiomyocytes and released cytokines, such as soluble-VCAM-1 (sVCAM-1) and very late antigen-4 (VLA-4) [73]. It was demonstrated that cyclosporine-A (CSA) improved the differentiation of mouse-derived embryonic stem cells (mESCs) into cardiomyocytes Flk1+/CXCR4+/VE-cadherin-, called FCV progenitor cells [74]. The results of this study indicated that even though the exact mechanism is still unknown, CSA plays a positive role in endogenous cardiac regeneration.

13.4 Current treatment strategies against atherosclerosis

13.4.1 Immunotherapy for plaque stabilization

Improved understanding of the inflammatory response has led to important advances in the treatment of diseases such as cryopyrin-associated periodic syndromes (CAPS), seropositive rheumatoid arthritis (RA), and atherosclerosis [75–77]. However, the inflammation cascades are coordinated by such many factors as activating macrophage, host-defense response, and an increase in oxidative stresses [78]. Regulating one or a few factors may not be appropriate to achieve the therapeutic effectiveness. Ongoing efforts to broaden the benefit-to-risk window of antiinflammatory therapy in atherosclerosis are to prevent the retention of atherogenic lipoproteins in addition to regulating the serum level of LDL [79].

13.4.1.1 Antiinflammatory therapy using monoclonal antibody (mAb)

The inflammation associated with leukocyte subpopulations and secretion of cytokines is recognized as to play a pivotal role in promoting atherosclerotic plaque growth and in subsequently destabilizing the atheroma [80–82]. Since the chronic inflammation obviously involves with the rupture of plaque, the modulation of the immune response is now viewed as a potential avenue of therapy [83–86]. The utilization of mAb has been widely established in a wide range of conditions such as cancers, the rejection of implanted transplants, and infectious disease [87–89]. These biotherapeutics have a higher potency than the conventional small-molecule therapeutics as the specificity at the target site and less dosing frequency could be enhanced [90,91].

In general, mAbs are well-tolerated and have some advantages such as the specificity for their target, no interaction with cytochrome p450, and other transport proteins in the body. This caused reduced potential for drug-drug interactions [92]. Circumstantial evidence demonstrates that eotaxin-2/CCL24 may play an essential role in the atherosclerotic progression [93]. In a subsequent study, the high level of induction of EO-2 circulating is associated with the presence of coronary artery disease (CAD) such as atherosclerosis [94]. In a recent study, it has been reported that blocking the EO-2 pathway by a newly developed monoclonal antibody, D8, may represent the inhibition of fatty streak formation and reduce the plaque size as employed to the ApoE−/− mouse model [95]. Likewise, various receptors causing an initiation of plaque formation would be recognized as a therapeutic targeting. It has been recently reported that BAFFR (B cell activation factor of the TNF family)-receptor might be a potential target to attenuate the atherosclerotic plaque [96]. BAFF is widely expressed by immune cells, primarily macrophages [97]. The induction B-cells by the interaction between BAFF and these receptors (BAFFR) could exacerbate the plaque instability. Anti-BAFFR monoclonal antibody has been developed and it ameliorated the progression of established atherosclerosis [96]. This mAbs were targeting B2 cells-depletion in atherosclerosis lesion, resulting in an attenuation of pathological progression.

However, the cytotoxic impacts of immune-modulation therapy on unintended tissue or target molecules are frequently found [98]. The toxicity related to mAbs can be dependent on the activity of the mAbs due to the ramification of interaction with target molecules (i.e., protein or receptor) or nonspecific targeting on target tissue (i.e., off-target) [99]. In addition, the major barrier is associated with a delivery route. The advanced knowledge for an appropriate delivery carrier or route of administration should be cumulated.

Taken together, the immune-modulation approach is effective, but it is still required to study the safety, site-specificity, stability, and delivery route.

13.4.1.2 Site-specific targeting mediated by pro-antibody design

As the antibody-therapeutics are bound to normal tissue, the patients have negative side effects including the increase in apoptotic pathway, on-target toxicities during the treatment, and reduced patient compliance [100]. Given these and other risks, several distinct strategies have been reported to enhance the selectivity of antibody-therapeutics [101].

Locally activated proteolytic enzymes are highly expressed at various pathological lesions such as neurodegenerative diseases, oncological diseases, and cardiovascular diseases [102–104]. Consequently, site-specific proteases are exploited as a targeting moiety [105,106]. In atherosclerotic plaque, it has been well-known that latent-matrix metalloproteases family (i.e., MMP-1, 8, and 9) has been activated by exocytic components such as reactive oxygen species and inflammatory cytokines [107]. The activated-MMP family could destabilize the plaque as the type-1 collagen matrix in the fibrous cap is degraded [108]. It has been recently reported that anti-VCAM-1-conjugated pro-antibody containing protease cleavable substrates was designed to render a specific binding affinity upon protease in atherosclerotic tissue [109]. The subsequent in vivo study demonstrated that the pro-antibody was efficiently activated by activated proteases in aorta tissue extracts from ApoE(−/−). No increase in the activation of pro-antibody was found as treated by the extracts from a normal mice. It indicated that the proteases (activation of MMP family) could be exploited to site-specifically target atherosclerotic lesion in vivo.

Newly developed microvessel in the plaque is a feature of angiogenesis induced by hypoxic condition [110,111]. The extradomain B of the ECM glycoprotein fibronectin (ED-B) was expressed in atherosclerotic plaque, but not in normal blood vessels of humans and ApoE(−/−) mice. The binding to this ED-B could be utilized for a targeting domain [112]. Since T effector cell proliferation increased with the deprivation of IL-2, which exacerbates the chronic inflammation at atherosclerotic site, subsequent administration of IL-2 could re-adjust the homeostatic balance of regulatory T cells (Treg) and T effector cells [77]. It was recently reported that the L19 antibody that specifically binds to ED-B was fused with interleukin-2 (IL-2). The Foxp3 and CTLA4 (Treg makers) were highly expressed after L19-IL2 treatment as compared to IL-2 only ($P < .03$) [113]. It indicated that IL-2 could be effectively delivered to the atherosclerotic plaque, resulting in the significant reduction of a plaque size mediated by the increase in Treg.

It is noteworthy that to achieve enhanced pharmacological efficacy, the immunosuppressive therapeutics could be functionalized with specific binding moieties, such as the protease specificity and a receptor-mediated binding affinity.

13.4.2 Induction of autophagy to treat atherosclerosis

Recent advances in characterizing the signaling cascades of MA revealed the machinery proteins that are particularly for autophagosome formation [24]. Especially, mammalian target of rapamycin (mTOR) plays a critical role in early and advanced atherosclerosis. Uptake of ox-LDL by macrophages triggered cytoplasmic accumulation of lipid debris in early atherosclerosis [114]. If the expression of autophagy increased through the inhibition of mTOR, the lipid debris was fused with autophagosome and subsequently with lysosomes, resulting in producing free cholesterol. The free cholesterols were effluxed from the cell. As compared to early stage, the cholesterols were trapped inside lysosomes of foam cells. This is mainly due to the increase in mTOR expression followed by the decrease in Atm expression. The prediction from these findings is that ATM-dependent inhibition of mTOR pathway at the molecular level could lead to activation of autophagy expression, resulting in the efflux of free cholesterol and rapid clearance by efferocytosis. The inactivation of mTOR pathway could provide a beneficial avenue for reduction of plaque size.

It has been recently reported that diet-induced fat accumulation and induction of atherosclerosis could be explained by Wip1-Atm-mTOR signaling cascades in macrophages [115]. In vivo data using Wip1-deficient mice showed that the transformation of macrophages into foam cells was suppressed. A noncanonical Atm-mTOR pathway which is governed by Wip1-dependent signaling cascades was considered as a novel target for the treatment of atherosclerosis. Although immune-modulating therapy with target specificity seems to have a promising future, clear evidence of the effectiveness of a variety of strategies to reduce immune rejection after i.v. injection has not been thoroughly elucidated yet.

13.4.3 Nanoparticles targeting to atherosclerosis

Drug delivery systems (DDS) for the therapeutics whose release was triggered by the external stimuli have been extensively investigated in the area of cancer and brain research [116,117], but rarely investigated for the treatment of atherosclerosis [118].

The narrowed blood vessel could serve as the potential stimuli by which the therapeutics are released from the carrier in a target-specific manner [119]. Obstruction of blood vessels due to formation of atherosclerotic plaque produced abnormally high shear stress. The study revealed that shear-activated microscale of nanoparticle (NP) aggregates was broken up into nanoscale particulates by high shear stress produced in the narrowed vascular sites. This strategy opened a new stimuli-responsive approach for the treatment of atherosclerosis with minimal side effects while maximizing drug efficacy.

It was demonstrated that specific enzymes present in atherosclerotic plaque could be an attractive option to achieve targeted drug delivery. NPs were designed by introducing the moiety labile to collagen IV (Col-IV), which abundantly presented in plaque lesion [120]. Specific accumulation of NPs at atherosclerotic lesions in $Ldlr^{-/-}$ mice model fed with high western diet for 12 weeks was achieved in Col IV-Ac2-26 NPs as compared to Ac2-26 NPs (without Col IV peptide, KLWVLPK peptide). However, the homing rates of Col IV-Ac2-26 NPs to spleen and liver were lower than those of NPs fabricated without Col IV peptide.

Taken together, the approaches based on enzymatically labile to varying degrees are very promising because it is free from surgery (i.e., angioplasty). However, it would not be a suitable method for the acute atherosclerotic occlusion due to the abundant collateral network and the infrequency of atherosclerosis.

13.5 Preparation and surface modification of nanofiber

Artificially fabricated biomaterials, such as polymeric porous matrix, micro-porous fiber, and nano-sized mean diameter of fibers, have been developed based on natural polymer, high-biocompatible synthetic polymer or composite (blending polymers), and evaluated for numerous biomaterial applications [121–125]. Nanofiber has been produced using various techniques, such as Conventional Electrospinning Process [126,127], Co-Axial Electrospinning Technique [128–130], and Phase Separation Method [131,132]. Among those techniques, the electrospinning method has been widely utilized due to cost-effectiveness and easiness in fiber-mat production. Electrospinning technique produces nonwoven three-dimensional matrix, whose prospects have been highlighted upon optimizing several factors, such as materials to be selected, mechanical property, biodegradability, and types of organs/tissues to regenerate [5].

One-dimensional nanoscale fiber has been recognized as a versatile scaffold for cells to grow, attach, and proliferate onto it due to its large surface to volume ratio [133]. In vivo outcomes of tissue engineering have elucidated how and based on what mechanism scaffolds interact with cells. Recently, free-standing membrane is introduced as a cell culture membrane reconstituting tissue-tissue interface to integrate with a curved substrate. The fluidic nature of the electrolyte collector in the free-standing nanofiber membrane enables flexible patterning and facile integration on complex substrates from a 2D flat surface to a 3D curved geometry [134]. The pros and cons of fabricated nanofibers were previously delineated in detail [13].

The interaction between scaffolds and cells is greatly influenced by the composition of the scaffold. Hence, the modification and composition of the surface texture have been a viable means to improve the adhesion rate of cells onto the surface. There were various methods available for the modification of the scaffold surface: (1) plasma treatment or wet chemical method for surface graft polymerization [135–138], (2) co-electrospinning with a drug molecule or a biological agent [139–142], and (3) utilization of aligned nanofiber [143,144].

13.5.1 Stem cell stability and viability in nanofiber

The stability issue of transplanted stem cell under the harsh conditions, such as ischemia and hypoxia, has greatly restricted the clinical usage of stem cell therapy. As described in Table 13.1, several strategies were introduced to enhance the stability of stem cell in vivo, which would pave the way to expand stem cell therapy (Table 13.2) [152].

13.5.1.1 Stem cell conditionings

Advanced tissue engineering via stem cell application seems to be a promising strategy to regenerate damaged tissue. However, transplanted cells will face critical barriers, such as low oxygen level, low interaction between artificial scaffold and native tissue, and lack of nutrients. Therefore, it would be integral for stem cells to be

Table 13.2 Various nanofiber scaffolds loaded with stem cells for advanced tissue engineering

Polymer	Cell sources	Description	Application	References
PLLA/PBLG/ collagen with *n*-HA	ADSCs	Cell source was seeded onto the electrospunned scaffold (PLLA/ PBLG/collagen) in the presence of nano-hydroxyapatite	Bone tissue engineering	[145]
PCL/gelatin	ADSCs	Nanofiber was made of PCL 10% and gelatin 10%, which were optimized to achieve proper tensile strength and homogeneity. As PCL was degraded gradually, the space for cell migration became larger	Cardiovascular tissue engineering	[146]
Collagen (Type 1)	hMSCs	As compared with tissue culture polystyrene (TCPS), 3D collagen nanofiber (Type-1) supports extracellular matrix where hMSCs can be differentiated into osteoblast-lineage	Bone tissue engineering	[147]
PCL	hMSCs	hMSCs were suitable cultured on PCL scaffold with TGF-β, indicating that hMSCs embedded into nanofibrous scaffolds were differentiated to a chondrocytic phenotype	Cartilage tissue engineering	[148]
PLGA	hMSCs	PLGA nanofiber scaffold can accommodate hMSCs, which were differentiated into chondrogenic and ostetogenic cells	Cartilage tissue engineering	[149]
PCL	Mouse-derived ES	PCL nanofiber scaffold and neural basal media (B27 supplement) rendered optimal differentiation of ES into neural tissue lineages, such as neurons, oligodendrocytes, and astrocytes	Neural tissue engineering	[150]
PLLA	hMSCs	The scaffold with MSCs accomplished antithrombogenic effect, opening the new era of novel engineered vascular grafts incorporated with stem cells	Vascular tissue engineering	[151]

ADSC, adipose-derived stem cells; *hMSCs*, bone marrow-derived human mesenchymal stem cells; *n-HA*, nano-hydroxyapatite particle; *PBLG*, poly-benzyl-L-glutamate; *PCL*, poly (ε-caprolactone); *PLGA*, poly(D,L-lactide-*co*-glycolide); *PLLA*, poly (L-lactic acid); *TGF-β*, transforming growth factor-β.

preconditioned before their transplantation. There are several techniques for pre- or postconditioning of stem cells: hypoxic treatment and stem cell aging.

Cardiovascular diseases, such as atherosclerosis and MI, are characterized with the low oxygen level, indicating that transplanted stem cells applied to cardiovascular diseases will be subjected to apoptotic pathway. Therefore, stem cell was pretreated with coercive hypoxia, in which the amount of hypoxia-inducible factor-1 (HIF-1) increased via stimulation of the PI3K/AKT pathway. Hypoxic-controlled MSCs showed a slower cell death rate and lowered caspase-3 activation, suggesting that preconditioning of stem cell could increase angiogenesis with a longer survival rate at the acute MI site [153].

The oxygenation treatment after transplantation of MSCs into rat myocardium (i.e., postconditioning) also enhanced endothelial NOS (NOS3) and subsequently improved MSCs engraftment [154], indicating that the pre- and postconditioning before/after transplantation of stem cells into disease lesions stimulate the survival rate of stem cells.

13.5.1.2 Genetic modification of stem cells

The genetic modification of such properties as antiinflammatory response, migration, and angiogenesis processes was used to boost the stability of stem cell [155]. As the inflammatory response at the transplanted-site affects the survival rate of stem cells, the efficacy of stem cell at targeted tissue was significantly attenuated by lowered antiinflammatory response. The modification of MSCs-TNFR by tumor necrosis factor receptor (TNFR) after 2 weeks of transplantation caused the reduction of the amount of inflammatory cytokines and proteins, such as TNF-α, IL-1β, and IL-6 [156]. The high secretion rate of growth factors, VEGF and bFGF, from transplanted stem cell stimulated the angiogenesis process, resulting in building up of myocardial vascular matrix [152,153].

The regulation of myocardial vascular matrix by MSCs transplantation enhanced a cardiac function [157]. It was also reported that overexpression of Nkx2.5 and GATA-4 in MSCs enhanced their survival rate in vivo as compared to bare-MSCs [158]. Both Nkx2.5 and GATA-4 are known as distinguished markers for cardiomyogenesis, playing an integral role in both upregulating P19 cells and hindering their apoptosis.

13.5.1.3 Addition of drugs or molecular biologics to stem cells

An addition of drug to stem cells seemed to be a viable approach to enhance the stability of stem cells in vivo. Statin, an inhibitor of cholesterol biosynthesis, enhanced the regulation rate of stem cell and its endothelial function [159]. Simvastatin and lovastatin were utilized to control two regulation pathways, such as PI3K/AKT and ERK1/2, achieving the positive enhancement of stem cell stability [160,161].

It would be beneficial for their biomedical applications to load nanofibers with molecular biologics, which could enhance therapeutic efficacies of transplanted cells. Biomolecules, such as proteins, genes, or small RNA, can provide stem cells with the biophysical environment in which they can be differentiated into the specific lineage [162].

Vascular endothelial growth factor (VEGF) along with the protective agents (dextran or bovine serum albumin (BSA)) was loaded in nanofiber electrospunned on poly (L-lactic acid-*co*-caprolactone) (PLCL) [163]. It was reported that proliferation ability of MSCs on different substrates, such as tissue culture-treated plate (TCP), PLCL and PLCL-VEGF, at specific time points (5, 10, and 20 days) enhanced cell population of PLCL-VEGF by 32.3% and 49.9%, respectively, as compared with TCP and PLCL nanofibers. Since VEGF plays a critical role in the differentiation of MSCs into cardiomyocytes, nanofibers containing biomolecules would have a positive outcome in cardiac tissue regeneration.

The more effective drugs or advanced technologies that immobilize and preserve a mixture of macromolecules will emerge as a suitable candidate in the near future. The thermostable mixture may contain sensitive biologicals that are secreted by stem cells in culture, such as cytokines, chemokines, growth factors, matricellular proteins, and enzymes, mRNAs and microRNAs.

13.5.2 Application of nanofibers for cardiovascular tissue engineering

As three-dimensional structure of nanofibers could closely simulate ECM, they can provide stem cells with a suitable environment to be differentiated into cardiovascular tissues. Along with the conventional nanofibers (i.e., based on biocompatible synthetic polymer), the blending of nanofibers with proteins, such as fibrinogen or gelatin or antibodies, has been generally considered as a next generation of tissue engineering applicable to the cardiovascular field.

13.5.2.1 Tubular nanofiber for treatment of vascular disease

Although autologous small-diameter blood vessel has been transplanted for the treatment of vascular diseases, the vascular tissue engineering field still suffers from the limited supply of artificial tissues available for patients [164,165]. Artificial blood vessels fabricated by synthetic polymers, such as poly ((L-lactic acid)-*co*-poly (epsilon-caprolactone) (P(LLA-CL 70:30)), polycaprolactone (PCL) coated with poly(dopamine), and poly(lactic-*co*-glycolic acid) (PLGA), have been regarded as an alternative means for the replacement of damaged blood vessels [166–168]. However, the practical use of artificial blood vessels fabricated by the electrospinning method still has some drawbacks due to the side effects, such as instantaneous thrombosis and initial thickening after the transplantation [164].

The design of vascular grafts based on nanofibers should consider the mechanical characteristics including stress/strain and elongation rate [169]. The dynamic environment of blood vessels, such as continuous blood flow and platelet adhesion by inflammatory cytokines, makes it difficult to apply nanofibers to artificial blood vessels. The elastic modulus of three-dimensional nanofibers made of polyethylene glycol dimethacrylate (PEGdma) blended with PEO (3.5 wt%) through singlet-electrospinning ranged from 2 to 15 kPa, whose magnitude is close to elasticity of the intima membrane and media layer [170].

To date, several strategies have been proposed to improve electrospunned artificial blood vessels. The nanofibers blend of natural polymers, in which collagen was mixed with other natural polymers like elastin, closely mimics the complex architecture of the blood vessel wall, offering a high-porosity and high-surface area [171]. Gelatin shell nanofibers improved the cell viability and biocompatibility, demonstrating their usefulness as an artificial vascular tissue and effectiveness in cardiac restoration [168,172].

13.5.2.2 Application of stem cell to MI

Stem cell therapy could represent the suitable means of MI treatment, replacing dead cells with new cells produced via either stem cell or progenitor cell differentiation. The biomedical techniques, such as ST-segment-elevation MI, Non-ST-segment elevation MI, cardiac angioplasty with stent implantation, coronary artery bypass-graft surgery (CABG), and antithrombotic therapy, have been frequently used in the treatment of MI lesion. However, since these techniques don't regenerate *de novo* cardiac muscle cells, MI could become even worse, unless the fundamental replacement of necrosis cells occurs [173,174]. Subsequently, stem cell could be considered as a novel way to reduce stent-induced restenosis, which relieved the symptoms of MI [151,175].

To date, most novel carriers for stem cells applied against heart diseases are produced in an injectable or implantable mode. Advanced carriers for delivering SPCs into the heart, such as self-assembling peptide, drug-eluting stent, hydrogel, and cardiac patch, have been developed and their properties were characterized [176–180].

The cell therapy approach for cardiovascular tissue regeneration has several limitations. The delivery of cells via coronary artery would be constrained due to the presence of narrowed or blocked coronary artery. There are two approaches to deliver stem cells into the MI site: Inject cells (1) through blood vessels, and (2) directly into myocardium. Since intravenous infusion of cells into whole body organs or tissues has been clinically restricted, the number of the cells that needed to be injected into the body should be large [176,181].

As compared to trans-vascular infusion, the direct injection of stem cell into the ventricular wall is considered as a novel approach, particularly given that the size of MSCs is large. For coronary artery occlusion, the direct injection of stem cell into myocardial tissues was as effective as those via the coronary artery route [182]. However, the survival rate of injected cells was low due to harsh conditions, in which a limited oxygen supply via blood stream is noticeable.

An injection of hydrogel has been considered as a novel approach to deliver stem cells into MI lesions without surgery [177,178,183]. However, the injected polymeric hydrogel could affect the expansion and stiffness of MI lesions. Injected cells will also be vulnerable under harsh environments, such as a hypoxia condition around MI lesions, which reduces cell viability. Decellularized organ ECM was emerged as a novel biomaterial scaffold to enhance the survival rate of cells in vivo, since it consisted of growth factors, offering an ideal and closely simulated organ geometry and three-dimensional structure. It was reported that decellularized-pocrine myocardial tissue injected into rat MI lesions formed a gelling scaffold, yielding sufficient cell adhesion in situ and in vivo [179].

As compared to the injection of cell sources, the implantation of cells loaded in a scaffold has improved cardiac function and vascular engraftment. When murine embryonic stem cell-derived cardiomyocyte (mESCDCs) was cultured onto the polyurethane (PU) scaffold with fibroblasts, the enriched sarcomere shape was formed onto a PU scaffold as a result of phenotype differentiation of mESCDCs with mouse embryonic fibroblasts (MEFs) [144]. This study supported that co-culture condition is integral for the biological response of stem cell when they are differentiated into cardiac muscle.

13.5.2.3 Nanofibers as cardiac patch for stem cell delivery

Both stem cells and progenitor cells can be delivered through a patch into MI lesions in the heart. The patch has several beneficial properties, such as high biocompatibility, controlled degradability, and flexibility. However, there are some drawbacks, such as necessity of surgery and ambiguity of stem cell effects [184,185].

Poly (glycerol sebacate)(PGS)/fibrinogen (core/shell) nanofibers produced by core-shell electrospinning were suitable as a cardiac patch for the treatment of MI. PGS/fibrinogen induced the larger number of cardiac markers than fibrinogen nanofiber. Also, it was feasible for PGS/fibrinogen to be utilized as a cardiac patch owing to the reasonable Young's modulus (3.28 ± 1.7 MPa), whose value is close to native myocardium [145]. In addition, cardiac patch made of type 1 collagen nanofibers seeded with hMSCs significantly enhanced myocardium functions via higher engraftment [186]. Nanofibers combined with various biocompatible hydrogels were introduced for the enhancement of their efficacy. For instance, short-fragmented nanofibers made of poly (glycerol sebacate) (PGS) were fabricated into core (PGS)-shell (PLLA) electrospinning [187]. PLLA (shell) was removed by immersing nanofibers into the mixture of hexane and DCM (1:2), resulting in short-fragmented PGS nanofibers. Several cardiac makers, such as actinin, troponin, myosin heavy chain, and connexin 43, were overexpressed from cardiomyocytes cultured on short-fragmented nanofibers. Because PGS are injectable, they need a minimally invasive technique for the treatment of MI.

Stimuli-sensitive hydrogel can also be utilized for the fabrication of nanofiber-hydrogel composite. PCL nanofibers electrospunned and mixed with hydrogel precursor (poly ethyleneglycol (PEG) hydrogel) were hardened through external stimuli, such as UV treatment [188]. Photo-patterned nanofibers demonstrated its usefulness in promoting the proliferation rate of mammalian cells incorporated within them. This study has expanded the utility of nanofiber-hydrogel to stem cell-based cell therapy as well as the sustained release of therapeutics agents, such as proteins and siRNA.

13.5.2.4 Stem cell-loaded stent for the treatment of atherosclerosis

Biomedical techniques to coat the metallic surface of stent with nanofibers can be applicable to various biomedical fields, such as esophageal stent, wide-necked aneurysm, tracheal regeneration, and cholangiocarcinoma treatment [189,190]. The stent coated with nanofibers was developed through electrospinning technique as shown in Fig. 13.1. Since in- and outside of the stent surface were coated with nanofiber,

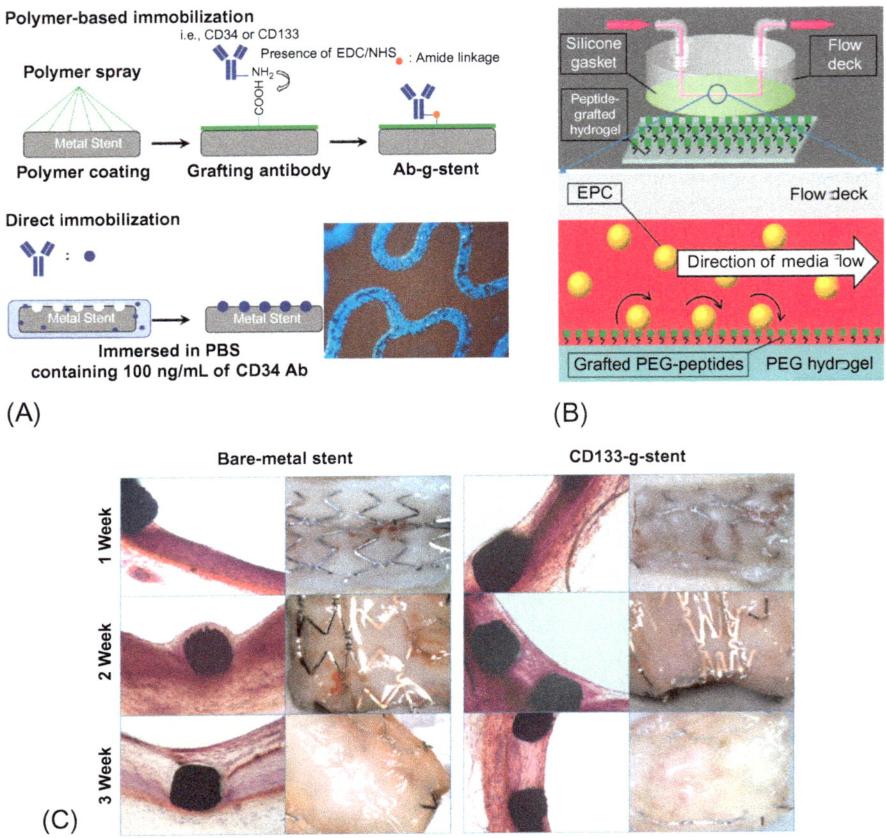

Fig. 13.1 (A) Schematic illustration of capture-antibody immobilization. Inset images from Ref. [191] demonstrated that CD34+ cells were captured by anti-CD34 antibodies. Anti-CD34 was immobilized using direct immobilization. (B) Evaluation of dynamic adhesion of circulating EPCs on Ab-attached stent was employed to examine EPCs binding to the pseudo-vascular wall simulator under shear fluid [192]. (C) Anti-CD133 was immobilized using polymer-based immobilization. CD133-attached stent showed rapid endothelialization in normal porcine coronary arteries after stent implantation [193].

nanofiber surface rather than bare-metal surface was exposed to blood flow when it was implanted. The advantages of this approach are: (1) the presence of the barrier between injured vascular site and bare-metal surface, (2) increase in surface area, and (3) reduction in in-stent restenosis. One minor drawback of this approach is that it requires the long-term observation to demonstrate or verify its efficiency.

The application of stem cells to the cardiovascular stent has been accentuated to clinical application in which the proper differentiation of stem cells on vascular disease lesion can prevent in-stent restenosis and thrombosis. For instance, immune responses were modulated by hMSCs, which play a critical role in regulating autoimmunity through the generation of regulatory T cells (Tregs) [194].

An advanced approach known as an electro-addressing technique was introduced to incorporate stem cells into biocompatible scaffolds using sodium alginate hydrogel [195]. Although the electro-addressing approach has several advantages in delivering a living organism, this approach has faced some difficulties due to the increase in the cellular cytotoxicity during the process [196]. One critical aspect along with increase in cytotoxicity is that hydrogel showed unfavorable mechanical property in terms of the expansion of stent in angioplasty. Therefore, the charged-nanofiber-gel (CNG) seems to be a promising alternative means to increase the mechanical property, guaranteeing the safety of incorporating cells during the process (Fig. 13.2).

As compared to externally delivered stem cells, endogenously derived stem cells significantly improved the efficacy of the cell-based therapy. Nanofibers containing biological therapeutics, such as antibody (CD34 or CD133), serve as a prototype to utilize endothelial progenitor cells for facilitating blood stream flow [197]. Recently, it was reported that re-endothelialization and neointima formation with the use of CD133-coated stent were not distinctively different from those with bare-metal stent [198]. These seemingly conflicted results can be explained by: (1) lack of population of EPCs in blood circulation, and (2) lack of efficacy due to unspecific binding. Since the conjugation of stent and antibodies, such as CD133, was highly dependent on the total surface of metallic stent, it is integral to thoroughly investigate based on long-term observation and make proper assessment of the cardiovascular stent coated with nanofibers containing a limited amount of antibodies.

13.6 Functionalized cardiovascular stents for treatment of atherosclerosis

Implantation of drug-eluted vascular stents (DESs) has shown positive outcomes as compared to bare-metal stent (BMS), resulting in decrease in the occurrence of in-stent restenosis [199]. Despite promising clinical outcomes, recent studies reported that DES can cause late stent thrombosis and lead to long-term failure, especially after stenting complex lesions [200]. This was because antiproliferative agents, such as sirolimus or paclitaxel, hindered not only SMCs migration, but also the re-endothelialization process of the substrate. That is why DES has been considered as "double-edged sword." Functionalized cardiovascular stents will be discussed (Table 13.3).

13.6.1 Rapid endothelialization by antibody-grafted cardiovascular stent

Recently, antibody-conjugated cardiovascular stent, which selectively captures circulating EPCs, has shown a remarkable improvement of re-endothelialization on metal surface (Fig. 13.1) [193,203,204]. This approach utilized circulating progenitor cells migrated on the surface of implanted devices, which are already coated with specific antibodies, such as antihuman CD133 or CD34 antibodies (Fig. 13.1A and C) [191,193,205]. To immobilize capturing antibodies on the metallic surface, various

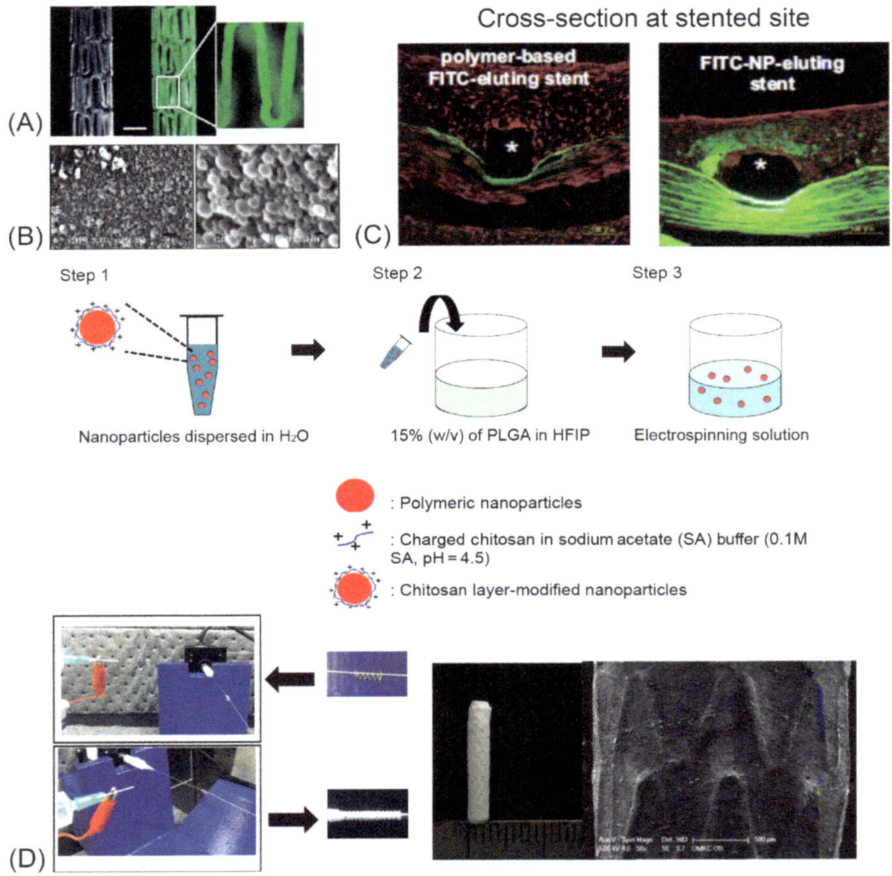

Fig. 13.2 Nanotechnology-based cardiovascular stent. (A) Nanoparticles-coated stent. *Green fluorescent* indicated FITC-NPs successfully coated on vascular stent. (B) SEM images of surface of vascular stent after the coating process. (C) In vivo deployment of NP-coated stent at coronary artery. NPs diffused fast and prevented neointima formation. (D) Nanofiber-coated vascular stent. Nanofibers were fabricated using electrospinning technique. Nanoparticles were encapsulated in nanofibers, resulting in sustained drug release.
(A), (B), and (C) from K. Nakano, K. Egashira, S. Masuda, K. Funakoshi, G. Zhao, S. Kimura, T. Matoba, K. Sueishi, Y. Endo, Y. Kawashima, K. Hara, H. Tsujimoto, R. Tominaga, K. Sunagawa, Formulation of nanoparticle-eluting stents by a cationic electrodeposition coating technology: efficient nano-drug delivery via bioabsorbable polymeric nanoparticle-eluting stents in porcine coronary arteries, JACC Cardiovasc. Interv. 2 (2009) 277–283 and (D) from B. Oh, C.H. Lee, Nanofiber for cardiovascular tissue engineering, Expert Opin. Drug Deliv. 10 (2013) 1565–1582.

Table 13.3 Delivery methods for stem cell therapy against atherosclerosis

	Description	Application	Significances	References
Injection	Injection of stem cells	Atherosclerotic renal artery stenosis (ARAS)	Improvement of medullary vascularization Increased expression of paracrine factors such as VEGF, FLK-1, and HIF-1α Additional intervention required to reduce oxidative stress	[63]
		Atherosclerotic vulnerable plaque	Downregulation of cytokines such as TNF-alpha, hs-CRP, and IL-6 in serum Upregulation of IL-10	[62]
		Prevention of progression of atherosclerosis	In vitro co-culture of endothelial cells with MSCs ameliorated stresses exerted by oxLDL and restored eNOS level In vivo infusion of MSCs alleviated endothelial dysfunction and plaque progression in $ApoE^{-/-}$	[61]
Ab-g-stent	Capture circulating progenitor cells	Rapid endothelialization on metallic surface	After deployment of CD34-g-stent in vivo, EPCs ($CD34^+$) adhered to CD34-g-stent much earlier and proliferated faster in the presence of anti-CD34 antibody Neointimal area in CD34-g-stent was much lower than that in control	[191]
			In vitro cell culture demonstrated that CD133-g-stent significantly promoted $CD133^+$ migration and proliferation In vivo arteriovenous shunt model showed CD133-g-stent captured $CD133^+$ from the blood stream within 6 h	[193]
Stentcoated with stem cells	Local delivery of stem cells	Localized delivery of AD-MSCs into atherosclerotic plaque	Sodium alginate hydrogels containing AD-MSCs were successfully deposited on metallic stent via EPD Robust hydrogels in the presence of fragmented nanofibers formed without any hindrance during EPD	[201]
		Prolonging delivery of paracrine factors (PFs) released by pMSCs	In vitro pMSCs seeded on NF sleeves continued to release PFs, inducing tubulogenesis of ECs In vivo deployment of pMSCs-impregnated NFs-coated stent showed no thrombotic occlusion or immune rejection	[202]

immobilization techniques, such as covalently binding to polymer basement and direct absorbing on metal surface, have been reported. The direct immobilization of capturing antibody on the metal stent has been considered an ideal approach, because it doesn't cause any risks from the polymers on the stent [191].

To evaluate dynamic adhesion of circulating EPCs on antibody-attached stent, the advanced model was employed to examine whether the incorporation of peptide ligands into poly(ethylene glycol) diacrylate (PEGDA) hydrogel had a positive effect on EPCs binding to the pseudo-vascular wall simulator under shear fluid [192]. Although the test samples were made of grafted PEG-peptides, such as RGD, REDV, and YIGSRG, a three-dimensional flow chamber design was useful to determine the efficacy of antibody-attached stent (Fig. 13.1B).

13.6.2 NP-coated cardiovascular stent

DES prepared with conventional dip-coating or spray-coating methods are widely used for angioplasty [199]. However, the therapeutics, such as sirolimus, used for DES deteriorated the endothelialization process on the surface of metallic stents, resulting in late restenosis [206].

Recent advances in nanotechnologies are poised to alter the development of implantable devices, such as bone replacement, dental implants, and cardiovascular stent [207–209]. The delivery of less invasive cardiovascular stent coated with NPs has been introduced (Fig. 13.2) [209,210]. NPs-coated stent (encapsulated with a fluorescent marker, FITC) has been formulated with an advanced technique called cation electrodeposition coating (Fig. 13.2A and B). In vitro cellular uptake study on cultured vascular SMCs demonstrated that NPs were taken up rapidly and efficiently as compared to FITC in the absence of NP [209]. Fluorescent images of cross-sections of stent deployment sites in vivo porcine coronary artery for 2 weeks showed that NPs had played a beneficial role in prolonged delivery of drug and its distribution into the stented coronary artery (Fig. 13.2C). Another study demonstrated the efficacy of NPs-coated stent delivering N-nitrosomelatonin (NOMela) [210]. Cardiovascular stent has been coated with NOMela nanoparticles using electrophoretic deposition (EPD) technique. The in vivo study on adult male Spraque-Dawley rats revealed that stent coated with poly (D,L-lactide-*co*-glycolide) (PLGA, 50:50)-NPs containing NOMela significantly reduced platelet aggregation as compared to those deployed with the control stent, resulting in alleviating neointima formation.

Comprehensively, the treatment for atherosclerosis has been evolved due to the benefits of using stem cells. Although various methods to deliver or capture stem cells have been introduced, it is still in initial stage. In-depth research on cell delivery method as the next generation therapeutics will be helpful in treating many patients struggled with CAD.

13.6.3 Surface coating of cardiovascular stents for stem cells

13.6.3.1 Electrophoretic deposition of stem cells in hydrogel

Hydrogels have been widely investigated for tissue engineering material, cosmetic purpose, and pharmaceutical ingredients due to its high biocompatibility [211–213].

Since the growth rate of implantable devices has substantially increased [214], development of metallic medical implants coated with hydrogels to alleviate inflammation would be a new era for researchers to rigorously investigate for future medicine. Among various implants, cardiovascular stents coated with hydrogels have been gradually introduced [215–217].

To coat the metallic surface of stent with homogenous hydrogel layer formed, various strategies have been developed. Conventional methods were either simple dip-coating approach or spray drying [217]. Natural polymers including chitosan and sodium alginate have intrinsic charges due to the presence of functional groups such as amine and carboxylic acid. It was developed that ITO glass was successfully coated with hydrogel using electrical potential [195,218,219]. The method indicated that the metallic surface can be coated with homogenous layer of hydrogels as polymer chains have shown proper electrical potential. In addition, three-dimensional (3D) patterning of human umbilical vein endothelial cells (HUVECs) deposited on ITO electrode by dielectrophoresis was reported. The study demonstrated that gelatin methacrylate (GelMA) was a promising hydrogel for use in cell encapsulation due to its high biocompatibility [220]. GelMA deposited on ITO electrode with HUVECs was further polymerized as exposed to UV treatment, resulting in maintaining cell viability and growth over 7 days.

Although deposition of hydrogel on the ITO electrode has been reported, proper scheme for metallic surface coating has not been developed. Recently, it has been developed that the metallic surface of cardiovascular stent can be coated with sodium alginate hydrogel containing adipose tissue-derived mesenchymal stem cells (AD-MSCs) using electrophoretic deposition method (EPD) (Fig. 13.3A) [201]. The study was aimed to develop the robust hydrogels mimicking "mud-and-straw" bird nest as cell delivery carrier for AD-MSCs. Mechanical property of hydrogels was tested using custom-built perfusive fluid channel, indicating that thiolated-sodium alginate in the presence of poly (ethylene glycol)-diacrylate (PEGDA) blended with fragmented nanofiber pieces showed substantial increase in physical strength against pseudo-perfusive flow (1 mL/s). The results from cell viability and proliferation rates demonstrated that AD-MSCs were mostly viable after the cells encapsulated in hydrogels and then proliferated well during the incubation for 7 days.

Apparently, stem cell encapsulation in hydrogels coated on metallic surface could be beneficial due to avoidance of autoimmune rejection against implantable device. Especially, cardiovascular stent coated with hydrogels containing stem cells could be the next generation of stent. The major drawbacks will be the requirement of appropriate modification on hydrogels due to its poor mechanical property and unpredictable behaviors of encapsulated stem cells in hydrogels.

13.6.3.2 Stem cell-impregnated nanofiber stent

Nanofibers have been extensively studied to coat surface of implantable devices in various fields such as esophageal stent, aneurysm, tracheal regeneration, cholangiocarcinoma, and cardiovascular stent [189,190,221–223]. Cardiovascular stent coated with nanofibers has not been frequently reported for clinical application due to its

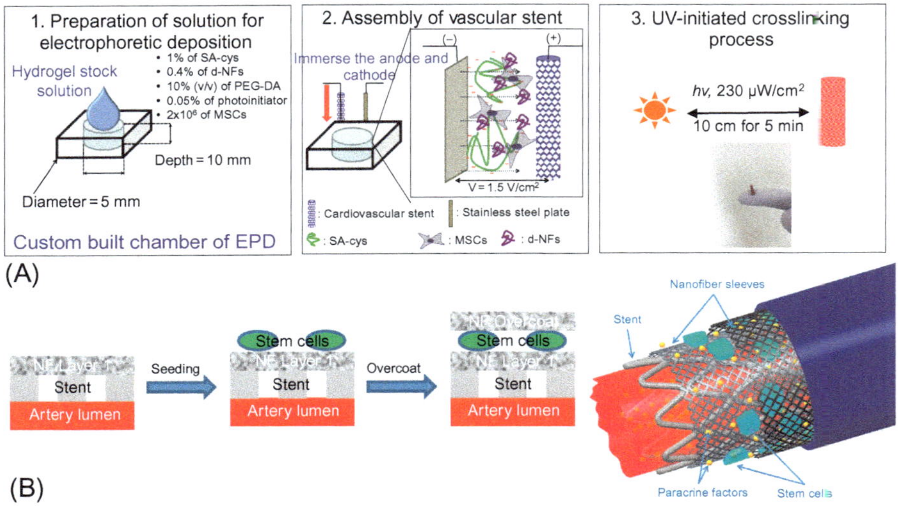

Fig. 13.3 Stem cell-coated cardiovascular stent. (A) Development of biomimicking robust hydrogel for the MSCs Carrier. Hydrogels containing fragmented nanofibers and MSCs were deposited on metallic stent using EPD technique [201]. (B) Stem cell-impregnated nanofiber stent sleeve. Paracrine factors released from MSCs freely pass through the nanofiber sleeves. Nanofiber sleeves acted as a barrier in which rapid wash by blood flow can be prevented [202].

small-diameter. However, coating with nanofibers would be beneficial since it creates (a) the barrier between injured plaque and blood flow, (b) increases surface area, and (c) accelerates endothelialization on nanofiber scaffolds (Fig. 13.2D).

It has been previously reported that the stent surface was fully coated with nanofibers via electrospinning [224]. According to the study, advanced technique has been developed that vascular stent was coated with nanofibers containing β-estradiol. Hormonal agent such as β-estradiol has been recognized as a promising agent to deal with ROS-related disease. It has been reported that β-estradiol-eluted stent had a potential to prevent restenosis in vitro because it enhanced EC proliferation. However, in vivo and early clinical study showed that it had no positive outcome due to the insufficient concentration of drug and complicated adventitial biology [225]. The results indicated that the concentration of β-estradiol increased with sustained release pattern for 3 months. In addition, the ECs proliferation increased as cultured on nanofibers scaffolds containing β-estradiol. Cell viability against excessive oxidative stress induced by hydrogen peroxide (H_2O_2) significantly increased.

As the significance of activated mast cells in atherosclerosis has been emphasized, the stabilization of mast cells is a promising approach against vascular remodeling after angioplasty [226]. It was continuously reported that mast cells might be stabilized when treated with exogenous NO molecules [227,228]. One of obvious problems associated with the use of ROS scavenger is that viable level of NO decreased [223,229,230]. It was recently reported that ROS scavenger (Edaravone) in a combination with NO donor drug (GSNO) successfully prevented further degranulation of activated mast cells. The % of β-hexosaminidase in the group treated with both drugs

(40% for 2 days activation) didn't increase as compared to the groups treated with only each drug (65% for Edaravone and 58% for GSNO). It indicated that the decrease in degranulation rate stabilizes plaque against external stress such as angioplasty.

Recent evidences indicate that complete cure for atherosclerosis will be stemmed from adventuring causal factors of atherosclerosis [231,232]. Immunological responses against cytokines such as IL-1b, IL-6, TNF-α, and IFN-γ participate in the progression of atherosclerosis and cause rapid disruption of atherosclerotic plaque, resulting in acute blockage of blood vessel [83]. The disruption of atherosclerotic plaque caused by immunological responses has been studied and it is concluded that acute disruption of atherosclerotic plaque was highly mediated by immunological complexity, meaning that the immunological stabilization can be a key factor to achieve a regression of atherosclerosis [83,233,234]. Immunoregulatory function governed by paracrine factors released from stem cells offers enhanced efficacy for atherosclerotic plaque stabilization and regression [55,235]. In addition, the prolonging delivery of paracrine factors at atherosclerotic lesions will enhance the reparative response, since paracrine factors can promote intrinsic repair mechanisms [236,237].

Angioplasty has been widely used for the treatment of atherosclerosis. However, delivering stem cells with cardiovascular stent is at early stage due to low biocompatible surface of metallic stent [238,239]. It has been recently developed that MSCs were impregnated on nanofiber-coated stent sleeves (Fig. 13.3B) [202]. MSCs were successfully seeded on nanofiber sleeves produced by electrospinning technique (PLGA nanofibers). In addition, seeded-MSCs were overcoated with another nanofiber sleeves, resulting in thorough protection of seeded-MSCs not only from perfused blood flow after the angioplasty, but also from being deteriorated by autoimmune rejections in the body.

The site-specific release of paracrine factors, such as VEGF and HGF from MSCs, would significantly alleviate the onset and progress of atherosclerosis. This approach guarantees in vivo and clinical studies to investigate the efficacy of stem cell-impregnated nanofiber stent in animal models and human volunteers. Although new coating technique could have brought various advantages beyond conventional-coating methods, the long-term observation is still necessary to validate its efficiency.

13.7 Conclusion

Nanofiber loaded with stem cells opened a new era to tissue regeneration with less immune rejection. It was demonstrated that nanofibers loaded with stem cells can be utilized for the treatment of cardiovascular diseases including atherosclerosis and cardiomyocyte regeneration.

Nanofiber approach for replacing impaired blood vessel has provided three-dimensional architecture, which mimics an ECM and could serve as a stem cell niche. In addition, they can serve as a physical barrier, which prevents monocyte adhesion and in-stent restenosis in the treatment of atherosclerosis. The modification of cell-carriers with biological cues, which provide rapid differentiation of stem cells into a specific lineage and protect stem cells under the harsh conditions (i.e., hypoxia), will significantly enhance therapeutic efficacy of transplanted cells.

To maximize the benefits of nanomaterial, it is essential to improve rapid regeneration of tissues on the implanted materials by which the immunogenic response and/or side effects can be alleviated. The stem cell approach may accomplish this task by reducing immune rejection and facilitating rapid differentiation in targeting lineage. The maintenance of biocompatibility of various cells loaded on nanofiber scaffold could furnish an integral array of physical and biochemical signals in a spatial and temporal manner. The preconditioning of nanofiber through the surface modification technique accelerates differentiation, attachment, and proliferation of stem cells.

As the precise mechanisms of differentiation of stem cell and its interaction with scaffold at microenvironment have been gradually elucidated, stem cell therapy via nanofibers will enhance its potential for the regeneration of cardiac tissues and subsequently for the treatment of cardiac diseases, such as atherosclerosis and MI.

References

[1] X. Zhang, Fundamentals of Fiber Science, first ed., DEStech Publications, Lancaster, PA, 2015.
[2] R. Langer, J.P. Vacanti, Tissue engineering, Science 260 (1993) 920–926.
[3] J. Venugopal, S. Low, A.T. Choon, S. Ramakrishna, Interaction of cells and nanofiber scaffolds in tissue engineering, J. Biomed. Mater. Res. B Appl. Biomater. 84 (2008) 34–48.
[4] W.J. Li, C.T. Laurencin, E.J. Caterson, R.S. Tuan, F.K. Ko, Electrospun nanofibrous structure: a novel scaffold for tissue engineering, J. Biomed. Mater. Res. 60 (2002) 613–621.
[5] M.M. Stevens, J.H. George, Exploring and engineering the cell surface interface, Science 310 (2005) 1135–1138.
[6] K. Abe, T. Yamashita, S. Takizawa, S. Kuroda, H. Kinouchi, N. Kawahara, Stem cell therapy for cerebral ischemia: from basic science to clinical applications, J. Cereb. Blood Flow Metab. 32 (2012) 1317–1331.
[7] M.D. Tibbetts, M.A. Samuel, T.S. Chang, A.C. Ho, Stem cell therapy for retinal disease, Curr. Opin. Ophthalmol. 23 (2012) 226–234.
[8] R. Soler, C. Fullhase, A. Hanson, L. Campeau, C. Santos, K.E. Andersson, Stem cell therapy ameliorates bladder dysfunction in an animal model of Parkinson disease, J. Urol. 187 (2012) 1491–1497.
[9] D.A. Cantu, P. Hematti, W.J. Kao, Cell encapsulating biomaterial regulates mesenchymal stromal/stem cell differentiation and macrophage immunophenotype, Stem Cells Transl. Med. 1 (2012) 740–749.
[10] W. Yang, S. Lee, Y.H. Jo, K.M. Lee, J.G. Nemeno, B.M. Nam, B.Y. Kim, I.J. Jang, H.N. Kim, T. Takebe, J.I. Lee, Effects of natural cartilaginous extracellular matrix on chondrogenic potential for cartilage cell transplantation, Transplant. Proc. 46 (2014) 1247–1250.
[11] M.S. Han, D.Y. Jung, C. Morel, S.A. Lakhani, J.K. Kim, R.A. Flavell, R.J. Davis, JNK expression by macrophages promotes obesity-induced insulin resistance and inflammation, Science 339 (2013) 218–222.
[12] C. Garris, M.J. Pittet, Therapeutically reeducating macrophages to treat GBM, Nat. Med. 19 (2013) 1207–1208.
[13] B. Oh, C.H. Lee, Nanofiber for cardiovascular tissue engineering, Expert Opin. Drug Deliv. 10 (2013) 1565–1582.

[14] K.J. Moore, I. Tabas, Macrophages in the pathogenesis of atherosclerosis, Cell 145 (2011) 341–355.
[15] P. Libby, P.M. Ridker, G.K. Hansson, Progress and challenges in translating the biology of atherosclerosis, Nature 473 (2011) 317–325.
[16] K.J. Moore, F.J. Sheedy, E.A. Fisher, Macrophages in atherosclerosis: a dynamic balance, Nat. Rev. Immunol. 13 (2013) 709–721.
[17] D.L. Laskin, Macrophages and inflammatory mediators in chemical toxicity: a battle of forces, Chem. Res. Toxicol. 22 (2009) 1376–1385.
[18] G. Chinetti-Gbaguidi, M. Baron, M.A. Bouhlel, J. Vanhoutte, C. Copin, Y. Sebti, B. Derudas, T. Mayi, G. Bories, A. Tailleux, S. Haulon, C. Zawadzki, B. Jude, B. Staels, Human atherosclerotic plaque alternative macrophages display low cholesterol handling but high phagocytosis because of distinct activities of the PPARgamma and LXRalpha pathways, Circ. Res. 108 (2011) 985–995.
[19] A.M. Choi, S.W. Ryter, B. Levine, Autophagy in human health and disease, N. Engl. J. Med. 368 (2013) 651–662.
[20] A. Baruch, N. van Bruggen, J.B. Kim, J.E. Lehrer-Graiwer, Anti-inflammatory strategies for plaque stabilization after acute coronary syndromes, Curr. Atheroscler. Rep. 15 (2013) 327.
[21] M. Ouimet, V. Franklin, E. Mak, X. Liao, I. Tabas, Y.L. Marcel, Autophagy regulates cholesterol efflux from macrophage foam cells via lysosomal acid lipase, Cell Metab. 13 (2011) 655–667.
[22] D.M. Schrijvers, G.R. De Meyer, W. Martinet, Autophagy in atherosclerosis: a potential drug target for plaque stabilization, Arterioscler. Thromb. Vasc. Biol. 31 (2011) 2787–2791.
[23] W. Martinet, I. De Meyer, S. Verheye, D.M. Schrijvers, J.P. Timmermans, G.R. De Meyer, Drug-induced macrophage autophagy in atherosclerosis: for better or worse? Basic Res. Cardiol. 108 (2013) 321.
[24] M.C. Maiuri, G. Grassia, A.M. Platt, R. Carnuccio, A. Ialenti, P. Maffia, Macrophage autophagy in atherosclerosis, Mediat. Inflamm. 2013 (2013) 584715.
[25] R. Virmani, A.P. Burke, F.D. Kolodgie, A. Farb, Vulnerable plaque: the pathology of unstable coronary lesions, J. Interv. Cardiol. 15 (2002) 439–446.
[26] X. Liao, J.C. Sluimer, Y. Wang, M. Subramanian, K. Brown, J.S. Pattison, J. Robbins, J. Martinez, I. Tabas, Macrophage autophagy plays a protective role in advanced atherosclerosis, Cell Metab. 15 (2012) 545–553.
[27] I. Tabas, Macrophage death and defective inflammation resolution in atherosclerosis, Nat. Rev. Immunol. 10 (2010) 36–46.
[28] R.D. Levit, N. Landazuri, E.A. Phelps, M.E. Brown, A.J. Garcia, M.E. Davis, G. Joseph, R. Long, S.A. Safley, J.D. Suever, A.N. Lyle, C.J. Weber, W.R. Taylor, Cellular encapsulation enhances cardiac repair, J. Am. Heart Assoc. 2 (2013) e000367.
[29] Q. Xu, The impact of progenitor cells in atherosclerosis, Nat. Clin. Pract. Cardiovasc. Med. 3 (2006) 94–101.
[30] Q. Xu, Stem cells and transplant arteriosclerosis, Circ. Res. 102 (2008) 1011–1024.
[31] Y. Hu, Z. Zhang, E. Torsney, A.R. Afzal, F. Davison, B. Metzler, Q. Xu, Abundant progenitor cells in the adventitia contribute to atherosclerosis of vein grafts in ApoE-deficient mice, J. Clin. Invest. 113 (2004) 1258–1265.
[32] Y. Hu, M. Mayr, B. Metzler, M. Erdel, F. Davison, Q. Xu, Both donor and recipient origins of smooth muscle cells in vein graft atherosclerotic lesions, Circ. Res. 91 (2002) e13–e20.
[33] Q. Xiao, F. Zhang, L. Lin, C. Fang, G. Wen, T.N. Tsai, X. Pu, D. Sims, Z. Zhang, X. Yin, B. Thomaszewski, B. Schmidt, M. Mayr, K. Suzuki, Q. Xu, S. Ye, Functional role of

matrix metalloproteinase-8 in stem/progenitor cell migration and their recruitment into atherosclerotic lesions, Circ. Res. 112 (2013) 35–47.
[34] F. Pelliccia, C. Cianfrocca, G. Rosano, G. Mercuro, G. Speciale, V. Pasceri, Role of endothelial progenitor cells in restenosis and progression of coronary atherosclerosis after percutaneous coronary intervention: a prospective study, J. Am. Coll. Cardiol. Interv. 3 (2010) 78–86.
[35] J. Liao, X. Chen, Y. Li, Z. Ge, H. Duan, Y. Zou, J. Ge, Transfer of bone-marrow-derived mesenchymal stem cells influences vascular remodeling and calcification after balloon injury in hyperlipidemic rats, J. Biomed. Biotechnol. 2012 (2012) 165296.
[36] P. Yan, C. Xia, C. Duan, S. Li, Z. Mei, Biological characteristics of foam cell formation in smooth muscle cells derived from bone marrow stem cells, Int. J. Biol. Sci. 7 (2011) 937–946.
[37] M.Y. Speer, H.Y. Yang, T. Brabb, E. Leaf, A. Look, W.L. Lin, A. Frutkin, D. Dichek, C.M. Giachelli, Smooth muscle cells give rise to osteochondrogenic precursors and chondrocytes in calcifying arteries, Circ. Res. 104 (2009) 733–741.
[38] Z. Tang, A. Wang, F. Yuan, Z. Yan, B. Liu, J.S. Chu, J.A. Helms, S. Li, Differentiation of multipotent vascular stem cells contributes to vascular diseases, Nat. Commun. 3 (2012) 875.
[39] E. Zengin, F. Chalajour, U.M. Gehling, W.D. Ito, H. Treede, H. Lauke, J. Weil, H. Reichenspurner, N. Kilic, S. Ergun, Vascular wall resident progenitor cells: a source for postnatal vasculogenesis, Development 133 (2006) 1543–1551.
[40] R.R. Pauly, A. Passaniti, M. Crow, J.L. Kinsella, N. Papadopoulos, R. Monticone, E.G. Lakatta, G.R. Martin, Experimental models that mimic the differentiation and dedifferentiation of vascular cells, Circulation 86 (1992). Iii68–73.
[41] Y. Tintut, Z. Alfonso, T. Saini, K. Radcliff, K. Watson, K. Bostrom, L.L. Demer, Multilineage potential of cells from the artery wall, Circulation 108 (2003) 2505–2510.
[42] E. Gonzalez-Rey, M.A. Gonzalez, N. Varela, F. O'Valle, P. Hernandez-Cortes, L. Rico, D. Buscher, M. Delgado, Human adipose-derived mesenchymal stem cells reduce inflammatory and T cell responses and induce regulatory T cells in vitro in rheumatoid arthritis, Ann. Rheum. Dis. 69 (2010) 241–248.
[43] G. Chamberlain, H. Smith, G.E. Rainger, J. Middleton, Mesenchymal stem cells exhibit firm adhesion, crawling, spreading and transmigration across aortic endothelial cells: effects of chemokines and shear, PLoS One 6 (2011). e25663.
[44] C.M. Dollery, P. Libby, Atherosclerosis and proteinase activation, Cardiovasc. Res. 69 (2006) 625–635.
[45] P. Van Lint, C. Libert, Matrix metalloproteinase-8: cleavage can be decisive, Cytokine Growth Factor Rev. 17 (2006) 217–223.
[46] J.B. Michel, O. Thaunat, X. Houard, O. Meilhac, G. Caligiuri, A. Nicoletti, Topological determinants and consequences of adventitial responses to arterial wall injury, Arterioscler. Thromb. Vasc. Biol. 27 (2007) 1259–1268.
[47] J.H. Campbell, O. Kocher, O. Skalli, G. Gabbiani, G.R. Campbell, Cytodifferentiation and expression of alpha-smooth muscle actin mRNA and protein during primary culture of aortic smooth muscle cells. Correlation with cell density and proliferative state, Arteriosclerosis 9 (1989) 633–643.
[48] S.S. Rensen, P.A. Doevendans, G.J. van Eys, Regulation and characteristics of vascular smooth muscle cell phenotypic diversity, Neth. Heart J. 15 (2007) 100–108.
[49] M.F. Pittenger, B.J. Martin, Mesenchymal stem cells and their potential as cardiac therapeutics, Circ. Res. 95 (2004) 9–20.

[50] M.F. Pittenger, A.M. Mackay, S.C. Beck, R.K. Jaiswal, R. Douglas, J.D. Mosca, M.A. Moorman, D.W. Simonetti, S. Craig, D.R. Marshak, Multilineage potential of adult human mesenchymal stem cells, Science 284 (1999) 143–147.
[51] E. Spaeth, A. Klopp, J. Dembinski, M. Andreeff, F. Marini, Inflammation and tumor microenvironments: defining the migratory itinerary of mesenchymal stem cells, Gene Ther. 15 (2008) 730–738.
[52] S. Wang, X. Qin, D. Sun, Y. Wang, X. Xie, W. Fan, D. Liang, X. Pei, F. Cao, Effects of hepatocyte growth factor overexpressed bone marrow-derived mesenchymal stem cells on prevention from left ventricular remodelling and functional improvement in infarcted rat hearts, Cell Biochem. Funct. 30 (2012) 574–581.
[53] J.W. Wright, J.W. Harding, The brain hepatocyte growth factor/c-Met receptor system: a new target for the treatment of Alzheimer's disease, J. Alzheimers Dis. 45 (4) (2015) 985–1000.
[54] K. Le Blanc, D. Mougiakakos, Multipotent mesenchymal stromal cells and the innate immune system, Nat. Rev. Immunol. 12 (2012) 383–396.
[55] L. Chen, E.E. Tredget, P.Y.G. Wu, Y. Wu, Paracrine factors of mesenchymal stem cells recruit macrophages and endothelial lineage cells and enhance wound healing, PLoS One 3 (2008) e1886.
[56] J. Kim, P. Hematti, Mesenchymal stem cell-educated macrophages: a novel type of alternatively activated macrophages, Exp. Hematol. 37 (2009) 1445–1453.
[57] K. Anton, D. Banerjee, J. Glod, Macrophage-associated mesenchymal stem cells assume an activated, migratory, pro-inflammatory phenotype with increased IL-6 and CXCL10 secretion, PLoS One 7 (2012). e35036.
[58] S. Adutler-Lieber, T. Ben-Mordechai, N. Naftali-Shani, E. Asher, D. Loberman, E. Raanani, J. Leor, Human macrophage regulation via interaction with cardiac adipose tissue-derived mesenchymal stromal cells, J. Cardiovasc. Pharmacol. Ther. 18 (2013) 78–86.
[59] Q.Z. Zhang, W.R. Su, S.H. Shi, P. Wilder-Smith, A.P. Xiang, A. Wong, A.L. Nguyen, C.W. Kwon, A.D. Le, Human gingiva-derived mesenchymal stem cells elicit polarization of m2 macrophages and enhance cutaneous wound healing, Stem Cells 28 (2010) 1856–1868.
[60] D.O. Freytes, J.W. Kang, I. Marcos-Campos, G. Vunjak-Novakovic, Macrophages modulate the viability and growth of human mesenchymal stem cells, J. Cell. Biochem. 114 (2013) 220–229.
[61] Y.L. Lin, S.F. Yet, Y.T. Hsu, G.J. Wang, S.C. Hung, Mesenchymal stem cells ameliorate atherosclerotic lesions via restoring endothelial function, Stem Cells Transl. Med. 4 (2015) 44–55.
[62] S.S. Wang, S.W. Hu, Q.H. Zhang, A.X. Xia, Z.X. Jiang, X.M. Chen, Mesenchymal stem cells stabilize atherosclerotic vulnerable plaque by anti-inflammatory properties, PLoS One 10 (2015). e0136026.
[63] B. Ebrahimi, A. Eirin, Z. Li, X.Y. Zhu, X. Zhang, A. Lerman, S.C. Textor, L.O. Lerman, Mesenchymal stem cells improve medullary inflammation and fibrosis after revascularization of swine atherosclerotic renal artery stenosis, PLoS One 8 (2013). e67474.
[64] S.S. Wang, J. Ren, Aging as an essential modifier for the efficacy in mesenchymal stem cell therapy through an inositol phosphate 6 kinase-inositol pyrophosphate 7-dependent mechanism, Stem Cell Res. Ther. 5 (2014) 43.
[65] N. Kotobuki, Y. Katsube, Y. Katou, M. Tadokoro, M. Hirose, H. Ohgushi, In vivo survival and osteogenic differentiation of allogeneic rat bone marrow mesenchymal stem cells (MSCs), Cell Transplant. 17 (2008) 705–712.
[66] X.P. Huang, Z. Sun, Y. Miyagi, H. McDonald Kinkaid, L. Zhang, R.D. Weisel, R.K. Li, Differentiation of allogeneic mesenchymal stem cells induces immunogenicity and limits their long-term benefits for myocardial repair, Circulation 122 (2010) 2419–2429.

[67] I. Chimenti, R.R. Smith, T.S. Li, G. Gerstenblith, E. Messina, A. Giacomello, E. Marban, Relative roles of direct regeneration versus paracrine effects of human cardiosphere-derived cells transplanted into infarcted mice, Circ. Res. 106 (2010) 971–980.
[68] S.W. Kim, D.W. Lee, L.H. Yu, H.Z. Zhang, C.E. Kim, J.M. Kim, T.H. Park, K.S. Cha, S.Y. Seo, M.S. Roh, K.C. Lee, J.S. Jung, M.H. Kim, Mesenchymal stem cells overexpressing GCP-2 improve heart function through enhanced angiogenic properties in a myocardial infarction model, Cardiovasc. Res. 95 (2012) 495–506.
[69] L. Yao, J. Heuser-Baker, O. Herlea-Pana, R. Iida, Q. Wang, M.H. Zou, J. Barlic-Dicen, Bone marrow endothelial progenitors augment atherosclerotic plaque regression in a mouse model of plasma lipid lowering, Stem Cells 30 (2012) 2720–2731.
[70] Y. Yao, J. Huang, Y. Geng, H. Qian, F. Wang, X. Liu, M. Shang, S. Nie, N. Liu, X. Du, J. Dong, C. Ma, Paracrine action of mesenchymal stem cells revealed by single cell gene profiling in infarcted murine hearts, PLoS One 10 (2015). e0129164.
[71] S. Bollini, N. Smart, P.R. Riley, Resident cardiac progenitor cells: at the heart of regeneration, J. Mol. Cell. Cardiol. 50 (2011) 296–303.
[72] J. Nussbaum, E. Minami, M.A. Laflamme, J.A. Virag, C.B. Ware, A. Masino, V. Muskheli, L. Pabon, H. Reinecke, C.E. Murry, Transplantation of undifferentiated murine embryonic stem cells in the heart: teratoma formation and immune response, FASEB J. 21 (2007) 1345–1357.
[73] K. Matsuura, A. Honda, T. Nagai, N. Fukushima, K. Iwanaga, M. Tokunaga, T. Shimizu, T. Okano, H. Kasanuki, N. Hagiwara, I. Komuro, Transplantation of cardiac progenitor cells ameliorates cardiac dysfunction after myocardial infarction in mice, J. Clin. Invest. 119 (2009) 2204–2217.
[74] M. Fujiwara, P. Yan, T.G. Otsuji, G. Narazaki, H. Uosaki, H. Fukushima, K. Kuwahara, M. Harada, H. Matsuda, S. Matsuoka, K. Okita, K. Takahashi, M. Nakagawa, T. Ikeda, R. Sakata, C.L. Mummery, N. Nakatsuji, S. Yamanaka, K. Nakao, J.K. Yamashita, Induction and enhancement of cardiac cell differentiation from mouse and human induced pluripotent stem cells with cyclosporin-A, PLoS One 6 (2011). e16734.
[75] T. Kubota, R. Koike, Cryopyrin-associated periodic syndromes: background and therapeutics, Mod. Rheumatol. 20 (2010) 213–221.
[76] I.B. McInnes, G. Schett, The pathogenesis of rheumatoid arthritis, N. Engl. J. Med. 365 (2011) 2205–2219.
[77] T.N. Dinh, T.S. Kyaw, P. Kanellakis, K. To, P. Tipping, B.H. Toh, A. Bobik, A. Agrotis, Cytokine therapy with interleukin-2/anti-interleukin-2 monoclonal antibody complexes expands CD4 + CD25 + Foxp3+ regulatory T cells and attenuates development and progression of atherosclerosis, Circulation 126 (2012) 1256–1266.
[78] I. Tabas, C.K. Glass, Anti-inflammatory therapy in chronic disease: challenges and opportunities, Science 339 (2013) 166–172.
[79] V. Brito, K. Mellal, S.G. Portelance, A. Perez, Y. Soto, D. deBlois, H. Ong, S. Marleau, A.M. Vazquez, Induction of anti-anti-idiotype antibodies against sulfated glycosaminoglycans reduces atherosclerosis in apolipoprotein E-deficient mice, Arterioscler. Thromb. Vasc. Biol. 32 (2012) 2847–2854.
[80] M. Adi, A. Arnon, E.-M. Michal, K. Gad, G. Jacob, Anti eotaxin-2 antibodies attenuate the initiation and progression of experimental atherosclerosis, World J. Cardiorespir. Dis. 03 (2013) 339.
[81] J. Sun, G.K. Sukhova, P.J. Wolters, M. Yang, S. Kitamoto, P. Libby, L.A. MacFarlane, J. Mallen-St Clair, G.P. Shi, Mast cells promote atherosclerosis by releasing proinflammatory cytokines, Nat. Med. 13 (2007) 719–724.

[82] R. Kodali, M. Hajjou, A.B. Berman, M.B. Bansal, S. Zhang, J.J. Pan, A.D. Schecter, Chemokines induce matrix metalloproteinase-2 through activation of epidermal growth factor receptor in arterial smooth muscle cells, Cardiovasc. Res. 69 (2006) 706–715.
[83] G.K. Hansson, P. Libby, The immune response in atherosclerosis: a double-edged sword, Nat. Rev. Immunol. 6 (2006) 508–519.
[84] E. Galkina, K. Ley, Immune and inflammatory mechanisms of atherosclerosis (*), Annu. Rev. Immunol. 27 (2009) 165–197.
[85] J. Kuiper, G.H. van Puijvelde, E.J. van Wanrooij, T. van Es, K. Habets, A.D. Hauer, T.J. van den Berkel, Immunomodulation of the inflammatory response in atherosclerosis, Curr. Opin. Lipidol. 18 (2007) 521–526.
[86] K. Lacek, K. Vercauteren, K. Grzyb, M. Naddeo, L. Verhoye, M.P. Slowikowski, S. Fafi-Kremer, A.H. Patel, T.F. Baumert, A. Folgori, G. Leroux-Roels, R. Cortese, P. Meuleman, A. Nicosia, Novel human SR-BI antibodies prevent infection and dissemination of HCV in vitro and in humanized mice, J. Hepatol. 57 (2012) 17–23.
[87] A.M. Scott, J.D. Wolchok, L.J. Old, Antibody therapy of cancer, Nat. Rev. Cancer 12 (2012) 278–287.
[88] S.L. Topalian, F.S. Hodi, J.R. Brahmer, S.N. Gettinger, D.C. Smith, D.F. McDermott, J.D. Powderly, R.D. Carvajal, J.A. Sosman, M.B. Atkins, P.D. Leming, D.R. Spigel, S.J. Antonia, L. Horn, C.G. Drake, D.M. Pardoll, L. Chen, W.H. Sharfman, R.A. Anders, J.M. Taube, T.L. McMiller, H. Xu, A.J. Korman, M. Jure-Kunkel, S. Agrawal, D. McDonald, G.D. Kollia, A. Gupta, J.M. Wigginton, M. Sznol, Safety, activity, and immune correlates of anti-PD-1 antibody in cancer, N. Engl. J. Med. 366 (2012) 2443–2454.
[89] S. Ansari, A. Moshaverinia, S.H. Pi, A. Han, A.I. Abdelhamid, H.H. Zadeh, Functionalization of scaffolds with chimeric anti-BMP-2 monoclonal antibodies for osseous regeneration, Biomaterials 34 (2013) 10191–10198.
[90] G. Casi, D. Neri, Antibody-drug conjugates: basic concepts, examples and future perspectives, J. Control. Release 161 (2012) 422–428.
[91] B.D. Choi, C.T. Kuan, M. Cai, G.E. Archer, D.A. Mitchell, P.C. Gedeon, L. Sanchez-Perez, I. Pastan, D.D. Bigner, J.H. Sampson, Systemic administration of a bispecific antibody targeting EGFRvIII successfully treats intracerebral glioma, Proc. Natl. Acad. Sci. U. S. A. 110 (2013) 270–275.
[92] A.L. Catapano, N. Papadopoulos, The safety of therapeutic monoclonal antibodies: implications for cardiovascular disease and targeting the PCSK9 pathway, Atherosclerosis 228 (2013) 18–28.
[93] R. Salcedo, H.A. Young, M.L. Ponce, J.M. Ward, H.K. Kleinman, W.J. Murphy, J.J. Oppenheim, Eotaxin (CCL11) induces in vivo angiogenic responses by human CCR3+ endothelial cells, J. Immunol. 166 (2001) 7571–7578.
[94] E. Emanuele, C. Falcone, A. D'Angelo, P. Minoretti, M.P. Buzzi, M. Bertona, D. Geroldi, Association of plasma eotaxin levels with the presence and extent of angiographic coronary artery disease, Atherosclerosis 186 (2006) 140–145.
[95] A. Mor, A. Afek, M. Entin-Meer, G. Karen, J. George, Anti eotaxin-2 antibodies attenuate the initiation and progression of experimental atherosclerosis, World J. Cardiovasc. Dis. 03 (2013) 339.
[96] T. Kyaw, P. Cui, C. Tay, P. Kanellakis, H. Hosseini, E. Liu, A.G. Rolink, P. Tipping, A. Bobik, B.H. Toh, BAFF receptor mAb treatment ameliorates development and progression of atherosclerosis in hyperlipidemic ApoE(-/-) mice, PLoS One 8 (2013). e60430.
[97] R. Brink, Regulation of B cell self-tolerance by BAFF, Semin. Immunol. 18 (2006) 276–283.
[98] T.T. Hansel, H. Kropshofer, T. Singer, J.A. Mitchell, A.J. George, The safety and side effects of monoclonal antibodies, Nat. Rev. Drug Discov. 9 (2010) 325–338.

[99] B. Gorovits, C. Krinos-Fiorotti, Proposed mechanism of off-target toxicity for antibody-drug conjugates driven by mannose receptor uptake, Cancer Immunol. Immunother. 62 (2013) 217–223.
[100] A. Hebar, P. Valent, E. Selzer, The impact of molecular targets in cancer drug development: major hurdles and future strategies, Expert. Rev. Clin. Pharmacol. 6 (2013) 23–34.
[101] Y. Bordon, Anticancer drugs: cracking the combination, Nat. Rev. Drug Discov. 12 (2013) 505.
[102] G.W. Huntley, Synaptic circuit remodelling by matrix metalloproteinases in health and disease, Nat. Rev. Neurosci. 13 (2012) 743–757.
[103] X.W. Cheng, G.P. Shi, M. Kuzuya, T. Sasaki, K. Okumura, T. Murohara, Role for cysteine protease cathepsins in heart disease: focus on biology and mechanisms with clinical implication, Circulation 125 (2012) 1551–1562.
[104] R.B. Singh, L. Hryshko, D. Freed, N.S. Dhalla, Activation of proteolytic enzymes and depression of the sarcolemmal Na+/K+-ATPase in ischemia-reperfused heart may be mediated through oxidative stress, Can. J. Physiol. Pharmacol. 90 (2012) 249–260.
[105] M.H. Chien, C.W. Lin, C.W. Cheng, Y.C. Wen, S.F. Yang, Matrix metalloproteinase-2 as a target for head and neck cancer therapy, Expert Opin. Ther. Targets 17 (2013) 203–216.
[106] A. Dufour, C.M. Overall, Missing the target: matrix metalloproteinase antitargets in inflammation and cancer, Trends Pharmacol. Sci. 34 (2013) 233–242.
[107] D.F. Ketelhuth, M. Back, The role of matrix metalloproteinases in atherothrombosis, Curr. Atheroscler. Rep. 13 (2011) 162–169.
[108] W. Peeters, W.E. Hellings, D.P. de Kleijn, J.P. de Vries, F.L. Moll, A. Vink, G. Pasterkamp, Carotid atherosclerotic plaques stabilize after stroke: insights into the natural process of atherosclerotic plaque stabilization, Arterioscler. Thromb. Vasc. Biol. 29 (2009) 128–133.
[109] O. Erster, J.M. Thomas, J. Hamzah, A.M. Jabaiah, J.A. Getz, T.D. Schoep, S.S. Hall, E. Ruoslahti, P.S. Daugherty, Site-specific targeting of antibody activity in vivo mediated by disease-associated proteases, J. Control. Release 161 (2012) 804–812.
[110] E. Marsch, J.C. Sluimer, M.J. Daemen, Hypoxia in atherosclerosis and inflammation, Curr. Opin. Lipidol. 24 (2013) 393–400.
[111] S. Parathath, S.L. Mick, J.E. Feig, V. Joaquin, L. Grauer, D.M. Habiel, M. Gassmann, L.B. Gardner, E.A. Fisher, Hypoxia is present in murine atherosclerotic plaques and has multiple adverse effects on macrophage lipid metabolism, Circ. Res. 109 (2011) 1141–1152.
[112] T. Dietrich, C. Perlitz, K. Licha, P. Stawowy, K. Atrott, M. Tachezy, H. Meyborg, C. Stocker, M. Grafe, E. Fleck, M. Schirner, K. Graf, ED-B fibronectin (ED-B) can be targeted using a novel single chain antibody conjugate and is associated with macrophage accumulation in atherosclerotic lesions, Basic Res. Cardiol. 102 (2007) 298–307.
[113] T. Dietrich, T. Hucko, C. Schneemann, M. Neumann, A. Menrad, J. Willuda, K. Atrott, D. Stibenz, E. Fleck, K. Graf, H.D. Menssen, Local delivery of IL-2 reduces atherosclerosis via expansion of regulatory T cells, Atherosclerosis 220 (2012) 329–336.
[114] A. Brichkina, D.V. Bulavin, WIP-ing out atherosclerosis with autophagy, Autophagy 8 (2012) 1545–1547.
[115] X. Le Guezennec, A. Brichkina, Y.F. Huang, E. Kostromina, W. Han, D.V. Bulavin, Wip1-dependent regulation of autophagy, obesity, and atherosclerosis, Cell Metab. 16 (2012) 68–80.
[116] S. Mura, J. Nicolas, P. Couvreur, Stimuli-responsive nanocarriers for drug delivery, Nat. Mater. 12 (2013) 991–1003.
[117] K. Park, Nanotechnology: what it can do for drug delivery, J. Control. Release 120 (2007) 1–3.

[118] P. Bawa, V. Pillay, Y.E. Choonara, L.C. du Toit, Stimuli-responsive polymers and their applications in drug delivery, Biomed. Mater. 4 (2009) 022001.
[119] N. Korin, M. Kanapathipillai, B.D. Matthews, M. Crescente, A. Brill, T. Mammoto, K. Ghosh, S. Jurek, S.A. Bencherif, D. Bhatta, A.U. Coskun, C.L. Feldman, D.D. Wagner, D.E. Ingber, Shear-activated nanotherapeutics for drug targeting to obstructed blood vessels, Science 337 (2012) 738–742.
[120] G. Fredman, N. Kamaly, S. Spolitu, J. Milton, D. Ghorpade, R. Chiasson, G. Kuriakose, M. Perretti, O. Farokhzad, I. Tabas, Targeted nanoparticles containing the proresolving peptide Ac2-26 protect against advanced atherosclerosis in hypercholesterolemic mice, Sci. Transl. Med. 7 (2015). 275ra220.
[121] Y.Z. Cai, G.R. Zhang, L.L. Wang, Y.Z. Jiang, H.W. Ouyang, X.H. Zou, Novel biodegradable three-dimensional macroporous scaffold using aligned electrospun nanofibrous yarns for bone tissue engineering, J. Biomed. Mater. Res. A 100 (2012) 1187–1194.
[122] G. Ma, D. Fang, Y. Liu, X. Zhu, J. Nie, Electrospun sodium alginate/poly(ethylene oxide) core–shell nanofibers scaffolds potential for tissue engineering applications, Carbohydr. Polym. 87 (2012) 737–743.
[123] E. Kijenska, M.P. Prabhakaran, W. Swieszkowski, K.J. Kurzydlowski, S. Ramakrishna, Electrospun bio-composite P(LLA-CL)/collagen I/collagen III scaffolds for nerve tissue engineering, J. Biomed. Mater. Res. B Appl. Biomater. 100 (2012) 1093–1102.
[124] M.E. Frohbergh, A. Katsman, G.P. Botta, P. Lazarovici, C.L. Schauer, U.G. Wegst, P.I. Lelkes, Electrospun hydroxyapatite-containing chitosan nanofibers crosslinked with genipin for bone tissue engineering, Biomaterials 33 (2012) 9167–9178.
[125] D. Kai, M.P. Prabhakaran, B. Stahl, M. Eblenkamp, E. Wintermantel, S. Ramakrishna, Mechanical properties and in vitro behavior of nanofiber-hydrogel composites for tissue engineering applications, Nanotechnology 23 (2012) 095705.
[126] S.E. Paramonov, H.W. Jun, J.D. Hartgerink, Self-assembly of peptide-amphiphile nanofibers: the roles of hydrogen bonding and amphiphilic packing, J. Am. Chem. Soc. 128 (2006) 7291–7298.
[127] J. Luo, Y.W. Tong, Self-assembly of collagen-mimetic peptide amphiphiles into biofunctional nanofiber, ACS Nano 5 (2011) 7739–7747.
[128] H.S. Koh, T. Yong, W.E. Teo, C.K. Chan, M.E. Puhaindran, T.C. Tan, A. Lim, B.H. Lim, S. Ramakrishna, In vivo study of novel nanofibrous intra-luminal guidance channels to promote nerve regeneration, J. Neural Eng. 7 (2010) 046003.
[129] J.-J. Liu, C.-Y. Wang, J.-G. Wang, H.-J. Ruan, C.-Y. Fan, Peripheral nerve regeneration using composite poly(lactic acid-caprolactone)/nerve growth factor conduits prepared by coaxial electrospinning, J. Biomed. Mater. Res. A 96A (2011) 13–20.
[130] H. Jiang, Y. Hu, P. Zhao, Y. Li, K. Zhu, Modulation of protein release from biodegradable core-shell structured fibers prepared by coaxial electrospinning, J. Biomed. Mater. Res. B Appl. Biomater. 79 (2006) 50–57.
[131] R. Chen, C. Huang, Q. Ke, C. He, H. Wang, X. Mo, Preparation and characterization of coaxial electrospun thermoplastic polyurethane/collagen compound nanofibers for tissue engineering applications, Colloids Surf. B: Biointerfaces 79 (2010) 315–325.
[132] M. Chen, M. Dong, R. Havelund, V.R. Regina, R.L. Meyer, F. Besenbacher, P. Kingshott, Thermo-responsive core—sheath electrospun nanofibers from poly (*N*-isopropylacrylamide)/polycaprolactone blends, Chem. Mater. 22 (2010) 4214–4221.
[133] Z. Ma, M. Kotaki, R. Inai, S. Ramakrishna, Potential of nanofiber matrix as tissue-engineering scaffolds, Tissue Eng. 11 (2005) 101–109.

[134] S.M. Park, D.S. Kim, Electrolyte-assisted electrospinning for a self-assembled, free-standing nanofiber membrane on a curved surface, Adv. Mater. 27 (2015) 1682–1687.
[135] H.S. Yoo, T.G. Kim, T.G. Park, Surface-functionalized electrospun nanofibers for tissue engineering and drug delivery, Adv. Drug Deliv. Rev. 61 (2009) 1033–1042.
[136] S. Turmanova, M. Minchev, K. Vassilev, G. Danev, Surface grafting polymerization of vinyl monomers on poly(tetrafluoroethylene) films by plasma treatment, J. Polym. Res. 15 (2008) 309–318.
[137] D. Yan, J. Jones, X.Y. Yuan, X.H. Xu, J. Sheng, J.C. Lee, G.Q. Ma, Q.S. Yu, Plasma treatment of electrospun PCL random nanofiber meshes (NFMs) for biological property improvement, J. Biomed. Mater. Res. A 101 (2013) 963–972.
[138] M. Zhang, Z. Wang, S. Feng, H. Xu, Q. Zhao, S. Wang, J. Fang, M. Qiao, D. Kong, Immobilization of anti-CD31 antibody on electrospun poly(varepsilon-caprolactone) scaffolds through hydrophobins for specific adhesion of endothelial cells, Colloids Surf. B: Biointerfaces 85 (2011) 32–39.
[139] W. Li, Y. Guo, H. Wang, D. Shi, C. Liang, Z. Ye, F. Qing, J. Gong, Electrospun nanofibers immobilized with collagen for neural stem cells culture, J. Mater. Sci. Mater. Med. 19 (2008) 847–854.
[140] T.G. Kim, T.G. Park, Biomimicking extracellular matrix: cell adhesive RGD peptide modified electrospun poly(D,L-lactic-*co*-glycolic acid) nanofiber mesh, Tissue Eng. 12 (2006) 221–233.
[141] J.R. Paletta, S. Bockelmann, A. Walz, C. Theisen, J.H. Wendorff, A. Greiner, S. Fuchs-Winkelmann, M.D. Schofer, RGD-functionalisation of PLLA nanofibers by surface coupling using plasma treatment: influence on stem cell differentiation, J. Mater. Sci. Mater. Med. 21 (2010) 1363–1369.
[142] C.K. Chan, S. Liao, B. Li, R.R. Lareu, J.W. Larrick, S. Ramakrishna, M. Raghunath, Early adhesive behavior of bone-marrow-derived mesenchymal stem cells on collagen electrospun fibers, Biomed. Mater. 4 (2009) 035006.
[143] J.A. Matthews, G.E. Wnek, D.G. Simpson, G.L. Bowlin, Electrospinning of collagen nanofibers, Biomacromolecules 3 (2002) 232–238.
[144] I.C. Parrag, P.W. Zandstra, K.A. Woodhouse, Fiber alignment and coculture with fibroblasts improves the differentiated phenotype of murine embryonic stem cell-derived cardiomyocytes for cardiac tissue engineering, Biotechnol. Bioeng. 109 (2012) 813–822.
[145] R. Ravichandran, J.R. Venugopal, S. Sundarrajan, S. Mukherjee, S. Ramakrishna, Precipitation of nanohydroxyapatite on PLLA/PBLG/collagen nanofibrous structures for the differentiation of adipose derived stem cells to osteogenic lineage, Biomaterials 33 (2012) 846–855.
[146] S. Heydarkhan-Hagvall, K. Schenke-Layland, A.P. Dhanasopon, F. Rofail, H. Smith, B.M. Wu, R. Shemin, R.E. Beygui, W.R. MacLellan, Three-dimensional electrospun ECM-based hybrid scaffolds for cardiovascular tissue engineering, Biomaterials 29 (2008) 2907–2914.
[147] Y.R. Shih, C.N. Chen, S.W. Tsai, Y.J. Wang, O.K. Lee, Growth of mesenchymal stem cells on electrospun type I collagen nanofibers, Stem Cells 24 (2006) 2391–2397.
[148] W.J. Li, R. Tuli, C. Okafor, A. Derfoul, K.G. Danielson, D.J. Hall, R.S. Tuan, A three-dimensional nanofibrous scaffold for cartilage tissue engineering using human mesenchymal stem cells, Biomaterials 26 (2005) 599–609.
[149] X. Xin, M. Hussain, J.J. Mao, Continuing differentiation of human mesenchymal stem cells and induced chondrogenic and osteogenic lineages in electrospun PLGA nanofiber scaffold, Biomaterials 28 (2007) 316–325.

[150] J. Xie, S.M. Willerth, X. Li, M.R. Macewan, A. Rader, S.E. Sakiyama-Elbert, Y. Xia, The differentiation of embryonic stem cells seeded on electrospun nanofibers into neural lineages, Biomaterials 30 (2009) 354–362.
[151] C.K. Hashi, Y. Zhu, G.Y. Yang, W.L. Young, B.S. Hsiao, K. Wang, B. Chu, S. Li, Antithrombogenic property of bone marrow mesenchymal stem cells in nanofibrous vascular grafts, Proc. Natl. Acad. Sci. U. S. A. 104 (2007) 11915–11920.
[152] Y.L. Tang, Y. Tang, Y.C. Zhang, K. Qian, L. Shen, M.I. Phillips, Improved graft mesenchymal stem cell survival in ischemic heart with a hypoxia-regulated heme oxygenase-1 vector, J. Am. Coll. Cardiol. 46 (2005) 1339–1350.
[153] Y.L. Tang, Q. Zhao, X. Qin, L. Shen, L. Cheng, J. Ge, M.I. Phillips, Paracrine action enhances the effects of autologous mesenchymal stem cell transplantation on vascular regeneration in rat model of myocardial infarction, Ann. Thorac. Surg. 80 (2005) 229–236.
[154] M. Khan, S. Meduru, R. Gogna, E. Madan, L. Citro, M.L. Kuppusamy, M. Sayyid, M. Mostafa, R.L. Hamlin, P. Kuppusamy, Oxygen cycling in conjunction with stem cell transplantation induces NOS3 expression leading to attenuation of fibrosis and improved cardiac function, Cardiovasc. Res. 93 (2012) 89–99.
[155] R. Mingliang, Z. Bo, W. Zhengguo, Stem cells for cardiac repair: status, mechanisms, and new strategies, Stem Cells Int. 2011 (2011) 310928.
[156] C. Bao, J. Guo, G. Lin, M. Hu, Z. Hu, TNFR gene-modified mesenchymal stem cells attenuate inflammation and cardiac dysfunction following MI, Scand. Cardiovasc. J. 42 (2008) 56–62.
[157] Y.Y. Du, S.H. Zhou, T. Zhou, H. Su, H.W. Pan, W.H. Du, B. Liu, Q.M. Liu, Immuno-inflammatory regulation effect of mesenchymal stem cell transplantation in a rat model of myocardial infarction, Cytotherapy 10 (2008) 469–478.
[158] X.-r. Gao, Y.-z. Tan, H.-j. Wang, Overexpression of Csx/Nkx2.5 and GATA-4 enhances the efficacy of mesenchymal stem cell transplantation after myocardial infarction, Circ. J. 75 (2011) 2683–2691.
[159] H. Xu, Y.J. Yang, T. Yang, H.Y. Qian, Statins and stem cell modulation, Ageing Res. Rev. 12 (2012) 1–7.
[160] Y.J. Yang, H.Y. Qian, J. Huang, J.J. Li, R.L. Gao, K.F. Dou, G.S. Yang, J.T. Willerson, Y.J. Geng, Combined therapy with simvastatin and bone marrow-derived mesenchymal stem cells increases benefits in infarcted swine hearts, Arterioscler. Thromb. Vasc. Biol. 29 (2009) 2076–2082.
[161] R. Xu, J. Chen, X. Cong, S. Hu, X. Chen, Lovastatin protects mesenchymal stem cells against hypoxia- and serum deprivation-induced apoptosis by activation of PI3K/Akt and ERK1/2, J. Cell. Biochem. 103 (2008) 256–269.
[162] V.F. Segers, R.T. Lee, Biomaterials to enhance stem cell function in the heart, Circ. Res. 109 (2011) 910–922.
[163] L. Tian, M. Prabhakaran, X. Ding, D. Kai, S. Ramakrishna, Emulsion electrospun vascular endothelial growth factor encapsulated poly(L-lactic acid-*co*-e-caprolactone) nanofibers for sustained release in cardiac tissue engineering, J. Mater. Sci. 47 (2012) 3272–3281.
[164] V.L. Roger, A.S. Go, D.M. Lloyd-Jones, E.J. Benjamin, J.D. Berry, W.B. Borden, D.M. Bravata, S. Dai, E.S. Ford, C.S. Fox, H.J. Fullerton, C. Gillespie, S.M. Hailpern, J.A. Heit, V.J. Howard, B.M. Kissela, S.J. Kittner, D.T. Lackland, J.H. Lichtman, L.D. Lisabeth, D.M. Makuc, G.M. Marcus, A. Marelli, D.B. Matchar, C.S. Moy, D. Mozaffarian, M.E. Mussolino, G. Nichol, N.P. Paynter, E.Z. Soliman, P.D. Sorlie, N. Sotoodehnia, T.N. Turan, S.S. Virani, N.D. Wong, D. Woo, M.B. TurnerAmerican

Heart Association Statistics Committee and Stroke Statistics Subcommittee, Heart disease and stroke statistics–2012 update: a report from the American Heart Association, Circulation 125 (2012) e2–e220.
[165] B.C. Isenberg, C. Williams, R.T. Tranquillo, Small-diameter artificial arteries engineered in vitro, Circ. Res. 98 (2006) 25–35.
[166] Z.X. Meng, Y.S. Wang, C. Ma, W. Zheng, L. Li, Y.F. Zheng, Electrospinning of PLGA/gelatin randomly-oriented and aligned nanofibers as potential scaffold in tissue engineering, Mater. Sci. Eng. C 30 (2010) 1204–1210.
[167] N. Thottappillil, P.D. Nair, Scaffolds in vascular regeneration: current status, Vasc. Health Risk Manag. 11 (2015) 79–91.
[168] Z. Ma, W. He, T. Yong, S. Ramakrishna, Grafting of gelatin on electrospun poly(caprolactone) nanofibers to improve endothelial cell spreading and proliferation and to control cell orientation, Tissue Eng. 11 (2005) 1149–1158.
[169] H. Wang, Y. Feng, M. Behl, A. Lendlein, H. Zhao, R. Xiao, J. Lu, L. Zhang, J. Guo, Hemocompatible polyurethane/gelatin-heparin nanofibrous scaffolds formed by a bilayer electrospinning technique as potential artificial blood vessels, Front. Chem. Sci. Eng. 5 (2011) 392–400.
[170] K. Wingate, W. Bonani, Y. Tan, S.J. Bryant, W. Tan, Compressive elasticity of three-dimensional nanofiber matrix directs mesenchymal stem cell differentiation to vascular cells with endothelial or smooth muscle cell markers, Acta Biomater. 8 (2012) 1440–1449.
[171] L. Buttafoco, N.G. Kolkman, P. Engbers-Buijtenhuijs, A.A. Poot, P.J. Dijkstra, I. Vermes, J. Feijen, Electrospinning of collagen and elastin for tissue engineering applications, Biomaterials 27 (2006) 724–734.
[172] R. Ravichandran, J.R. Venugopal, S. Sundarrajan, S. Mukherjee, S. Ramakrishna, Poly(glycerol sebacate)/gelatin core/shell fibrous structure for regeneration of myocardial infarction, Tissue Eng. A 17 (2011) 1363–1373.
[173] F. Van de Werf, D. Ardissino, A. Betriu, D.V. Cokkinos, E. Falk, K.A. Fox, D. Julian, M. Lengyel, F.J. Neumann, W. Ruzyllo, C. Thygesen, S.R. Underwood, A. Vahanian, F.W. Verheugt, W. WijnsTask Force on the Management of Acute Myocardial Infarction of the European Society of Cardiology, Management of acute myocardial infarction in patients presenting with ST-segment elevation. The Task Force on the Management of Acute Myocardial Infarction of the European Society of Cardiology, Eur. Heart J 24 (2003) 28–66.
[174] Q.D. Wang, J. Pernow, P.O. Sjoquist, L. Ryden, Pharmacological possibilities for protection against myocardial reperfusion injury, Cardiovasc. Res. 55 (2002) 25–37.
[175] R.K. Binder, A.A. Khattab, Acute myocardial infarction—the role of drug-eluting stents in treatment strategies, Eur. Cardiol. 7 (2011) 113–116.
[176] A.A. Kocher, M.D. Schuster, M.J. Szabolcs, S. Takuma, D. Burkhoff, J. Wang, S. Homma, N.M. Edwards, S. Itescu, Neovascularization of ischemic myocardium by human bone-marrow-derived angioblasts prevents cardiomyocyte apoptosis, reduces remodeling and improves cardiac function, Nat. Med. 7 (2001) 430–436.
[177] A.A. Rane, K.L. Christman, Biomaterials for the treatment of myocardial infarction a 5-year update, J. Am. Coll. Cardiol. 58 (2011) 2615–2629.
[178] H. Das, N. Abdulhameed, M. Joseph, R. Sakthivel, H.Q. Mao, V.J. Pompili, Ex vivo nanofiber expansion and genetic modification of human cord blood-derived progenitor/stem cells enhances vasculogenesis, Cell Transplant. 18 (2009) 305–318.
[179] J.M. Singelyn, J.A. DeQuach, S.B. Seif-Naraghi, R.B. Littlefield, P.J. Schup-Magoffin, K.L. Christman, Naturally derived myocardial matrix as an injectable scaffold for cardiac tissue engineering, Biomaterials 30 (2009) 5409–5416.

[180] J.M. Singelyn, P. Sundaramurthy, T.D. Johnson, P.J. Schup-Magoffin, D.P. Hu, D.M. Faulk, J. Wang, K.M. Mayle, K. Bartels, M. Salvatore, A.M. Kinsey, A.N. Demaria, N. Dib, K.L. Christman, Catheter-deliverable hydrogel derived from decellularized ventricular, J. Am. Coll. Cardiol. 59 (2012) 751–763.

[181] S.H. Lee, P.L. Wolf, R. Escudero, R. Deutsch, S.W. Jamieson, P.A. Thistlethwaite, Early expression of angiogenesis factors in acute myocardial ischemia and infarction, N. Engl. J. Med. 342 (2000) 626–633.

[182] R. Lin, J. Cai, C. Nathan, X. Wei, S. Schleidt, R. Rosenwasser, L. Iacovitti, Neurogenesis is enhanced by stroke in multiple new stem cell niches along the ventricular system at sites of high BBB permeability, Neurobiol. Dis. 74 (2015) 229–239.

[183] Y.L. Tang, W. Zhu, M. Cheng, L. Chen, J. Zhang, T. Sun, R. Kishore, M.I. Phillips, D.W. Losordo, G. Qin, Hypoxic preconditioning enhances the benefit of cardiac progenitor cell therapy for treatment of myocardial infarction by inducing CXCR4 expression, Circ. Res. 104 (2009) 1209–1216.

[184] J. Jin, S.I. Jeong, Y.M. Shin, K.S. Lim, H.S. Shin, Y.M. Lee, H.C. Koh, K.-S. Kim, Transplantation of mesenchymal stem cells within a poly(lactide-*co*-Œμ-caprolactone) scaffold improves cardiac function in a rat myocardial infarction model, Eur. J. Heart Fail. 11 (2009) 147–153.

[185] H. Piao, J.S. Kwon, S. Piao, J.H. Sohn, Y.S. Lee, J.W. Bae, K.K. Hwang, D.W. Kim, O. Jeon, B.S. Kim, Y.B. Park, M.C. Cho, Effects of cardiac patches engineered with bone marrow-derived mononuclear cells and PGCL scaffolds in a rat myocardial infarction model, Biomaterials 28 (2007) 641–649.

[186] D. Simpson, H. Liu, T.H. Fan, R. Nerem, S.C. Dudley Jr., A tissue engineering approach to progenitor cell delivery results in significant cell engraftment and improved myocardial remodeling, Stem Cells 25 (2007) 2350–2357.

[187] R. Ravichandran, J.R. Venugopal, S. Sundarrajan, S. Mukherjee, R. Sridhar, S. Ramakrishna, Minimally invasive injectable short nanofibers of poly(glycerol sebacate) for cardiac tissue engineering, Nanotechnology 23 (2012) 385102.

[188] H.J. Lee, Y.H. Park, W.-G. Koh, Fabrication of nanofiber microarchitectures localized within hydrogel microparticles and their application to protein delivery and cell encapsulation, Adv. Funct. Mater. 23 (2012) 591–597.

[189] J.J. Yoo, C. Kim, C.W. Chung, Y.I. Jeong, D.H. Kang, 5-aminolevulinic acid-incorporated poly(vinyl alcohol) nanofiber-coated metal stent for application in photodynamic therapy, Int. J. Nanomedicine 7 (2012) 1997–2005.

[190] K. Kuraishi, H. Iwata, S. Nakano, S. Kubota, H. Tonami, M. Toda, N. Toma, S. Matsushima, K. Hamada, S. Ogawa, W. Taki, Development of nanofiber-covered stents using electrospinning: in vitro and acute, J. Biomed. Mater. Res. B Appl. Biomater. 88 (2009) 230–239.

[191] Z. Yu, H. Zhu, S. Lu, X. Yang, Accelerated endothelialization with a polymer-free sirolimus-eluting antibody-coated stent, J. Mater. Sci. Mater. Med. 24 (2013) 2601–2609.

[192] W.J. Seeto, Y. Tian, E.A. Lipke, Peptide-grafted poly(ethylene glycol) hydrogels support dynamic adhesion of endothelial progenitor cells, Acta Biomater. 9 (2013) 8279–8289.

[193] J. Li, D. Li, F. Gong, S. Jiang, H. Yu, Y. An, Anti-CD133 antibody immobilized on the surface of stents enhances endothelialization, Biomed. Res. Int. 2014 (2014) 902782.

[194] O. DelaRosa, W. Dalemans, E. Lombardo, Mesenchymal stem cells as therapeutic agents of inflammatory and autoimmune diseases, Curr. Opin. Biotechnol. 23 (2012) 978–983.

[195] Y. Cheng, C.-Y. Tsao, H.-C. Wu, X. Luo, J.L. Terrell, J. Betz, G.F. Payne, W.E Bentley, G.W. Rubloff, Electroaddressing functionalized polysaccharides as model biofilms for interrogating cell signaling, Adv. Funct. Mater. 22 (2012) 519–528.

[196] A. Townsend-Nicholson, S.N. Jayasinghe, Cell electrospinning: a unique biotechnique for encapsulating living organisms for generating active biological microthreads/scaffolds, Biomacromolecules 7 (2006) 3364–3369.

[197] J. Aoki, P.W. Serruys, H. van Beusekom, A.T.L. Ong, E.P. McFadden, G. Sianos, W.J. van der Giessen, E. Regar, P.J. de Feyter, H.R. Davis, S. Rowland, M.J.B. Kutryk, Endothelial progenitor cell capture by stents coated with antibody against CD34: the HEALING-FIM (Healthy Endothelial Accelerated Lining Inhibits Neointimal Growth-First In Man) Registry, J. Am. Coll. Cardiol. 45 (2005) 1574–1579.

[198] A. Sedaghat, J.M. Sinning, K. Paul, G. Kirfel, G. Nickenig, N. Werner, First in vitro and in vivo results of an anti-human CD133-antibody coated coronary stent in the porcine model, Clin. Res. Cardiol. 102 (2013) 1–13.

[199] K. Bonaventura, A.W. Leber, C. Sohns, M. Roser, L.H. Boldt, F.X. Kleber, W. Haverkamp, M. Dorenkamp, Cost-effectiveness of paclitaxel-coated balloon angioplasty and paclitaxel-eluting stent implantation for treatment of coronary in-stent restenosis in patients with stable coronary artery disease, Clin. Res. Cardiol. 101 (2012) 573–584.

[200] P.S. Teirstein, Drug-eluting stent restenosis: an uncommon yet pervasive problem, Circulation 122 (2010) 5–7.

[201] B. Oh, R.B. Melchert, C.H. Lee, Biomimicking robust hydrogel for the mesenchymal stem cell carrier, Pharm. Res. 32 (2015) 3213–3227.

[202] C.W. Hwang, P.V. Johnston, G. Gerstenblith, R.G. Weiss, G.F. Tomaselli, V.E Bogdan, A. Panigrahi, A. Leszczynska, Z. Xia, Stem cell impregnated nanofiber stent sleeve for on-stent production and intravascular delivery of paracrine factors, Biomaterials 52 (2015) 318–326.

[203] S. Petersen, A. Strohbach, R. Busch, S.B. Felix, K.P. Schmitz, K. Sternberg, Site-selective immobilization of anti-CD34 antibodies to poly(l-lactide) for endovascular implant surfaces, J. Biomed. Mater. Res. B Appl. Biomater. 102 (2014) 345–355.

[204] P. Benvenuto, M.A. Neves, C. Blaszykowski, A. Romaschin, T. Chung, S.R. Kim, M. Thompson, Adlayer-mediated antibody immobilization to stainless steel for potential application to endothelial progenitor cell capture, Langmuir 31 (2015) 5423–5431.

[205] C.L. Song, Q. Li, Y.P. Yu, G. Wang, J.P. Wang, Y. Lu, J.C. Zhang, H.Y. Diao, J.G. Liu, Y.H. Liu, J. Liu, Y. Li, D. Cai, B. Liu, Study of novel coating strategy for coronary stents: simutaneous coating of VEGF and anti-CD34 antibody, Rev. Bras. Cir. Cardiovasc. 30 (2015) 159–163.

[206] W.H. Maisel, Unanswered questions—drug-eluting stents and the risk of late thrombosis, N. Engl. J. Med. 356 (2007) 981–984.

[207] A.P. Tomisa, M.E. Launey, J.S. Lee, M.H. Mankani, U.G. Wegst, E. Saiz, Nanotechnology approaches to improve dental implants, Int. J. Oral Maxillofac. Implants 25 (Suppl) (2011) 25–44 (discussion 45–29).

[208] K. Chaudhury, V. Kumar, J. Kandasamy, S. RoyChoudhury, Regenerative nanomedicine: current perspectives and future directions, Int. J. Nanomedicine 9 (2014) 4153–4167.

[209] K. Nakano, K. Egashira, S. Masuda, K. Funakoshi, G. Zhao, S. Kimura, T Matoba, K. Sueishi, Y. Endo, Y. Kawashima, K. Hara, H. Tsujimoto, R. Tominaga, K. Sunagawa, Formulation of nanoparticle-eluting stents by a cationic electrodeposition coating technology: efficient nano-drug delivery via bioabsorbable polymeric nanoparticle-eluting stents in porcine coronary arteries, J. Am. Coll. Cardiol. Interv. 2 (2009) 277–283.

[210] Y. Liu, W. Wang, G. Acharya, Y.B. Shim, E.S. Choi, C.H. Lee, Advanced stent coating for drug delivery and in vivo biocompatibility, J. Nanopart. Res. 15 (2014) 1–16.
[211] J. Zhu, R.E. Marchant, Design properties of hydrogel tissue-engineering scaffolds, Expert Rev. Med. Devices 8 (2011) 607–626.
[212] M.E. Parente, A. Ochoa Andrade, G. Ares, F. Russo, A. Jimenez-Kairuz, Bioadhesive hydrogels for cosmetic applications, Int. J. Cosmet. Sci. 37 (2015) 511–518.
[213] N.A. Peppas, P. Bures, W. Leobandung, H. Ichikawa, Hydrogels in pharmaceutical formulations, Eur. J. Pharm. Biopharm. 50 (2000) 27–46.
[214] P. Thevenot, W. Hu, L. Tang, Surface chemistry influence implant biocompatibility, Curr. Top. Med. Chem. 8 (2008) 270–280.
[215] L. Indolfi, F. Causa, C. Giovino, F. Ungaro, F. Quaglia, P.A. Netti, Microsphere-integrated drug-eluting stents: PLGA microsphere integration in hydrogel coating for local and prolonged delivery of hydrophilic antirestenosis agents, J. Biomed. Mater. Res. A 97 (2011) 201–211.
[216] H. Zhong, O. Matsui, K. Xu, T. Ogi, J. Sanada, Y. Okamoto, Y. Tabata, Y. Takuwa, Gene transduction into aortic wall using plasmid-loaded cationized gelatin hydrogel-coated polyester stent graft, J. Vasc. Surg. 50 (2009) 1433–1443.
[217] Y. Nakayama, K. Ji-Youn, S. Nishi, H. Ueno, T. Matsuda, Development of high-performance stent: gelatinous photogel-coated stent that permits drug delivery and gene transfer, J. Biomed. Mater. Res. 57 (2001) 559–566.
[218] X.-W. Shi, C.-Y. Tsao, X. Yang, Y. Liu, P. Dykstra, G.W. Rubloff, R. Ghodssi, W.E. Bentley, G.F. Payne, Electroaddressing of cell populations by Co-deposition with calcium alginate hydrogels, Adv. Funct. Mater. 19 (2009) 2074–2080.
[219] X. Yang, E. Kim, Y. Liu, X.-W. Shi, G.W. Rubloff, R. Ghodssi, W.E. Bentley, Z. Pancer, G.F. Payne, In-film bioprocessing and immunoanalysis with electroaddressable stimuli-responsive polysaccharides, Adv. Funct. Mater. 20 (2010) 1645–1652.
[220] J. Ramon-Azcon, S. Ahadian, R. Obregon, G. Camci-Unal, S. Ostrovidov, V. Hosseini, H. Kaji, K. Ino, H. Shiku, A. Khademhosseini, T. Matsue, Gelatin methacrylate as a promising hydrogel for 3D microscale organization and proliferation of dielectrophoretically patterned cells, Lab Chip 12 (2012) 2959–2969.
[221] C.G. Park, M.H. Kim, M. Park, J.I. Lee, S.H. Lee, J.H. Park, K.H. Yoon, Y.B. Choy, Polymeric nanofiber coated esophageal stent for sustained delivery of an anticancer drug, Macromol. Res. 19 (2011) 1210–1216.
[222] D.N. Heo, J.B. Lee, M.S. Bae, Y.S. Hwang, K.H. Kwon, I.K. Kwon, Development of nanofiber coated indomethacin-eluting stent for tracheal regeneration, J. Nanosci. Nanotechnol. 11 (2011) 5711–5716.
[223] B. Oh, C.H. Lee, Nanofiber-coated drug eluting stent for the stabilization of mast cells, Pharm. Res. 31 (2014) 2463–2478.
[224] B. Oh, C.H. Lee, Advanced cardiovascular stent coated with nanofiber, Mol. Pharm. 10 (2013) 4432–4442.
[225] S.K. Ryu, E. Mahmud, S. Tsimikas, Estrogen-eluting stents, J. Cardiovasc. Transl. Res. 2 (2009) 240–244.
[226] E. Kritikou, J. Kuiper, P.T. Kovanen, I. Bot, The impact of mast cells on cardiovascular diseases, Eur. J. Pharmacol. 778 (2015) 103–115.
[227] M. Iikura, T. Takaishi, K. Hirai, H. Yamada, M. Iida, T. Koshino, Y. Morita, Exogenous nitric oxide regulates the degranulation of human basophils and rat peritoneal mast cells, Int. Arch. Allergy Immunol. 115 (1998) 129–136.
[228] B.J. Davis, B.F. Flanagan, A.M. Gilfillan, D.D. Metcalfe, J.W. Coleman, Nitric oxide inhibits IgE-dependent cytokine production and Fos and Jun activation in mast cells, J. Immunol. 173 (2004) 6914–6920.

[229] J.I. Noor, T. Ikeda, Y. Ueda, T. Ikenoue, A free radical scavenger, edaravone, inhibits lipid peroxidation and the production of nitric oxide in hypoxic-ischemic brain damage of neonatal rats, Am. J. Obstet. Gynecol. 193 (2005) 1703–1708.

[230] E.J. Swindle, D.D. Metcalfe, The role of reactive oxygen species and nitric oxide in mast cell-dependent inflammatory processes, Immunol. Rev. 217 (2007) 186–205.

[231] M. Gomez, J. Vila, R. Elosua, L. Molina, J. Bruguera, J. Sala, R. Masia, M.I. Covas, J. Marrugat, M. Fito, Relationship of lipid oxidation with subclinical atherosclerosis and 10-year coronary events in general population, Atherosclerosis 232 (2014) 134–140.

[232] D.M. Shih, Z. Wang, R. Lee, Y. Meng, N. Che, S. Charugundla, H. Qi, J. Wu, C. Pan, J.M. Brown, T. Vallim, B.J. Bennett, M. Graham, S.L. Hazen, A.J. Lusis, Flavin containing monooxygenase 3 exerts broad effects on glucose and lipid metabolism and atherosclerosis, J. Lipid Res. 56 (2015) 22–37.

[233] G.K. Hansson, Regulation of immune mechanisms in atherosclerosis, Ann. N. Y. Acad. Sci. 947 (2001) 157–165 (discussion 165–166).

[234] K.Y. Chyu, J. Nilsson, P.K. Shah, Immune mechanisms in atherosclerosis and potential for an atherosclerosis vaccine, Discov. Med. 11 (2011) 403–412.

[235] Z. Yang, M.W. von Ballmoos, D. Faessler, J. Voelzmann, J. Ortmann, N. Diehm, W. Kalka-Moll, I. Baumgartner, S. Di Santo, C. Kalka, Paracrine factors secreted by endothelial progenitor cells prevent oxidative stress-induced apoptosis of mature endothelial cells, Atherosclerosis 211 (2010) 103–109.

[236] P.R. Baraniak, T.C. McDevitt, Stem cell paracrine actions and tissue regeneration, Regen. Med. 5 (2010) 121–143.

[237] M. Gnecchi, Z. Zhang, A. Ni, V.J. Dzau, Paracrine mechanisms in adult stem cell signaling and therapy, Circ. Res. 103 (2008) 1204–1219.

[238] N.J. Hallab, K.J. Bundy, K. O'Connor, R.L. Moses, J.J. Jacobs, Evaluation of metallic and polymeric biomaterial surface energy and surface roughness characteristics for directed cell adhesion, Tissue Eng. 7 (2001) 55–71.

[239] M. Martinesi, S. Bruni, M. Stio, C. Treves, T. Bacci, F. Borgioli, Biocompatibility evaluation of surface-treated AISI 316L austenitic stainless steel in human cell cultures, J. Biomed. Mater. Res. A 80 (2007) 131–145.

Nitric oxide donor delivery

14

S.A. Omar, A. de Belder
Brighton and Sussex University Hospital, Brighton, United Kingdom

14.1 Introduction

Ischemic heart disease is the leading cause of death worldwide [1]. Although coronary bypass surgery was the mainstay for the treatment of coronary artery disease, in 1977 the first balloon angioplasty performed in the cardiac catheter lab on humans would change the future management of this disease [2]. From earlier balloon angioplasty models preformed in animals, it was clear that there was a degree of local injury and there were concerns regarding vessel and lesion recoil. Following the implantation of the first bare metal coronary stent in humans in 1987 [3], it was thought that a definitive and easily deliverable treatment had been established that obviated these short comings of balloon angioplasty. However, it soon became clear that vessel injury leads to dysfunction of the endothelial layer, a dysfunction that persists for up to 4–8 weeks, in turn leading to [4,5] negative vessel remodeling [6]. Thus, it results in an exaggerated healing response and intimal hyperplasia causing in-stent restenosis in 20%–30% of cases [7].

In 2001, drug-eluting stents (DES) revolutionized percutaneous coronary intervention (PCI) by reducing the incidence of in-stent restenosis (ISR) seen in bare metal stents (BMS) by 70%–80% [8,9]. The first-generation DESs mitigated excessive proliferation and rapid healing through the use of drugs with an anti-inflammatory function such as sirolimus [10] or drugs that were used as anti-cancer drugs such as paclitaxel [11]. The initial optimism was dampened when impaired re-endothelialization and delayed healing unmasked the specter of late in-stent thrombosis (LST) [12–14], a complication that was not previously seen with BMSs. This mandates the use of dual antiplatelet agents for a minimum of 6 months to a year and a single agent life long as the discontinuation of these agents is associated with the incidence of LST [15]. However, the use of dual antiplatelet therapy (DAPT) increases the risk of bleeding and delays any form of operative intervention, be it cardiac or non-cardiac surgery. Furthermore, considering that nearly a tenth of patients undergoing PCI will have an indication requiring permanent anticoagulation in addition to the mandatory DAPT will evidently increase their risk of bleeding even further [16].

Hence, it is clear that DESs have only solved one problem to raise another. It would be ideal if an alternative to anti-proliferative agents were found that would better modulate the endothelial cellular response to injury. Thus, nitric oxide (NO) donor stents provide an attractive substitute to DES. To grasp why this may be the case, an understanding of the pathophysiology of atherosclerosis, vessel injury healing, and intimal hyperplasia and the pivotal role that NO plays in these process is essential.

Functionalized Cardiovascular Stents. https://doi.org/10.1016/B978-0-08-100496-8.00015-9

14.2 Nitric oxide

Since identifying NO as being endothelial-derived relaxing factor (EDRF) ([17]), it has been shown to play a pivotal role as a messenger and signaling molecule, activating soluble guanylyl cyclase (sGC) and generating cyclic guanosine monophosphate (cGMP) [18]; making NO fundamental in the maintenance of numerous physiological pathways.

Nitric oxide is formed through an oxygen-dependent enzymatic process where the guanidino nitrogen atoms of L-arginine are oxidized to form NO and L-citrulline [19], a process facilitated by a family of enzymes named nitric oxide synthase (NOS) (Formula 1) comprising three distinct isoforms; neuronal (nNOS/NOS1), inducible (iNOS/NOS2), and endothelial (eNOS/NOS3) [20].

Formula 1

$$2\,\mathrm{L-arginine} + 3\,\mathrm{NADPH} + 1\,\mathrm{H}^{+} + 4\,\mathrm{O}_2 \rightleftharpoons 2\,\mathrm{citrulline} + 2\,\mathrm{NO} + 4\,\mathrm{H}_2\mathrm{O} + 3\,\mathrm{NADP}^{+}$$

Due to its extremely short half-life, NO is mostly shuttled within the circulation either bound to proteins in the form of s-nitrosothiols (s-nitroalbumin, s-nitrosoglutathione (GSNO), or s-nitrosocystine) at relatively high concentrations [21] or as a salt in the form of nitrate or nitrite [22].

14.2.1 S-nitrosothiols

S-nitrosothiols (RSNO) are produced through the addition of a nitroso group (RNO) to the sulfur atom of cysteine [23] (Formula 2). They are present in human plasma primarily as s-nitrosoproteins [24]. Although all proteins can undergo a process of s-nitrosation, s-nitrosoalbumin is the commonest circulating form as a consequence of the abundance of albumin and its possession of a free cysteine group that is readily accessible to nitrogen oxides [24,25]. Another endogenous s-nitrosthiol (although present in smaller quantities), GSNO, has a more potent relaxant effect on the vasculature [26]. Their effects are mediated through the liberation of NO as demonstrated in Formula 3.

Formula 2

$$\mathrm{RSH} + \mathrm{HONO} \Rightarrow \mathrm{RSNO} + \mathrm{H}_2\mathrm{O}$$

Formula 3

$$2\,\mathrm{RSNO} \Rightarrow \mathrm{RSSR} + 2\,\mathrm{NO}$$

S-nitrosothiols have been shown to convey cardioprotective effects via the transnitrosation of many proteins, including mitochondrial complex I [27].

14.2.2 Inorganic nitrite

Inorganic nitrite, a simple ion (NO_2^-), provides the largest directly accessible circulating pool of NO [22]. Endogenously produced inorganic nitrite (formed via the oxidation of NO derived from the L-arginine NOS pathway) constitutes 70% of its stored and circulating volume [28]. The remaining 30% is acquired through dietary intake via the nitrate-nitrite-NO entero-salivary circuit [29]. Contrary to the long held assumption that nitrite had no biological function other than being a mere byproduct of NO metabolism [30], the demonstration of an atrial-to-venous gradient under normal physiological conditions suggested that the anion is in fact biologically active and is reduced to NO in the vascular bed [31,32].

14.3 Nitric oxide and vascular function

Nitric oxide plays a key role in vascular biology, vascular endothelial function, regulating tone [33] inhibiting the adhesion of platelets [34] and leukocyte [35], down-regulating vascular smooth muscle cell (VSMC) proliferation [36], and their synthesis of protein and collagen [37].

14.3.1 Endothelial dysfunction

Endothelial dysfunction (directly affected by NO bioavailability) is the primary step in the development of atherosclerotic disease [38]. Furthermore, it has been shown to be an independent predictor of cardiac events and adverse sequel [39]. Many of the drugs used to mitigate the progression of atherosclerotic disease and/or improve clinical outcomes such as statins, angiotensin-converting enzyme inhibitors (ACEi), and b-blockers have been shown to improve endothelial function and the bioavailability of NO [40–44]. Conversely, the decline in NO bioavailability through either reduced production or increased consumption has been implicated in multiple cardiovascular disease processes [45,46].

14.3.2 Platelet aggregation

Platelets play an important role in the physiology of hemostasis [47,48] and normal healing ([49]), but also play key roles in the pathophysiology of acute coronary syndrome (ACS) [50,51] and stent thrombosis [52,53]. Platelets are anuclear cell fragments found within the circulation and continuously interacting with the vascular endothelium [54]; their activation remains inhibited via the function of vascular endothelium, which produces inhibitors such as prostacyclin [55] and NO [56,57].

In addition to direct NO inhibition of platelets, it has been shown that circulating donors of NO demonstrated the same effect. GSNO has been shown to selectively inhibit platelet function in the human forearm vasculature [58] and this antiplatelet property was confirmed in the coronary circulation of patients with atherosclerosis [59]. Nitrite has also been shown to inhibit platelet function and aggregation in a dose-dependent manner [60,61].

14.3.3 Intimal hyperplasia

Vascular injury usually leads to an appropriate healing response. However, the de-endothelialization caused by balloon angioplasty and stent implantation and the promotion of platelet deposition that follows can precipitate an exaggerated healing process, leading to incongruous remodeling and intimal hyperplasia (IH), which in turn causes the development of plaques and flow limiting disease [62,63].

It is the release of growth factors and cytokines by platelets and leukocytes that stimulate the proliferation of VSMCs and the formation of neointima causing restenosis [64,65]. Nitric Oxide inhibits the function of platelets, leukocytes, and VSMCs migration; thus, there is evidence that reduction in NO bioavailability is responsible for this inappropriate healing and its presence may mitigate it [66,67].

14.4 Localized NO delivery

Thus, it should not be surprising that there has been interest in developing vascular stents that donate NO providing targeted release of the messenger with a range of delivery kinetics by using a variety of coatings including polymers [68–71].

Nitric oxide has an extremely short half-life. There are two ways in which a stent would allow for localized NO delivery. Firstly, in keeping with current DES technology, a polymer coating would be impregnated with a nitric oxide donor. The choice of NO donor is limited by the rate and duration at which they release NO. Secondly, utilizing the endogenously circulating NO donors to release NO when they come into contact with the implanted device.

Animal studies confirmed that sodium nitroprusside (SNP) (an organic nitrate that acts as a direct NO donor [72]) impregnated polyurethane polymer stents released NO in a controlled manner for up to 28 days following arterial injury [73].

14.4.1 Exogenous NO donors

Organic nitrate and nitrite have been used as NO donors for 150 years. Their medicinal use is limited by bioavailability profile, systemic absorption and metabolism, therapeutic half-life, and development of tolerance [74,75].

There are numerous NO donor pro-drugs that have been developed to date. The familiar organic nitrates and nitrite [such as glyceryl trinitrate (GTN), isosorbide mononitrate (ISMN), Isosorbide dinitrate (ISDN), amyl nitrite and nicorandil] would not be suitable for impregnation of the stent as they require metabolism in order to release NO [75]. In addition, they have a short half-life, lack selectivity, and develop tolerance [75]. Other organic nitrates, such as SNP, have a more favorable activity profile spontaneously liberating NO without the need for prior metabolism [75].

Molsidomine, another direct nitric oxide donor [76], was used in the MEDCOR trial, but failed to demonstrate a significant difference in any of the primary or secondary endpoints, between the cohort receiving the drug and those that received placebo [77]. However, there are new and novel synthetic NO donors constantly produced, such as MK-8150 [78], which may offer even better delivery profiles.

14.4.2 Endogenous NO donors

Stent properties and NO delivery using SNOs.

S-nitrosothiols are viewed as stable carriers of NO able to buffer its extra- and intra-cellular concentrations with the ability of releasing NO via the interaction with numerous reducing agents and transitional metal ions [79]. It has been revealed that Fe^{2+}, Cu^{2+}, Ag^{+}, and Hg^{2+} induce the release of NO from both endogenous and synthetic RSNOs [80–82]. On the other hand, Co^{2+}, Ni^{2+}, and Zn^{+} were reported to lack the ability to liberate NO from RSNO [83]. It has been suggested that the release of NO at the stent blood interface would mitigate the occurrence of in-stent thrombosis and restenosis [84,85].

One study determined the feasibility of using stents developed from specific metals and metal alloys that would encourage the release of NO from RSNO [86]. In this study, S-nitroso-N-acetyl-D-penicillinamine (SNAP), a synthetic RSNO commonly utilized to replicate the physiological behavior of endogenous RSNOs in vitro [87], was used. The study investigated metal ions that could possibly be released from currently manufactured BMSs (such as Pt^{2+}, Fe^{2+}, Fe^{3+}, Mg^{2+}, Mn^{2+}, Ni^{2+}, Co^{2+}, Cu^{2+}, and Zn^{2+}); in addition, wires composed of the typical materials used in the manufacturing of stents were also tested [86].

The quantity of NO released via the interaction of the various salts or metal wires with SNAP was measured by using chemiluminescence. The study demonstrated that Fe^{2+}, Cu^{2+}, Co^{2+}, Ni^{2+}, and Zn^{2+} all induced the release of NO from SNAP; on the other hand, Fe^{3+}, Pt^{2+}, Mg^{2+}, and Mn^{2+} did not [86]. These findings are helpful in directing future intravascular stent design where the utilization of the endogenous circulating pool of NO carriers is desirable.

The study then used zinc wires (as a potent liberator of NO) and platinum wires (as a control due to its poor NO release profile) in rat vasculature. The platinum wires were prothrombotic and exhibited abnormal healing with excessive cellular and fibrin deposition; and the zinc wires showed minimal cellular coverage and no thrombosis [86]. These differences were noted in periods of time as short as 2 h and as long as 6 months post-implantation.

And as many of these transition metals and their alloys have been shown to be bioabsorbable, such as zinc and magnesium, they have undergone development as base materials for BMSs [86,88].

A host of proteins and enzymes are capable of reducing nitrite and liberating NO. These include hemoglobin [89], myoglobin [90], neuroglobin [91], cytoglobin ([92]), mitochondrial proteins [93], xanthine oxidase [94], aldehyde oxidase [95], cytochrome P450 [96], carbonic anhydrase [97], and NOS [98]. Many of these enzymes are located within the cells comprising the vascular bed or within the circulation itself. So it is conceivable that a stent can be designed that is either impregnated with nitrite or with one of these enzymes that would facilitate the reduction of circulating nitrite to NO.

14.5 Nitric oxide donor stents (the evidence)

The TiNOX trial compared titanium BMS with identical titanium stents coated with nitric oxide (so-called bioactive stents—BAS). The trial demonstrated a significant reduction of restenosis in the BAS arm (33% vs. 15%) and significant reduction in major adverse cardiac events (MACE) at 6 months (27% vs. 7%; $P = .02$). However, it must be noted that the latter was mainly driven by the need for TLR (23% vs. 7%; $P = .07$) [99].

In the context of an acute myocardial infarction (AMI), the titanium-coated-nitric-oxide stents (TITANOX) were compared to paclitaxel-eluting stents (PES). At 2-years follow-up, the primary endpoints of composite MI, TLR, or death from a cardiac cause were significantly lower in the TITANOX group when equated with the PES group (11.2% vs. 21.8%, $P = .004$), which in turn was driven by a reduction in MI and cardiac death (5.1 vs. 15.6%, $P < .001$; and 0.9 vs. 4.7%, $P = .02$), respectively. In addition, there was significantly reduced incidence of stent thrombosis (ST) in the TITANOX group (0.5% vs. 6.2%, $P = .001$) [100]. The increased incidence of ST in the PES group may in part be explained by the findings of the TITAX-OCT study, which showed better healing and endothelialization of the TITANOX (BAS) stents and greater incidence of incomplete endothelialization in the PES group [101].

When comparing the use of titanium nitric oxide-coated bioactive stents (BAS) with second-generation everolimus drug-eluting stents (EES) in the setting of patients presenting within the setting of acute coronary syndrome including ST elevation myocardial infarction (STEMI) and non-ST elevation myocardial infarction (NSTEMI), the BASE ACS trial demonstrated that the former was statistically non-inferior to the latter for the major outcome of MACE or target lesion revascularization (TLR) in both a time frame of 1-year and 5-years follow-up [102,103]. At 5-year follow-up, the comparators for BAS vs. EES for the primary endpoint of MACE and secondary endpoints of non-fatal MI and TLR were 14.4% vs. 17.8%, $P < .001$; 5.9% vs. 9.7%, $P = .028$; 8.3% vs. 9.9%, $P = .58$, respectively [102]. Furthermore, following 1 year the cases of ischemia-driven TLR and definite stent thrombosis were fewer in the BAS cohort when compared to the EES cohort (1.8% vs. 5%, $P = .028$; and 1.1% vs. 3.8%, $P = .015$, respectively). These finding are suggestive of a better safety profile of BAS when compared to EES in the long term, although the study was underpowered to exclude a type I statistical error [102]. Interestingly, patients with diabetes who represented about 16.9% of the trial cohort exhibited a greater preponderance for ISR and need for TLR [102]. This may reflect the extremely abnormal endothelial function that this cohort of patients exhibits.

14.6 The future

Made up of a backbone of polylactic acids, bioabsorbable vascular scaffolds (BVSs) are fully reabsorbed by the vascular endothelium at 4 years [104]. The use of BVSs represents a novel and exciting advancement in the strategies available to us in the treatment of coronary artery disease by overcoming the restrictions caused by the long-term constraints imposed by the use of metallic stents. These limitations include disrupting normal vascular response, restricting future surgical options, impinging on side branches, and necessitating the lifelong use of an antiplatelet agent.

Although the above study was conducted in a porcine model, the pilot human study demonstrated that BVSs performed favorably against EESs both via intracoronary imaging and outcomes [105]; the encouraging data was replicated in a randomized multicenter trial (ABSORB II) comparing EESs and BVS in elective cases [106] and in patients presenting with STEMI [107]. Scaffolds, however, did show a high rate of acute stent thrombosis in the first 2 years. Currently, the scaffolds are impregnated with everolimus. However, impregnating the scaffolds with an NO donor may prove to be more promising as this more closely simulates normal vascular function and may, as described above, encourage a more normal process of healing and reduce the incidence of acute thrombosis. It is interesting to note that in one study (conducted in a rodent model) used scaffolds coated with zinc (a potent liberator of circulating NO as described above) demonstrated that low doses of Zn had a positive effect on VSMCs behavior in the healing process [108].

14.7 Conclusion

Nitric oxide is a key messenger in the regulation of numerous physiological processes, playing a pivotal role in normal vascular function. The diminution in the bioavailability of NO through reduced production or increased consumption has many adverse effects: loss of natural endothelial function, promotion of abnormal platelet aggregation, and amplifying the development of atherosclerosis. Thus, when treating coronary artery disease with percutaneous revascularization, the benefit of using a platform that enhances local NO is a highly desirable concept, mitigating many of the above-mentioned problems: improving endothelial function, inhibiting platelet aggregation, and supressing intimal hyperplasia. These benefits lead to a diminution of stent thrombosis and in-stent restenosis, translating into a reduction in MACE and TLR.

Boosting local NO delivery can be achieved via the use of a stent that is impregnated with an NO donor. This donor could either be a manufactured synthetic drug that releases NO in a constant and controlled manner, or it could be a naturally occurring physiological donor that spontaneously provides NO on coming in contact with enzymes and proteins found in the circulation or in the vascular bed. An alternative strategy would be to impregnate the stent with an enzyme or protein that would liberate NO from circulating carriers of NO such as s-nitrosothiols, nitrate, and nitrite.

A scaffold that would provide radial support and mimic the normal vessel dynamic physical response in addition to its biological one (by delivering NO) would be the ideal platform by which to treat established coronary atherosclerotic disease. If this platform were to disappear over time, leaving behind a vessel free of atherosclerotic disease, a repaired endothelium with restored function, and physiologically normal levels of NO, then surely this is the perfect manner in which to treat coronary artery disease.

References

[1] WHO, Top 10 causes of death. http://www.who.int/mediacentre/factsheets/fs310/en/, 2014.

[2] A. Gruentzig, Results from coronary angioplasty and implications for the future, Am. Heart J. 103 (1982) 779–783.

[3] U. Sigwart, J. Puel, V. Mirkovitch, F. Joffre, L. Kappenberger, Intravascular stents to prevent occlusion and restenosis after transluminal angioplasty, N. Engl. J. Med. 316 (1987) 701–706.

[4] R.M. Saroyan, M.P. Roberts, J.T. Light Jr., I.L. Chen, M.Y. Vaccarella, D.J. Bang, P. Kvamme, S. Singh, S.V. Scalia, M.D. Kerstein, et al., Differential recovery of prostacyclin and endothelium-derived relaxing factor after vascular injury, Am. J. Physiol. 262 (1992) 1449–1457.

[5] F.F. Weidinger, J.M. Mclenachan, M.I. Cybulsky, J.B. Gordon, H.G. Rennke, N.K. Hollenberg, J.T. Fallon, P. Ganz, J.P. Cooke, Persistent dysfunction of regenerated endothelium after balloon angioplasty of rabbit iliac artery, Circulation 81 (1990) 1667–1679.

[6] P.W. Serruys, P. de Jaegere, F. Kiemeneij, C. Macaya, W. Rutsch, G. Heyndrickx, H. Emanuelsson, J. Marco, V. Legrand, P. Materne, et al., A comparison of balloon-expandable-stent implantation with balloon angioplasty in patients with coronary artery disease. Benestent Study Group, N. Engl. J. Med. 331 (1994) 489–495.

[7] D.L. Fischman, M.B. Leon, D.S. Baim, R.A. Schatz, M.P. Savage, I. Penn, K. Detre, L. Veltri, D. Ricci, M. Nobuyoshi, et al., A randomized comparison of coronary-stent placement and balloon angioplasty in the treatment of coronary artery disease. Stent Restenosis Study Investigators, N. Engl. J. Med. 331 (1994) 496–501.

[8] A.J. Kirtane, A. Gupta, S. Iyengar, J.W. Moses, M.B. Leon, R. Applegate, B. Brodie, E. Hannan, K. Harjai, L.O. Jensen, S.J. Park, R. Perry, M. Racz, F. Saia, J.V. Tu, R. Waksman, A.J. Lansky, R. Mehran, G.W. Stone, Safety and efficacy of drug-eluting and bare metal stents: comprehensive meta-analysis of randomized trials and observational studies, Circulation 119 (2009) 3198–3206.

[9] M.C. Morice, P.W. Serruys, J.E. Sousa, J. Fajadet, E. Ban Hayashi, M. Perin, A. Colombo, G. Schuler, P. Barragan, G. Guagliumi, F. Molnar, R. FaloticoRAVEL Study Group, A randomized comparison of a sirolimus-eluting stent with a standard stent for coronary revascularization, N. Engl. J. Med. 346 (2002) 1773–1780.

[10] J.W. Moses, M.B. Leon, J.J. Popma, P.J. Fitzgerald, D.R. Holmes, C. O'Shaughnessy, R.P. Caputo, D.J. Kereiakes, D.O. Williams, P.S. Teirstein, J.L. Jaeger, R.E. Kuntz, S. Investigators, Sirolimus-eluting stents versus standard stents in patients with stenosis in a native coronary artery, N. Engl. J. Med. 349 (2003) 1315–1323.

[11] G.W. Stone, S.G. Ellis, D.A. Cox, J. Hermiller, C. O'Shaughnessy, J.T. Mann, M. Turco, R. Caputo, P. Bergin, J. Greenberg, J.J. Popma, M.E. Russell, T.-I. Investigators, One-year clinical results with the slow-release, polymer-based, paclitaxel-eluting TAXUS stent: the TAXUS-IV trial, Circulation 109 (2004) 1942–1947.

[12] J. Daemen, P. Wenaweser, K. Tsuchida, L. Abrecht, S. Vaina, C. Morger, N. Kukreja, P. Juni, G. Sianos, G. Hellige, R.T. van Domburg, O.M. Hess, E. Boersma, B. Meier, S. Windecker, P.W. Serruys, Early and late coronary stent thrombosis of sirolimus-eluting and paclitaxel-eluting stents in routine clinical practice: data from a large two-institutional cohort study, Lancet 369 (2007) 667–678.

[13] I. Iakovou, T. Schmidt, E. Bonizzoni, L. Ge, G.M. Sangiorgi, G. Stankovic, F. Airoldi, A. Chieffo, M. Montorfano, M. Carlino, I. Michev, N. Corvaja, C. Briguori, U. Gerckens, E. Grube, A. Colombo, Incidence, predictors, and outcome of thrombosis after successful implantation of drug-eluting stents, JAMA 293 (2005) 2126–2130.

[14] G. Nakazawa, A.V. Finn, M. Joner, E. Ladich, R. Kutys, E.K. Mont, H.K. Gold, A.P. Burke, F.D. Kolodgie, R. Virmani, Delayed arterial healing and increased late stent thrombosis at culprit sites after drug-eluting stent placement for acute myocardial infarction patients: an autopsy study, Circulation 118 (2008) 1138–1145.

[15] E.P. Mcfadden, E. Stabile, E. Regar, E. Cheneau, A.T. Ong, T. Kinnaird, W.O. Suddath, N.J. Weissman, R. Torguson, K.M. Kent, A.D. Pichard, L.F. Satler, R. Waksman, P.W. Serruys, Late thrombosis in drug-eluting coronary stents after discontinuation of antiplatelet therapy, Lancet 364 (2004) 1519–1521.
[16] A.E. May, T. Geisler, M. Gawaz, Individualized antithrombotic therapy in high risk patients after coronary stenting. A double-edged sword between thrombosis and bleeding, Thromb. Haemost. 99 (2008) 487–493.
[17] L.J. Ignarro, G.M. Buga, K.S. Wood, R.E. Byrns, G. Chaudhuri, Endothelium-derived relaxing factor produced and released from artery and vein is nitric oxide, Proc. Natl. Acad. Sci. U. S. A. 84 (1987) 9265–9269.
[18] F. Murad, Shattuck lecture. Nitric oxide and cyclic GMP in cell signaling and drug development, N. Engl. J. Med. 355 (2006) 2003–2011.
[19] S. Moncada, Nitric oxide: discovery and impact on clinical medicine, J. R. Soc. Med. 92 (1999) 164–169.
[20] C. Villanueva, C. Giulivi, Subcellular and cellular locations of nitric oxide synthase isoforms as determinants of health and disease, Free Radic. Biol. Med. 49 (2010) 307–316.
[21] E. Nagababu, J.M. Rifkind, Routes for formation of S-nitrosothiols in blood, Cell Biochem. Biophys. 67 (2013) 385–398.
[22] S.A. Omar, A.J. Webb, Nitrite reduction and cardiovascular protection, J. Mol. Cell. Cardiol. 73 (2014) 57–69.
[23] B.C. Smith, M.A. Marletta, Mechanisms of S-nitrosothiol formation and selectivity in nitric oxide signaling, Curr. Opin. Chem. Biol. 16 (2012) 498–506.
[24] J.S. Stamler, O. Jaraki, J. Osborne, D.I. Simon, J. Keaney, J. Vita, D. Singel, C.R. Valeri, J. Loscalzo, Nitric oxide circulates in mammalian plasma primarily as an S-nitroso adduct of serum albumin, Proc. Natl. Acad. Sci. U. S. A. 89 (1992) 7674–7677.
[25] J.S. Stamler, S-nitrosothiols in the blood: roles, amounts, and methods of analysis, Circ. Res. 94 (2004) 414–417.
[26] B. Gaston, J. Reilly, J.M. Drazen, J. Fackler, P. Ramdev, D. Arnelle, M.E. Mullins, D.J. Sugarbaker, C. Chee, D.J. Singel, et al., Endogenous nitrogen oxides and bronchodilator S-nitrosothiols in human airways, Proc. Natl. Acad. Sci. U. S. A. 90 (1993) 10957–10961.
[27] L.S. Burwell, S.M. Nadtochiy, A.J. Tompkins, S. Young, P.S. Brookes, Direct evidence for S-nitrosation of mitochondrial complex I, Biochem. J. 394 (2006) 627–634.
[28] P. Rhodes, A.M. Leone, P.L. Francis, A.D. Struthers, S. Moncada, P.M. Rhodes, The L-arginine: nitric oxide pathway is the major source of plasma nitrite in fasted humans, Biochem. Biophys. Res. Commun. 209 (1995) 590–596.
[29] S. Lidder, A.J. Webb, Vascular effects of dietary nitrate (as found in green leafy vegetables and beetroot) via the nitrate-nitrite-nitric oxide pathway, Br. J. Clin. Pharmacol. 75 (2013) 677–696.
[30] T. Lauer, M. Preik, T. Rassaf, B.E. Strauer, A. Deussen, M. Feelisch, M. Kelm, Plasma nitrite rather than nitrate reflects regional endothelial nitric oxide synthase activity but lacks intrinsic vasodilator action, Proc. Natl. Acad. Sci. U. S. A. 98 (2001) 12814–12819.
[31] E. Cicinelli, L.J. Ignarro, L.M. Schonauer, M.G. Matteo, P. Galantino, N. Falco, Different plasma levels of nitric oxide in arterial and venous blood, Clin. Physiol. 19 (1999) 440–442.
[32] M.T. Gladwin, J.H. Shelhamer, A.N. Schechter, M.E. Pease-Fye, M.A. Waclawiw, J.A. Panza, F.P. Ognibene, R.O. Cannon 3rd, Role of circulating nitrite and S-nitrosohemoglobin in the regulation of regional blood flow in humans, Proc. Natl. Acad. Sci. U. S. A. 97 (2000) 11482–11487.

[33] R.F. Furchgott, J.V. Zawadzki, The obligatory role of endothelial cells in the relaxation of arterial smooth muscle by acetylcholine, Nature 288 (1980) 373–376.
[34] M.W. Radomski, R.M. Palmer, S. Moncada, An L-arginine/nitric oxide pathway present in human platelets regulates aggregation, Proc. Natl. Acad. Sci. U. S. A. 87 (1990) 5193–5197.
[35] P.M. Bath, D.G. Hassall, A.M. Gladwin, R.M. Palmer, J.F. Martin, Nitric oxide and prostacyclin. Divergence of inhibitory effects on monocyte chemotaxis and adhesion to endothelium in vitro, Arterioscler. Thromb. 11 (1991) 254–260.
[36] T. Scott-Burden, P.M. Vanhoutte, Regulation of smooth muscle cell growth by endothelium-derived factors, Tex. Heart Inst. J. 21 (1994) 91–97.
[37] V. Kolpakov, D. Gordon, T.J. Kulik, Nitric oxide-generating compounds inhibit total protein and collagen synthesis in cultured vascular smooth muscle cells, Circ. Res. 76 (1995) 305–309.
[38] K.V. Fitch, E. Stavrou, S.E. Looby, L. Hemphill, M.R. Jaff, S.K. Grinspoon, Associations of cardiovascular risk factors with two surrogate markers of subclinical atherosclerosis: endothelial function and carotid intima media thickness, Atherosclerosis 217 (2011) 437–440.
[39] Y. Kitta, J.E. Obata, T. Nakamura, M. Hirano, Y. Kodama, D. Fujioka, Y. Saito, K. Kawabata, K. Sano, T. Kobayashi, T. Yano, K. Nakamura, K. Kugiyama, Persistent impairment of endothelial vasomotor function has a negative impact on outcome in patients with coronary artery disease, J. Am. Coll. Cardiol. 53 (2009) 323–330.
[40] F. de Nigris, F.P. Mancini, M.L. Balestrieri, R. Byrns, C. Fiorito, S. Williams-Ignarro, A. Palagiano, E. Crimi, L.J. Ignarro, C. Napoli, Therapeutic dose of nebivolol, a nitric oxide-releasing beta-blocker, reduces atherosclerosis in cholesterol-fed rabbits, Nitric Oxide 19 (2008) 57–63.
[41] L.J. Ignarro, R.E. Byrns, K. Trinh, M. Sisodia, G.M. Buga, Nebivolol: a selective beta(1)-adrenergic receptor antagonist that relaxes vascular smooth muscle by nitric oxide- and cyclic GMP-dependent mechanisms, Nitric Oxide 7 (2002) 75–82.
[42] S. John, M.P. Schneider, C. Delles, J. Jacobi, R.E. Schmieder, Lipid-independent effects of statins on endothelial function and bioavailability of nitric oxide in hypercholesterolemic patients, Am. Heart J. 149 (2005) 473.
[43] J.C. Larosa, S.M. Grundy, D.D. Waters, C. Shear, P. Barter, J.C. Fruchart, A.M. Gotto, H. Greten, J.J. Kastelein, J. Shepherd, N.K. WengerTreating to New Targets (TNT) Investigators, Intensive lipid lowering with atorvastatin in patients with stable coronary disease, N. Engl. J. Med. 352 (2005) 1425–1435.
[44] S. Yusuf, P. Sleight, J. Pogue, J. Bosch, R. Davies, G. Dagenais, Effects of an angiotensin-converting-enzyme inhibitor, ramipril, on cardiovascular events in high-risk patients. The Heart Outcomes Prevention Evaluation Study Investigators, N. Engl. J. Med. 342 (2000) 145–153.
[45] L.J. Ignarro, Nitric oxide as a unique signaling molecule in the vascular system: a historical overview, J. Physiol. Pharmacol. 53 (2002) 503–514.
[46] C. Napoli, L.J. Ignarro, Nitric oxide and pathogenic mechanisms involved in the development of vascular diseases, Arch. Pharm. Res. 32 (2009) 1103–1108.
[47] R.M. Hardisty, The role of platelets in haemostasis, Br. J. Haematol. 17 (1969) 299.
[48] G.R. Sambrano, E.J. Weiss, Y.W. Zheng, W. Huang, S.R. Coughlin, Role of thrombin signalling in platelets in haemostasis and thrombosis, Nature 413 (2001) 74–78.
[49] H.C. de Boer, C. Verseyden, L.H. Ulfman, J.J. Zwaginga, I. Bot, E.A. Biessen, T.J. Rabelink, A.J. van Zonneveld, Fibrin and activated platelets cooperatively guide stem cells to a vascular injury and promote differentiation towards an endothelial cell phenotype, Arterioscler. Thromb. Vasc. Biol. 26 (2006) 1653–1659.

[50] M. Farstad, The role of blood platelets in coronary atherosclerosis and thrombosis, Scand. J. Clin. Lab. Invest. 58 (1998) 1–10.
[51] V. Fuster, P.M. Steele, J.H. Chesebro, Role of platelets and thrombosis in coronary atherosclerotic disease and sudden death, J. Am. Coll. Cardiol. 5 (1985) 175B–184B.
[52] M. Gawaz, F.J. Neumann, I. Ott, A. May, S. Rudiger, A. Schomig, Changes in membrane glycoproteins of circulating platelets after coronary stent implantation, Heart 76 (1996) 166–172.
[53] F.J. Neumann, M. Gawaz, I. Ott, A. May, G. Mossmer, A. Schomig, Prospective evaluation of hemostatic predictors of subacute stent thrombosis after coronary Palmaz-Schatz stenting, J. Am. Coll. Cardiol. 27 (1996) 15–21.
[54] K. Kaushansky, Historical review: megakaryopoiesis and thrombopoiesis, Blood 111 (2008) 981–986.
[55] B.B. Weksler, C.W. Ley, E.A. Jaffe, Stimulation of endothelial cell prostacyclin production by thrombin, trypsin, and the ionophore A 23187, J. Clin. Invest. 62 (1978) 923–930.
[56] S. Moncada, M.W. Radomski, R.M. Palmer, Endothelium-derived relaxing factor. Identification as nitric oxide and role in the control of vascular tone and platelet function, Biochem. Pharmacol. 37 (1988) 2495–2501.
[57] M.W. Radomski, R.M. Palmer, S. Moncada, Endogenous nitric oxide inhibits human platelet adhesion to vascular endothelium, Lancet 2 (1987) 1057–1058.
[58] A.J. de Belder, R. Macallister, M.W. Radomski, S. Moncada, P.J. Vallance, Effects of S-nitroso-glutathione in the human forearm circulation: evidence for selective inhibition of platelet activation, Cardiovasc. Res. 28 (1994) 691–694.
[59] E.J. Langford, A.S. Brown, R.J. Wainwright, A.J. de Belder, M.R. Thomas, R.E. Smith, M.W. Radomski, J.F. Martin, S. Moncada, Inhibition of platelet activity by S-nitrosoglutathione during coronary angioplasty, Lancet 344 (1994) 1458–1460.
[60] G.L. Apostoli, A. Solomon, M.J. Smallwood, P.G. Winyard, M. Emerson, Role of inorganic nitrate and nitrite in driving nitric oxide-cGMP-mediated inhibition of platelet aggregation in vitro and in vivo, J. Thromb. Haemost. 12 (2014) 1880–1889.
[61] M. Kadan, S. Doganci, V. Yildirim, G. Ozgur, G. Erol, K. Karabacak, F. Avcu, In vitro effect of sodium nitrite on platelet aggregation in human platelet rich plasma—preliminary report, Eur. Rev. Med. Pharmacol. Sci. 19 (2015) 3935–3939.
[62] R. Kornowski, G.S. Mintz, K.M. Kent, A.D. Pichard, L.F. Satler, T.A. Bucher, M.K. Hong, J.J. Popma, M.B. Leon, Increased restenosis in diabetes mellitus after coronary interventions is due to exaggerated intimal hyperplasia A serial intravascular ultrasound study, Circulation 95 (1997) 1366–1369.
[63] M.W. Liu, G.S. Roubin, S.B. King 3rd, Restenosis after coronary angioplasty. Potential biologic determinants and role of intimal hyperplasia, Circulation 79 (1989) 1374–1387.
[64] M.A. Costa, D.I. Simon, Molecular basis of restenosis and drug-eluting stents, Circulation 111 (2005) 2257–2273.
[65] T.F. Luscher, J. Steffel, F.R. Eberli, M. Joner, G. Nakazawa, F.C. Tanner, R. Virmani, Drug-eluting stent and coronary thrombosis: biological mechanisms and clinical implications, Circulation 115 (2007) 1051–1058.
[66] J.W. Fischer, S. Hawkins, A.W. Clowes, Pharmacologic inhibition of nitric oxide synthases and cyclooxygenases enhances intimal hyperplasia in balloon-injured rat carotid arteries, J. Vasc. Surg. 40 (2004) 115–122.
[67] R.D. Rudic, E.G. Shesely, N. Maeda, O. Smithies, S.S. Segal, W.C. Sessa, Direct evidence for the importance of endothelium-derived nitric oxide in vascular remodeling, J. Clin. Invest. 101 (1998) 731–736.

[68] O.F. Bertrand, R. Sipehia, R. Mongrain, J. Rodes, J.C. Tardif, L. Bilodeau, G. Cote, M.G. Bourassa, Biocompatibility aspects of new stent technology, J. Am. Coll. Cardiol. 32 (1998) 562–571.

[69] D.S. Ettenson, E.R. Edelman, Local drug delivery: an emerging approach in the treatment of restenosis, Vasc. Med. 5 (2000) 97–102.

[70] S.K. Pulfer, D. Ott, D.J. Smith, Incorporation of nitric oxide-releasing crosslinked polyethyleneimine microspheres into vascular grafts, J. Biomed. Mater. Res. 37 (1997) 182–189.

[71] D.J. Smith, D. Chakravarthy, S. Pulfer, M.L. Simmons, J.A. Hrabie, M.L. Citro, J.E. Saavedra, K.M. Davies, T.C. Hutsell, D.L. Mooradian, S.R. Hanson, L.K. Keefer, Nitric oxide-releasing polymers containing the [N(O)NO]- group, J. Med. Chem. 39 (1996) 1148–1156.

[72] J.N. Bates, M.T. Baker, R. Guerra Jr., D.G. Harrison, Nitric oxide generation from nitroprusside by vascular tissue. Evidence that reduction of the nitroprusside anion and cyanide loss are required, Biochem. Pharmacol. 42 (suppl.) (1991) S157–S165.

[73] J.H. Yoon, C.J. Wu, J. Homme, R.J. Tuch, R.G. Wolff, E.J. Topol, A.M. Lincoff, Local delivery of nitric oxide from an eluting stent to inhibit neointimal thickening in a porcine coronary injury model, Yonsei Med. J. 43 (2002) 242–251.

[74] L.J. Ignarro, C. Napoli, J. Loscalzo, Nitric oxide donors and cardiovascular agents modulating the bioactivity of nitric oxide: an overview, Circ. Res. 90 (2002) 21–28.

[75] S.A. Omar, E. Artime, A.J. Webb, A comparison of organic and inorganic nitrates/nitrites, Nitric Oxide 26 (2012) 229–240.

[76] E. Barbato, A. Herman, E. Benit, L. Janssens, J. Lalmand, E. Hoffer, P. Chenu, A. Guedes, L. Missault, B. Pirenne, F. Cardinal, S. Vercauteren, W. Wijns, Double-blind parallel placebo-controlled study to evaluate the effect of molsidomine on the endothelial dysfunction in patients with stable angina pectoris undergoing percutaneous coronary intervention: the MEDCOR Trial, J. Cardiovasc. Transl. Res. 7 (2014) 226–231.

[77] E. Barbato, A. Herman, E. Benit, L. Janssens, J. Lalmand, E. Hoffer, P. Chenu, A. Guedes, L. Missault, B. Pirenne, F. Cardinal, S. Vercauteren, W. Wijns, Long-term effect of molsidomine, a direct nitric oxide donor, as an add-on treatment, on endothelial dysfunction in patients with stable angina pectoris undergoing percutaneous coronary intervention: results of the MEDCOR trial, Atherosclerosis 240 (2015) 351–354.

[78] C.D. Knox, P.J. de kam,, K. Azer, P. Wong, A.G. Ederveen, D. Shevell, C. Morabito, A.G. Meehan, W. Liu, T. Reynders, J.F. Denef, A. Mitselos, D. Jonathan, D.E. Gutstein, K. Mitra, S.Y. Sun, M.M. Lo, D. Cully, A. Ali, Discovery and clinical evaluation of MK-8150, a novel nitric oxide donor with a unique mechanism of nitric oxide release, J. Am. Heart Assoc. 5 (2016).

[79] R.J. Singh, N. Hogg, J. Joseph, B. Kalyanaraman, Mechanism of nitric oxide release from S-nitrosothiols, J. Biol. Chem. 271 (1996) 18596–18603.

[80] A.R. Butler, S. Elkins-Daukes, D. Parkin, D.L. Williams, Direct NO group transfer from S-nitrosothiols to iron centres, Chem. Commun. (Camb.) (2001) 1732–1733.

[81] A.P. Dicks, D.L. Williams, Generation of nitric oxide from S-nitrosothiols using protein-bound Cu^{2+} sources, Chem. Biol. 3 (1996) 655–659.

[82] R.S. Wright, G.S. Reeder, C.A. Herzog, R.C. Albright, B.A. Williams, D.L. Dvorak, W.L. Miller, J.G. Murphy, S.L. Kopecky, A.S. Jaffe, Acute myocardial infarction and renal dysfunction: a high-risk combination, Ann. Intern. Med. 137 (2002) 563–570.

[83] N. Naghavi, A. de Mel, O.S. Alavijeh, B.G. Cousins, A.M. Seifalian, Nitric oxide donors for cardiovascular implant applications, Small 9 (2013) 22–35.

[84] A.W. Carpenter, M.H. Schoenfisch, Nitric oxide release: Part II. Therapeutic applications, Chem. Soc. Rev. 41 (2012) 3742–3752.

[85] P.S. Fleser, V.K. Nuthakki, L.E. Malinzak, R.E. Callahan, M.L. Seymour, M.M. Reynolds, S.I. Merz, M.E. Meyerhoff, P.J. Bendick, G.B. Zelenock, C.J. Shanley, Nitric oxide-releasing biopolymers inhibit thrombus formation in a sheep model of arteriovenous bridge grafts, J. Vasc. Surg. 40 (2004) 803–811.

[86] C.W. Mccarthy, R.J. Guillory 2nd, J. Goldman, M.C. Frost, Transition-metal-mediated release of nitric oxide (NO) from S-nitroso-N-acetyl-D-penicillamine (SNAP): potential applications for endogenous release of NO at the surface of stents via corrosion products, ACS Appl. Mater. Interfaces 8 (2016) 10128–10135.

[87] E. Nisoli, E. Clementi, C. Paolucci, V. Cozzi, C. Tonello, C. Sciorati, R. Bracale, A. Valerio, M. Francolini, S. Moncada, M.O. Carruba, Mitochondrial biogenesis in mammals: the role of endogenous nitric oxide, Science 299 (2003) 896–899.

[88] D. Persaud-Sharma, A. Mcgoron, Biodegradable magnesium alloys: a review of material development and applications, J. Biomimetics Biomater. Tissue Eng. 12 (2012) 25–39.

[89] K. Cosby, K.S. Partovi, J.H. Crawford, R.P. Patel, C.D. Reiter, S. Martyr, B.K. Yang, M.A. Waclawiw, G. Zalos, X. Xu, K.T. Huang, H. Shields, D.B. Kim-Shapiro, A.N. Schechter, R.O. Cannon 3rd, M.T. Gladwin, Nitrite reduction to nitric oxide by deoxyhemoglobin vasodilates the human circulation, Nat. Med. 9 (2003) 1498–1505.

[90] S. Shiva, Z. Huang, R. Grubina, J. Sun, L.A. Ringwood, P.H. Macarthur, X. Xu, E. Murphy, V.M. Darley-Usmar, M.T. Gladwin, Deoxymyoglobin is a nitrite reductase that generates nitric oxide and regulates mitochondrial respiration, Circ. Res. 100 (2007) 654–661.

[91] M. Tiso, J. Tejero, S. Basu, I. Azarov, X. Wang, V. Simplaceanu, S. Frizzell, T. Jayaraman, L. Geary, C. Shapiro, C. Ho, S. Shiva, D.B. Kim-shapiro, M.T. Gladwin, Human neuroglobin functions as a redox-regulated nitrite reductase, J. Biol. Chem. 286 (2011) 18277–18289.

[92] E. Fordel, L. Thijs, L. Moens, S. Dewilde, Neuroglobin and cytoglobin expression in mice. Evidence for a correlation with reactive oxygen species scavenging, FEBS J. 274 (2007) 1312–1317.

[93] A.V. Kozlov, K. Staniek, H. Nohl, Nitrite reductase activity is a novel function of mammalian mitochondria, FEBS Lett. 454 (1999) 127–130.

[94] B.L. Godber, J.J. Doel, G.P. Sapkota, D.R. Blake, C.R. Stevens, R. Eisenthal, R. Harrison, Reduction of nitrite to nitric oxide catalyzed by xanthine oxidoreductase, J. Biol. Chem. 275 (2000) 7757–7763.

[95] T.K. Kundu, M. Velayutham, J.L. Zweier, Aldehyde oxidase functions as a superoxide generating NADH oxidase: an important redox regulated pathway of cellular oxygen radical formation, Biochemistry 51 (2012) 2930–2939.

[96] C.E. Immoos, J. Chou, M. Bayachou, E. Blair, J. Greaves, P.J. Farmer, Electrocatalytic reductions of nitrite, nitric oxide, and nitrous oxide by thermophilic cytochrome P450 CYP119 in film-modified electrodes and an analytical comparison of its catalytic activities with myoglobin, J. Am. Chem. Soc. 126 (2004) 4934–4942.

[97] R. Aamand, T. Dalsgaard, F.B. Jensen, U. Simonsen, A. Roepstorff, A. Fago, Generation of nitric oxide from nitrite by carbonic anhydrase: a possible link between metabolic activity and vasodilation, Am. J. Physiol. Heart Circ. Physiol. 297 (2009) H2068–H2074.

[98] C. Gautier, E. van Faassen, I. Mikula, P. Martasek, A. Slama-Schwok, Endothelial nitric oxide synthase reduces nitrite anions to NO under anoxia, Biochem. Biophys. Res. Commun. 341 (2006) 816–821.

[99] S. Windecker, R. Simon, M. Lins, V. Klauss, F.R. Eberli, M. Roffi, G. Pedrazzini, T. Moccetti, P. Wenaweser, M. Togni, D. Tuller, R. Zbinden, C. Seiler, J. Mehilli, A. Kastrati, B. Meier, O.M. Hess, Randomized comparison of a titanium-nitride-oxide-coated stent with a stainless steel stent for coronary revascularization: the TiNOX trial, Circulation 111 (2005) 2617–2622.

[100] P.P. Karjalainen, A. Ylitalo, M. Niemela, K. Kervinen, T. Makikallio, M. Pietila, J. Sia, P. Tuomainen, K. Nyman, K.E. Airaksinen, Two-year follow-up after percutaneous coronary intervention with titanium-nitride-oxide-coated stents versus paclitaxel-eluting stents in acute myocardial infarction, Ann. Med. 41 (2009) 599–607.
[101] T. Lehtinen, K.E. Airaksinen, A. Ylitalo, P.P. Karjalainen, Stent strut coverage of titanium-nitride-oxide coated stent compared to paclitaxel-eluting stent in acute myocardial infarction: TITAX-OCT study, Int. J. Cardiovasc. Imaging 28 (2012) 1859–1866.
[102] P.P. Karjalainen, W. Nammas, A. Ylitalo, B. de Bruyne, J. Lalmand, A. de Belder, F. Rivero-Crespo, K. Kervinen, J.K. Airaksinen, Long-term clinical outcome of titanium-nitride-oxide-coated stents versus everolimus-eluting stents in acute coronary syndrome: final report of the BASE ACS trial, Int. J. Cardiol. 222 (2016) 275–280.
[103] P.P. Karjalainen, M. Niemela, J.K. Airaksinen, F. Rivero-Crespo, H. Romppanen, J. Sia, J. Lalmand, B. de Bruyne, A. Debelder, M. Carlier, W. Nammas, A. Ylitalo, C.M. Hess, B.-A.S. Investigators, A prospective randomised comparison of titanium-nitride-oxide-coated bioactive stents with everolimus-eluting stents in acute coronary syndrome: the BASE-ACS trial, EuroIntervention 8 (2012) 306–315.
[104] Y. Onuma, P.W. Serruys, L.E. Perkins, T. Okamura, N. Gonzalo, H.M. Garcia-Garcia, E. Regar, M. Kamberi, J.C. Powers, R. Rapoza, H. van Beusekom, W. van der Giessen, R. Virmani, Intracoronary optical coherence tomography and histology at 1 month and 2, 3, and 4 years after implantation of everolimus-eluting bioresorbable vascular scaffolds in a porcine coronary artery model: an attempt to decipher the human optical coherence tomography images in the ABSORB trial, Circulation 122 (2010) 2288–2300.
[105] P.W. Serruys, Y. Onuma, D. Dudek, P.C. Smits, J. Koolen, B. Chevalier, B. de Bruyne, L. Thuesen, D. Mcclean, R.J. van Geuns, S. Windecker, R. Whitbourn, I. Meredith, C. Dorange, S. Veldhof, K.M. Hebert, K. Sudhir, H.M. Garcia-Garcia, J.A. Ormiston, Evaluation of the second generation of a bioresorbable everolimus-eluting vascular scaffold for the treatment of de novo coronary artery stenosis: 12-month clinical and imaging outcomes, J. Am. Coll. Cardiol. 58 (2011) 1578–1588.
[106] B. Chevalier, Y. Onuma, A.J. van Boven, J.J. Piek, M. Sabate, S. Helqvist, A. Baumbach, P.C. Smits, R. Kumar, L. Wasungu, P.W. Serruys, Randomised comparison of a bioresorbable everolimus-eluting scaffold with a metallic everolimus-eluting stent for ischaemic heart disease caused by de novo native coronary artery lesions: the 2-year clinical outcomes of the ABSORB II trial, EuroIntervention 12 (2016) 1102–1107.
[107] R. Chakraborty, S. Patra, S. Banerjee, A. Pande, A. Khan, P.C. Mandol, D. Ghosh, S.K. De, S.S. Das, R. Nag, Outcome of everolimus eluting bioabsorbable vascular scaffold (BVS) compared to non BVS drug eluting stent in the management of ST-segment elevation myocardial infarction (STEMI)—a comparative study, Cardiovasc. Revasc. Med. 17 (2016) 151–154.
[108] J. Ma, N. Zhao, D. Zhu, Bioabsorbable zinc ion induced biphasic cellular responses in vascular smooth muscle cells, Sci. Rep. 6 (2016) 26661.

Immobilization of peptides on cardiovascular stent

15

F. Boccafoschi, L. Fusaro, M. Cannas
University of Oriental Piedmont (UPO), Novara, Italy

15.1 Introduction: Cardiovascular materials and biocompatibility

Cardiovascular diseases are the primary cause of death in developed countries and, in particular, arterial bypass graft remains the primary surgery for patients with advanced cardiovascular disease [1], bringing tremendous financial burden in every country [2]. Cardiovascular implants find wide clinical applications, making a great contribution to the treatment of cardiovascular diseases. Although several commercial cardiovascular implants are approved by FDA and millions of patients benefit from these products, unfortunately, there are still not ideal device which can help patients recover completely by the way of a perfect physiological function of the substitutes or a perfect tissue regeneration [3].

Concerning endovascular stent, the intervention process induces the inflammatory response and a limited hemocompatibility of the surface and the release of toxic ions such as Cr, Ni, and Co from a bare metal stent (BMS) results in a restenosis, which means the failure of the implant requiring additional surgeries.

Drug-eluting stent (DES) was considered as the breakthrough in stent design. However, they performed not so well in long-term clinical studies [4], also considering that the released drug from the stent surface delays endothelialization and triggers late thrombosis.

In order to overcome unexpected interaction between surface and tissues, surface modification has been introduced for cardiovascular devices to allow independent tailoring of surface and bulk properties [5].

Three are the key issues to focus on in order to optimize the use of the endovascular stent:

- Blood compatibility
- Compatible inflammatory reaction
- Re-endothelialization.

The stable bond on the stent' surface of peptides able to guide these issues is the main goal of endovascular stent enrichment.

In blood vessels, blood compatibility is naturally guaranteed by the presence of the endothelial layer.

Functionalized Cardiovascular Stents. https://doi.org/10.1016/B978-0-08-100496-8.00016-0

Concerning endovascular stent, blood compatibility is to be evaluated according to ISO-10993, thus considering several parameters such as thrombogenicity, hemolysis, and the inflammatory response (complement activation).

15.1.1 Endothelium, thrombosis, and inflammatory response

Endothelial cells naturally constitute a thin layer that covers the interior surface of blood vessels, forming an interface between circulating blood in the lumen and the rest of the vessel wall. The endothelium is able to respond to physical and chemical signals by the production of factors that regulate vascular tone, inflammation, thrombosis, and fibrinolysis. In physiological conditions, endothelium remains thromboresistant to the circulating blood. The antithrombotic nature of endothelial cells is attributed to different molecules including: prostacyclins [6,7]; heparan sulfates [8]; protein C-thrombomodulin complexes [9,10]; and plasminogen activators [11,12]. When artificial biomaterials are implanted in the human body, a series of thrombotic and inflammatory reactions occur similarly to the exposition of blood to the ECM of a disrupted vessel. In fact, when biomaterials come in contact with blood, the anticoagulant properties of the endothelium are abrogated and it becomes prothrombotic by virtue of adhesive factors such as von Willebrand factor [13], tissue factor (TF) [14,15], fibronectin [16], thrombospondin, and coagulation factor binding and activation [17,18]. Activation of the endothelium generates intravascular signals that increase the expression of procoagulant proteins such as TF, cytokines, and surface adhesion molecules, thereby promoting the recruitment of leukocytes to the inflammation area. This source of TF can sustain the production of thrombin, the deposition of platelets, and the conversion of fibrinogen to fibrin that entraps the red blood corpuscles, leading to thrombus formation. Moreover, reactive oxygen species (ROS) are released from the vessel wall and leukocytes oxidize hemoglobin and induce red blood cells lysis [19]. In addition to regulating thrombosis, endothelial cells direct biological responses such as leukocyte traffic to inflammatory sites in response to chemotactic factors. Upon activation, endothelial cells express immunologically relevant surface molecules such as adhesion molecules of the immunoglobulin gene superfamily as well as cytokines and growth factors. Secreted cytokines such as interleukin-1 (IL-1), tumor necrosis factor (TNFa), and interferon (IFNg) promote activation of neutrophils and trigger an acute inflammatory cascade with consequent endothelium injury. These events alter the morphology of the cellular membrane, the cytoskeleton structure, and the cell-matrix organization acting on molecule permeability [20]. It has been shown that cytokine-dependent stimulation of endothelial cells promotes activation of leukocytes (in particular monocytes) mediated by platelet-activating factor (PAF) [21] as well as cell adhesion molecules on PECAM, ICAM-1, P-Selectin, E-Selectin, and VCAM-1 [22,23]. In addition, ECs express molecules involved in complement regulation as well as receptors for a number of complement system proteins [24–26] (Fig. 15.1).

Finally, hemolysis may result in impairment of oxygen-carrying capacity of red blood cells (RBC). Hemolysis occurs when the RBC comes in contact with the material or its degradation products formed due to the shear stress generated because of relative motion between blood and the material surface.

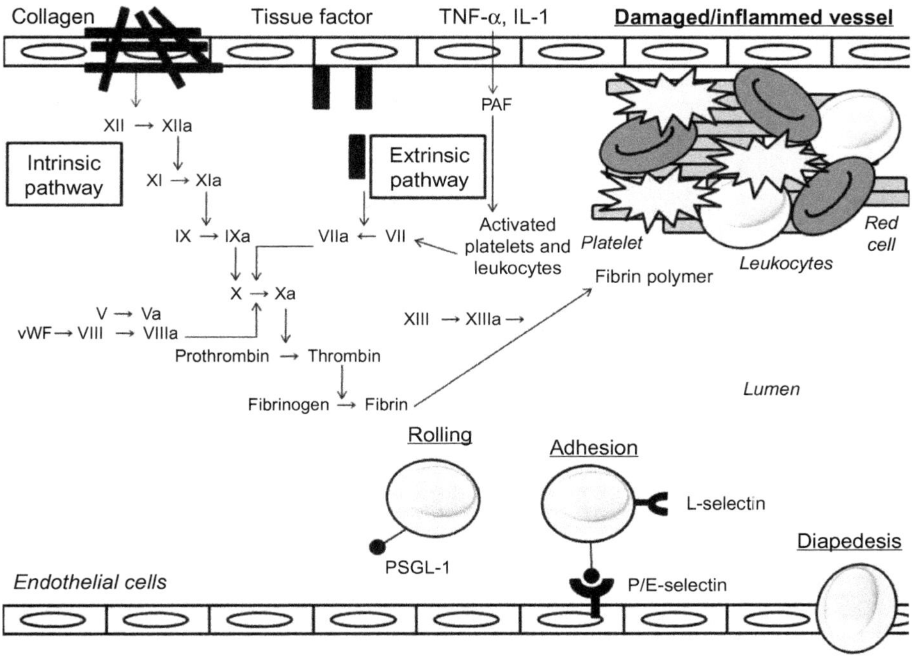

Fig. 15.1 Activation of endothelium after vessel injury with related intravascular signals that lead to a thrombus formation and a recruitment of leucocytes in the injured site.
From F. Boccafoschi, C Mosca, M Cannas, Cardiovascular biomaterials: when the inflammatory response helps to efficiently restore tissue functionality?, J. Tissue Eng. Regen. Med. 8(4) (2014) 253–267.

15.2 Metals and alloys for endovascular stent

Metals have been used for more than a century in biomedical field [27]. In majority of cardiovascular applications, stainless steel, cobalt chromium (CoCr) alloys, and titanium (Ti) alloys are widely used [28]. The basic properties of these common metallic cardiovascular biomaterials are strength, stiffness, corrosion resistance, and blood compatibility. Endovascular stents can be typically classified into three types based on their function and physical characters, namely, BMSs, DES, and bioresorbable stents.

In the beginning, stainless steel used for implants contained vanadium, but it has been replaced with the advent of 18% Cr and 8% Ni alloy making it stronger for applications. Soon, addition of molybdenum (Mo) and reduction of carbon (C) made it corrosion-resistant (316L) and suitable for blood-contacting devices like stent. Stainless steel is still the gold standard material for stent application in order to provide mechanical support to diseased arteries [29].

Recently, cobalt (Co)-based alloys have gained entry for production of stents, although Co-based alloys were used in medicine since 1937. The use of Co-based alloys is highly preferred in coronary stent manufacturing because coronary interventionist demands for thinner struts which can be easily achieved by using the Co-alloys. Further, due to nonferromagnetic characteristic and a consistence denser than stainless steel, these properties make cobalt alloy a feasible material for coronary stents.

Titanium-based alloys (Ti-alloys) have wide acceptance and usage since 1970. The most commonly used Ti-alloys are commercially pure titanium (CP-Ti) and 5Ti-6Al-4V (titanium-aluminum-vanadium). One of the remarkable features of titanium is light weight [30]. Moreover, Ti-alloys are known for their excellent tensile strength and pitting corrosion resistance suitable for cardiovascular applications. Another interesting feature of titanium alloys is shape memory effect possessed by the nickel-titanium (nitinol) alloys widely utilized for producing self-expanding memory stents [31,32]. Although BMS have excellent mechanical characteristics, they failed because of serious limitations such as stent thrombosis which requires prolonged antiplatelet therapy and mismatch of the stent to the vessel size. Moreover, the metallic stents impair the vessel geometry and obstruct side branches.

In order to rectify the complications present in BMS, DESs have been developed [33]. The DES were further classified into polymer-free stents and metallic stents with polymer carrier to hold and release the drug. DESs basically consist of three parts: stent platform, coating, and drug. Some DES release Paclitaxel, Tacrolimus, or Sirolimus as antiproliferative agents, in order to avoid the vessel lumen restenosis, which leads to secondary surgeries. Unfortunately, the polymer used as a vehicle for drug delivery may induce vessel irritation, endothelial dysfunction, vessel hypersensitivity, and chronic inflammation at the stent site.

Bioadsorbable stents have been developed to overcome the mentioned issues related to the use of polymers as drug carriers. Bioabsorbable stents stay for a limited period and promote healing of the blood vessel. The main purpose of the stent is to assist the arterial remodeling and this may take 6–12 months. This may overcome the need of unnecessary medication and also avoid late stent thrombosis. However, material for biodegradable stents is expected to meet some basic demands as it should be biocompatible and also its degradation products of the material must also be biocompatible. Finally, it should be able to stay in the place for several months before its complete bioabsorption and also its radial force of the resultant stent must be enough for scaffolding effect during the arterial remodeling period [34]. Based on these requirements, two metallic elements including iron and magnesium have been explored for this application [29]. Magnesium alloy stent is the first metallic bioabsorbable stent implanted in humans. Clinical evaluation conducted by Heublein and colleagues demonstrated higher degradation rates for Mg alloy from 60 to 90 days. Moreover, the stent was well-integrated with both endothelial and smooth muscle cells indicating its overall biocompatibility [35].

15.3 Immobilization of peptides: The grafting technique

The use of functional groups on material surface to form covalent bonds between the substrate and bioactive molecules, such as peptides, is a classical approach for constructing surface chemical modification. The reaction is specific and binding effect is stable. The main approach, in order to activate functional groups on biomaterial surface, is surface modification. Two types of modification are applied, chemical modification, mainly through silanization process, and physical modification, in particular with the use of plasma technique [36].

15.3.1 Silanization

Silanization is a low-cost and effective covalent coating method to modify material surface that is rich in hydroxyl groups, such as titanium, hydroxyapatite, and many other metal oxide surfaces. Binding of silicon-based molecules on metal surfaces occurs because hydroxyl groups on metal surfaces can link steadily to silicon atoms. There are many types of commercially available silane-coupling agents, which are easy to react with hydroxylated surface and introduce active groups (e.g., amino group and carboxyl group) to the surface. Silanized surface can easily be modified by further grafting. Holmberg et al. modified titanium disks with 3-(chloropropyl)-triethoxysilane (CPTES), and then grafting GL13K, an antimicrobial peptide, through amine alkylation of peptides amino groups to propyl group of the silane and through nucleophilic aliphatic substitution of the chloride atom [37]. Godoy-Gallardo et al. silanized a titanium surface with 3-aminopropyl-triethoxysilane (APTES), modified the amino groups with Iodoacetic acid N-hydroxysuccinimide ester, and bound hLf1-11 peptide, an antibacterial peptide, through amine alkylation on the position with iodine atom [38,39]. Although the silanization is simple and effective, the reaction conditions such as concentration of the silane and reaction time must be carefully controlled to prevent forming thick polymerized silane network on the surface. Otherwise, the bond between silane and the surface can also be subjected to hydrolysis in some conditions [40].

15.3.2 Plasma etching and grafting

Plasma is a gas composed by ionized atoms or molecules and free electrons. The electrons are at high temperature, and therefore high energy, and can interact with neutral molecules that can be injected in plasma atmosphere, forming free radicals and dissociating complex molecules, in order to form reactive molecules that are not stable in normal conditions.

Plasma can be used not only to prepare coatings on biomaterials, but is also able to conduct various noncoating surface modification processes, for example, plasma etching and plasma grafting. Plasma etching is a simple plasma-surface treatment method. During the sputtering process, a negative voltage (about 1 to several kVs) is applied to the substrate and an argon plasma is generated by rfGD or ECR. The ions are accelerated towards the substrate by the applied electric field. Since the energy is not very high, the argon ions cannot go very deeply into the substrate and a big portion of their energy is transferred to the surface atoms via elastic and inelastic collisions with the materials. Some surface atoms will acquire enough energy and escape from the substrate into the vacuum chamber. After the first layer of atoms has been sputtered off, the underlying layers will be exposed and gradually etched. With sufficient sputtering time, surface contamination will be cleaned off. [41]. This process can be used as a pretreatment for subsequent implantation and deposition and is helpful to improve surface activity for bioinert polymers, with less influence on surface topography than chemical etching process [42]. For example, Ye et al. pretreated a titanium alloy surface by H_2O plasma, before surface silanization [43]. Also, Sevilla et al. cleaned titanium disks with plasma before silanization and peptide grafting [44].

Plasma grafting is used to modify surface chemical properties of biomaterials by grafting specific chemical groups on the surface, to improve biocompatibility of the material or linking chemically active molecules, in order to subsequently bind bioactive molecules to materials' surface. Zhang et al. modified 316 L stainless steel with silicon oxycarbide (SiCOH), in order to improve endothelialization and anticoagulation properties of the metallic coronary stent [45].

Puleo et al. treated a titanium alloy surface with plasma, grafting on it a polymeric film of allyl amine, exploiting the free amino group to bind proteins such as bone morphogenetic protein 4 [46].

15.3.3 Peptide immobilization

Silanization and plasma grafting are both techniques that chemically activate metal surface. Peptides cannot be directly bound to metal surface with silanization [47], and their chemical structure would not resist to plasma's working conditions [48]. In order to chemically graft peptides, the reaction environment temperature needs to be between 0°C and room temperature, and in water solution with pH near 7. For these reasons, there are only few methods to immobilize peptides to activated metal surfaces, and these methods are strictly correlated to the type of the active molecule exposed on material surface. One of the methods used for chemically bind peptides to free amino and carboxylic groups on materials' surface is amidation reaction with water-soluble carbodiimides [49,50]. Amidation technique uses the carboxylic groups of the aminoacids of a substrate to bind active molecules that contain at least one reactive amino group. To avoid the *cross-link* of internal amino and carboxylic groups in the peptide of interest, the *grafting* reaction is performed through the following steps:

(1) 1-Ethyl-3-(3-dimethylaminopropyl)carbodiimide (EDC) binds to carboxyl groups of extracellular matrix (ECM) proteins, which makes carboxyl group more reactive and allows the binding of N-hydroxysuccinimide (NHS), which binds more firmly than EDC to carboxyl group, but is more susceptible to reaction with other amines, such as those of the peptides.
(2) The amino group of the peptide reacts with the carbon of the amide group, forming in turn an amide and taking away the NHS.

If the surface presents free amino groups, for example, after silanization, there are two possible solutions for peptide immobilization. The first one involves the reaction of the amino groups with succinic anhydride, one of the two carboxyl groups of the anhydride form an amide with NH_2, while the other one remains available for an amidation reaction with peptides NH_2.

Another possible approach is to bind a molecule, such as iodoacetic acid, which presents a carboxylic acid, in order to bind the amino group, and an alkyl halide, which allows nucleophilic aliphatic substitution of the nucleophilic amino group of the peptide on halogen-bound carbon [38,39].

15.3.4 Metal-binding peptides

The harshness of surface treatments required to bind peptides on metal biodevices pushed many scientists to study alternative methods for peptide immobilization. Sano e Shiba succeeded in selecting a 12-mer peptide, called titanium binding peptide 1 (TBP-1), containing a short motif (RKLPDA, arginine-lysine-leucine-proline-aspartic acid-alanine) that can electrostatically titanium oxide surfaces, generating an ionic bond. The ionic bond is more stable than idrophobic interaction involved in protein surface coating, the classical method used to improve metal stents' biocompatibility [51]. This study led to a high number of researches focused on finding peptides that can bind to metal stent surfaces. Sano et al. studied TBP-1 affinity for other metal surfaces, finding that TBP-1 binds efficiently also with silver and silicon [52]. Estephan et al. found another 12-mer peptide (SVSVGMKPSPRP, serine-valine-serine-valine-glicine-methionine-lysine-proline-serine-proline-arginine-proline), not specific to a particular metal surface, but a good linker for the functionalization of a wide range of metallic and mineral materials, such as titanium, titanium grade 5 alloys (Ti_6Al_4V), hydroxyapatite, cobalt, and silica [53].

Starting from these researches, several authors used these surface-binding peptides for immobilization of bioactive molecules. Meyer et al. proposed to attach the 12-mer titanium binding peptide (TiBP) to the RGD sequence. The authors demonstrated better adhesion and proliferation of endothelial cells on titanium functionalized with the peptide-RGD binder [54].

In 2013, Yazici H et al. employed the same approach to bind phage display-selected TiBP and Arg-Gly-Asp-Ser (RGDS) sequence to improve fibroblast cell adhesion on commercial grade Ti surface [55].

15.4 Guiding the tissue regeneration: Surface modification of cardiovascular stents

Depending on the seeked bioactivity, the grafting technique allows the binding of different types of molecules to the material surface, as growth factors, hormones, antibiotics, anticoagulant molecules, etc.

The main challenge for endovascular stent is to reestablish the endothelial layer useful to maintain the adequate hemodynamic balance. For this reason, many of the literature reports several approach able to promote a rapid re-endothelialization at the implant site and to improve the hemocompatibility of the used materials.

Two are the possible strategies: (1) mimicking a self-healing system with accelerated repair of the damaged endothelium by supporting the homing of stem cells or progenitor cells to the site of injury and (2) avoid blood clot when in contact with the implant, until the endothelial layer is reestablished.

Endothelial cells (ECs) are end-differentiated cells which are not capable of cell dividing and expansion [56]. For these reasons, and also because of a limited long-term efficacy of biomolecule-modified implants, and the time-, labor-, and cost-consuming procedures of cell seeding onto implants, the use of adult ECs for re-endothelialization was unsuccessful in several studies [57]. In 1997, Asahara and colleagues discovered endothelial progenitor cells (EPC), opening new frontiers to cardiovascular stent re-endothelialization [58].

EPCs are the relatively small population of $CD34^+$ circulating mononuclear cells in the circulatory system available as two forms, namely, early EPC and late EPC. Based on these two forms, EPC have been exploited in two different ways: one way is to construct and immobilize the early EPCs at the site of injury which will secrete angiogenic cytokines, which will flourish the resident ECs and the late EPCs. Another way is to construct the surface with late EPCs which in turn promotes neoangiogenesis and repairs the damaged site by their native ability to proliferate at high rate [59,60].

Altough the exact signaling pathways involved in EPC homing remain to be tested, some growth factors, cytokines and chemokines such as vascular endothelial growth factor (VEGF), stromal-derived factor-1 (SDF-1), nerve growth factor (NGF), granulocyte colony-stimulating factor (G-CSF), hypoxia-inducible factor (HIF), and brain-derived neurotrophic factor are known to play a role in the repair of vascular injury and neovascolarization [61].

A recent study utilizing NGF-bound vascular grafts showed significant immobilization of EPC and a similar preparation using SDF-1/heparin found to recruit both EPCs and smooth muscle progenitor cells tackling the two important issues namely, endothelization and remodeling of blood vessels [62,63].

EPC capture technology is the way through which circulating EPC is captured by using anti-$CD34^+$ that was impregnated on the surface of stents. Genous R-Stent is the first medical device utilizing this technology [64]. One of the studies postulated that this EPC capture technology was feasible and safe for primary percutaneous coronary intervention for ST segment elevation myocardial infarction (STEMI) without the incidence of late stent restenosis [65]. In another independent trial, coronary stenting with the Genous resulted in good clinical outcomes and low incidences of repeat revascularization and stent thrombosis [66]. However, some recent evaluation came in contrast to the above findings, where they reported higher risk of restenosis while using Genous compared to DESs [67]. To add further, Genous stent used in a population of elderly patients resulted in a significantly higher target vessel failure rates compared with younger patients. Moreover, target lesion revascularization rates were higher with increasing age and there was no difference in stent thrombosis [58].

Apart from Genous R-Stent, other devices with similar functions are designed by different strategies. For instance, DNA-aptamers with a high affinity to EPCs were identified and isolated by Jan Hoffmann and colleagues [69]. Studies demonstrated that the rapid adhesion of EPCs to aptamer-enriched implants can be useful to promote endothelial wound healing and to prevent the neointimal hyperplasia, which may lead to second surgeries because of the lumen occlusion [70].

Recently, Joo Myung Lee and colleagues compared endothelialization and neointimal formation of two EPC capture stents with $CD34^+$ antibodies or vascular endothelial-cadherin, respectively, coated on BMSs. It seemed that VE-cadherin resulted more suitable target molecule than $CD34^+$ to be used on EPC-capturing stents [71].

Literature reports other inspiring researches which show as the peptide sequence Arg-Glu-Asp-Val (REDV) mediates the adhesion and migration of the endothelial cells via the integrin $\alpha_4\beta_1$ subunit, while smooth muscle cells and platelets adhesion is not regulated by the presence of the REDV peptide sequence [72–74].

Moreover, a very recent study demonstrated the efficacy of the use of immobilized CD47 in inhibiting early inflammatory and thrombotic events that contribute to the pathophysiology of arterial injury poststent angioplasty. In particular, using in vitro and in vivo models, the immunomodulatory activity and the platelet inhibition have been studied, indicating CD47 as one of the elected molecules to develop innovative surface-modified endovascular stents [75].

15.5 Future trends/conclusion

With a worth of about $20.7 billion, cardiovascular biomaterials are projected to be a predominant category of biomaterials market in the next years. The main issue associated with the development of cardiovascular biomaterials is to maintain an adequate blood compatibility, allowing a functional tissue regeneration. Thus, blood compatibility is a major subject and several surface modifications are adopted to circumvent secondary undesirable effects and to develop innovative biocompatible cardiovascular biomaterials.

Moreover, there is a need to develop materials that mimic the properties of the native cardiovascular tissues. This can be achieved by producing composite materials that combine the properties of both natural and synthetic materials. The main goal will be to optimize the tissue regeneration after the implant, firstly obtaining a perfectly integrated vascular prosthesis, and then allowing in time the regeneration of the native tissue. In this context, developing surface modification strategies, such as the enrichment with specific peptides able to contain the inflammatory response while inducing a functional endothelialization, should indicate the promising way for the next generation for cardiovascular repair.

References

[1] C. Quint, M. Arief, A. Muto, A. Dardik, L.E. Niklason, Allogeneic human tissue-engineered blood vessel, J. Vasc. Surg. 55 (3) (2012) 790–798.
[2] P.A. Heidenreich, J.G. Trogdon, O.A. Khavjou, J. Butler, K. Dracup, M.D. Ezekowitz, E.A. Finkelstein, Y. Hong, S.C. Johnston, A. Khera, D.M. Lloyd-Jones, S.A. Nelson, G. Nichol, D. Orenstein, P.W. Wilson, Y.J. Woo, Forecasting the future of cardiovascular disease in the United States: a policy statement from the American Heart Association, Circulation 123 (8) (2011) 933–944.
[3] W. Sapirstein, S. Alpert, T.J. Callahan, The role of clinical trials in the Food and Drug Administration approval process for cardiovascular devices, Circulation 89 (4) (1994) 1900–1902.
[4] D.M. Martin, F.J. Boyle, Drug-eluting stents for coronary artery disease: a review, Med. Eng. Phys. 33 (2) (2011) 148–163.
[5] S.R. Meyers, M.W. Grinstaff, Biocompatible and bioactive surface modifications for prolonged in vivo efficacy, Chem. Rev. 112 (3) (2012) 1615–1632.

[6] A. Bhagyalakshmi, J.A. Frangos, Mechanism of shear-induced prostacyclin production in endothelial cells, Biochem. Biophys. Res. Commun. 158 (1989) 31–37.
[7] J.A. Frangos, S.G. Eskin, L.V. McIntire, C.L. Ives, Flow effects on prostacyclin production by cultured human endothelial cells, Science 227 (1985) 1477–1479.
[8] J.A. Marcum, J.B. McKenney, R.D. Rosenberg, Acceleration of thrombin-antithrombin complex formation in rat hindquarters via heparin-like molecules bound to the endothelium, J. Clin. Invest. 74 (1984) 341–350.
[9] A.M. Malek, R. Jackman, R.D. Rosenberg, S. Izumo, Endothelial expression of thrombomodulin is reversibly regulated by fluid shear stress, Circ. Res. 74 (1994) 852–860.
[10] Y. Takada, F. Shinkai, S. Kondo, S. Yamamoto, H. Tsuboi, R. Korenaga, J. Ando, Fluid shear stress increases the expression of thrombomodulin by cultured human endothelial cells, Biochem. Biophys. Res. Commun. 205 (1994) 1345–1352.
[11] S.L. Diamond, S.G. Eskin, L.V. McIntire, Fluid flow stimulates tissue plasminogen activator secretion by cultured human endothelial cells, Science 243 (1989) 1483–1485.
[12] E.G. Levin, L. Santell, K.G. Osborn, The expression of endothelial tissue plasminogen activator in vivo: a function defined by vessel size and anatomic location, J. Cell Sci. 110 (1997) 139–148.
[13] M. Galbusera, C. Zoja, R. Donadelli, S. Paris, M. Morigi, A. Benigni, M. Figliuzzi, G. Remuzzi, A. Remuzzi, Fluid shear stress modulates von Willebrand factor release from human vascular endothelium, Blood 90 (1997) 1558–1564.
[14] E.F. Grabowski, A.J. Reininger, P.G. Petteruti, O. Tsukurov, R.W. Orkin, Shear stress decreases endothelial cell tissue factor activity by augmenting secretion of tissue factor pathway inhibitor, Arterioscler. Thromb. Vasc. Biol. 21 (2001) 157–162.
[15] A.D. Westmuckett, C. Lupu, S. Roquefeuil, T. Krausz, V.V. Kakkar, F. Lupu, Fluid flow induces upregulation of synthesis and release of tissue factor pathway inhibitor in vitro, Arterioscler. Thromb. Vasc. Biol. 20 (2000) 2474–2482.
[16] T. Pompe, F. Kobe, K. Salchert, B. Jorgensen, J. Oswald, C. Werner, Fibronectin anchorage to polymer substrates controls the initial phase of endothelial cell adhesion, J. Biomed. Mater. Res. A 67 (2003) 647–657.
[17] G.P. Clagett, R.W. Colman, J. Hirsh, V.J. Marder, E.W. Salzman, Artificial devices in clinical practice, in: Hemostasis and Thrombosis: Basic Principles and Clinical Practice, J.B. Lippincott Co, Philadelphia, 1987, pp. 1348.
[18] M.A. Gimbrone, Vascular endothelium: nature's blood container, Vascular Endothelium in Hemostasis and Thrombosis, Churchill Livingstone, Edinburgh, 1986.
[19] P. Saha, J. Humphries, B. Modarai, K. Mattock, M. Waltham, C.E. Evans, A. Ahmad, A.S. Patel, S. Premaratne, O.T. Lyons, A. Smith, Leukocytes and the natural history of deep vein thrombosis: current concepts and future directions, Arterioscler. Thromb. Vasc. Biol. 31 (2011) 506–512.
[20] M. Pate, V. Damarla, D.S. Chi, S. Negi, G. Krishnaswamy, Endothelial cell biology: role in the inflammatory response, Adv. Clin. Chem. 52 (2010) 109–130.
[21] G.A. Zimmerman, T.M. McIntyre, M. Mehra, S.M. Prescott, Endothelial cell-associated platelet-activating factor: a novel mechanism for signaling intercellular adhesion, J. Cell Biol. 110 (1990) 529–540.
[22] R.K. Andrews, M.C. Berndt, Platelet physiology and thrombosis, Thromb. Res. 114 (2004) 447–453.
[23] C. Garlanda, E. Dejana, Heterogeneity of endothelial cells. Specific markers, Arterioscler. Thromb. Vasc. Biol. 17 (1997) 1193–1202.
[24] K.K. Hamilton, Z. Ji, S. Rollins, B.H. Stewart, P.J. Sims, Regulatory control of the terminal complement proteins at the surface of human endothelial cells: neutralization of a C5b-9 inhibitor by antibody to CD59, Blood 76 (1990) 2572–2577.

[25] E.A. Lidington, D.O. Haskard, J.C. Mason, Induction of decay-accelerating factor by thrombin through a protease-activated receptor 1 and protein kinase C-dependent pathway protects vascular endothelial cells from complement-mediated injury, Blood 96 (2000) 2784–2792.
[26] F. Tedesco, F. Fischetti, M. Pausa, A. Dobrina, R.B. Sim, M.R. Daha, Complement-endothelial cell interactions: pathophysiological implications, Mol. Immunol. 36 (1999) 261–268.
[27] W.A. Lane, Some remarks on the treatment of fractures, Br. Med. J. 1 (1895) 861.
[28] H. Hermawan, D. Ramdan, J.R.P. Djuansjah, Metals for Biomedical Applications, in: Biomedical Engineering—From Theory to Applications, 2011, p. 498.
[29] M. Moravej, D. Mantovani, Biodegradable metals for cardiovascular stent application: interests and new opportunities, Int. J. Mol. Sci. 12 (7) (2011) 4250–4270.
[30] E.A. Brandes, G.B. Brook, Smithells Metals Reference Book, seventh ed., Butterworth-Heinemann, Oxford, UK, 1998.
[31] H.J. Rack, J.I. Qazi, Titanium alloys for biomedical applications, Mater. Sci. Eng. C 26 (8) (2006) 1269–1277.
[32] D. Stoekel, A. Pelton, T. Duering, Self-expanding nitinol stents—material and design considerations, Eur. Radiol. 14 (2) (2004) 292–301.
[33] R. Waksman, Biodegradable stents: they do their job and disappear, J. Invasive Cardiol. 18 (2) (2006) 70–74.
[34] S. Saito, New horizon of bioabsorbable stent, Catheter. Cardiovasc. Interv. 66((4) (2005) 595–596.
[35] B. Heublein, R. Rohde, V. Kaese, M. Niemeyer, W. Hartung, A. Haverich, Biocorrosion of magnesium alloys: a new principle in cardiovascular implant technology? Heart 89 (6) (2003) 651–656.
[36] Z.Y. Qiu, C. Chen, X.M. Wang, I.S. Lee, Advances in the surface modification techniques of bone-related implants for last 10 years, Regen. Biomater. 1 (1) (2014) 67–79.
[37] K.V. Holmberg, M. Abdolhosseini, Y. Li, X. Chen, S.U. Gorr, C. Aparicio, Bio-inspired stable antimicrobial peptide coatings for dental applications, Acta Biomater. 9 (9) (2013) 8224–8231.
[38] M. Godoy-Gallardo, C. Mas-Moruno, K. Yu, J.M. Manero, F.J. Gil, J.N. Kizhakkedathu, D. Rodriguez, Antibacterial properties of hLf1-11 peptide onto titanium surfaces: a comparison study between silanization and surface initiated polymerization, Biomacromolecules 16 (2) (2015) 483–496.
[39] M. Godoy-Gallardo, C. Mas-Moruno, K. Yu, J.M. Manero, F.J. Gil, J.N. Kizhakkedathu, D. Rodriguez, Antibacterial properies of hLf1-11 peptide onto titanium surfaces: a comparison study between silanization and surface initiated polymerization, Biomacromolecules 16 (2) (2015) 483–496.
[40] S.R. Wasserman, Y.T. Tao, G.M. Whitesides, Structure and reactivity of alkylsiloxane monolayers formed by reaction of alkyltrichlorosilanes on silicon substrates, Langmuir 5 (1989) 1074–1087.
[41] P.K. Chu, J.Y. Chen, L.P. Wang, N. Huang, Plasma-surface modification of biomaterials, Mater. Sci. Eng. R 36 (2002) 143–206.
[42] S.W. Ha, R. Hauert, K.H. Ernst, Surface analysis of chemically-etched and plasma-treated polyetheretherketone (PEEK) for biomedical applications, Surf. Coat. Technol. 96 (1997) 293–299.
[43] S.H. Ye, C.A. Johnson Jr., J.R. Woolley, T.A. Snyder, L.J. Gamble, W.R. Wagner, Covalent surface modification of a titanium alloy with a phosphorylcholine-containing copolymer for reduced thrombogenicity in cardiovascular devices, J. Biomed. Mater. Res. 91 (1) (2009) 18–28.

[44] P. Sevilla, M. Godoy, E. Salvagni, D. Rodríguez, F.J. Gil, Biofunctionalization of titanium surfaces for osseintegration process improvement, J. Phys. Conf. Ser. 252 (2010).
[45] Q. Zhang, Y. Shen, C. Tang, X. Wu, Q. Yu, G. Wang, Surface modification of coronary stents with SiCOH plasma nanocoatings for improving endothelialization and anticoagulation, J. Biomed. Mater. Res. B Appl. Biomater. 103 (2) (2015) 464–472.
[46] D.A. Puleo, R.A. Kissling, M.S. Sheu, A technique to immobilize bioactive proteins, including bone morphogenetic protein-4 (BMP-4), on titanium alloy, Biomaterials 23 (9) (2002) 2079–2087.
[47] M.C. Porté-Durrieu, F. Guillemot, S. Pallu, C. Labrugère, B. Brouillaud, R. Bareille, J. Amédée, N. Barthe, M. Dard, Ch. Baquey, Cyclo-(DfKRG) peptide grafting onto Ti-6Al-4V physical characterization and interest towards human osteoprogenitor cells adhesion, Biomaterials 25 (19) (2004) 4837–4846.
[48] L. Karam, C. Jama, P. Dhulster, N.E. Chihib, Study of surface interactions between peptides, materials and bacteria for setting up antimicrobial surfaces and active food packaging, J. Mater. Environ. Sci. 4 (5) (2013) 798–821.
[49] D.L. Taylor, J.J. Thevarajah, D.K. Narayan, P. Murphy, M.M. Mangala, S. Lim, R. Wuhrer, C. Lefay, M.D. O'Connor, M. Gaborieau, P. Castignolles, Real-time monitoring of peptide grafting onto chitosan films using capillary electrophoresis, Anal. Bioanal. Chem 407 (9) (2015) 2543–2555.
[50] G. Vidal, T. Blanchi, A.J. Mieszawska, R. Calabrese, C. Rossi, P. Vigneron, J.L. Duval, D.L. Kaplan, C. Egles, Enhanced cellular adhesion on titanium by silk functionalized with titanium binding and RGD peptides, Acta Biomater. 9 (1) (2013) 4935–4943.
[51] K. Sano, K. Shiba, A hexapeptide motif that electrostatically binds to the surface of titanium, J. Am. Chem. Soc. 125 (47) (2003) 14234–14235.
[52] K. Sano, H. Sasaki, K. Shiba, Specificity and biomineralization activities of Ti-binding peptide-1 (TBP-1), Langmuir 21 (7) (2005) 3090–3095.
[53] E. Estephan, J. Dao, M.B. Saab, I. Panayotov, M. Martin, C. Larroque, C. Gergely, F.J. Cuisinier, S. Levallois, VSVGMKPSPRP: a broad range adhesion peptide, Biomed. Tech. (Berl.) 57 (6) (2012) 481–489.
[54] S.R. Meyers, P.T. Hamilton, E.B. Walsh, D.J. Kenan, Grinstaff MW, Endothelialization of titanium surfaces, Adv. Mater. 19 (18) (2007) 2492–2498.
[55] H. Yazici, H. Fong, B. Wilson, E.E. Oren, F.A. Amos, H. Zhang, J.S. Evans, M.L. Snead, M. Sarikaya, C. Tamerler, Biological response on a titanium implant-grade surface functionalized with modular peptides, Acta Biomater. 9 (2) (2013) 5341–5352.
[56] F. Otsuka, A.V. Finn, S.K. Yazdani, M. Nakano, F.D. Kolodgie, R. Virmani, The importance of the endothelium in atherothrombosis and coronary stenting, Nat. Rev. Cardiol. 9 (8) (2012) 439–453.
[57] G.A. Villalona, B. Udelsman, D.R. Duncan, E. McGillicuddy, R.F. Sawh-Martinez, N. Hibino, C. Painter, T. Mirensky, B. Erickson, T. Shinoka, C.K. Breuer, Cell-seeding techniques in vascular tissue engineering, Tissue Eng. Part B Rev. 16 (3) (2010) 341–350.
[58] T. Asahara, J.M. Isner, Endothelial progenitor cells for vascular regeneration, J. Hematother. Stem Cell Res. 11 (2) (2002) 171–178.
[59] M. Adali, G. Ziemer, H.P. Wendel, Induction of EPC homing on biofunctionalized vascular grafts for rapid in vivo self-endothelialization—a review of current strategies, Biotechnol. Adv. 28 (1) (2010) 119–129.
[60] J. Hur, C.-H. Yoon, H.S. Kim, J.H. Choi, H.J. Kang, K.K. Hwang, B.H. Oh, M.M. Lee, Y.B. Park, Characterization of two types of endothelial progenitor cells and their different contributions to neovasculogenesis, Arterioscler. Thromb. Vasc. Biol. 24 (2) (2004) 288–293.

[61] A.J. Melchiorri, B.N. Nguyen, J.P. Fisher, Mesenchymal stem cells: roles and relationships in vascularization, Tissue Eng. Part B Rev. 20 (3) (2014) 218–228.
[62] J. Yu, A. Wang, Z. Tang, J. Henry, B. Li-Ping Lee, Y. Zhu, F. Yuan, F. Huang, S. Li, The effect of stromal cell-derived factor-1α/heparin coating of biodegradable vascular grafts on the recruitment of both endothelial and smooth muscle progenitor cells for accelerated regeneration, Biomaterials 33 (2012) 8062–8074.
[63] W. Zeng, W. Yuan, L. Li, et al., The promotion of endothelial progenitor cells recruitment by nerve growth factors in tissue-engineered blood vessels, Biomaterials 31 (7) (2010) 1636–1645.
[64] M. Klomp, M.A.M. Beijk, R.J. de Winter, Genous endothelial progenitor cell-capturing stent system: a novel stent technology, Expert Rev. Med. Dev. 6 (4) (2009) 365–375.
[65] M. Co, E. Tay, C.H. Lee, K.K. Poh, A. Low, J. Lim, I.H. Lim, Y.T. Lim, H.C. Tan, Use of endothelial progenitor cell capture stent (Genous Bio-Engineered R Stent) during primary percutaneous coronary intervention in acute myocardial infarction: intermediate- to long-term clinical follow-up, Am. Heart J. 155 (1) (2008) 128–132.
[66] S. Silber, P. Damman, M. Klomp, M.A. Beijk, M. Grisold, E.E. Ribeiro, H. Suryapranata, J. Wójcik, K. Hian Sim, J.G. Tijssen, R.J. de Winter, Clinical results after coronary stenting with the Genous bio-engineered R stent: 12-month outcomes of the e-HEALING (Healthy Endothelial Accelerated Lining Inhibits Neointimal Growth) worldwide registry, EuroIntervention 6 (7) (2011) 819–825.
[67] R. Sethi, C.H. Lee, Endothelial progenitor cell capture stent, safety and effectiveness, J. Interv. Cardiol. 25 (5) (2012) 493–500.
[68] P. Damman, A. Iñiguez, M. Klomp, M. Beijk, P. Woudstra, S. Silber, E.E. Ribeiro, H. Suryapranata, K.H. Sim, J.G. Tijssen, R.J. de Winter, Coronary stenting with the Genous Bio-Engineered R Stent in elderly patients, Circ. J. 75 (11) (2011) 2590–2597.
[69] J. Hoffmann, A. Paul, M. Harwardt, J. Groll, T. Reeswinkel, D. Klee, M. Moeller, H. Fischer, T. Walker, T. Greiner, G. Ziemer, H.P. Wendel, Immobilized DNA aptamers used as potent attractors for porcine endothelial precursor cells, J. Biomed. Mater. Res. A 84 (3) (2008) 614–621.
[70] P. Qi, W. Yan, Y. Yang, Y. Li, Y. Fan, J. Chen, Z. Yang, Q. Tu, N. Huang, Immobilization of DNA aptamers via plasma polymerized allylamine film to construct an endothelial progenitor cell-capture surface, Colloids Surf. B Biointerfaces 126 (2015) 70–79.
[71] J.M. Lee, W. Choe, B.K. Kim, W.W. Seo, W.H. Lim, C.K. Kang, S. Kyeong, K.D. Eom, H.J. Cho, Y.C. Kim, J. Hur, H.M. Yang, H.J. Cho, Y.S. Lee, H.S. Kim, Comparison of endothelialization and neointimal formation with stents coated with antibodies against CD34 and vascular endothelial-cadherin, Biomaterials 33 (35) (2012) 8917–8927.
[72] M.I. Castellanos, A.S. Zenses, A. Grau, J.C. Rodríguez-Cabello, F.J. Gil, J.M. Manero, M. Pegueroles, Biofunctionalization of REDV elastin-like recombinamers improves endothelialization on CoCr alloy surfaces for cardiovascular applications, Colloids Surf. B Biointerfaces 127 (2015) 22–32.
[73] H. Ceylan, A.B. Tekinay, M.O. Guler, Selective adhesion and growth of vascular endothelial cells on bioactive peptide nanofiber functionalized stainless steel surface, Biomaterials 32 (34) (2011) 8797–8805.
[74] Y. Wei, Y. Ji, L.L. Xiao, Q.K. Lin, J.P. Xu, K.F. Ren, J. Ji, Surface engineering of cardiovascular stent with endothelial cell selectivity for in vivo re-endothelialisation, Biomaterials 34 (11) (2013) 2588–2599.
[75] J.B. Slee, I.S. Alferiev, C. Nagaswami, J.W. Weisel, R.J. Levy, I. Fishbein, S.J. Stachelek, Enhanced biocompatibility of CD47-functionalized vascular stents, Biomaterials 87 (2016) 82–92.

Further reading

[1] Biological evaluation of medical devices. Part 1. Evaluation and testing within a risk management process, http://www.fda.gov/downloads/MedicalDevices/DeviceRegulationandGuidance/GuidanceDocuments/UCM348890.pdf.

[2] F. Boccafoschi, C. Mosca, M. Cannas, Cardiovascular biomaterials: when the inflammatory response helps to efficiently restore tissue functionality? J. Tissue Eng. Regen. Med. 8 (4) (2014) 253–267.

[3] M.W. Curtis, B. Russel, Cardiac tissue engineering, J. Cardiovasc. Nurs. 24 (2) (2009) 87–92.

Immobilization of antibodies on cardiovascular stents

16

I.B. O'Connor, J.G. Wall
National University of Ireland, Galway, Ireland

16.1 Introduction

One of the main treatments used to reopen blocked coronary arteries is angioplasty, followed by deployment of a stent to provide radial strength and keep the vessel open [1]. Metals such as stainless steel, cobalt-chromium, platinum, and titanium alloys have traditionally been preferred in stent manufacturing due to their mechanical strength and their biological and chemical inertness that reduces the possibility of a foreign body response in vivo [2]. The use of such nonnative materials can also result in negative clinical consequences due to their limited biocompatibility, however, leading to the occurrence of thrombosis or intimal hyperplasia. Many original stents were associated with a high frequency of acute stent thrombosis, caused by the interaction of the device surface with surrounding cells, proteins and extracellular matrix materials, and/or intimal hyperplasia due to excessive growth of tissues surrounding the metal [3,4]. The focus of stent design shifted therefore in the 1980s and 1990s from bioinert, bare metal stents (BMS) to bioactive metal-based devices such as drug-eluting stents (DES) that released antiproliferative and antiinflammatory drugs like paclitaxel or sirolimus in order to inhibit restenosis [5]. While the use of DES successfully reduced smooth muscle cell proliferation and inhibited neointimal growth, numerous studies identified delayed re-endothelialization and arterial healing in patients with DES [6]: stent re-endothelialization is typically almost complete within 3–4 months of BMS implantation in the coronary circulation but this took much longer following DES implantation, leading to a phenomenon of late in-stent thrombosis and a requirement for patients to continue long-term anticlotting therapy after stent placement. With the increasing recognition of the critical role of forming a healthy endothelial layer on the stented luminal surface to avoid thrombosis, the emphasis in stent design moved again to biofunctionalized devices which would accelerate formation of a native or native-like endothelium as the optimal available surface for blood contact [7]. As the coating of stents with appropriate endothelial cells in vitro remains particularly challenging and costly, this has led to the emergence of stents with tethered surface moieties, particularly antibodies, designed to capture circulating endothelial (EC) and endothelial progenitor (EPC) cells in vivo [8,9]. The acceleration of surface re-endothelialization is the dominant application of stent-immobilized antibodies by some distance. The technology has made a breakthrough into clinical practice and remains a major focus of the biofunctionalization of stents with biomolecules to reshape their cell and protein interactions and to improve their efficacies in vivo [10].

Functionalized Cardiovascular Stents. https://doi.org/10.1016/B978-0-08-100496-8.00017-2

In this review, we describe the use of antibody and antibody fragment molecules in stent-biofunctionalization applications, highlight limitations in current approaches and technologies, and discuss likely future directions in this field.

16.2 The use of antibodies in stent functionalization

16.2.1 Immunoglobulin structure and target binding

The typical IgG immunoglobulin molecule has a highly modular structure that derives from its dual role in recognition and destruction in the immune system. It is a large (~150 kDa) Y-shaped glycoprotein with dimensions of the order of 15 nm × 7 nm × 3.5 nm [11]. The stem of the molecule is largely conserved between antibodies of different binding specificities and functions in mediating elimination of the assorted targets that are bound in the highly diverse binding clefts located at the ends of the arms of the protein [12]. The molecule is a heterodimer, composed of two identical heavy chain and two identical light chains as shown in Fig. 16.1. Each chain consists of distinct constant or variable domains of approximately 12.5 kDa in size that fold into a typical immunoglobulin-like fold, each stabilized by an intra-domain disulfide bridge between two conserved cysteine residues. Heavy chains contain C_H1, C_H2, and C_H3 constant domains as well as a V_H variable domain, while light chains consist of a single C_L constant domain and a V_L variable domain [11]. The heavy and light chains are also held together by interchain disulfide bridges. Within

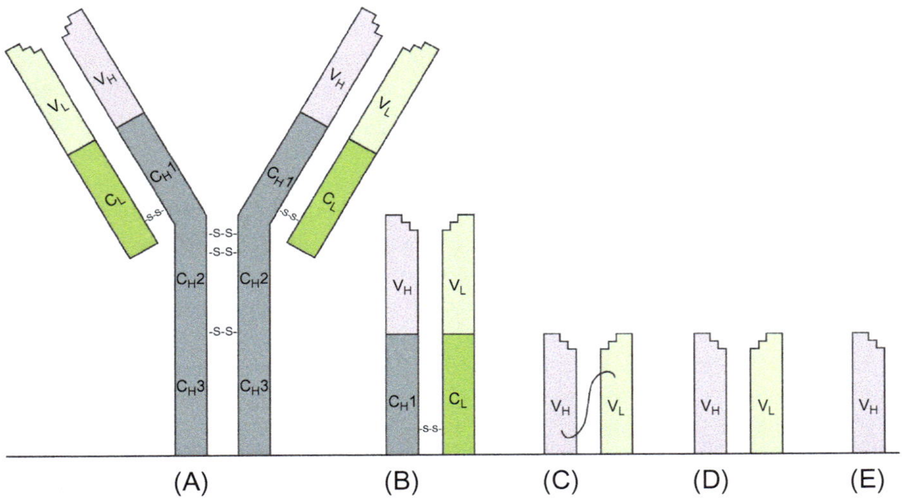

Fig. 16.1 Structure of typical IgG immunoglobulin and derived antigen-binding fragments. (A) Whole IgG molecule. Heavy chains are in *gray*, light chains in *green*. Variable domains are in *lighter shading* on each chain. (B) Fab fragment. (C) Single-chain Fv (scFv) fragment. (D) Fv fragment. (E) V_H domain.

the variable domains of both heavy and light chains, three hypervariable loops, termed complementarity determining regions (CDRs), protrude from the otherwise relatively conserved domain frameworks to form the highly specific binding pockets in which antigen molecules are bound. These CDRs are highly diverse between antibodies in both amino acid sequence and length and undergo further "affinity maturation" upon exposure to antigen through a process of mutation and selection, typically yielding molecules with affinities for target in the nM range [13]. Meanwhile, the Fc stem of the molecule is made up entirely of heavy chain constant domains and undergoes post-translational addition of glycan chains that are critical for interaction with Fc-receptor-bearing effector cells such as phagocytes in the immune response.

The overall antibody structure, therefore, is one of a highly conserved stem that mediates conserved functions such as interaction with the complement cascade or Fc-receptor containing cytotoxic cells, linked to two identical arms with rotational flexibility about the central hinge region to enable their terminal binding pockets to interact with specific ligands [14]. This spatial separation of the recognition and elimination functions enables antibody populations to combine the enormous conformational diversity required for target binding with the structural conservation necessary to interact with common, highly conserved antigen-elimination effector processes. The modular structure also lends itself to immobilization of antibody molecules on material surfaces via the conserved Fc domains to impart new binding functions to such materials in vivo.

16.2.2 Antibody fragments: Design, expression, and engineering

16.2.2.1 Antibody fragment expression and engineering

Conventional monoclonal antibody (mAb) generation relied on immunizing laboratory animals with antigens of interest and fusing the resultant antibody-producing B cells with tumor cells to create immortalized antibody-secreting cell lines that could be scaled up to produce, e.g., murine monoclonal antibodies in the laboratory [15]. The relatively high cost of mammalian cell culture, and the incompatibility of some of the steps with human antibody generation, led researchers to investigate different and cheaper production platforms, resulting in the development of a number of approaches to producing mAb-derived fragments that broadly retained the antigen-binding properties of their source antibodies. The emergence in the late 1980s of small, monoclonal antibody-derived fragments that retained the binding properties of their parent antibodies [16,17] opened up the field of antibody engineering to a myriad of applications—both in vitro and in vivo—in which antibody-binding specificity can be exploited, but interactions with immune effector functions are unnecessary or even undesirable.

While the fragmentation of antibodies can be achieved through proteolysis of the parent IgG using pepsin or papain to yield a bivalent $F(ab')_2$ or monovalent Fab arms, respectively, recombinant DNA technology provides a facility to amplify and clone antibody V_H and V_L domains in various combinations and structural formats to recreate the parent immunoglobulin-binding pocket and retain its binding strength and specificity in a smaller molecular package. Most common among these were the Fab antibody arm and Fv (Fig. 16.1), the latter containing only the antibody V_H and V_L domains, and now the single-chain Fv (scFv) [18,19]. In scFvs, the antibody V_H and V_L domains are covalently linked by a short, flexible peptide linker of typically 15–20 amino acids that overcomes the reduced interaction energy frequently observed between V_H and V_L domains in Fvs due to the absence of the C_H1-C_L disulfide bridge that holds them together in the whole antibody molecule (Fig. 16.1). While smaller antibody variants such as single V_H domains have also been demonstrated to be capable of binding antigen [20], the scFv is the smallest fragment in common use due to its typical retention of the affinity and specificity of its parent monoclonal antibody and its compatibility with recombinant expression systems such as *Escherichia coli* [21].

An important advantage of recombinant antibody fragments in general is that they can be produced relatively easily in bacterial or yeast expression platforms that are more robust, cheaper, less labor-intensive, and less susceptible to contamination than mammalian cell-based systems [22]. *E. coli,* in particular, is a very well-characterized expression system for recombinant proteins but is incapable of efficiently producing large monoclonal antibodies in active form due to their size, multiple disulfide bridges, and Fc glycosylation. While yields of Fab′ antibody fragments of g/L have been reported in high cell density *E. coli* cultures [23], this is 2–3 orders of magnitude higher than yields routinely associated with fragments in closed system, shake flask cultures [24,25]. Nevertheless, increasingly routine optimization approaches such as strain engineering, manipulation of cell physiology and expression parameters, and more complex engineering of polypeptide folding pathways in the expressing cells [19,26,27], many of which were pioneered with nonantibody recombinant proteins [28,29], now allow ongoing improvements in expression and increased yields of fragments.

The use of routine cloning and expression hosts such as *E. coli* and yeasts like *Saccharomyces cerevisiae* and *Pichia pastoris* also facilitates targeted engineering of the antibody fragments for improved expression and/or to modify their application-relevant properties. While the smaller size of the fragments (scFvs have approximate dimensions 5 nm × 4 nm × 4 nm) lends them to applications in drug delivery [30] and imaging [31] in vivo due to their greater tissue penetration or in in vitro immobilization-based scenarios [32], with higher binding pocket densities achievable on support scaffolds for increased detection or capture sensitivities [33], one of the main advantages of recombinant fragments over whole antibodies is the potential to modify their properties by protein engineering. A powerful arsenal of targeted [34] or random [35] molecular techniques allows enhancement of binding affinities or elimination of binding target cross-reactivities, while fusion protein expression enables the production of antibody fragments with novel "effector functions," such as fused cytotoxic drugs, which add activities to the binding moieties in vivo [36–38]. In the field of immobilization or surface modification of materials such as stents, the addition of flexible peptide tags to antibody fragments may facilitate their enhanced attachment by achieving covalent attachment for long-term stability and reduced leaching and ensure the correct orientation of fragments on the surface for increased accessibility of binding pockets.

Other well-characterized antibody fragment conformations include multivalent molecules such as diabodies and triabodies [39,40], which overcome the lower affinities of monovalent Fabs, Fvs, and scFvs resulting from their reduced avidities [39,41,42], and bispecific molecules that combine, e.g., cell targeting and T cell recruitment for in vivo applications [43,44]. Though these engineered moieties find roles in niche applications, their advantages over the basic scFv structure in surface functionalization applications such as stent modification are limited. Instead, the major objective of antibody engineering for stent functionalization, after structure minimization, is the isolation and production of human rather than animal antibodies to minimize the potential for immunological complications in vivo.

16.2.2.2 Isolation of antibodies of desired binding specificities

As outlined in Section 16.2.2.1, the use of the revolutionary hybridoma technology pioneered in the 1970s is increasingly giving way to in vitro antibody isolation approaches that obviate the need both for animal immunization and for expensive mammalian cell culture approaches. Such approaches typically involve amplification of (human) V_H and V_L antibody genes and their assembly into large combinatorial collections of antibody fragments—effectively "immortalizing" antibody genes in recombinant expression platforms rather than the antibody-secreting B cells [45]. These antibody fragment libraries are usually displayed on cells, viruses, or ribosomes in a format in which molecules can be screened in vitro to rapidly isolate binders of a ligand of interest [46]. Incorporation of a mutagenic step into the procedure allows an iterative cycle of generation of diversity followed by selection of higher affinity binders to be carried out in a process that mimics affinity maturation [47]. Combined with the use of human antibody genes derived from, e.g., peripheral blood, in the initial stages of the procedure, the technology presents a powerful approach to relatively rapidly generate high affinity, human antibody fragments with specificity for almost any ligand, and without the need for immunization [48].

The most common combinatorial antibody library approach is phage display, in which antibody fragments are expressed on the surface of bacteriophages such as the filamentous M13 phage that infects *E. coli* cells. Fragments are genetically fused to the pIII minor capsid protein-encoding gene to achieve their display in 1–3 copies at the phage tip [49] and libraries of up to 10^{10} phage-displayed antibody fragments are routinely generated for screening on immobilized antigens [50,51]. Repetitive cycles of ligand-binding selection of scFv-displaying phage particles and infection of *E. coli* cells for amplification (and to increase antibody diversity) can yield high affinity, highly specific binders in only 2–4 weeks [48,52]. Antibody fragments can also be raised against whole cells [53] or cell surface-derived proteins [50] identified using in silico tools [54] to isolate cell-binding moieties for characterization. The isolated antibody fragments are then expressed in soluble form in *E. coli* for characterization and exploitation. Critically for end applications such as stent functionalization, the resultant antibody molecules are human in origin and therefore less likely to give rise to immunogenicity problems than those from hybridoma-type platforms.

Antibody fragment libraries can also be displayed on bacterial, yeast, or mammalian cell surfaces by genetic fusion to naturally occurring cell surface proteins [55]. These include agglutinin protein Aga2p in yeast cells [56] and β-barrel outer membrane proteins in *E. coli* [57], with the former better suited to folding and soluble display of larger, more complex proteins such as antibody molecules as yeasts have a eukaryotic protein folding machinery, greater capacity for formation of complex disulfide bond patterns, and can carry out *N*-linked glycosylation [58]. In vitro antibody fragment display systems such as ribosome display have also been well-established and exploited in antibody isolation in recent years. Here, mRNA transcripts are translated in the absence of a stop codon, resulting in linking of the ligand-binding phenotype with its encoding genotype via the formation of stable complexes of mRNA, ribosome, and synthesized polypeptides [59]. As library sizes are not limited by cell transformation efficiency because the approach is cell-free, the technique can be used to generate extremely large antibody fragment libraries with increased potential to isolate high affinity antibodies against diverse targets of interest [60–63]. An additional impact of the powerful mutagenic and selection molecular tools used to increase combinatorial diversity in recombinant libraries is that naïve antibody collections can be used to isolate fragments in the nanomolar affinity range typically obtained from immune libraries—allowing both the creation of libraries from nonimmunized human donors and the isolation of fragments with highly varied binding specificities from a single, diverse antibody collection.

16.2.2.3 Future perspectives of antibodies in stent functionalization

As an estimated 97% of amino acids in antibodies are not involved in making antigen contacts [11], minimization of the ligand-binding moiety has clear potential to allow the effective packing density of binding pockets to be increased on surfaces—thus potentially improving the detection sensitivity for low concentration analytes [33,64] or capture efficiency of low abundance cells on antibody-modified surfaces. Similarly, the absence of nonantigen-interacting domains in stent-tethered antibody fragments can avoid undesirable interactions with Fc-receptor-bearing immune effector cells on device surfaces and reduce nonspecific interactions in vivo. Furthermore, the ability to engineer recombinant antibodies for controlled, oriented, and stable attachment to surfaces or devices [65] is a significant advance on the use of physisorbed molecules of which as few as one tenth of proteins may be functional. Nevertheless, while antibody engineering and recombinant expression techniques have become standard R&D tools over the past three decades, their impact has yet to become apparent in biomedical device or materials functionalization pipelines. It may be that the potential to isolate and express *human*-derived antibody fragments, however, proves the critical consideration that persuades manufacturers to embrace combinatorial library and recombinant expression platforms for next-generation medical device and stent functionalization. Antibody-coated stents currently used in clinical practice employ mouse [66] or occasionally rabbit [9] rather than human antibodies. While the use of animal molecules overcomes previous technical bottlenecks associated with isolation of their human counterparts [67], the potential to elicit a human antimurine antibody (HAMA)-type response in vivo is clear [68]. As HAMA responses would rapidly destroy an antibody-coated surface designed for, e.g., cell capture or drug delivery, the use of recombinant technologies to create immune-tolerated human molecules will clearly be preferable for stent coating. The adaptation of human-derived or -based antibody molecules has been the norm in other clinical sectors for some time: of 11 new antibodies approved for clinical use as far back as 2006–09, six were fully human and four others had been "humanized" to reduce their potential immunogenicity in vivo [69].

16.3 Protein-stent linking approaches

16.3.1 Overview

The most commonly used method of attaching proteins to solid supports for many applications is simple physisorption [14] (Fig. 16.2). Antibodies are routinely adsorbed onto diverse surfaces such as polystyrene, polyvinylidene fluoride (PVDF), or nitrocellulose in standard in vitro techniques like enzyme immunoassays and immunoblotting. The main advantages of this approach are its simplicity, low cost, relatively high antibody-binding capacity, and the fact that it involves no additional manipulation of the antibody. For a more complex application such as stent biofunctionalization with

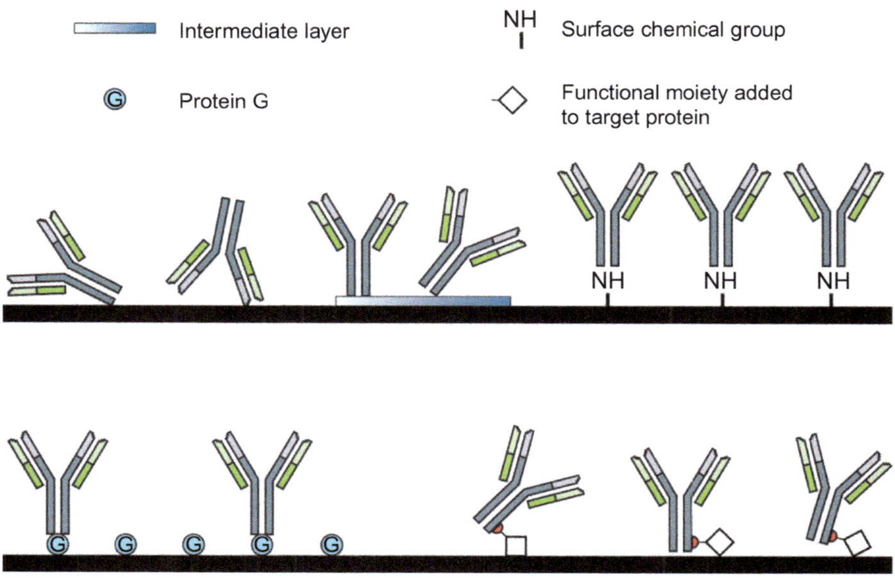

Fig. 16.2 Immobilization of antibody molecules on solid support [45]. Antibody heavy chain is *gray*, *light chain green*. Constant domains are in *darker shading*; variable domains are *lighter*. Upper level, Left: Physisorbed antibodies; Centre: antibodies tethered via intermediate "sticky" layer; Right: antibodies attached via specific chemical moiety, e.g., terminal NH_4 or COOH groups; Lower level, Left: attachment via antibody-binding protein, e.g., Protein G or Protein A; Right: antibodies with engineered motif for specific interaction with (chemically activated) surface.

antibodies, however, the disadvantages of such an approach are considerable: many proteins denature when adsorbed onto a solid surface, resulting in a disordered or inaccessible binding site in up to 90% of adsorbed molecules [70]; as attachment occurs via nonsite-specific interactions, antibody molecules are randomly oriented on the substrate and binding pockets may be inaccessible to the target ligand or cell (Fig. 16.2); protein adsorption is poorly reproducible between analyses; the reversible nature of the noncovalent attachment leads to reduced stability of the bioengineered stent over extended usage due to antibody desorption; and deposition of multiple layers of protein is common, leading to reduced sensitivity and higher antibody production costs [71]. Ideally, tethering of antibodies onto stent surfaces should yield molecules that are correctly oriented on the surface, stably (preferably covalently) attached, arranged in a well-structured monolayer, and conformationally and functionally unmodified by the immobilization procedure, as discussed in Section 16.3.2.

16.3.2 Ensuring orientation and/or stability of immobilized antibodies

16.3.2.1 Covalent attachment of antibodies

Covalent attachment of antibodies to the stent structure can be used to overcome the fundamental problem of protein leaching from the support over time—but not, in many cases, that of antibody orientation [14]. At the simplest level of covalent attachment, the terminal amino or carboxyl group in any protein molecule can be coupled via relatively simple chemistries to a variety of engineered supports, such as aldehyde- and epoxy-activated substrates, resulting in highly stable display of the protein of interest [30]. Cross-linking of solvent-exposed amino acids with similar chemistries can also occur, however, as can cross-linking within or between the protein molecules themselves, which can result in incorrectly oriented and/or inaccessible binding pockets in the immobilized antibodies [72]. Covalent tethering of antibodies to solid supports typically exploits one of three functionalities in the antibody molecule: lysine amino acids, cysteine residues, and carbohydrate moieties, which occur in the Fc stem [73]. The use of the ε-NH_3^+ approach with surface-exposed lysines cannot ensure the correct orientation of the antibody molecules on supports as approximately 40–50 such amino acid residues may be solvent-exposed and accessible for modification in standard immunoglobulin structures (Fig. 16.2). In the case of cysteine amino acids, the typical IgG immunoglobulin contains approximately 32 cysteine residues that are involved in domain stabilization (one S-S bond in each C or V region) or interchain covalent disulfide bonds that attach the heavy and light chains in the heterodimeric antibody molecule (see Fig. 16.1). The occurrence of unpaired cysteine residues that are accessible for attachment to stent surfaces is, therefore, unusual, though Fab′ antibody fragments in which a single cysteine is available for reaction can be generated enzymatically by pepsin and acid treatment [74]. Alternatively, Fab′ fragments can be created recombinantly, though modified scFv-type fragments are more likely to be preferred in recombinant expression formats. The 2–5 glycan chains that occur in the Fc stem of antibody molecules and function in immune effector functions in vivo may also be exploited as they meet the combined requirements of ensuring stable, covalent attachment of molecules and orienting the binding pocket away from the stent surface by virtue of their location in the binding pocket-distal stem of the antibody structure. A well-established chemistry of oxidation of the carbohydrate groups to generate aldehydes in the Fc stem then allows specific reaction with amine or hydrazide groups that must be introduced into the stent surface [75,76]. A somewhat similar, recently described technology for covalent, oriented attachment of antibodies exploits the histidine-rich nature of antibody Fc regions to directionally immobilize IgGs on heterofunctional metal chelate-glyoxyl supports [77].

16.3.2.2 Use of intermediate, antibody-binding layer

Another approach to directionally immobilize antibodies, albeit typically not covalently, is via an intermediate antibody-binding layer. This can be achieved by the use of broad specificity protein-binding molecules, antibody-specific binding partners, or recombinant interaction motifs engineered into the antibody molecules. Generic protein-binding layers such as mussel adhesive proteins (MAPs) can be easily coated onto stent surfaces such as stainless steel and titanium to provide a "sticky" layer to which antibodies can be readily attached [78] (Fig. 16.2). This approach has been successfully utilized to coat anti-CD34 antibodies on 316L stainless steel to mediate EPC capture [79]. A greater degree of specificity for antibodies can be added by the use of a MAP-Protein A (see below) fusion protein in which the MAP layer can attach to polymeric and metal surfaces to tether the antibody-binding Protein A domain, which then captures antibodies that retain their antigen binding upon immobilization [80]. Protein A and Protein G, meanwhile, are naturally occurring proteins secreted by *Staphylococcus aureus* that bind antibody Fc regions and are commonly used in antibody purification in vitro [81]. Due to the location of their target motifs in the antibody stem, immobilization via these binding partners is likely to correctly orient the antibody layer on surfaces, as demonstrated in the case of anti-VEGFR-2 monoclonal antibodies tethered using Protein G [82]. Antibodies can also be immobilized in an oriented manner on supports by the use of short peptides that recognize exposed hydrophobic patches at the bottom of the antibody Fc stem [83].

The use of intermediate, antibody-capture layers avoids the need for engineering of antibodies to mediate their surface tethering and they are broadly compatible with molecules of diverse binding specificities as they typically exploit conserved structural or sequence motifs in the conserved Fc stem of the molecule. The interaction between capture protein and antibody is typically reversible, however, and, more critically, the precoating of bacterial or other nonhuman proteins on the stent has the potential to capture antibodies of varying binding specificities from the circulation or to elicit a highly destructive immune response against the stent protein layer.

16.3.2.3 Immobilization of engineered antibody fragments

The relative ease of recombinant isolation and expression of antibody fragments of desired binding specificities (see Section 16.2.2.1) also extends to their production with engineered recombinant tags to mediate their controlled attachment to surfaces. scFvs can be produced with a tag such as a leucine-zipper motif that interacts with a partner motif on the support matrix to form a correctly oriented heterodimeric pair [84] or with various peptide motifs that are recognized by surface-bound antibodies [85] (Fig. 16.2). Short (typically 4–8 amino acid residue) peptide tags containing, e.g., cysteine residues, can also be added at surface-exposed locations on antibody fragments to mediate oriented immobilization of the purified protein molecules via appropriate chemical groups introduced into the stent surface [86]. In an approach that mimics oriented immobilization of whole antibody molecules via their Fc-glycan chains, we [65,87] and others [88] have also expressed glycosylated antibody fragments in *E. coli* for chemical modification and correctly oriented, covalent immobilization on aminated stent surfaces.

16.4 Applications of stent-immobilized antibodies

16.4.1 Overview

The dominant application of antibodies in stent functionalization lies in mediating E(P)C attachment to promote endothelialization in vivo and prevent thrombosis in the period in which the stent surface has not yet been completely re-endothelialized. Efforts to accelerate endothelialization of the stent surface gained considerable focus with the recognition that antiproliferative drugs released from DES delayed re-endothelialization. Attempts to seed biomaterial surfaces with EPCs and/or ECs in vitro and to create an artificial endothelium on stent surfaces are complex and require an initial biopsy to obtain a patient's own cells to seed the graft, expansion, and growth of the cells to confluence on the graft, and re-implantation into the patient [89]. The work is technically demanding, time-consuming, expensive, prone to bacterial contamination, and not readily compatible with stent manufacturing practices [90], which has led to a clear preference for biofunctionalization of stent surfaces with E(P)C-capturing antibodies in an attempt to promote endothelial cell capture and accelerate formation of a new endothelial layer in vivo.

16.4.2 Stent re-endothelialization

EPCs have been the focus of intense study over the past two decades due to their ability to attain EC characteristics in vitro [91] and their potential, therefore, to drive surface endothelialization for improved biocompatibility. Numerous studies have demonstrated their ability to decrease neointimal formation and accelerate re-endothelialization upon recruitment to sites of vascular damage. As EPCs deposit extracellular matrix to a greater extent than ECs, these are typically preferred in materials engineering and functionalization-type applications. While there is still no universally accepted definition of EPCs, a number of cell surface markers with varying degrees of specificity for EPCs have been identified, foremost among which are CD34, CD31, CD105, CD133, and CD144, which play a key role in endothelium lining [92]. A number of studies have indicated that EPCs in the bone marrow and immediately after their migration into the systemic circulation are positive for CD34, CD133, and VEGFR2, whereas EPCs circulating in the bloodstream no longer express CD133 [93].

Antibodies have particular potential to accelerate re-endothelialization by capture of circulating EPCs or ECs and effectively immobilizing them on the stent surface. The hydrophobicity and complexity of these membrane-spanning proteins in general leads to considerable difficulties in their purification or recombinant expression [94,95]. While this can clearly create a bottleneck in the isolation of binding antibodies, antibodies against a number of potential EPC and EC markers, either in their full-length state or extracellular domains, have been isolated. The most extensively characterized of these are CD34, CD133, VEGFR2, CD144, and anti-CD105 molecules, which are discussed in more detail below.

16.4.2.1 CD34-binding antibodies

CD34 (http://www.uniprot.org/uniprot/P28906) is a highly glycosylated protein with a single transmembrane (TM) region and an extracellular domain of 259 amino acids. While it has been used in a number of investigations as a marker to capture EPC populations, it has also been demonstrated that over 99% of cells bound by CD34-binding antibodies are not EPCs, as defined by expression of CD133 and VEGFR-2 [96]. Purified mouse anti-CD34 monoclonal antibody has, however, been found to interact with CD34-expressing hematopoietic precursor cells and endothelial cells, and not with normal peripheral lymphocytes, monocytes, erythrocytes, granulocytes, or platelets [97]. Ex vivo and animal studies have demonstrated that stents coated with CD34 antibodies, which specifically targeted circulating EPCs, achieved complete endothelial coverage of struts in as little as 48 h postimplantation [98]. Stents with an immobilized murine IgG2a antihuman CD34 antibody were developed by Orbus Medical Technologies and have been commercially available since 2010 (http://www.orbusneich.com). The first clinical investigations of this anti-CD34 Genous technology indicated its safety and feasibility in treating *de novo* coronary artery disease [8] and in long-term promotion of significant late regression of neointimal hyperplasia. Subsequent randomized clinical trials comparing the stent with BMS [99,100] and DES [101] led to development of a next-generation COMBO stent which combines EPC capture using anti-CD34 antibodies with eluting sirolimus to reduce in-stent late luminal loss [102].

Other researchers have combined EPC capture, mediated by immobilized anti-CD34 antibodies, with the supply of growth factors to promote the differentiation of EPCs into ECs. Incorporation of VEGF and basic fibroblast growth factor in the form of conjugated microparticles led to improved EPC adhesion and proliferation on a polymeric coating on a vascular stent [103]. Others have reported new coating-by-layer functionalization approaches on 316L stainless steel [104,105] and titanium [106] with an anti-CD34 antibody and VEGF in an approach which could be utilized with many antibody-growth factor combinations. In all antibody immobilization analyses, of course, the method of immobilization is critical in determining the orientation and functionality of the surface-bound capture molecules. Petersen and co-workers demonstrated the importance of optimizing orientation and surface density of antibodies on stents (approximately 1 μg/cm^2 of anti-CD34 mAb in the Genous stent [107]) by using a site-specific immobilization approach. Similar levels of coated antibody were achieved in this study via physisorption or Fc-mediated reaction of the antibody with amino-activated PLLA-coated stent surfaces, but 2–3-fold higher binding activities (of CD34$^+$ cells) were measured in the latter molecules [108]. Similarly, others have demonstrated that CD34-binding antibody immobilized on coated stainless steel via its Fc stem promoted better EC attachment in vitro and reduced blood coagulation than the same surface functionalized by (nonorientation-specific) glutaraldehyde treatment [109]. Immobilization of chitosan in combination with loading of an anti-CD34 antibody has also been demonstrated to accelerate re-endothelialization in a commercial sirolimus-eluting stent (Cypher, Cordis, Miami, Florida, United States) [107,110].

16.4.2.2 CD133-binding antibodies

CD133 (http://www.uniprot.org/uniprot/O43490) is proposed to play a role in apical plasma membrane organization of epithelial cells and is a 5-TM protein with three extracellular topological domains. Less attention has been paid to creation of anti-CD133 stents than in the case of CD34, probably largely due to its less well-understood association with circulating EPCs [93,111]. In addition, in our experience at least, recombinant expression of CD133 or component domains has proven challenging, though the native molecule is commercially available from a number of suppliers for antibody isolation. A recent comparison of the cell types bound by anti-CD34 and anti-CD133 antibodies found that while the latter selectively captured hematopoietic stem cells (HSCs) which differentiate into ECs, anti-CD34 antibodies bound, in addition to HSCs, hematopoietic progenitor cells that differentiate into vascular ECs and immune cells, thereby promoting smooth muscle cell growth and potentially leading to thrombosis, inflammation, and rejection [112]. A further comparison of the efficacies of anti-CD34 and anti-CD133 antibodies in EPC capture also concluded that the latter were preferable as they mediated more rapid endothelialization and a more effective inhibition of in-stent restenosis [113]. CD133-binding antibodies have been successfully covalently attached to polymeric coatings on bare metal stents, leading to EPC capture from circulation and significantly improved cell growth compared to BMS over a short (6 h) time period, with no difference in endothelial function noted over 6 months [114]. A more in-depth comparison of multiple antibodies against CD133 and CD34, incorporating evaluation of coating densities, functionalities of coated molecules, and cell specificities, is required before more general conclusions can be drawn about the relative efficacies of the two capture approaches in E(P)C binding.

16.4.2.3 VEGFR2-binding antibodies

Vascular endothelial growth factor receptor 2 (VEGFR2) is the primary mediator of the effects of VEGF, which is involved in the growth and maintenance of vascular structures. VEGF has been reported to accelerate re-endothelialization of damaged arteries in the rat carotid artery [115], while its elution from stents promotes re-endothelialization by stimulating EPC capture and maturation [116,117]. VEGFR2 contains an exposed chain of seven repeated Ig-like domains (http://www.uniprot.org/uniprot/P35968) and it has been produced recombinantly as a fusion product and in component domains for biochemical and functional characterization studies [118,119]. A number of groups have isolated VEGFR2-binding antibodies for EPC capture. While it has been argued that VEGFR2 may be a better target than CD34 or CD133 for EPC capture, many reports with these antibodies to date have focused on their immobilization, e.g., by exploiting Fc-binding protein G [82] or via engineered motifs in recombinant antibody fragments [120], and their efficacy in EPC capture and surface re-endothelialization remains to be investigated in detail. Reports of the successful use of immobilized VEGF to capture EPCs [121] and even of positive effects on re-endothelialization of upregulation of VEGF expression [122] provide indications of the potential usefulness of this VEGFR2-capture approach, however.

16.4.2.4 Other antibodies used in re-endothelialization

Human vascular endothelial (VE)-cadherin, also known as CD144 (http://www.uniprot.org/uniprot/P33151), is an endothelial cell marker found in veins, arteries, capillary, and large vessels and reported to be expressed on late EPC and differentiated ECs, but not on early EPCs or other leukocytes. Coating of a stainless steel stent with rabbit polyclonal antihuman VE-cadherin antibodies has been demonstrated to capture EPCs and reported to significantly reduce neointimal formation and accelerate re-endothelialization in vivo [9]. A comparison between anti-CD34- and anti-CD144-coated stents also indicated that the latter are more effective at capturing outgrowth endothelial cells (OECs) and mature ECs in vitro and in vivo [123], while a recent rabbit study indicated a superior performance of an anti-VE-cadherin-coated zwitterionic stent over BMS in reducing neointimal formation, reducing platelet adhesion, and promoting endothelial healing [124].

CD105 (or endoglin) is a homodimeric, single-TM glycoprotein that binds transforming growth factor β receptor in ECs (http://www.uniprot.org/uniprot/P17813). A comparison of the performance of anti-CD104-coated stainless steel stents with BMS and sirolimus-eluting stents revealed improved endothelialization in the antibody-coated stents after 14 days in porcine coronary arteries, while neointimal formation and stenosis were also decreased [125]. Subsequent investigation also revealed the antiendoglin-coated stent to behave similarly to an anti-CD34-coated stent, with no difference in the neointima area, percent area stenosis, or percentage of re-endothelialization detected between the two devices [126].

16.4.3 Other applications of stent-immobilized antibodies

In a novel application, DNA-binding antibodies have been tethered onto cardiovascular stents to achieve localized gene delivery [127]. Studies in a porcine model demonstrated delivery of a green fluorescent protein-encoding plasmid and expression of the reporter protein in the left anterior descending coronary artery, while in vivo delivery of nitric oxide synthase cDNA via the same approach was reported to inhibit restenosis [128]. In a similar approach, other groups have tethered adenovirus-binding antibodies on coronary stents for gene delivery. While the technique realized transfection of arterial smooth muscle cells in porcine coronary arteries, neointimal transduction levels were more than 17% of total neointimal cells [129] and the approach does not appear to have been developed further over the past decade.

16.5 Future perspectives

The current state of the art in stent functionalization using antibodies is dominated by the use of antibody molecules to capture endothelial or endothelial progenitor cells for acceleration of re-endothelialization of the inserted device and minimization of the risk of thrombosis in vivo. Foremost among developments has been the realization of a number of clinical products with CD34-binding antibodies, designed to capture EPCs from the circulation. Meanwhile, a variety of additional antibodies against both CD34 and other proposed E(P)C marker proteins have been isolated and characterized in R&D settings, with some progressing through animal studies and along the product development pipeline. The bottleneck of identifying a surface marker unique to ECs or EPCs remains, however, with no currently accepted "ideal" marker (i.e., specific for and highly expressed on E(P)Cs) for use in endothelial cell identification and capture. This is a significant limitation of antibody-stent biofunctionalization as it limits the potential to capture only desired cell types to coat the stent luminal surface. Genomic investigation of surface protein expression on EPCs, ECs, and other physiologically relevant cell types such as smooth muscle cells may help to identify improved markers, though a compromise between specificity and expression level on target cells is likely to prove necessary.

In the interim, carefully designed comparison of the cell binding of current antibody candidates will be instructive in understanding their potential to mediate endothelialization. Many current studies may be limited not only by their use of small numbers of antibodies, but also by the reported changes in the expression of recognized EPC "markers" in cell lines due to senescence upon ongoing culture. This has important potential consequences for the ability to extrapolate in vitro results to in vivo settings—or to isolate and characterize antibodies of clinical significance using available cell culture approaches. Again, a more detailed evaluation of differential expression of protein targets both in vitro and in isolated cells in vivo will improve the ability to successfully translate technological advances in antibody isolation into clinically useful products.

One area that seems certain to grow in importance in the stent-biofunctionalization sector over the coming years is the use of recombinant antibody fragments. While fragments are advantageous in material functionalization applications for the many reasons discussed above, not least their lower production costs and higher yields compared with monoclonal antibodies, their defining advantage in stent modification is their human, i.e., nonimmunogenic, origin. It is unlikely that entirely new stent products containing nonhuman antibodies will continue to be developed and progress into the clinic over the coming decade. Recombinant antibody technology instead allows the relatively simple isolation of human-derived antibody fragments against practically any molecule of interest. Whole antibodies can be reconstituted from these fragments to re-generate classical IgG-type molecules, though the use of the smaller, scFv-type fragments are advantageous in many applications in which effector functions are unnecessary or undesirable. The ease with which recombinant fragments can be engineered to achieve highly specific, covalent (stable) and correctly oriented (accessible for cell capture) attachment to stent surfaces constitutes another important advantage over the use of whole antibody molecules. They can also be affinity-matured in vitro to increase their binding strength which may benefit EPC capture applications in which potentially as low as 0.01% of circulating cells in the blood may be the target cells.

Other than re-endothelialization, there are few reported applications of antibody use to modify stent properties in vivo. Gene delivery has been investigated to increase localized concentrations of growth factors or NO, but the approaches have not been extensively developed, due largely to the existence of simpler approaches to achieve the same goals. It seems likely, therefore, that the predominant use of antibodies in adding function to cardiac stents in the near future will continue to be in the area of promoting surface endothelialization, albeit with increased use of recombinant fragments to overcome existing limitations of noncovalent immobilization, poor orientation, and the nonhuman origin of currently preferred antibodies.

Acknowledgments

Work on biomaterials and protein engineering in JGW's laboratory is supported by European Union Project Grant FP7-PEOPLE-2012-IAPP-324514; Science Foundation Ireland Grant 13/RC/2073, co-funded under the European Regional Development Fund; and NUI Galway College of Science studentship to IBOC.

References

[1] N.A. Scott, Restenosis following implantation of bare metal coronary stents: pathophysiology and pathways involved in the vascular response to injury, Adv. Drug Deliv. Rev. 58 (2006) 358–376.

[2] L.L. Hench, I. Thompson, Twenty-first century challenges for biomaterials, J. R. Soc. Interface 7 (2010) S379–S391.

[3] J.M. Anderson, A. Rodriguez, D.T. Chang, Foreign body reaction to biomaterials, Semin. Immunol. 20 (2008) 86–100.

[4] M. Kastellorizios, N. Tipnis, D.J. Burgess, Foreign body reaction to subcutaneous implants, Adv. Exp. Med. Biol. 865 (2015) 93–108.

[5] M.C. Chen, Y. Chang, C.T. Liu, W.Y. Lai, S.F. Peng, Y.W. Hung, H.W. Tsai, H.W. Sung, The characteristics and in vivo suppression of neointimal formation with sirolimus-eluting polymeric stents, Biomaterials 30 (2009) 79–88.

[6] T. Inoue, K. Croce, T. Morooka, M. Sakuma, K. Node, D.I. Simon, Vascular inflammation and repair: implications for re-endothelialization, restenosis, and stent thrombosis, JACC Cardiovasc. Interv. 4 (2011) 1057–1066.

[7] Y. Weng, J. Chen, Q. Tu, Q. Li, M.F. Maitz, N. Huang, Biomimetic modification of metallic cardiovascular biomaterials: from function mimicking to endothelialization in vivo, Interface Focus 2 (2012) 356–365.

[8] J. Aoki, P.W. Serruys, H. Van Beusekom, A.T. Ong, E.P. McFadden, G. Sianos, W.J. Van Der Giessen, E. Regar, P.J. De Feyter, H.R. Davis, S. Rowland, M.J. Kutryk, Endothelial progenitor cell capture by stents coated with antibody against CD34: the HEALING-FIM (Healthy Endothelial Accelerated Lining Inhibits Neointimal Growth-First In Man) Registry, J. Am. Coll. Cardiol. 45 (2005) 1574–1579.

[9] W.-H. Lim, W.-W. Seo, W. Choe, C.-K. Kang, J. Park, H.-J. Cho, S. Kyeong, J. Hur, H.-M. Yang, H.-J. Cho, Y.-S. Lee, H.-S. Kim, Stent coated with antibody against vascular endothelial-cadherin captures endothelial progenitor cells, accelerates re-endothelialization, and reduces neointimal formation, Arterioscler. Thromb. Vasc. Biol. 31 (2011) 2798–2805.

[10] A.J. Melchiorri, N. Hibino, T. Yi, Y.U. Lee, T. Sugiura, S. Tara, T. Shinoka, C. Breuer, J.P. Fisher, Contrasting biofunctionalization strategies for the enhanced endothelialization of biodegradable vascular grafts, Biomacromolecules 16 (2015) 437–446.

[11] E. Kabat, Antibody complementarity and antibody structure, J. Immunol. 141 (1988) 25S–36S.

[12] H.W. Schroeder, L. Cavacini, Structure and function of immunoglobulins, J. Allergy Clin. Immunol. 125 (2010) S41–S52.

[13] B. Li, A.E. Fouts, K. Stengel, P. Luan, M. Dillon, W.-C. Liang, B. Feierbach, R.F. Kelley, I. Hötzel, In vitro affinity maturation of a natural human antibody overcomes a barrier to in vivo affinity maturation, MAbs 6 (2014) 437–445.

[14] A. Srivastava, I.B. O'Connor, A. Pandit, J.G. Wall, Polymer-antibody fragment conjugates for biomedical applications, Prog. Polym. Sci. 39 (2014) 308–329.

[15] G. Kohler, C. Milstein, Continuous cultures of fused cells secreting antibody of predefined specificity, Nature 256 (1975) 495–497.

[16] R.E. Bird, K.D. Hardman, J.W. Jacobson, S. Johnson, B.M. Kaufman, S.M. Lee, T. Lee, S.H. Pope, G.S. Riordan, M. Whitlow, Single-chain antigen-binding proteins, Science 242 (1988) 423–426.

[17] A. Skerra, A. Plückthun, Assembly of a functional immunoglobulin Fv fragment in *Escherichia coli*, Science 240 (1988) 1038–1041.

[18] X. Hu, R. O'Dwyer, J.G. Wall, Cloning, expression and characterisation of a single-chain Fv antibody fragment against domoic acid in *Escherichia coli*, J. Biotechnol. 120 (2005) 38–45.

[19] X. Hu, L. O'Hara, S. White, E. Magner, M. Kane, J.G. Wall, Optimisation of production of a domoic acid-binding scFv antibody fragment in *Escherichia coli* using molecular chaperones and functional immobilisation on a mesoporous silicate support, Protein Expr. Purif. 52 (2007) 194–201.

[20] J. Cheng, X. Wang, Z. Zhang, H. Huang, Construction and expression of a reshaped VH domain against human CD28 molecules, Prep. Biochem. Biotechnol. 32 (2002) 239–251.

[21] Z.A. Ahmad, S.K. Yeap, A.M. Ali, W.Y. Ho, N.B. Alitheen, M. Hamid, scFv antibody: principles and clinical application, Clin. Dev. Immunol. 2012 (2012). 980250.

[22] N. Ferrer-Miralles, J. Domingo-Espin, J.L. Corchero, E. Vazquez, A. Villaverde, Microbial factories for recombinant pharmaceuticals, Microb. Cell Fact. 8 (2009) 17.

[23] P. Carter, R.F. Kelley, M.L. Rodrigues, B. Snedecor, M. Covarrubias, M.D. Velligan, W.L. Wong, A.M. Rowland, C.E. Kotts, M.E. Carver, et al., High level *Escherichia coli* expression and production of a bivalent humanized antibody fragment, Biotechnology 10 (1992) 163–167.

[24] W. Chen, L. Hu, A. Liu, J. Li, F. Chen, X. Wang, Expression and characterization of single-chain variable fragment antibody against staphylococcal enterotoxin A in *Escherichia coli*, Can. J. Microbiol. 60 (2014) 737–743.

[25] M.A. Thiel, D.J. Coster, C. Mavrangelos, H. Zola, K.A. Williams, An economical 20 litre bench-top fermenter, Protein Expr. Purif. 26 (2002) 14–18.

[26] K. Intachai, P. Singboottra, N. Leksawasdi, W. Kasinrerk, C. Tayapiwatana, B. Butr-Indr, Enhanced production of functional extracellular single chain variable fragment against HIV-1 matrix protein from *Escherichia coli* by sequential simplex optimization, Prep. Biochem. Biotechnol. 45 (2015) 56–68.

[27] O. Kolaj, S. Spada, S. Robin, J.G. Wall, Use of folding modulators to improve heterologous protein production in *Escherichia coli*, Microb. Cell Fact. 8 (2009) 9.

[28] R. O'Dwyer, R. Razzaque, X. Hu, S.K. Hollingshead, J.G. Wall, Engineering of cysteine residues leads to improved production of a human dipeptidase enzyme in *E. coli*, Appl. Biochem. Biotechnol. 159 (2009) 178–190.

[29] S. Robin, D.M. Togashi, A.G. Ryder, J.G. Wall, Trigger factor from the psychrophilic bacterium *Psychrobacter frigidicola* is a monomeric chaperone, J. Bacteriol. 191 (2009) 1162–1168.

[30] A. Srivastava, C. Cunningham, A. Pandit, J.G. Wall, Improved gene transfection efficacy and cytocompatibility of multifunctional polyamidoamine-cross-linked hyaluronan particles, Macromol. Biosci. 15 (2015) 682–690.

[31] S. Kaur, G. Venktaraman, M. Jain, S. Senapati, P.K. Garg, S.K. Batra, Recent trends in antibody-based oncologic imaging, Cancer Lett. 315 (2012) 97–111.

[32] L. Torrance, A. Ziegler, H. Pittman, M. Paterson, R. Toth, I. Eggleston, Oriented immobilisation of engineered single-chain antibodies to develop biosensors for virus detection, J. Virol. Methods 134 (2006) 164–170.

[33] X. Hu, S. Spada, S. White, S. Hudson, E. Magner, J.G. Wall, Adsorption and activity of a domoic acid binding antibody fragment on mesoporous silicates, J. Phys. Chem. B 110 (2006) 18703–18709.

[34] L.K. Denzin, E.W. Voss Jr., Construction, characterization, and mutagenesis of an anti-fluorescein single chain antibody idiotype family, J. Biol. Chem. 267 (1992) 8925–8931.

[35] R.A. Irving, A.A. Kortt, P.J. Hudson, Affinity maturation of recombinant antibodies using *E. coli* mutator cells, Immunotechnology 2 (1996) 127–143.

[36] H. Cheng, X. Ye, X. Chang, R. Ma, X. Cong, Y. Niu, M. Zhang, K. Liu, H. Cui, J. Sang, Construction, expression, and function of 6B11ScFv-mIL-12, a fusion protein that attacks human ovarian carcinoma, Med. Oncol. 32 (2015) 130.

[37] N. Simon, A. Antignani, R. Sarnovsky, S.M. Hewitt, D. Fitzgerald, Targeting a cancer-specific epitope of the epidermal growth factor receptor in triple-negative breast cancer, J. Natl. Cancer Inst. 108 (2016) djw028.

[38] L.H. Wang, C.W. Ni, Y.Z. Lin, L. Yin, C.B. Jiang, C.T. Lv, Y. Le, Y. Lang, C.Y. Zhao, K. Yang, B.H. Jiao, J. Yin, Targeted induction of apoptosis in glioblastoma multiforme cells by an MRP3-specific TRAIL fusion protein in vitro, Tumour Biol. 35 (2014) 1157–1168.

[39] O. Dolezal, L.A. Pearce, L.J. Lawrence, A.J. McCoy, P.J. Hudson, A.A. Kortt, ScFv multimers of the anti-neuraminidase antibody NC10: shortening of the linker in single-chain Fv fragment assembled in V(L) to V(H) orientation drives the formation of dimers, trimers, tetramers and higher molecular mass multimers, Protein Eng. 13 (2000) 565–574.

[40] P. Holliger, P.J. Hudson, Engineered antibody fragments and the rise of single domains, Nat. Biotechnol. 23 (2005) 1126–1136.

[41] F. Le Gall, S.M. Kipriyanov, G. Moldenhauer, M. Little, Di-, tri- and tetrameric single chain Fv antibody fragments against human CD19: effect of valency on cell binding, FEBS Lett. 453 (1999) 164–168.

[42] W. Scheuer, M. Thomas, P. Hanke, J. Sam, F. Osl, D. Weininger, M. Baehner, S. Seeber, H. Kettenberger, J. Schanzer, U. Brinkmann, K.M. Weidner, J. Regula, C. Klein, Anti-tumoral, anti-angiogenic and anti-metastatic efficacy of a tetravalent bispecific antibody (TAvi6) targeting VEGF-A and angiopoietin-2, MAbs 8 (2016) 562–573.

[43] J.U. Schmohl, M.K. Gleason, P.R. Dougherty, J.S. Miller, D.A. Vallera, Heterodimeric bispecific single chain variable fragments (scFv) killer engagers (BiKEs) enhance NK-cell activity against CD133+ colorectal cancer cells, Target. Oncol. 11 (2016) 353–361.

[44] M. Xu, H. Jin, Z. Chen, W. Xie, Y. Wang, M. Wang, J. Zhang, D.O. Acheampong, A novel bispecific diabody targeting both vascular endothelial growth factor receptor 2 and epidermal growth factor receptor for enhanced antitumor activity, Biotechnol. Prog. 32 (2016) 294–302.

[45] M.A. Wronska, I.B. O'Connor, M.A. Tilbury, A. Srivastava, J.G. Wall, Adding Functions to Biomaterial Surfaces through Protein Incorporation, Adv. Mater. 28 (2016) 5485–5508.

[46] J. McCafferty, D. Schofield, Identification of optimal protein binders through the use of large genetically encoded display libraries, Curr. Opin. Chem. Biol. 26 (2015) 16–24.

[47] J.G. Jung, G.M. Jeong, S.S. Yim, K.J. Jeong, Development of high-affinity single chain Fv against foot-and-mouth disease virus, Enzyme Microb. Technol. 84 (2016) 50–55.

[48] C.M. Hammers, J.R. Stanley, Antibody phage display: technique and applications, J. Invest. Dermatol. 134 (2014). e17.

[49] N.V. Tikunova, V.V. Morozova, Phage display on the base of filamentous bacteriophages: application for recombinant antibodies selection, Acta Nat. 1 (2009) 20–28.

[50] C. Cunningham, A. Srivastava, E. Collin, S. Grad, M. Alini, A. Pandit, J.G. Wall, Isolation and characterisation of a recombinant antibody fragment that binds NCAM1-expressing intervertebral disc cells, PLoS One 8 (2013). e83678.

[51] T. Huovinen, M. Syrjanpaa, H. Sanmark, T. Seppa, S. Akter, L.M. Khan, U. Lamminmaki, The selection performance of an antibody library displayed on filamentous phage coat proteins p9, p3 and truncated p3, BMC Res. Notes 7 (2014) 661.

[52] S. Spada, J.T. Pembroke, J.G. Wall, Isolation of a novel *Thermus thermophilus* metal efflux protein that improves *Escherichia coli* growth under stress conditions, Extremophiles 6 (2002) 301–308.

[53] E. Pavoni, P. Vaccaro, A.M. Anastasi, O. Minenkova, Optimized selection of anti-tumor recombinant antibodies from phage libraries on intact cells, Mol. Immunol. 57 (2014) 317–322.

[54] K.A. Power, S. Grad, J. Rutges, L.B. Creemers, M.H.P. Van Rijen, P. O'Gaora, J.G. Wall, M. Alini, A. Pandit, W.M. Gallagher, Identification of cell surface specific markers to target human nucleus pulposus cells: expression of carbonic anhydrase 12 varies with age and degeneration, Arthritis Rheum. 63 (2011) 3876–3886.

[55] K.D. Miller, N.B. Pefaur, C.L. Baird, Construction and screening of antigen targeted immune yeast surface display antibody libraries, Curr. Protoc. Cytom. 45 (2008) 4.7:4.7.1–4.7.30.

[56] T.F. Chen, S. De Picciotto, B.J. Hackel, K.D. Wittrup, Engineering fibronectin-based binding proteins by yeast surface display, in: E.K. Amy (Ed.), Methods in Enzymology, Academic Press, Amsterdam, 2013 (Chapter 14).

[57] E. Van Bloois, R.T. Winter, H. Kolmar, M.W. Fraaije, Decorating microbes: surface display of proteins on Escherichia coli, Trends Biotechnol. 29 (2011) 79–86.

[58] G. Chao, W.L. Lau, B.J. Hackel, S.L. Sazinsky, S.M. Lippow, K.D. Wittrup, Isolating and engineering human antibodies using yeast surface display, Nat. Protoc. 1 (2006) 755–768.

[59] C. Zahnd, P. Amstutz, A. Plückthun, Ribosome display: selecting and evolving proteins in vitro that specifically bind to a target, Nat. Methods 4 (2007) 269–279.

[60] A. Rothe, A. Nathanielsz, F. Oberhauser, S.E. Von Pogge, A. Engert, P.J. Hudson, B.E. Power, Ribosome display and selection of human anti-cD22 scFvs derived from an acute lymphocytic leukemia patient, Biol. Chem. 389 (2008) 433–439.

[61] B. Schimmele, A. Plückthun, Identification of a functional epitope of the Nogo receptor by a combinatorial approach using ribosome display, J. Mol. Biol. 352 (2005) 229–241.

[62] H. Zhou, B. Zhou, H. Ma, C. Carney, K.D. Janda, Selection and characterization of human monoclonal antibodies against Abrin by phage display, Bioorg. Med. Chem. Lett. 17 (2007) 5690–5692.

[63] L. Zhou, W.P. Mao, J. Fen, H.Y. Liu, C.J. Wei, W.X. Li, F.Y. Zhou, Selection of scFvs specific for the HepG2 cell line using ribosome display, J. Biosci. 34 (2009) 221–226.

[64] Z. Shen, H. Yan, F.F. Parl, R.L. Mernaugh, X. Zeng, Recombinant antibody piezo-immunosensors for the detection of cytochrome P450 1B1, Anal. Chem. 79 (2007) 1283–1289.

[65] M.J. Hortigüela, L. Aumailley, A. Srivastava, C. Cunningham, S. Anandakumar, S. Robin, A. Pandit, X. Hu, J.G. Wall, Engineering recombinant antibodies for polymer biofunctionalization, Polym. Adv. Technol. 26 (2015) 1394–1401.

[66] M. Klomp, M.A.M. Beijk, R.J. De Winter, Genous™ endothelial progenitor cell-capturing stent system: a novel stent technology, Expert Rev. Med. Devices 6 (2009) 365–375.

[67] A. Nissim, Y. Chernajovsky, Historical development of monoclonal antibody therapeutics, Handb. Exp. Pharmacol. 181 (2008) 3–18.

[68] L.G. Presta, Engineering of therapeutic antibodies to minimize immunogenicity and optimize function, Adv. Drug Deliv. Rev. 58 (2006) 640–656.

[69] G. Walsh, Biopharmaceutical benchmarks 2010, Nat. Biotechnol. 28 (2010) 917–924.

[70] I.-H. Cho, E.-H. Paek, H. Lee, J.Y. Kang, T.S. Kim, S.-H. Paek, Site-directed biotinylation of antibodies for controlled immobilization on solid surfaces, Anal. Biochem. 365 (2007) 14–23.

[71] Y. Jung, J.Y. Jeong, B.H. Chung, Recent advances in immobilization methods of antibodies on solid supports, Analyst 133 (2008) 697–701.

[72] G. Hermanson, Bioconjugate Techniques, Academic Press-Elsevier, San Diego, 2008.

[73] M.P. Deonarain, G. Yahioglu, I. Stamati, J. Marklew, Emerging formats for next-generation antibody drug conjugates, Expert Opin. Drug Discov. 10 (2015) 463–481.

[74] K.D. Fowers, J. Callahan, P. Byron, J.I. Kopecek, Preparation of Fab' from murine IgG2a for thiol reactive conjugation, J. Drug Target. 9 (2001) 281–294.

[75] P. Batalla, M. Fuentes, V. Grazu, C. Mateo, R. Fernandez-Lafuente, J.M. Guisan, Oriented covalent immobilization of antibodies on physically inert and hydrophilic support surfaces through their glycosidic chains, Biomacromolecules 9 (2008) 719–723.

[76] J.P. Gering, L. Quaroni, G. Chumanov, Immobilization of antibodies on glass surfaces through sugar residues, J. Colloid Interface Sci. 252 (2002) 50–56.

[77] P. Batalla, J.M. Bolívar, F. Lopez-Gallego, J.M. Guisan, Oriented covalent immobilization of antibodies onto heterofunctional agarose supports: a highly efficient immunoaffinity chromatography platform, J. Chromatogr. A 1262 (2012) 56–63.

[78] B.H. Choi, Y.S. Choi, D.S. Hwang, H.J. Cha, Facile surface functionalization with glycosaminoglycans by direct coating with mussel adhesive protein, Tissue Eng. Part C Methods 18 (2012) 71–79.

[79] M. Yin, Y. Yuan, C. Liu, J. Wang, Combinatorial coating of adhesive polypeptide and anti-CD34 antibody for improved endothelial cell adhesion and proliferation, J. Mater. Sci. Mater. Med. 20 (2009) 1513–1523.
[80] C.S. Kim, Y.S. Choi, W. Ko, J.H. Seo, J. Lee, H.J. Cha, A mussel adhesive protein fused with the BC domain of protein A is a functional linker material that efficiently immobilizes antibodies onto diverse surfaces, Adv. Funct. Mater. 21 (2011) 4101–4108.
[81] A. Grodzki, E. Berenstein, Antibody purification: affinity chromatography—protein A and protein G sepharose, in: C. Oliver, M.C. Jamur (Eds.), Immunocytochemical Methods and Protocols, Humana Press, New York, 2010.
[82] B.D. Markway, O.J. McCarty, U.M. Marzec, D.W. Courtman, S.R. Hanson, M.T. Hinds, Capture of flowing endothelial cells using surface-immobilized anti-kinase insert domain receptor antibody, Tissue Eng. Part C Methods 14 (2008) 97–105.
[83] C.-W. Tsai, S.-L. Jheng, W.-Y. Chen, R.-C. Ruaan, Strategy of Fc-recognizable peptide ligand design for oriented immobilization of antibody, Anal. Chem. 86 (2014) 2931–2938.
[84] K. Nakanishi, T. Sakiyama, Y. Kumada, K. Imamura, H. Imanaka, Recent advances in controlled immobilization of proteins onto the surface of the solid substrate and its possible application to proteomics, Curr. Proteomics 5 (2008) 161–175.
[85] J.W. Slootstra, D. Kuperus, A. Plückthun, R.H. Meloen, Identification of new tag sequences with differential and selective recognition properties for the anti-FLAG monoclonal antibodies M1, M2 and M5, Mol. Divers. 2 (1997) 156–164.
[86] M.J. Hortigüela, J.G. Wall, Improved detection of domoic acid using covalently immobilised antibody fragments, Mar. Drugs 11 (2013) 881–895.
[87] X. Hu, M.J. Hortigüela, S. Robin, H. Lin, Y. Li, A.P. Moran, W. Wang, J.G. Wall, Covalent and oriented immobilization of scFv antibody fragments via an engineered glycan moiety, Biomacromolecules 14 (2013) 153–159.
[88] F. Schwarz, W. Huang, C. Li, B.L. Schulz, C. Lizak, A. Palumbo, S. Numao, D. Neri, M. Aebi, L.X. Wang, A combined method for producing homogeneous glycoproteins with eukaryotic N-glycosylation, Nat. Chem. Biol. 6 (2010) 264–266.
[89] D.N. Prater, J. Case, D.A. Ingram, M.C. Yoder, Working hypothesis to redefine endothelial progenitor cells, Leukemia 21 (2007) 1141–1149.
[90] M. Avci-Adali, N. Perle, G. Ziemer, H.P. Wendel, Current concepts and new developments for autologous in vivo endothelialisation of biomaterials for intravascular applications, Eur. Cell. Mater. 21 (2011) 157–176.
[91] T. Asahara, T. Murohara, A. Sullivan, M. Silver, R. van der Zee, T. Li, B. Witzenbichler, G. Schatteman, J.M. Isner, Isolation of putative progenitor endothelial cells for angiogenesis, Science 275 (1997) 964–966.
[92] G.J. Padfield, D.E. Newby, N.L. Mills, Understanding the role of endothelial progenitor cells in percutaneous coronary intervention, J. Am. Coll. Cardiol. 55 (2010) 1553–1565.
[93] M. Hristov, W. Erl, P.C. Weber, Endothelial progenitor cells: mobilization, differentiation, and homing, Arterioscler. Thromb. Vasc. Biol. 23 (2003) 1185–1189.
[94] F.J. Fernandez, M.C. Vega, Technologies to keep an eye on: alternative hosts for protein production in structural biology, Curr. Opin. Struct. Biol. 23 (2013) 365–373.
[95] Y. He, K. Wang, N. Yan, The recombinant expression systems for structure determination of eukaryotic membrane proteins, Protein Cell 5 (2014) 658–672.
[96] M. Peichev, A.J. Naiyer, D. Pereira, Z. Zhu, W.J. Lane, M. Williams, M. Oz, D.J. Hicklin, L. Witte, M.A. Moore, S. Rafil, Expression of VEGFR-2 and AC133 by circulating human CD34(+) cells identifies a population of functional endothelial precursors, Blood 95 (2000) 952–958.

[97] F. Timmermans, J. Plum, M.C. Yöder, D.A. Ingram, B. Vandekerckhove, J. Case, Endothelial progenitor cells: identity defined? J. Cell. Mol. Med. 13 (2009) 87–102.
[98] M.A. Beijk, P. Damman, M. Klomp, P. Woudstra, S. Silber, M. Grisold, E.E. Ribeiro, H. Suryapranata, J. Wojcik, K.H. Sim, J.G. Tijssen, A.R. De Winter, Twelve-month clinical outcomes after coronary stenting with the Genous Bio-engineered R Stent in patients with a bifurcation lesion: from the e-HEALING (Healthy Endothelial Accelerated Lining Inhibits Neointimal Growth) registry, Coron. Artery Dis. 23 (2012) 201–207.
[99] M. Bystron, P. Cervinka, R. Spacek, M. Kvasnak, J. Jakabcin, M. Cervinkova, P. Kala, P. Widimsky, Randomized comparison of endothelial progenitor cells capture stent versus cobalt-chromium stent for treatment of ST-elevation myocardial infarction. Six-month clinical, angiographic, and IVUS follow-up, Catheter. Cardiovasc. Interv. 76 (2010) 627–631.
[100] H.J. Duckers, T. Soullie, P. Den Heijer, B. Rensing, R.J. De Winter, M. Rau, H. Mudra, S. Silber, E. Benit, S. Verheye, W. Wijns, P.W. Serruys, Accelerated vascular repair following percutaneous coronary intervention by capture of endothelial progenitor cells promotes regression of neointimal growth at long term follow-up: final results of the Healing II trial using an endothelial progenitor cell capturing stent (Genous R stent), EuroIntervention 3 (2007) 350–358.
[101] M.A. Beijk, M. Klomp, N.J. Verouden, N. Van Geloven, K.T. Koch, J.P. Henriques, J. Baan, M.M. Vis, E. Scheunhage, J.J. Piek, J.G. Tijssen, R.J. De Winter, Genous endothelial progenitor cell capturing stent vs. the Taxus Liberte stent in patients with de novo coronary lesions with a high-risk of coronary restenosis: a randomized, single-centre, pilot study, Eur. Heart J. 31 (2010) 1055–1064.
[102] M. Haude, S.W. Lee, S.G. Worthley, S. Silber, S. Verheye, S. Erbs, M.A. Rosli, R. Botelho, I. Meredith, K.H. Sim, P.R. Stella, H.C. Tan, R. Whitbourn, S. Thambar, A. Abizaid, T.H. Koh, P. Den Heijer, H. Parise, E. Cristea, A. Maehara, R. Mehran, The REMEDEE trial: a randomized comparison of a combination sirolimus-eluting endothelial progenitor cell capture stent with a paclitaxel-eluting stent, JACC Cardiovasc. Interv. 6 (2013) 334–343.
[103] H. Xu, K.T. Nguyen, E.S. Brilakis, J. Yang, E. Fuh, S. Banerjee, Enhanced endothelialization of a new stent polymer through surface enhancement and incorporation of growth factor-delivering microparticles, J. Cardiovasc. Transl. Res. 5 (2012) 519–527.
[104] C.L. Song, Q. Li, Y.P. Yu, G. Wang, J.P. Wang, Y. Lu, J.C. Zhang, H.Y. Diao, J.G. Liu, Y.H. Liu, J. Liu, Y. Li, D. Cai, B. Liu, Study of novel coating strategy for coronary stents: simutaneous coating of VEGF and anti- CD34 antibody, Rev. Bras. Cir. Cardiovasc. 30 (2015) 159–163.
[105] C.L. Song, Q. Li, J.C. Zhang, J.P. Wang, X. Xue, G. Wang, Y.F. Shi, H.Y. Diao, B. Liu, Study of a novel coating strategy for coronary stents: evaluation of stainless metallic steel coated with VEGF and anti-CD34 antibody in vitro, Eur. Rev. Med. Pharmacol. Sci. 20 (2016) 311–316.
[106] L. Shihui, C. Junying, C. Cheng, N. Huang, The nanofabrication of CD34 antibody-VEGF-heparin on titanium surface via layer-by-layer assembly for biofunctionalization, in: 3rd International Nanoelectronics Conference (INEC), 2010, p. 1357.
[107] G. Nakazawa, J.F. Granada, C.L. Alviar, A. Tellez, G.L. Kaluza, M.Y. Guilhermier, S. Parker, S.M. Rowland, F.D. Kolodgie, M.B. Leon, R. Virmani, Anti-CD34 antibodies immobilized on the surface of sirolimus-eluting stents enhance stent endothelialization, JACC Cardiovasc. Interv. 3 (2010) 68–75.
[108] S. Petersen, C. Häcker, G. Turan, S. Knödler, A. Brodehl, A. Drynda, S. Kische, B. Frerich, R. Birkemeyer, H. Ince, B. Vollmar, K.-P. Schmitz, K. Sternberg, Implications

for the biofunctionalization of drug-eluting devices at the example of a site-selective antibody modification for drug eluting stents, BioNanoMaterials 16 (2015) 275.
[109] Y. Yuan, M. Yin, J. Qian, C. Liu, Site-directed immobilization of antibodies onto blood contacting grafts for enhanced endothelial cell adhesion and proliferation, Soft Matter 7 (2011) 7207–7216.
[110] F. Yang, S.C. Feng, X.J. Pang, W.X. Li, Y.H. Bi, Q. Zhao, S.X. Zhang, Y. Wang, B. Feng, Combination coating of chitosan and anti-CD34 antibody applied on sirolimus-eluting stents can promote endothelialization while reducing neointimal formation, BMC Cardiovasc. Disord. 12 (2012) 96.
[111] A. Tan, D. Goh, Y. Farhatnia, G. Natasha, J. Lim, S.-H. Teoh, J. Rajadas, M.S. Alavijeh, A.M. Seifalian, An anti-CD34 antibody-functionalized clinical-grade POSS-PCU nanocomposite polymer for cardiovascular stent coating applications: a preliminary assessment of endothelial progenitor cell capture and hemocompatibility, PLoS One 8 (2013) e77112.
[112] S. Zhang, F. Zhang, B. Feng, Q. Fan, F. Yang, D. Shang, J. Sui, H. Zhao, Hematopoietic stem cell capture and directional differentiation into vascular endothelial cells for metal stent-coated chitosan/hyaluronic acid loading CD133 antibody, Tissue Eng. Part A 21 (2015) 1173–1183.
[113] X. Wu, T. Yin, J. Tian, C. Tang, J. Huang, Y. Zhao, X. Zhang, X. Deng, Y. Fan, D. Yu, G. Wang, Distinctive effects of CD34- and CD133-specific antibody-coated stents on re-endothelialization and in-stent restenosis at the early phase of vascular injury, Regen. Biomater. 2 (2015) 87–96.
[114] J. Li, D. Li, F. Gong, S. Jiang, H. Yu, Y. An, Anti-CD133 antibody immobilized on the surface of stents enhances endothelialization, Biomed. Res. Int. 2014 (2014). 902782.
[115] T. Asahara, C. Bauters, C. Pastore, M. Kearney, S. Rossow, S. Bunting, N. Ferrara, J.F. Symes, J.M. Isner, Local delivery of vascular endothelial growth factor accelerates reendothelialization and attenuates intimal hyperplasia in balloon-injured rat carotid artery, Circulation 91 (1995) 2793–2801.
[116] M. Lahtinen, P. Blomberg, G. Baliulis, F. Carlsson, H. Khamis, V. Zemgulis, In vivo h-VEGF165 gene transfer improves early endothelialisation and patency in synthetic vascular grafts, Eur. J. Cardiothorac. Surg. 31 (2007) 383–390.
[117] N. Swanson, K. Hogrefe, Q. Javed, A.H. Gershlick, In vitro evaluation of vascular endothelial growth factor (VEGF)-eluting stents, Int. J. Cardiol. 92 (2003) 247–251.
[118] W. Cao, H. Li, J. Zhang, D. Li, D.O. Acheampong, Z. Chen, M. Wang, Periplasmic expression optimization of VEGFR2 D3 adopting response surface methodology: antiangiogenic activity study, Protein Expr. Purif. 90 (2013) 55–66.
[119] R. Di Stasi, D. Diana, D. Capasso, R. Palumbo, A. Romanelli, C. Pedone, R. Fattorusso, L.D. D'Andrea, VEGFR1(D2) in drug discovery: expression and molecular characterization, Biopolymers 94 (2010) 800–809.
[120] A. Foerster, I. Hołowacz, G.B. Sunil Kumar, S. Anandakumar, J.G. Wall, M. Wawrzyńska, M. Paprocka, A. Kantor, H. Kraskiewicz, S. Olsztyńska-Janus, S.J. Hinder, D. Bialy, H. Podbielska, M. Kopaczyńska, Stainless steel surface functionalization for immobilization of antibody fragments for cardiovascular applications, J. Biomed. Mater. Res. A 104 (2016) 821–832.
[121] S. Takabatake, K. Hayashi, C. Nakanishi, H. Hao, K. Sakata, M.A. Kawashiri, T. Matsuda, M. Yamagishi, Vascular endothelial growth factor-bound stents: application of in situ capture technology of circulating endothelial progenitor cells in porcine coronary model, J. Interv. Cardiol. 27 (2014) 63–72.

[122] X. Ma, B. Hibbert, M. Mcnulty, T. Hu, X. Zhao, F.D. Ramirez, T. Simard, J.S. De Belleroche, E.R. O'Brien, Heat shock protein 27 attenuates neointima formation and accelerates reendothelialization after arterial injury and stent implantation: importance of vascular endothelial growth factor up-regulation, FASEB J. 28 (2014) 594–602.

[123] J.M. Lee, W. Choe, B.K. Kim, W.W. Seo, W.H. Lim, C.K. Kang, S. Kyeong, K.D. Eom, H.J. Cho, Y.C. Kim, J. Hur, H.M. Yang, Y.S. Lee, H.S. Kim, Comparison of endothelialization and neointimal formation with stents coated with antibodies against CD34 and vascular endothelial-cadherin, Biomaterials 33 (2012) 8917–8927.

[124] H. Tang, Q. Wang, X. Wang, J. Zhou, M. Zhu, T. Qiao, C. Liu, C. Mao, M. Zhou, Effect of a novel stent on re-endothelialization, platelet adhesion, and neointimal formation, J. Atheroscler. Thromb. 23 (2016) 67–80.

[125] S. Cui, J.H. Liu, X.T. Song, G.L. Ma, B.J. Du, S.Z. Lv, L.J. Meng, Q.S. Gao, K. Li, A novel stent coated with antibodies to endoglin inhibits neointimal formation of porcine coronary arteries, Biomed. Res. Int. 2014 (2014). 428619.

[126] S. Cui, X.T. Song, C. Ding, L.J. Meng, S.Z. Lv, K. Li, Comparison of reendothelialization and neointimal formation with stents coated with antibodies against endoglin and CD34 in a porcine model, Drug Des. Dev. Ther. 9 (2015) 2249–2256.

[127] X. Jin, L. Mei, C. Song, L. Liu, X. Leng, H. Sun, D. Kong, R.J. Levy, Immobilization of plasmid DNA on an anti-DNA antibody modified coronary stent for intravascular site-specific gene therapy, J. Gene Med. 10 (2008) 421–429.

[128] L.H. Zhang, T. Luo, C. Zhang, P. Luo, X. Jin, C.X. Song, R.L. Gao, Anti-DNA antibody modified coronary stent for plasmid gene delivery: results obtained from a porcine coronary stent model, J. Gene Med. 13 (2011) 37–45.

[129] B.D. Klugherz, C. Song, S. Defelice, X. Cui, Z. Lu, J. Connolly, J.T. Hinson, R.L. Wilensky, R.J. Levy, Gene delivery to pig coronary arteries from stents carrying antibody-tethered adenovirus, Hum. Gene Ther. 13 (2002) 443–454.

Index

Note: Page numbers followed by *f* indicate figures, *t* indicate tables and *s* indicate schemes.

R

CPI Antony Rowe
Chippenham, UK
2018-02-16 10:25